SUBSURFACE HYDROGEN ENERGY STORAGE

SUBSURFACE HYDROGEN ENERGY STORAGE

Current Status, Prospects, and Challenges

Edited by

ACHINTA BERA

Department of Petroleum Engineering, School of Energy Technology, Pandit Deendayal Energy University, Gandhinagar, Gujarat, India

SUNIL KUMAR

RINOLYST, SciGlyph Exploration Private Limited, Patna, Bihar, India

ELSEVIER

Elsevier
Radarweg 29, PO Box 211, 1000 AE Amsterdam, Netherlands
125 London Wall, London EC2Y 5AS, United Kingdom
50 Hampshire Street, 5th Floor, Cambridge, MA 02139, United States

ISBN: 978-0-443-24071-3

For Information on all Elsevier publications
visit our website at https://www.elsevier.com/books-and-journals

Publisher: Megan Ball
Acquisitions Editor: Peter Adamson
Editorial Project Manager: Ashi Jain
Production Project Manager: Erragounta Saibabu Rao
Cover Designer: Mark Rogers

Typeset by MPS Limited, Chennai, India

Dedication

This book is dedicated to our parents and family members, whose unwavering support, love, and guidance have been the cornerstone of our personal and professional endeavors. Their belief in us and encouragement to pursue our passions have shaped us into the individuals we are today. We also dedicate this book to the tireless efforts of scientists, engineers, and innovators worldwide, whose dedication to sustainable energy solutions inspires us to push the boundaries of knowledge and technology. Your relentless pursuit of a better future motivates us to contribute meaningfully to the journey toward a sustainable world for generations to come.

Contents

4. Hydrogen surface storage

Shilpa Singh

5. Navigating hydrogen storage and transport networks: grid connectivity and storage site logistics

Gourav Kumar Rath, Gaurav Pandey, Ajay Mittal, Sakshi Singh, Swagata Sharma, Dharmendra Kumar Gupta, Sanket Joshi and Nadezhda Molokitina

6. Subsurface underground hydrogen storage

Jagruti Thakur and Ahmed Elberry

7. Physics of hydrogen flow in porous media: two-phase (or multiphase) flow and transport of hydrogen in porous geological storage

Birger Hagemann, Sebastian Hogeweg and Julia Michelsen

8. Physiochemical parametric considerations for optimal underground hydrogen storage

Nawal Noshad, Shwetank Krishna, Gerhard Thonhauser, Syahrir Ridha, Aymn Abdulrahman and Suhaib Umer Ilyas

9. Geochemical and geomechanical synergies during duringgies underground hydrogen storage schemes

Prashant Jadhawar and Hosein Doryani

10. Microbial considerations/aspects of underground hydrogen storage

Tausif Ahmad, Narendra Kumar and Muhammad Mubashir

List of contributors

Aymn Abdulrahman Chemical Engineering Department, University of Jeddah, Jeddah, Kingdom of Saudi Arabia

Tausif Ahmad Physical Science and Engineering Division, King Abdullah University of Science and Technology (KAUST), Thuwal, Saudi Arabia

Achinta Bera Department of Petroleum Engineering, School of Energy Technology, Pandit Deendayal Energy University, Gandhinagar, Gujarat, India

Hosein Doryani School of Engineering, University of Aberdeen, Aberdeen, United Kingdom

Ahmed Elberry Faculty of Science (HIMS), University of Amsterdam, Amsterdam, The Netherlands; TNO Energy and Materials Transition, Amsterdam, The Netherlands

Omar Elkhatib ChampionX USA Inc., Sugar Land, TX, United States

Ramesh Guduru (Kumar) Department of Mechanical Engineering, School of Technology, Pandit Deendayal Energy University, Gandhinagar, Gujarat, India

Dharmendra Kumar Gupta Department of Sustainability, School of Advanced Engineering, UPES, Dehradun, Uttarakhand, India

Birger Hagemann Department Reservoir Engineering, Clausthal University of Technology, Institute of Subsurface Energy Systems, Clausthal-Zellerfeld, Germany

Sebastian Hogeweg Department Reservoir Engineering, Clausthal University of Technology, Institute of Subsurface Energy Systems, Clausthal-Zellerfeld, Germany

Suhaib Umer Ilyas Chemical Engineering Department, University of Jeddah, Jeddah, Kingdom of Saudi Arabia

Prashant Jadhawar School of Engineering, University of Aberdeen, Aberdeen, United Kingdom

Sanket Joshi Amity Institute of Microbial Technology, Amity University, Jaipur, Rajasthan, India

Shwetank Krishna Chair of Drilling and Completion Engineering, Department Geo Energy, Montanuniversität Leoben, Leoben, Austria

Sandeep D. Kulkarni Deysarkar Centre of Excellence in Petroleum Engineering, IIT Kharagpur, Kharagpur, India

Narendra Kumar Ali I. Al-Naimi Petroleum Engineering Research Center, King Abdullah University of Science and Technology, Thuwal, Saudi Arabia

Sunil Kumar RINOLYST, SciGlyph Exploration Private Limited, Patna, Bihar, India

Anooja Sara Mathew Deysarkar Centre of Excellence in Petroleum Engineering, IIT Kharagpur, Kharagpur, India

Julia Michelsen Department Reservoir Engineering, Clausthal University of Technology, Institute of Subsurface Energy Systems, Clausthal-Zellerfeld, Germany

Ajay Mittal Department of Chemical Engineering, Energy Cluster, School of Advanced Engineering, UPES, Dehradun, Uttarakhand, India

Nadezhda Molokitina Earth Cryosphere Institute, Tyumen Scientific Center SB, RAS, Tyumen, Russia

Soubhik Mondal Deysarkar Centre of Excellence in Petroleum Engineering, IIT Kharagpur, Kharagpur, India

Muhammad Mubashir Water Technologies Innovation Institute & Research Advancement, Saudi Water Authority, WTIIRA-SWA, Jubail, Saudi Arabia

Vishnu Chandrasekharan Nair Deysarkar Centre of Excellence in Petroleum Engineering, IIT Kharagpur, Kharagpur, India

Nawal Noshad Department of Chemical Engineering, University of Gujrat, Gujrat, Pakistan

Gaurav Pandey Department of Petroleum Engineering and Earth Science, Energy Cluster, School of Advanced Engineering, UPES, Dehradun, Uttarakhand, India; Earth Cryosphere Institute, Tyumen Scientific Center SB, RAS, Tyumen, Russia

Ricurve Patel Department of Mechanical Engineering, School of Technology, Pandit Deendayal Energy University, Gandhinagar, Gujarat, India

Ranjan Phukan Department of Petroleum Technology, Dibrugarh University, Dibrugarh, Assam, India

Devanshi Raman Department of Chemical Engineering, National Institute of Technology Hamirpur, Hamirpur, Himachal Pradesh, India

Gourav Kumar Rath Department of Petroleum Engineering and Earth Science, Energy Cluster, School of Advanced Engineering, UPES, Dehradun, Uttarakhand, India

Arshad Raza Department of Petroleum Engineering, King Fahd University of Petroleum and Minerals, Dhahran, Saudi Arabia

Mohamed Gamal Rezk Center for Integrative Petroleum Research, College of Petroleum Engineering and Geosciences, King Fahd University of Petroleum and Minerals, Dhahran, Saudi Arabia

Syahrir Ridha Institute of Hydrocarbon Recovery, Universiti Teknologi PETRONAS, Seri Iskandar, Perak Darul Ridzuan, Malaysia

Rahul Saha Department of Chemical Engineering, National Institute of Technology Hamirpur, Hamirpur, Himachal Pradesh, India

Swagata Sharma Department of Petroleum Engineering and Earth Science, Energy Cluster, School of Advanced Engineering, UPES, Dehradun, Uttarakhand, India

Robin Singh Department of Chemistry, School of Energy Technology, Pandit Deendayal Energy University, Gandhinagar, Gujarat, India

Sakshi Singh Department of Petroleum Engineering and Earth Science, Energy Cluster, School of Advanced Engineering, UPES, Dehradun, Uttarakhand, India

Shilpa Singh Department of Physics and Electronics, St. Xavier's College, Ahmedabad, Ahmedabad, India

Jagruti Thakur Department of Energy Technology, KTH Royal Institute of Technology, Stockholm, Sweden

Gerhard Thonhauser Chair of Drilling and Completion Engineering, Department Geo Energy, Montanuniversität Leoben, Leoben, Austria

Rakesh Kumar Vij Department of Petroleum Engineering, School of Energy Technology, Pandit Deendayal Energy University, Gandhinagar, Gujarat, India

About the editors

Dr. Achinta Bera is currently working as an Assistant Professor of Petroleum Engineering at the School of Energy Technology, Pandit Deendayal Energy University, Gandhinagar, India. He received his PhD in petroleum engineering from the Indian Institute of Technology (Indian School of Mines), Dhanbad, in 2014. After that, he worked as a Postdoctoral Fellow at the University of Alberta, Canada, and Khalifa University, Abu Dhabi from 2014 to 2018. Dr. Bera mainly focuses on the topics related to carbon capture, utilization, and storage (CCUS) and subsurface hydrogen storage, which directly indicate global decarbonization. As a petroleum engineer, Dr. Bera's main aim is to develop technology that can help the current petroleum industry to reduce greenhouse gas emissions. His research works related to CCUS address the protocol to optimize the process of carbon storage. He is an active reviewer of more than 40 reputable international journals such as *Industrial and Engineering Chemistry Research*, *Journal of Petroleum Science and Engineering*, *Energy & Fuels*, *SPE Journal*, *Applied Energy*, *Fuel*, and *Chemical Engineering Science*. Based on his contribution in energy research, Dr. Bera has been listed as one among the 2% scientist published by Stanford University, USA, in 2021 and 2022. He is also serving as an Editor of some reputed journals.

Dr. Sunil Kumar is a seasoned professional currently serving as the Head of RINOLYST and Chief Innovation Officer at SciGlyph Exploration Private Limited, Patna, India. He received his PhD in petroleum engineering from the Indian Institute of Technology (Indian School of Mines), Dhanbad, in 2017. After that, he worked as a Postdoctoral Associate at the University of Calgary, Canada (2017–19), and later at Universiti Teknologi PETRONAS, Malaysia (2019–22). His expertise covers hydrocarbon recovery, surface and interfacial science, complex fluids, flow in porous media, underground hydrogen storage, carbon sequestration, and decarbonization, contributing significantly to research in energy sciences. Dr. Kumar has authored several articles in refereed international journals, book chapters, and conference articles. He also serves as a reviewer for reputable international journals, including *Nature Communication*, *Renewable and Sustainable Energy Reviews*, and *Fuel*. Dr. Kumar is currently leading efforts in decarbonization, sustainability, and advancing hydrogen technology across its entire life cycle, from production to end-use applications.

Preface

Welcome to the realm of possibilities, innovations, and challenges in the exciting domain of hydrogen energy storage. *Subsurface Hydrogen Energy Storage: Current Status, Prospects, and Challenges* embarks on a journey through the intricate landscapes of subsurface hydrogen storage technologies, aiming to enlighten, inspire, and catalyze advancements in sustainable energy solutions. This book represents a collective effort of dedicated researchers, engineers, and industry experts who share a common vision of leveraging hydrogen's potential for a cleaner and greener future. It is a testament to their unwavering commitment to unraveling the complexities, overcoming hurdles, and charting pathways toward a robust hydrogen economy. The narrative begins with a foundational exploration of underground hydrogen storage, laying a solid groundwork for the readers to delve deeper into subsequent chapters. We traverse the multifaceted terrain of hydrogen as an energy carrier, unraveling its production methods, surface storage intricacies, and the pivotal role of transportation networks in ensuring seamless energy transitions. Central to our discourse is the captivating realm of subsurface hydrogen storage, where geological, hydrogeological, physiochemical, and geochemical forces converge to shape storage efficiencies and operational dynamics. From microscopic microbial considerations to macroscopic experimental studies and cutting-edge modeling techniques, each chapter unfolds a tapestry of insights, challenges, and promising strategies. Navigating through regulatory frameworks, safety imperatives, and strategic imperatives, we spotlight not just the technical intricacies but also the strategic imperatives of embracing hydrogen storage on a global scale. Real-world case studies illuminate success stories, cautionary tales, and invaluable lessons learned from field applications, enriching our collective knowledge and inspiring future endeavors. As editors, along with our esteemed contributors, we aspire to ignite curiosity, spark dialogs, and catalyze actionable insights that transcend the pages of this book. Our goal is to empower readers—researchers, policymakers, industry leaders, and students—with the knowledge, vision, and motivation to drive transformative changes in the energy landscape. We invite you to immerse yourself in the pages of *Subsurface Hydrogen Energy Storage: Current Status, Prospects, and Challenges* and embark on a transformative odyssey toward a sustainable energy future. Let us unite in our quest to harness the boundless potential of hydrogen, turning aspirations into achievements and challenges into opportunities. Together, let us write the next chapter in the remarkable journey of hydrogen energy storage.

Achinta Bera
Sunil Kumar

Acknowledgments

We extend our heartfelt gratitude to all the authors, including professors and their research teams, whose generous contributions in crafting the chapters of this book have been invaluable. Their dedication and expertise have played a pivotal role in completing this comprehensive work.

We are deeply appreciative of the global community committed to addressing climate change and advancing sustainability. Your proactive efforts inspire us all to strive for a greener and more sustainable future.

Special thanks to Elsevier and all the members of the *Energy with Purpose initiative, Elsevier*, for their invaluable support and trust in this project. The opportunity to contribute to the field of sustainable energy is both inspiring and rewarding.

A sincere thanks to Dr. Gyan Prakash Satyam for his creative vision in designing the captivating front cover photo of the book, adding a distinctive touch to its presentation.

We extend our appreciation to the dedicated members of the Elsevier sales and marketing team, whose efforts are instrumental in disseminating knowledge to a global audience.

Our gratitude also extends to the Department of Petroleum Engineering, School of Energy Technology, Pandit Deendayal Energy University, Gandhinagar, Gujarat, India, and exceptional team members of RINOLYST and GreenSapien at SciGlyph Exploration Private Limited for their unwavering support in various capacities throughout this project. Your contributions have been invaluable and have played a significant role in shaping this book.

The invaluable support, sacrifices, and love from our family members have been instrumental in completing this book. Their unwavering assistance and timely help made it possible to bring this book to fruition. Without their constant support and encouragement, this endeavor would have been impossible to accomplish. We are deeply grateful for their contributions throughout this journey.

Finally, we express our gratitude to all the anticipated readers of this book from different corners of the world. Your interest and engagement are paramount to us. We eagerly await your feedback and insights, which inspire us to continue pursuing innovative and sustainable solutions for a better tomorrow, in the hope of creating a more sustainable world for generations to come.

Summary

Subsurface Hydrogen Energy Storage: Current Status, Prospects, and Challenges presents a comprehensive and authoritative exploration of the pivotal role of hydrogen storage technologies in the ongoing global shift toward sustainable energy systems. Authored by leading experts in the field, this book offers profound insights into the current landscape, future potential, and the hurdles that demand attention.

The journey unfolds with a profound analysis of hydrogen's significance as a versatile energy carrier, underlining the urgency of transitioning toward cleaner energy amid pressing global climate concerns. It meticulously dissects various hydrogen production methodologies, spanning from renewable sources to conventional fossil fuels integrated with carbon capture technologies, accentuating their relevance in shaping sustainable energy ecosystems.

Progressing forward, this book navigates through surface and subsurface storage methodologies, placing emphasis on the distinct advantages of subsurface storage within porous geological formations such as aquifers and depleted hydrocarbon reservoirs. It meticulously examines the intricate physics of hydrogen flow within porous media, unraveling the complexities of two-phase flow dynamics and the paramount influence of reservoir properties on storage efficacy.

Deeper insights are uncovered in subsequent chapters that delve into the physiochemical, geochemical, and mineralogical nuances pivotal for the successful implementation of underground hydrogen storage systems. Addressing microbial challenges within subsurface environments, this book articulates robust strategies for effective mitigation and operational resilience.

A substantial segment is dedicated to elucidating experimental methodologies, advanced modeling, and simulation approaches crucial for unraveling the intricacies of subsurface hydrogen storage dynamics. Encompassing fluid dynamics, geomechanics, geochemical reactions, and microbial interactions, these chapters serve as pillars supporting optimized storage operations.

Beyond the technical realm, the book scrutinizes the broader landscape of government policies, regulatory frameworks, international collaborations, and public perceptions that profoundly influence the trajectory of hydrogen storage technology adoption. Real-world case studies provide tangible insights into ongoing projects worldwide, encapsulating operational intricacies, technological innovations, and best practices shaping the industry.

Concluding with a holistic analysis of prospects and challenges, the book advocates strategic advancements in technology, stringent safety protocols, and comprehensive workforce training as catalysts propelling the global transition toward sustainable energy paradigms.

Subsurface Hydrogen Energy Storage: Current Status, Prospects, and Challenges stands as an indispensable beacon for researchers, engineers, policymakers, and industry stakeholders across diverse engineering domains. It not only offers a roadmap for advancing hydrogen storage technologies but also contributes significantly to the broader mission of fostering sustainable energy ecosystems on a global scale.

CHAPTER

1

Introduction to underground hydrogen storage

Achinta Bera[1] *and Sunil Kumar*[2]

[1]Department of Petroleum Engineering, School of Energy Technology, Pandit Deendayal Energy University, Gandhinagar, Gujarat, India [2]RINOLYST, SciGlyph Exploration Private Limited, Patna, Bihar, India

1.1 Importance of hydrogen as an energy carrier

Hydrogen stands out as a crucial energy carrier due to its remarkable properties and potential contributions to addressing environmental challenges toward sustainable energy transition. With an energy content of about 142 MJ/kg, hydrogen possesses a significantly higher energy density than traditional fossil fuels such as gasoline (around 46 MJ/kg) and even natural gas (approximately 53 MJ/kg). This high energy density translates hydrogen to a greater amount of energy per unit mass, making it an attractive option for various energy-intensive applications (Acar & Dincer, 2018; Muhammed et al., 2023). Furthermore, hydrogen's role in fuel cells showcases its impressive efficiency. Fuel cells can convert hydrogen into electricity with an efficiency of up to 60%, significantly higher than internal combustion engines found in conventional vehicles, which typically have efficiencies around 20%–30%. This superior efficiency not only reduces energy waste but also leads to lower overall emissions during energy conversion processes. Hydrogen's clean combustion properties are also noteworthy. When hydrogen is used as a fuel, whether in fuel cells or combustion engines, the only byproduct produced is water vapor (Andersson & Grönkvist, 2019; Ishimoto et al., 2020). This contrasts sharply with fossil fuels, which emit pollutants such as carbon dioxide (CO_2), nitrogen oxides (NOx), and particulate matter (PM) during combustion, contributing to air pollution and climate change toward global warming. Moreover, hydrogen offers versatility in storage and transport, crucial for integrating renewable energy sources like wind and solar into the grid effectively (Andersson & Grönkvist, 2019; Ishimoto et al., 2020). Hydrogen can be stored for long periods without significant losses, allowing for seasonal energy storage and buffering intermittent renewable energy generation. Its ability to be transported over long distances, whether through

Subsurface Hydrogen Energy Storage
DOI: https://doi.org/10.1016/B978-0-443-24071-3.00002-9

pipelines or in liquid form, makes it a viable option for global energy distribution networks. However, the scientific data and results highlight hydrogen's immense potential as an energy carrier, offering high energy density, superior efficiency in energy conversion, clean combustion with zero greenhouse gas emissions, and versatile storage and transport capabilities. These attributes the position of hydrogen as a key player in sustainable energy transitions and the shift toward a low-carbon energy future.

Hydrogen emerges as a compelling choice for energy storage due to its impressive specific energy capacity of 142 MJ/kg and its emission of environment-friendly combustion byproducts (Acar & Dincer, 2018; Muhammed et al., 2023). Currently, the main production methods of hydrogen involve natural gas steam methane reforming (NGSMR) and water electrolysis utilizing fossil fuel feedstock (blue and gray hydrogen) and renewable sources based on electrolyzer capacity (green hydrogen) with energy efficiencies 65%−85% and 55%−75%, respectively (Buttler & Spliethoff, 2018; Lee et al., 2019; Ozturk & Dincer, 2021). Some NGSMR approaches for blue hydrogen production are considered as clean energy sources since the resulting CO_2 can be sequestered and other NGSMR processes leading to gray hydrogen production are deemed environmentally harmful due to formation of impurities like CO, CO_2, CH_4, Ar, and N_2 (Muhammed et al., 2022). As an energy carrier, hydrogen possesses a low density ($\sim$0.089 kg/m^3) under standard conditions; necessitates large-scale volumes of hydrogen beyond the storage capacity of surface-based facilities. The capability to effectively store energy on the scale of GWh to TWh (Hashemi et al., 2021; Le Gallo, 2024) enables the balancing of consumer demand and during peak seasons. Fig. 1.1 illustrates the various applications of hydrogen as an energy carrier across different sectors and production of hydrogen from various sources.

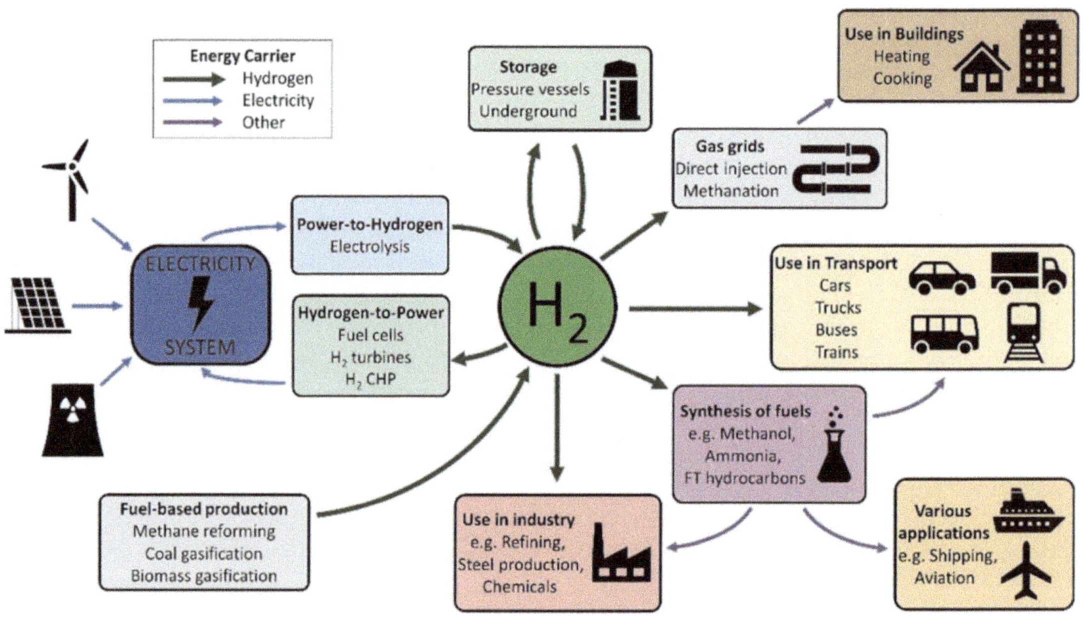

FIGURE 1.1 Utilization of hydrogen as an energy carrier in different industries (Quarton et al., 2020).

In recent times, there has been a growing shift away from fossil fuels toward sustainable, renewable energy sources. While hydro, wind, and solar power plants offer viable alternatives, their reliance on environmental conditions has historically threatened their reliability, presenting a hurdle in transitioning from fossil fuels to renewable energy sources (Bilgen et al., 2004; Ebhota & Jen, 2020; Herzog et al., 2001; Rahman et al., 2022). One emerging solution is hydrogen, which could help combat the widespread use of conventional and environmentally harmful energy sources. However, the advancement of hydrogen production is hindered by inadequate storage capacity. The implementation of underground hydrogen storage is thus crucial for ensuring a consistent energy supply and stabilizing the electrical grid (Gür, 2018; Matos et al., 2019; Preuster et al., 2017; Rasul et al., 2022). Consequently, challenges are encountered during transitioning energy harnessing from fossil fuels to renewable sources. To mitigate the energy supply-demand problems, extensive research works are urgently needed to identify and explore economically feasible hydrogen energy storage methods. From a large-scale production perspective, hydrogen meets the challenging criteria for replacing fossil fuels as the primary energy carrier (Andersson & Grönkvist, 2019; Ishimoto et al., 2020). However, despite being a promising clean energy carrier for the future, hydrogen faces limitations due to insufficient infrastructural advancements, hindering its global commercial adoption (Dunn, 2002; Sazali, 2020). The characteristics of hydrogen in the context of underground storage raise numerous uncertainties that require careful exploration. Although hydrogen is the most abundant element in the universe, constituting less than 0.14% of Earth naturally (Keçebaş & Kayfeci, 2019; Santhanam et al., 2017), its highly reactive nature necessitates binding to other elements to exist as a stable compound. Nonetheless, a significant challenge for the global advancement of hydrogen production is the lack of storage capacity for potential seasonal uses (Olabi et al., 2021; Osman et al., 2022).

Hydrogen is currently stored using a variety of methods, including high-pressure gas cylinders, cryogenic tanks, and adsorption onto materials with high specific surface areas and interstitial sites within a host metal. Additionally, it can be chemically stored, bonded in covalent or ionic compounds, or through the oxidation of reactive metals with water (Anson et al., 2004; Germain et al., 2006; Hirscher & Panella, 2005). However, the challenge of storing substantial volumes of hydrogen as an additional energy source for the long term is significant. The primary solution for storing such huge quantities of hydrogen lies within subsurface geological formations, including depleted hydrocarbon reservoirs, saline aquifers, or salt or rock mine caverns (Epelle et al., 2022; Małachowska et al., 2022; Tarkowski, 2019). Hydrogen, with its remarkable ability to permeate reservoir rocks, presents a fascinating aspect of geology and energy storage. Its interaction with the rock formations can lead to notable transformations in the pore structure, thus impacting various aspects of reservoir dynamics. Through processes such as mineral dissolution and precipitation, hydrogen's reactivity contributes to changes in the geological composition, ultimately influencing factors like injectivity and storage capacity within the reservoirs. This intricate interplay between hydrogen and the geological environment underscores the complexity of energy storage systems and the importance of understanding subsurface processes for effective resource management. Current research suggests that hydrogen loss during storage is minimal, with less than 2% attributed to diffusion, and the most substantial losses arising from residual trapping during the cycling process, geochemical reactions, and caprock leakage (Ershadnia et al., 2023; Sambo et al., 2022). Thus similar to storing other gases such as methane and carbon dioxide, a fundamental aspect of developing an underground storage repository for hydrogen is gaining a comprehensive understanding of fluid migration through the reservoir.

During the hydrogen storage process, mineralogical changes occur within reservoir rocks, altering their pore structure and flow characteristics (Ershadnia et al., 2023; Sambo et al., 2022). These modifications depend on the in situ conditions of the reservoirs, such as pressure and temperature, as well as the initial pore and mineralogical characteristics of the rock (Lankof & Tarkowski, 2020; Zivar et al., 2021). Thus a comprehensive dataset detailing the changes in properties of various depleted gas reservoir rocks upon interaction with hydrogen is crucial. However, there exists a significant knowledge gap, particularly concerning alterations in the mineralogical, chemical, and petrophysical properties of reservoir rocks upon hydrogen interaction. Additionally, reservoir microbes catalyze new reactions, converting hydrogen into other gases (e.g., methane and hydrogen sulfide), which can impede the production of pure hydrogen (Ali et al., 2023; Dopffel et al., 2021). These interactions have significant implications for hydrogen injection, storage capacity, and safety. Although rock–fluid interactions during hydrogen storage have received minimal attention, similar interactions have been extensively studied in other underground reservoir applications like oil and gas reserve characterization.

1.2 Understanding hydrogen: properties and characteristics

Hydrogen, the simplest chemical element, is a colorless, odorless, tasteless, and flammable gas. Its atom comprises a proton with a positive charge in the nucleus and an electron with a negative charge orbiting it (Keçebaş & Kayfeci, 2019). Under typical conditions, hydrogen exists as diatomic molecules, each consisting of two atoms, denoted as H_2. Its notable chemical property is its ability to combine with oxygen to form water (H_2O), hence its name derived from Greek, meaning "maker of water" (Kommula, 2021). It exists abundantly in water bodies, ice caps, rivers, lakes, and the atmosphere. It is also a component of numerous carbon compounds found in both living organisms and petroleum. Considering its prevalence in various compounds, hydrogen compounds might surpass carbon compounds in number (Idriss et al., 2015; Keçebaş & Kayfeci, 2019). Hydrogen has three known isotopes, with mass numbers 1, 2, and 3, respectively. The most common isotope, known as protium or hydrogen-1 (symbol H or ^{1}H), has a single proton in its nucleus. Deuterium, or heavy hydrogen (symbol D or ^{2}H), constitutes a minute portion of hydrogen, with a nucleus containing one proton and one neutron. Tritium (symbol T or ^{3}H), the mass 3 isotope, contains one proton and two neutrons and makes up an extremely small fraction of hydrogen (Idriss et al., 2015; Keçebaş & Kayfeci, 2019). The distinct naming of these isotopes reflects their differing properties.

The remarkably low melting and boiling points of hydrogen arise from the feeble forces of attraction between its molecules. Evidence of these weak intermolecular forces is observed when hydrogen gas expands from high to low pressure at room temperature, causing its temperature to increase, contrary to most other gases whose temperature decreases under such conditions (Idriss et al., 2015; Keçebaş & Kayfeci, 2019). This phenomenon, in line with thermodynamic principles, suggests that repulsive forces surpass attractive forces among hydrogen molecules at room temperature; otherwise, the expansion would lead to cooling. However, at temperature below −68.6°C, attractive forces become dominant, resulting in the cooling of hydrogen upon expansion (Zgonnik, 2020). This cooling effect becomes particularly pronounced at temperatures below that of liquid nitrogen (−196°C), a phenomenon exploited to achieve the liquefaction of hydrogen gas itself. The physical and chemical properties of hydrogen are very important to understand for its underground storage. Table 1.1 illustrates the physical and chemical properties of hydrogen, highlighting its lowest density among gases, as well as its colorless and odorless nature. Often addressed as the clean fuel of the future, hydrogen is produced from water and reverts to water upon oxidation. It is ubiquitous in living organisms and water molecules, forming bonds with carbon and oxygen atoms. Notably, hydrogen is the most abundant element in the universe (Idriss et al., 2015; Keçebaş & Kayfeci, 2019). While concerning the large-scale storage of hydrogen for energy transition toward green aspect, underground storage is one of the main options to mitigate the problem of storing huge amount of hydrogen. Fig. 1.2 depicts various options for storing 100 TWh of hydrogen, revealing that depleted hydrocarbon reservoirs offer the highest storage capacity, while long pipelines and gasometer tanks require significant infrastructure for the same volume. Salt caverns and aquifers also necessitate multiple sites for storing equivalent amounts of hydrogen.

TABLE 1.1 Physical and chemical properties of hydrogen.

Physical properties	Chemical properties
Chemical symbol: H	Electron configuration: $1s^1$
Atomic number: 1	Oxidation states: +1, −1
Atomic mass: 1.008 g/mol	Electronegativity: 2.20 (Pauling scale)
Phase at room temp.: Gas	Ionization energy: 1312 kJ/mol
Density at standard temperature and pressure (STP): 0.08988 g/L	First ionization energy: 1312 kJ/mol
Melting point: 14.01 K (−259.14°C or −434.45°F)	Heat of vaporization: 0.904 kJ/mol
Boiling point: 20.28 K (−252.87°C or −423.17°F)	Heat of fusion: 0.117 kJ/mol
Color: Colorless	Standard heat of formation: 0 kJ/mol
Odor: Odorless	Combustion reaction: $2H_2(g) + O_2(g) \rightarrow 2H_2O(g)$
Taste: Tasteless	Reactivity: Reacts with halogens to form hydrogen halides (e.g., HCl)
Solubility: Insoluble in water and most organic solvents	Reacts with metals to form metal hydrides (e.g., LiH)
State: Diatomic	Electron configuration: $1s^1$
Flame color: Pale blue	Oxidation states: +1, −1

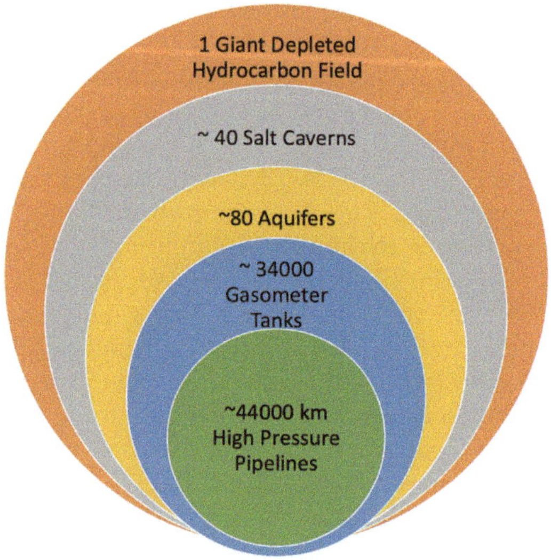

FIGURE 1.2 Approximate volumes of different media require storing 100 TWh of hydrogen energy.

1.3 Hydrogen as an energy carrier

Hydrogen emerges as a versatile energy source with the potential to significantly reduce greenhouse gas emissions. Produced from various domestic/green resources, it holds the promise of nearly zero emissions. When utilized, hydrogen yields electrical power through fuel cells, emitting only water vapor and warm air. Its applications span both stationery and transportation energy sectors, offering avenues for growth (Mazloomi & Gomes, 2012; Züttel et al., 2010).

Functioning as a potent and transportable energy carrier, hydrogen facilitates electricity production, industrial operations, and transportation. Unlike fossil fuels, its combustion results solely in water, mitigating harmful emissions. Hence, it stands as an appealing fuel option for the future (Mazloomi & Gomes, 2012; Züttel et al., 2010). As an energy carrier, hydrogen must be produced from other substances. It can be derived or separated from diverse sources such as water, fossil fuels, or biomass, serving as a versatile energy or fuel source (Mazloomi & Gomes, 2012; Züttel et al., 2010). Presently, hydrogen is utilized in various industrial processes, serves as rocket fuel, and powers fuel cells for electricity generation and vehicle propulsion. Some operators of natural gas-fired power plants are exploring hydrogen as a supplement or substitute for natural gas (Mills, 2018; Olaniyi et al., 2023). The hydrogen market, valued at approximately $183 billion in 2022, is anticipated to witness a compound annual growth rate of around 5.5% from 2023 to 2032, driven by sustainable and energy-efficient technologies. The ongoing energy transition and aggressive initiatives aimed at promoting energy-efficient fuels present lucrative prospects for hydrogen adoption as an alternative to conventional fuels. Both developed and developing nations worldwide are intensifying efforts to promote clean energy solutions and accelerate climate control initiatives. However, as depicted in Fig. 1.3, the global hydrogen market, segmented region-wise, indicates consumption patterns as of 2019. Green hydrogen emerges as a viable substitute for traditional fossil fuels in transportation,

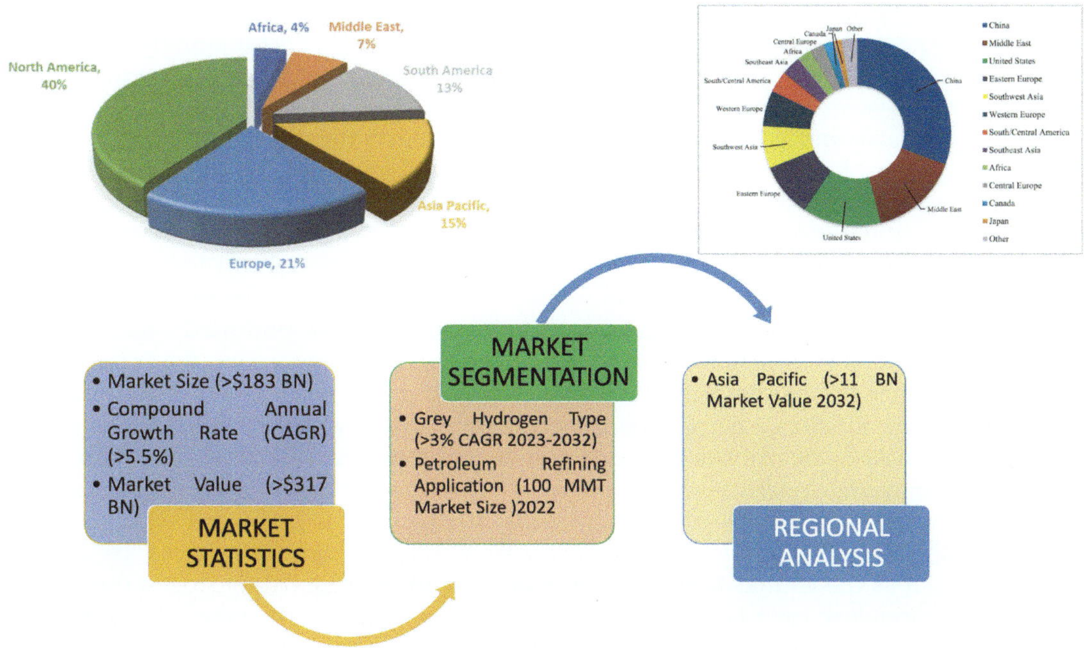

FIGURE 1.3 Global hydrogen market and consumption as of 2019.

which substantially contributes to greenhouse gas emissions. It also holds promise in industrial applications, including ammonia, methanol, and steel production, sectors heavily reliant on fossil fuels (Liu et al., 2022; Reddy et al., 2023). The key advantages of hydrogen utilization include (1) emission-free energy usage, with water being the sole by-product; (2) capability to produce other gases and liquid fuels; and (3) adaptability of existing infrastructure, such as gas transport and storage, for hydrogen application.

Hydrogen, especially when derived from clean electricity sources, is widely regarded as a pivotal facilitator in the global shift toward sustainable energy and an essential element in achieving economies with net-zero emissions. Despite its current limited usage primarily in certain industrial sectors and its comparatively higher cost compared to fossil fuel alternatives, the global hydrogen market is gaining momentum for valid reasons. Hydrogen serves as a significant energy carrier, facilitating the transfer of energy between different locations via different transportation modes (Sigfusson, 2007). Its potential extends to driving the decarbonization of entire industrial sectors and combating climate change, whether as an e-fuel in transportation or as a means of emission reduction in steel production and various other industries (Hosseini & Wahid, 2020; Sigfusson, 2007). Moreover, hydrogen can address energy accessibility challenges by enabling energy distribution to regions typically reliant on imports. Competitive energy production in regions rich in renewables, such as wind, can be converted into hydrogen or other molecules and transported to importing regions, thereby broadening energy access. The utilization of hydrogen as stored energy holds the potential to enhance the stability of increasingly volatile power grids. Even during periods of low wind or sunlight, reliable sources of electricity are crucial to prevent blackouts. This exemplifies how collaborative efforts across borders can translate grand ideas into tangible actions. Encouragingly, numerous countries are actively supporting these initiatives and have set ambitious goals in the field of hydrogen (Hosseini & Wahid, 2020; Sigfusson, 2007). The focus now shifts toward scaling up the industry and economically producing hydrogen from renewables, particularly "green" hydrogen. Encouragingly, the costs of renewable electricity required for green hydrogen production are declining, alongside prices for the electrolyzers essential for splitting water into hydrogen and oxygen.

The shift toward a hydrogen-integrated energy system necessitates a holistic approach encompassing various facets of infrastructure, technology, policy, regulation, behavior, and investment. This comprehensive transformation signifies a profound "technological transition" from one sociotechnical paradigm to another. Understanding the intricacies of sociotechnical change becomes pivotal in laying the groundwork for a successful hydrogen energy transition. Efficiency across energy, economic, and environmental domains underscore the significance of prioritizing sectors like industry and transportation within this transition. Enhancing the efficiency of low-carbon hydrogen production emerges as a key strategy, showcasing a country's commitment to advancing the hydrogen energy agenda through sector coupling. Drawing inspiration from the strategies employed by leading economies such as the United Kingdom and Germany in their global hydrogen transitions, a proposed framework for a country's hydrogen energy transition is delineated in Fig. 1.4, offering valuable insights for effective implementation and adaptation.

Creating a hydrogen-integrated energy system requires comprehensive and integrated changes across infrastructure, technology, policies, regulations, behaviors, habits, and investment patterns. These broad transformations represent a "technological transition" from one sociotechnical system to another. Understanding the dynamics of sociotechnical change is crucial for establishing a framework for the hydrogen energy transition. Efficiency in energy, economics, and the environment underscores the importance of prioritizing the industry and transportation sectors in a country's hydrogen energy transition.

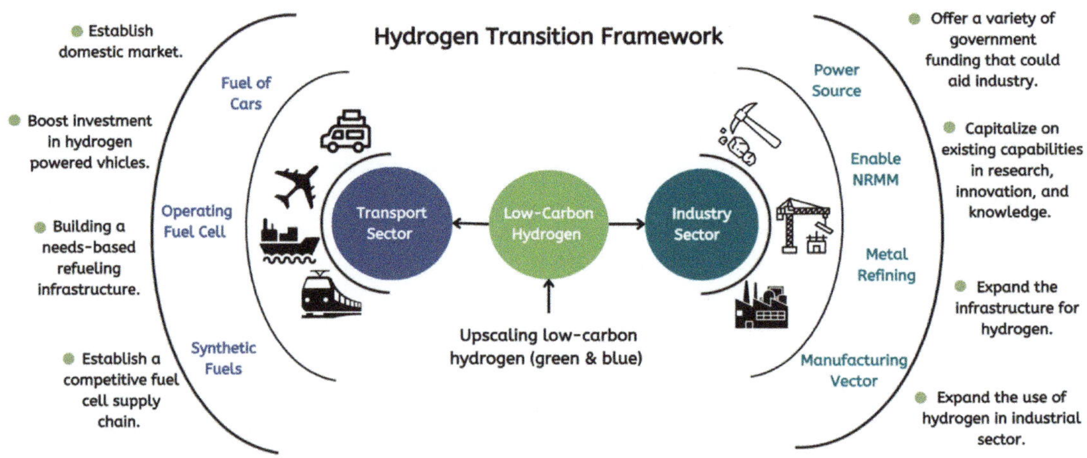

FIGURE 1.4 Hydrogen energy transition framework for the transportation and industry sectors (Ibrahim et al., 2022).

By improving the efficiency of low-carbon hydrogen production, a country can showcase its capability in achieving the hydrogen energy transition through sector coupling. Drawing insights from the global hydrogen transition strategies of leading economies such as the United Kingdom and Germany, a proposed framework for a country's hydrogen energy transition is illustrated in Fig. 1.4. Creating a hydrogen-integrated energy system requires comprehensive and integrated changes across infrastructure, technology, policies, regulations, behaviors, habits, and investment patterns. These broad transformations represent a "technological transition" from one sociotechnical system to another. Understanding the dynamics of sociotechnical change is crucial for establishing a framework for the hydrogen energy transition. Efficiency in energy, economics, and the environment underscores the importance of prioritizing the industry and transportation sectors in a country's hydrogen energy transition. By improving the efficiency of low-carbon hydrogen production, a country can showcase its capability in achieving the hydrogen energy transition through sector coupling. Drawing insights from the global hydrogen transition strategies of leading economies such as the United Kingdom and Germany, a proposed framework for a country's hydrogen energy transition is illustrated in Fig. 1.4.

1.4 Energy storage: need and challenges

Energy storage stands as a central issue in the transition toward a sustainable and decarbonized economy. Among the numerous challenges confronting renewable energy production—such as wind, solar, and tidal power, one critical hurdle is ensuring the availability of electricity generated from these intermittent sources when needed, mirroring the current reliability of fossil fuel-derived energy. This challenge manifests in various forms. Renewable energy production is susceptible to weather and seasonal variations, leading to a mismatch between supply and end-user demand. Additionally, limitations within national electricity grids can pose obstacles, particularly when renewable energy generation sites are far away from major demand centers (Hosseini & Abdul Wahid, 2014; Marks-Bielska et al., 2020). Energy storage emerges as a solution to address these challenges, sparking increased interest among investors, developers, and businesses seeking to integrate storage solutions into their portfolios or operations. Despite the ongoing momentum in the energy transition, energy storage hurdles persist, with technology advancements lagging in this domain. While energy generation from renewable sources is on the rise, energy storage capabilities are not keeping pace (Hosseini & Abdul Wahid, 2014; Marks-Bielska et al., 2020). Yet, what is the value of cheaply generating energy if it lacks the means to store for peak demand usage? In the 21st century, there are challenges concerning batteries and storage that demand attention in the years ahead for an effective energy transition as swiftly as possible (Hosseini & Abdul Wahid, 2014; Marks-Bielska et al., 2020). However, the following concerns should be taken care while large-scale energy storage concept will be implemented.

1. High cost of implementation: The high cost of implementation remains a significant barrier. Despite declining costs in recent years, batteries still demand substantial investment from many companies. Hence, integrating thermal energy storage with

efficient district cooling and heating systems becomes crucial to mitigate initial costs and establish an effective energy system for any facility.

2. Lack of standardization in storage systems: The lack of standardization in storage systems poses challenges. Without effective standardization, each manufacturer develops its own batteries, leading to compatibility issues and potential need for replacements as projects evolve. Ensuring scalability and ease of expansion is essential, not merely focusing on initial efficiency.
3. Outdated regulatory policy and market design: Outdated regulatory policies and market designs hinder the progress. Like with any emerging technology, regulation often lags behind until its necessity is well-established. Energy storage confronts this challenge, with regulation primarily tailored to the industrial level. Clear regulation for residential battery use may not materialize until batteries are fully integrated into the residential market.

1.5 Introduction to underground hydrogen storage

In the quest for a sustainable and decarbonized energy future, the exploration of alternative energy storage solutions is paramount. Among the emerging technologies, the understanding of geological hydrogen storage for seasonal uses stands out as a promising avenue. As the world continues to grapple with climate change and the urgent need to reduce greenhouse gas emissions, harnessing hydrogen's potential as a clean and abundant energy carrier becomes increasingly critical. Geological hydrogen storage offers a ground-breaking approach to address the intermittent nature of renewable energy sources and the challenges associated with seasonal energy demands (Lankof & Tarkowski, 2020; Zivar et al., 2021). By utilizing subsurface formations such as depleted oil and gas reservoirs, salt caverns, or aquifers, hydrogen can be safely stored during periods of high renewable energy generation and subsequently released during periods of high energy demand. This unique storage solution provides a reliable and flexible means of meeting energy needs while significantly reducing reliance on carbon-intensive power generation. The prologue sets the stage for an in-depth exploration of the understanding of geological hydrogen storage for seasonal uses toward decarbonization. Through an analysis of various storage technologies, including their design, capacity, efficiency, and safety measures, a clearer understanding of the potential and limitations of geological hydrogen storage emerges. Moreover, it is important to examine the integration of renewable energy sources, such as solar and wind power, with hydrogen storage to create a synergistic and sustainable energy system. Environmental impacts and considerations hold a central place within this study, as the work investigates the carbon footprint, land use, and potential risks associated with geological hydrogen storage (Lankof & Tarkowski, 2020; Zivar et al., 2021). Additionally, the economic viability and market potential of this technology are thoroughly explored, examining the costs, potential revenue streams, and investment opportunities to ensure a sustainable and commercially viable transition to hydrogen-based energy systems. By providing a comprehensive analysis of the understanding of geological hydrogen storage for seasonal uses toward decarbonization, this work aims to inform policymakers, researchers, and industry stakeholders. It presents insights, challenges, and recommendations for future research, policy reforms, and strategic decision-making to foster the adoption of this transformative technology and expedite the transition to a low-carbon energy future. The journey of understanding geological hydrogen storage for seasonal uses toward decarbonization starts here, as we unlock the potential of hydrogen to reshape our energy landscape and forge a path toward a more sustainable, resilient, and carbon-neutral world. By delving into the understanding of geological hydrogen storage for seasonal uses toward decarbonization, this report aims to contribute to the knowledge base and pave the way for informed decision-making, policy formulation, and the adoption of sustainable energy solutions on the path to a greener future.

The imperative transition to a sustainable, low-carbon energy future is becoming increasingly critical in light of climate change and environmental concerns. As the world seeks alternative sources of clean energy, the role of hydrogen has gained significant attention. Hydrogen, as a versatile and zero-emission fuel, holds immense potential to decarbonize various sectors, including transportation, industry, and power generation. However, the intermittent nature of renewable energy sources poses a challenge to its widespread adoption. Therefore the significance of understanding geological hydrogen storage for seasonal uses toward decarbonization is crucial for implementation (Lankof & Tarkowski, 2020; Zivar et al., 2021). Geological hydrogen storage offers a promising solution to overcome the intermittency of renewable energy and ensure a reliable supply of clean energy throughout the year. By utilizing existing underground formations, such as depleted oil and gas reservoirs, salt caverns, or aquifers, hydrogen can be stored during periods of high renewable energy production and released during times of high demand. This approach not only facilitates the integration of intermittent renewable but also provides a means to balance energy supply and demand on a seasonal scale.

1.6 Geological considerations for underground storage

Different geological formations can be used for subsurface hydrogen storage and their potentials for hydrogen storage are not equal. Salt caverns, depleted hydrocarbon reservoirs, aquifers, abandoned mine shafts, and engineered hard rock caverns can be used for large-scale hydrogen storage. Fig. 1.5 shows the options for subsurface hydrogen storage. Based on the size and storage capacity and requirement for uses selecting the suitable option for subsurface hydrogen storage is one of the important aspects for energy transition. In this regards, Fig. 1.6 depicts the underground storage in different scenarios with respect to usability and storage capacity. It is clear from Fig. 1.6 that depleted hydrocarbon reservoirs and saline aquifers are the best options for subsurface hydrogen storage in large-scale as well as for long-term use. However, each storage option comes with its own set of pros and cons, making the selection process complex. The screening of storage options involves intricate evaluations before implementation in real-world scenarios.

1.6.1 Salt caverns

Salt caverns, artificial cavities carved into geological salt deposits, typically reach depths of 500–1500 m. Constructing such caverns involves two main steps: excavating the salt

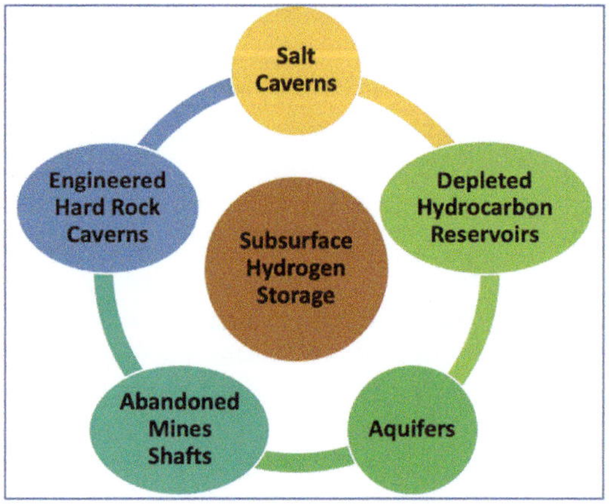

FIGURE 1.5 Different subsurface hydrogen storage options for large-scale.

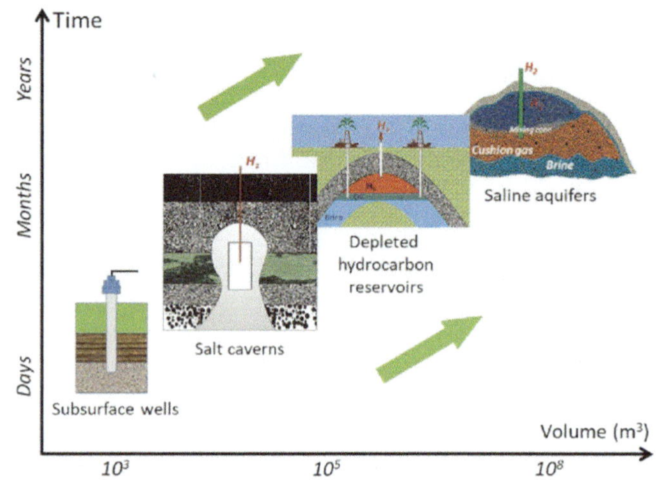

FIGURE 1.6 Comparison of underground storage in different scenarios with respect to usability and storage capacity (Chen et al., 2023).

and subsequently injecting water to dissolve it. The resulting brine, a mixture of water and salt, is then extracted, creating ample space for a sizable, dense cavern suitable for pressurized hydrogen storage. Pertinent factors for hydrogen storage in salt caverns include the sealing properties of evaporite minerals (such as anhydrous gypsum, gypsum, and rock salt), the mechanical stability of the cavern induced by injection, and the ability to redistribute stress through viscoplastic deformation (Epelle et al., 2022; Małachowska et al., 2022; Tarkowski, 2019). Salt caverns have been successfully employed for storing hydrocarbons, including natural gas, liquid hydrocarbons like crude oil or refined products, and liquefied petroleum gas, with over 1900 such caverns in operation globally. The creation of salt caverns involves solution mining, wherein fresh water or seawater is injected into a deep well. Typically, 8–10 volumes of water are injected to create one volume of cavern (Epelle et al., 2022). This process, also known as leaching, can be conducted through direct or reverse leaching methods. Direct leaching involves injecting fresh water through central tubing and withdrawing saturated brine through the annular space, while reverse leaching reverses this flow pattern (Epelle et al., 2022). The shape of the cavern is controlled by tubing placement, water flow rate adjustment, and the selection of leaching method (Małachowska et al., 2022). Cavern dimensions are tailored to the owner's requirements and salt layer characteristics, with common practices including cavern heights exceeding 300 m and diameters ranging from 50 to 100 m. Sizes vary widely, from 50,000 m^3 to over 1,000,000 m^3, and constructing a 500,000 m^3 cavern typically takes 6 months to 2 years. Operating pressure depends on cavern depth, with depths exceeding 1000 m allowing for pressures exceeding 200 bar. Maintaining a minimum pressure is essential to prevent salt damage and cavern closure due to excessive viscoplastic behavior, known as "salt creep." A portion of the gas inventory, known as cushion gas, remains at minimum pressure, constituting approximately one-third of the total gas.

Salt caverns have been utilized for hydrogen storage since the 1970s, with the first hydrogen cavern commissioned near Teesside, United Kingdom, in 1972 with three constructed caverns for a volume of $\sim 2\ Mm^3$ at 400 m depth. Since then, additional six caverns have been constructed in the United States (Moss Bluff, Clemens Dome, Spindle top) and the United Kingdom (Teesside), primarily serving the chemical or petrochemical industry. However, recent projects aim to utilize hydrogen as an energy carrier, reflecting the growing interest in hydrogen for its high energy density. Technical feasibility studies and laboratory tests demonstrate the suitability of salt caverns for hydrogen storage, with salt exhibiting impermeability to hydrogen. Aboveground and well equipment has proven adequate for hydrogen storage purposes (Schlichtenmeyer & Klafki, 2016).

1.6.2 Storage perspective of salt cavern

Salt cavern storage is excellent for short-term to medium-term variations in energy demand since it allows for many injection-reproduction cycles per year and extremely fast production rates. These caverns are, however, spatially limited owing to the scarcity of appropriate evaporitic deposits with storage capacities ranging from a few tens of thousands to one million cubic meters and energy contents ranging from several hundred giga tonnes per hour. A recovery method suitable for medium-term and short-term storage is being implemented in countries such as Denmark, Turkey, Canada, Romania, Germany, and Poland where the project is under progress (Sambo et al., 2022). Fig. 1.7 illustrates the major worldwide locations of salt formations.

1.6.3 Depleted hydrocarbon reservoirs

Porous rocks, whether in depleted fields or aquifers, have served as natural gas storage sites for over a century. The first underground storage for natural gas was established in the United States in 1915 (Londe, 2021). Unlike salt caverns, porous rock storages utilize existing voids within the geological structure. Porosity is commonly found in sedimentary rocks like sandstones or fractured carbonates, where gas is trapped due to the structure's low density. Typically, the structure comprises an anticline covered by an impermeable caprock, such as clay or claystone. The stored gas volume is carefully managed to ensure it never reaches the "spill point" of the structure. Wells are drilled for gas injection and

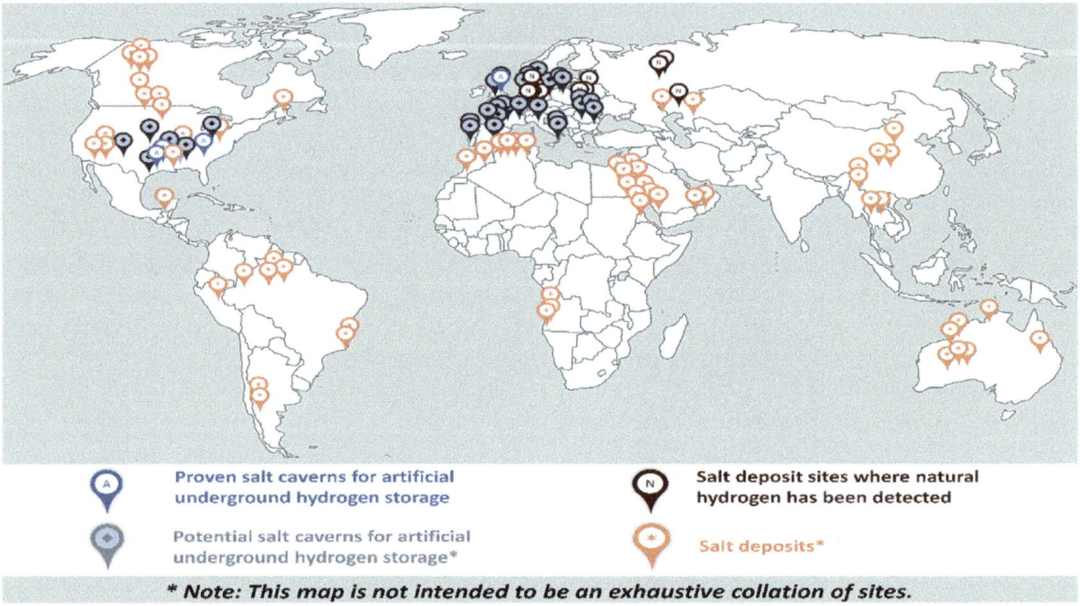

FIGURE 1.7 Occurrence of salt formations worldwide (Epelle et al., 2022). Map lines delineate study areas and do not necessarily depict accepted national boundaries.

withdrawal, with additional wells positioned above and around the structure to monitor system integrity. Operating pressures, often reaching up to 200 bar, are determined based on theoretical calculations and historical data. Similar to salt caverns, a portion of the injected gas remains in the storage as "cushion gas," comprising CO_2, N_2, and CH_4 (Kanaani et al., 2022; Zamehrian & Sedaee, 2022).

Storing hydrogen in porous rock typically involves blending hydrogen with natural gas (primarily methane). This hydrogen–methane mix can be utilized directly for energy production or separated at the wellhead using technologies like pressure swing adsorption. Various criteria guide the selection of depleted fields or aquifers for hydrogen storage. Recommended depths range from 500 to 2500 m, with trap structures ideally featuring a minimum height of 20 m. Effective porosity should exceed 5% for carbonates and 10% for sandstones, while hydraulic conductivity should surpass 10 mD for carbonates and 50 mD for sandstones. Fluid analysis, including salinity, pH, ion composition, and bacteria content, is crucial for predicting microbiological reactions. CO_2, sulfurous, or iron-rich fluids should be avoided. In aquifers, the geological requirements entail good reservoir properties and porous, permeable formations conducive to gas storage (Okoroafor et al., 2022; Perera, 2023). However, aquifer storage tends to be more expensive due to limited geological knowledge compared to depleted oil and gas reservoirs, as observed in countries like Spain, Canada, and Poland, where projects are underway.

Depleted oil and gas reservoirs require 50%–60% cushion gas, compared to 80% for aquifers. However, the remaining gas in depleted reservoirs can affect the purity of injected hydrogen, potentially compromising gas integrity during withdrawal (Okoroafor et al., 2022;

Perera, 2023). Both aquifer and depleted hydrocarbon storage options are ideal for seasonal gas storage due to their cycling flexibility, with ongoing projects in countries like the United Kingdom and Poland (Okoroafor et al., 2022; Perera, 2023). Each site requires optimization for well placement, operational design, and cost considerations to ensure proper ranking.

1.6.4 Storage perspectives of depleted hydrocarbon reservoir

For longer-term storage (e.g., months to year), such as supplying energy to heat homes during the colder months, a large-scale capacity is needed, which only the porous saline aquifers and depleted hydrocarbon fields can provide (Kanaani & Sedaee, 2022; Salmachi et al., 2024). Subsurface porous material, on the other hand, might be a potential site for hydrogen storage. Because of their high porosity, appropriate permeability, well-formed traps, and massive cavernous volumes, depleted oilfields have enormous potential for large-scale hydrogen storage (Okoroafor et al., 2022; Perera, 2023). They offer hydrogen storage capacities, which are several times larger than those that can be held in salt caverns. Furthermore, their inclusion can also provide a more diverse (independent of location) and viable solution for large-scale subsurface hydrogen storage. Fig. 1.8 shows the power duration and output for different storage systems.

1.6.5 Salt caverns and depleted hydrocarbon reservoirs storage capacity

The capacity of a storage site is a crucial consideration in selecting the appropriate storage medium for hydrogen. Salt caverns offer limited capacity due to their smaller sizes, whereas depleted hydrocarbon reservoirs and aquifers boast practically unlimited capacity (Sambo et al., 2022). Moreover, aquifers and hydrocarbon reservoirs are deemed more cost-effective over the long term compared to salt caverns and are more widespread globally. Another significant factor is hydrogen loss during storage. Although experimental data on underground hydrogen storage are limited, studies suggest that depleted hydrocarbon reservoirs and aquifers experience considerable losses due to diffusion and dispersion compared to salt caverns (Sambo et al., 2022). Additionally, maintaining pressure

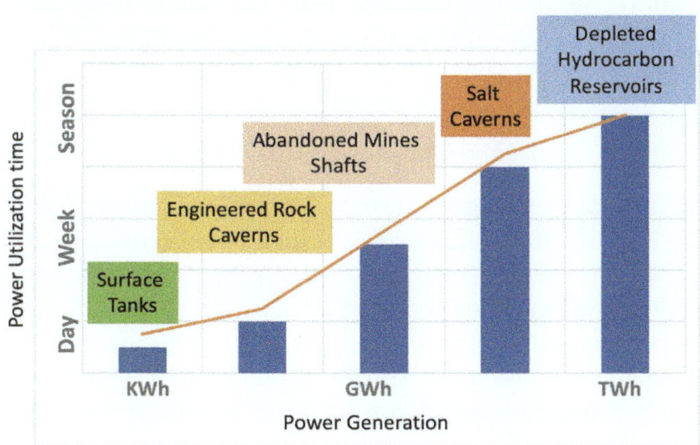

FIGURE 1.8 The scalability of hydrogen storage in various confining media of subsurface storage of hydrogen provides the maximum leverage.

requires a baseline of unrecoverable cushion gas, which is higher for depleted hydrocarbon reservoirs and aquifers than for salt caverns under typical conditions (Małachowska et al., 2022; Sambo et al., 2022). This underscores the importance of optimized production and injection strategies to mitigate losses and ensure consistent hydrogen delivery rates, particularly challenging in depleted hydrocarbon reservoirs. These factors indicate that while salt caverns offer advantages for short-term underground hydrogen storage, depleted hydrocarbon reservoirs and aquifers are better suited for the mid to long term (Okoroafor et al., 2022; Perera, 2023). However, Fig. 1.2 illustrates the volumes of different media required for storing 10 TWh of hydrogen energy, providing a visual comparison of storage requirements across different mediums.

1.6.6 Aquifers

The combination of porous rock formations, sealed by impermeable mudstones or evaporites, provides an ideal subsurface environment for trapping and storing gases like methane, natural gas, and CO_2, alongside formation brine or water. These formations, which have securely contained gases for millennia, offer promising opportunities for large-scale interseasonal hydrogen storage in the TWh range, encompassing both saline aquifers and gas fields. Noteworthy experience in hydrogen storage within saline aquifers was gained during the aquifer storage of town gas from the 1950s to the 1970s. Town gas, derived from coal gasification, typically comprised around 50%–60% hydrogen, 30% methane, and 20% carbon dioxide and carbon monoxide. Aquifer storage of town gas was undertaken in various locations, including France (Beynes), Czech Republic (Lobodice), and Germany (Engelborstel, Bad Lauchstaedt). Despite decades of commercial operation, no containment failures were reported at these town gas storage sites, although there may have been some gas composition changes due to bio-geochemical reactions within the reservoirs. Notably, observations of H_2S generation, possibly from abiotic pyrite reduction, and methane generation from methanogenic bacteria within the reservoir were made. Although hydrogen storage within porous media has attracted increasing interest, only two pilot studies have injected and recovered hydrogen to date, namely green methane projects in Austria and Argentina. These projects involve injecting a mixture of natural gas with 10% hydrogen, produced from renewable energy sources, into sandstone reservoirs to produce green methane via coupled CO_2/H_2 injection into reservoirs hosting methanogenic bacteria.

1.6.7 Abandoned mines shafts

Abandoned mines worldwide stand as vast expanses, harboring unique features and infrastructure that lie dormant, awaiting revitalization. In Europe alone, an estimated 30,000 flooded mines present both a challenge and an opportunity. Repurposing these dormant sites is not only essential for unlocking potential in other sectors but also for addressing the legacy of their past activities. Repurposing these spaces can mitigate environmental impacts, reduce greenhouse gas emissions from electricity production in their respective regions, and pave the way for new job opportunities. Underground hydrogen storage emerges as a promising solution, offering a host of advantages that align with these objectives. The extensive underground cavities and intricate networks of tunnels and chambers within abandoned mines offer ample space for large-scale hydrogen storage, eliminating the need for constructing new facilities. Additionally, the geological stability of these underground spaces ensures the integrity and longevity of the storage infrastructure, making abandoned mines a compelling option for sustainable energy storage solutions.

Furthermore, underground hydrogen storage in abandoned mines presents several safety advantages. The inherent geological stability of these sites reduces the risk of external disturbances or natural disasters, providing a secure environment for hydrogen storage. The sturdy construction of mines offers a solid foundation, enhancing the safety of the storage infrastructure. Enclosed underground spaces, equipped with ventilation systems and advanced gas monitoring technologies, enable early detection of potential hydrogen leaks, facilitating swift response and mitigation measures. While enclosed environments minimize accident risks, ongoing monitoring remains crucial to ensure the safety of equipment, hydrogen stacks, and personnel both within and outside the storage area. Infrastructure requirements for underground hydrogen storage resemble those of conventional sites, encompassing hydrogen production systems, gas compressors, and injection systems interconnected via pipelines to prevent hydrogen embrittlement or leaks. Injection wells play a key role in facilitating controlled hydrogen injection into underground reservoirs. However, infrastructure differences primarily stem from the enclosed nature of the reservoirs, necessitating facilities like shafts or tunnels to link production and distribution systems, communication networks, power supplies, and transportation routes for personnel and equipment. These measures collectively bolster the safety and reliability of underground hydrogen storage in abandoned mines.

1.7 Technical requirements and challenges

The unique characteristics of hydrogen introduce uncertainties in the context of underground gas storage, necessitating thorough exploration. As discussed before, despite being the most abundant element in the universe, hydrogen comprises less than 0.14% of Earth naturally and exhibits extreme reactivity, requiring it to bind with other elements to form stable compounds (Bosu & Rajamohan, 2023; Godula-Jopek, 2015; Züttel, 2004). Given these challenges, various methods are employed for hydrogen storage, including high-pressure gas cylinders, cryogenic tanks, adsorption on materials with large specific surface areas, interstitial sites in host metals, and chemical bonding in covalent or ionic compounds, as well as oxidation of reactive metals with water. During the hydrogen storage process, mineralogical changes occur within the reservoir rock, altering its pore structure and flow characteristics. These changes depend on in situ reservoir conditions (e.g., pressure and temperature) and the initial pore and mineralogical characteristics of the rocks (Małachowska et al., 2022; Sambo et al., 2022). Therefore akin to storing other gases like methane and carbon dioxide, a comprehensive understanding of fluid migration through the reservoir is essential in developing an underground storage solution for hydrogen. Factors such as the tightness of the reservoir, geochemistry and mineralogy, working gas, cushion gas, and effectiveness of the process mainly influence the performance of the storage process. Fig. 1.9 elucidates the impacts of different factors on underground hydrogen storage operation.

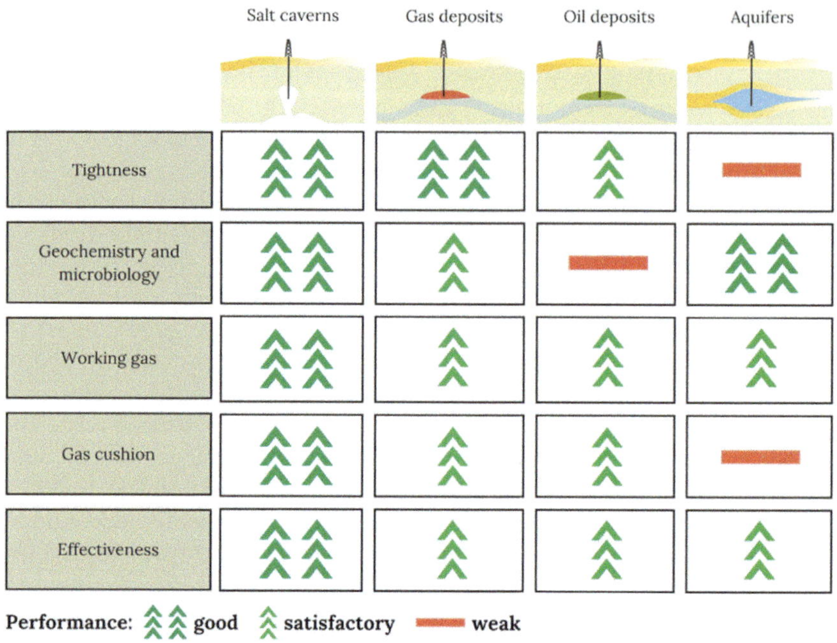

FIGURE 1.9 Requirements and comparison of selected operational parameters of geological structures suitable for underground hydrogen storage (UHS) (Quintos Fuentes & Santos, 2023).

1.7.1 Technical challenges

The technical challenges associated with underground hydrogen storage are considerable. One major obstacle is the lack of infrastructural advancements hindering the global commercial adoption of hydrogen (Kumari & Ranjith, 2023; Navaid et al., 2023). While hydrogen, from a large-scale production perspective, holds promise as a replacement for fossil fuels as the primary energy carrier, the long-term storage of large volumes of hydrogen presents difficulties. Storing hydrogen in underground reservoirs necessitates a comprehensive understanding of fluid migration through the reservoir. Furthermore, hydrogen's unique reactivity with reservoir rock can lead to notable alterations in the rock's pore structure via mineral dissolution/precipitation, impacting hydrogen injectivity and storage capacity in reservoirs (Kumari & Ranjith, 2023; Navaid et al., 2023). Ultimately, a detailed comprehension of fluid migration through the reservoir is crucial for the development of underground storage solutions for hydrogen.

1.7.2 Economic challenges

The underground hydrogen storage encounters associated economic challenges for large-scale prospect. The development and deployment of underground hydrogen storage require significant investments in research and infrastructure. The cost of hydrogen production, transportation, and storage is currently higher than that of fossil fuels, making it economically unfeasible (Ma et al., 2023; Tarkowski & Uliasz-Misiak, 2022). The cost of underground hydrogen storage hinges on several factors, including transportation, monitoring, storage, and injection expenses. Moreover, the overall cost is influenced by the location and characteristics of the geological storage site. Potential leakage points within the site, particularly in porous rock formations, can escalate storage costs (Epelle et al., 2022; Tarkowski & Uliasz-Misiak, 2022). For instance, in scenarios involving salt caverns, rigorous tightness tests should be conducted before licensing each cavern to ensure its suitability for hydrogen storage. Earlier conceptual studies have indicated that large-scale underground hydrogen storage in deposits is relatively economical compared to alternative storage technologies. Additionally, key capital costs associated with underground hydrogen storage include gas cost (cushion gas and hydrogen), well capital cost (new well cost including new well drilling and wellhead equipment installation, and retrofitting well cost), compressor capital cost (total hours of operation, compressor size, capitol cost per compressor, cost of electricity, water requirement, water and cooling cost, etc.). Storage methods like fast-cycling salt caverns and seasonal storage in depleted gas fields have varying cost ranges per kilogram of hydrogen. Developing new storage sites requires significant investment, with estimates in the hundreds of millions to billions of dollars range. Green hydrogen is favored due to zero CO_2 emissions, but "blue" hydrogen is also important for reducing emissions rapidly. Projected costs for "blue" hydrogen vary depending on factors like natural gas and CO_2 prices. By 2050, significant production volumes of green hydrogen are expected at competitive price points, assuming technical advancements and cost reductions are achieved.

1.7.3 Future market

Globally, approximately 120 million tonnes of hydrogen are produced annually, with two-thirds being pure hydrogen and the remaining one-third being mixed with other gases. China stands out as the leading producer and consumer of hydrogen, generating nearly 24 million tonnes of pure hydrogen each year, representing close to one-third of the world's dedicated production. This surge in clean hydrogen investment is fueled by national initiatives and support programs, resulting in significant growth in recent years. By November 2021, global announcements for hydrogen projects slated for completion by 2030 amounted to a staggering $160 billion in investment, with half of these funds earmarked for green hydrogen production utilizing renewable energy sources and electrolysis, as illustrated in Fig. 1.10 (Hydrogen Hydrogen Council, 2021). Hydrogen versatility enables its utilization across a spectrum of end-use applications. It serves as a conventional electricity storage option for re-electrification purposes, while also finding application in fuel cell electric vehicles (FCEVs), serving as a feedstock and heating fuel in various industries such as chemical, petrochemical, and steel, and contributing to the "greening" of natural gas by blending hydrogen into the natural gas grid (Messaoudani et al., 2016; Ozturk & Dincer, 2021). Numerous industries are actively engaged in advancing this transformative journey.

1.7.3.1 Electricity industry

Operating a vast electricity transport and distribution grid, particularly in regions rich in renewable electricity potential and with connections to large energy imports like

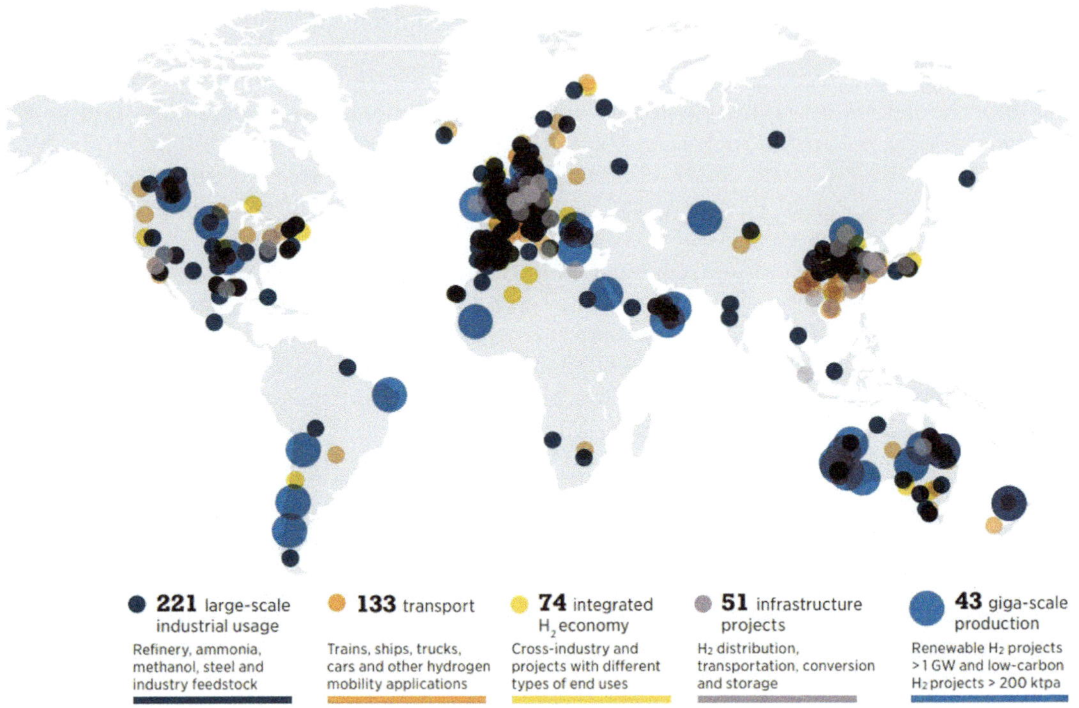

FIGURE 1.10 Clean hydrogen projects and investment as of November 2021 (Hydrogen Hydrogen Council, 2021). Map lines delineate study areas and do not necessarily depict accepted national boundaries.

offshore wind farms, holds significant promise for growth, provided it gains public acceptance. In the long term, the integration of hydrogen large-scale storage will become essential. However, in the short term, it could alleviate the pressure on grid expansion by offering ancillary services (Gür, 2018; Matos et al., 2019).

1.7.3.2 Mobility

A hydrogen refueling infrastructure tailored for hydrogen-powered fuel cell vehicles like cars and buses is projected to take shape across Europe, commencing around 2015. Initial adoption is expected to concentrate in densely populated regions such as megacities or urban clusters, with a gradual expansion into rural areas over time (Apostolou & Xydis, 2019; Greene et al., 2020). Germany, for instance, has plans to establish 400 hydrogen refueling stations by end of 2024 and aims for approximately 1000 stations by 2030.

1.7.3.3 Chemical industry

In selected European regions, the hydrogen industry has established robust networks of pipelines (Bünger et al., 2016) seamlessly integrated with large-scale storage facilities. These pipelines facilitate direct connections between individual production or end-use sites through point-to-point links, as evidenced by the connection of the Brunsbüttel chemical cluster with the refinery in Heide in northern Germany. Moreover, sophisticated multikilometer grids interconnect numerous production and user locations, such as those in Nordrhein-Westfalen and Leuna-Bitterfeld. Over time, there is significant potential for the hydrogen grid to merge with the existing natural gas grid (Baufumé et al., 2013; Bünger et al., 2016).

1.7.3.4 Natural gas industry

The natural gas industry has strategically deployed a dense pipeline grid across extensive regions of Europe, providing a robust framework to seamlessly integrate synthetic methane gas and limited quantities of hydrogen into the existing infrastructure. This flexibility enables the accommodation of varying natural gas flow rates and maximum admixture rates for hydrogen, thereby enhancing the efficiency of the electricity sector. Analytical calculations have been conducted to assess the additional demand for hydrogen supplied from renewable electricity through electrolysis, particularly for industrial purposes replacing current hydrogen production from fossil primary energy sources. Benchmarking exercises have been instrumental in evaluating the surplus electricity stored as hydrogen and forecasting future hydrogen demand. For example, in Germany, it was found that transitioning the surplus electricity storage requirement to hydrogen from electrolysis could significantly contribute to meeting the transport fuel demand for light-duty FCEVs by 2050. Evaluation of specific hydrogen costs is based on unit costs, encompassing operating costs and annualized capital expenditures. Typically, specific hydrogen costs range between 4 and 6 €/kg hydrogen for most applications, except in the electricity market, where costs are higher due to additional investment outlays for the re-electrification unit and lower round-trip efficiency of the system as shown in Fig. 1.11. Major cost components include

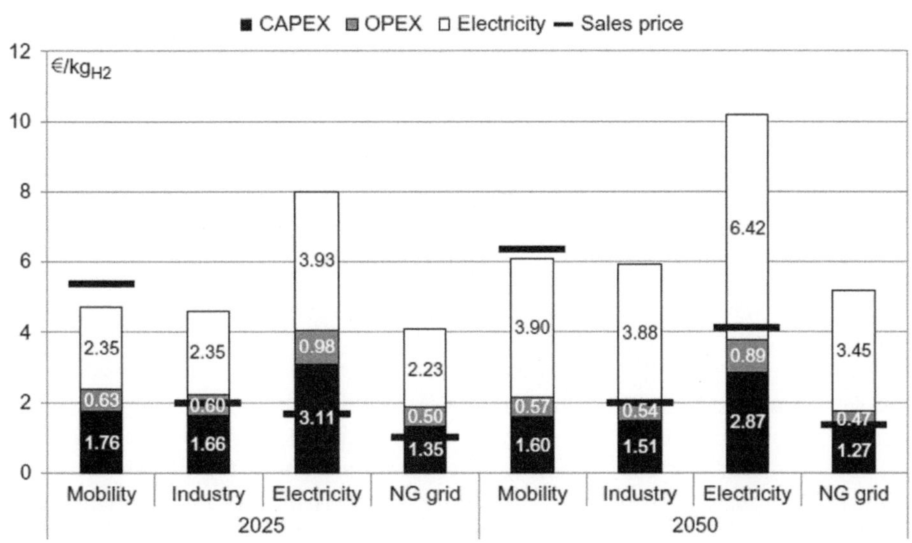

FIGURE 1.11 Comparison of hydrogen production costs with allowable hydrogen sale prices for hydrogen for (1) transport, (2) industry, (3) re-electrification, and (4) use in the natural gas grid in 2025 and 2050 (Simon et al., 2015).

annualized capital expenditures primarily from electrolysis investments, with minimal investments in underground storage and other equipment, as well as electricity costs from electrolysis operation. Fixed operation and maintenance (O&M) costs have limited influence on economic outcomes, and this cost structure applies uniformly across all hydrogen applications.

1.8 Regulatory and safety aspects and risks

Hydrogen storage reliability is a critical factor in the development of the hydrogen economy. The successful integration of hydrogen as a sustainable energy source depends on the availability of reliable and safe storage technologies (Andrews & Shabani, 2012; Ehteshami & Chan, 2014). Recent years have witnessed significant advancements in hydrogen storage technologies, leading to notable improvements in their reliability. However, challenges remain in ensuring the long-term safety and reliability of hydrogen storage systems.

Overall, hydrogen storage reliability is a function of several factors, including the design of the storage system, the materials used, the operating conditions, and the maintenance and inspection protocols. The reliability of different hydrogen storage technologies varies, with some technologies proving more reliable than others (Hassan et al., 2021; Moradi & Groth, 2019). To ensure the reliability of hydrogen storage systems, a range of strategies can be employed, including rigorous testing and evaluation during the development phase, implementing safety and maintenance protocols during operation, and investing in research and development to continuously improve the technology.

However, ensuring the reliability of hydrogen storage is pivotal for integrating hydrogen as a sustainable energy source successfully. Despite significant advancements in hydrogen storage technologies, there remains a need for continued efforts to enhance their long-term safety and reliability (Hassan et al., 2021; Moradi & Groth, 2019). Continued research, development, and thorough testing are essential to achieve this objective. Large-scale hydrogen storage is indispensable to address future energy and environmental challenges, with hydrogen being primarily utilized for industrial decarbonization, carbon-free mobility, and power generation. Initial demand for underground storage will arise for industrial and mobility decarbonization within the next few years, with underground pure hydrogen storage for power generation likely to follow after 2040 (Hassan et al., 2021; Moradi & Groth, 2019). In the interim, the gas and power sectors will adopt methane-hydrogen blends, with storage mainly in porous rock reservoirs, gradually increasing hydrogen content to potentially exceed 20%.

Salt caverns are anticipated to fulfill a significant portion of underground storage requirements, followed by porous rock reservoirs, particularly for methane-hydrogen blends. As technology advances, mined caverns may become integral in international trading for ammonia or liquid hydrogen, or where salt formations are unavailable for pure hydrogen storage. While salt caverns and porous rock reservoirs are poised for commercialization, ongoing research aims to deepen our understanding and establish standards (Lankof & Tarkowski, 2020; Zivar et al., 2021). While specific demonstrations may be necessary to address localized concerns, no major obstacles have been identified overall.

TABLE 1.2 Characteristics and risks of underground hydrogen storage in different geological formations (Muhammed et al., 2022).

Geological storage type	Storage capacity	Discharge rate	Initial cost	Operational cost	Seismic risk	Chemical reaction	Leakage risks	Usability purpose	Cushion gas requirement
Salt caverns	Moderate high	High	Very high	Low	Low	Low	Low	Frequent	Low
Aquifer	High	Low	Moderate	Moderate	High	High	Moderate	Seasonal	High
Depleted reservoirs	Very high	Moderate	Moderate	Moderate	Moderate	Moderate	High	Seasonal	High
Abandoned mines shaft	High	Moderate	Moderate	Moderate	Moderate	Average	Very high	Seasonal	Average

The future holds vast potential for shaping the energy landscape, with a crucial area of exploration being the development of optimized operational strategies for seasonal hydrogen storage. This entails studying hydrogen's dynamic behavior within geological formations and identifying efficient injection and extraction techniques to maximize storage capacities and minimize losses (Lankof & Tarkowski, 2020; Zivar et al., 2021).

Additionally, investigating the potential synergies between hydrogen storage and other renewable energy technologies can provide better insights. Exploring hybrid storage systems that combine geological hydrogen storage with battery storage or power-to-gas technologies can offer more comprehensive and flexible solutions for decarbonization (Llera et al., 2018; Li et al., 2021). Furthermore, evaluating the scalability and long-term sustainability of geological hydrogen storage is essential. Future research can be directed toward evaluating the long-term performance and stability of storage sites. Additionally, exploring the potential to repurpose existing infrastructure, such as natural gas pipelines, for hydrogen transportation, could also be a significant area of focus (Gondal & Sahir, 2012; Haeseldonckx & D'haeseleer, 2007). Lastly, considering the socioeconomic and policy dimensions of geological hydrogen storage is crucial. Analyzing the economic viability, job creation potential, and policy frameworks required to support the widespread deployment of hydrogen storage can guide decision-makers in making informed choices and accelerating the transition to a decarbonized future. This can provide better judgment and valuable insights for stakeholders, enabling them to make informed decisions and prioritize actions toward leveraging geological hydrogen storage for seasonal uses in the pursuit of decarbonization goals. However, Table 1.2 shows summary of the characteristics and risks of underground hydrogen storage for different geological formations. The summary provides a quick overview of the key attributes of each storage type, including storage capacity, discharge rate, cost, risk factors, and usability purposes.

1.9 Chapter summary and conclusions

Setting the stage for urgent sustainable energy solutions amid global warming and fossil fuel depletion, the introduction underscores hydrogen's pivotal role as a clean energy carrier. It emphasizes the critical need for refining storage methodologies to unlock hydrogen's full potential. Providing a comprehensive overview of the book's structure, the focus is on advancing subsurface hydrogen storage technologies to pave the way for a sustainable energy future. Chapter 2 delves into the pivotal role of hydrogen as a versatile and clean energy carrier, especially in mitigating the intermittency of renewable sources. It extensively discusses various hydrogen types, including green, gray, and blue hydrogen, and their pivotal roles across stationary and mobile applications. By analyzing production processes like electrolysis, the chapter underscores hydrogen's potential to revolutionize energy systems toward sustainability while addressing economic and technical feasibility aspects. Chapter 3 meticulously explores diverse hydrogen production methods, meticulously comparing renewable and nonrenewable approaches. It evaluates the efficiencies, costs, and technical complexities inherent in each method, accentuating the vital role of plant location concerning energy resource accessibility and subsurface geological conditions. Chapter 4 intricately unpacks hydrogen surface storage techniques, classifying them into physical and material-based paradigms. This analysis brings to light the nuanced advantages and challenges inherent in each storage method, highlighting the economic viability and technical feasibility of large-scale storage within geological formations such as aquifers and depleted hydrocarbon reservoirs. Chapter 5 thoroughly investigates the intricate domain of hydrogen transportation from production hubs to storage facilities, prioritizing safety and environmental sustainability. Through meticulous comparisons of transportation modes like pipelines, trucks, tankers, and ships, the chapter elucidates critical factors such as cost, efficiency, and scalability crucial for sustainable hydrogen logistics. Chapter 6 zeroes in on subsurface underground hydrogen storage (UHS), delineating its superiority over surface-based storage systems. By delving into diverse UHS types like storage in caverns, aquifers, and depleted hydrocarbon reservoirs, the chapter addresses pivotal challenges such as hydrogen purity, storage depth, cushion gas selection, and reservoir engineering intricacies. Chapter 7 delves deeply into the nuanced physics of hydrogen flow within porous media during UHS operations. By dissecting two-phase flow phenomena alongside interfacial tension dynamics, relative permeability nuances, and hysteresis effects, the chapter offers profound insights into optimizing storage efficiency while considering hydrogen's distinct properties compared to other gases. Chapter 8 rigorously analyzes the physiochemical intricacies underpinning successful UHS operations, meticulously considering rock–fluid interactions, hydrogen migration dynamics, and optimal storage capacity determination. By delving into reservoir properties, fluid characteristics, and their cascading impacts on secure and efficient hydrogen storage, the chapter forms a comprehensive understanding of UHS viability. Chapter 9 delves into the critical realm of geochemical and mineralogical considerations within UHS, meticulously unpacking hydrogen's reactivity vis-a-vis rock-forming minerals and its consequential impacts on storage dynamics. Through a rigorous exploration of potential chemical reactions, flow alterations, and in situ migration challenges across varied reservoir types, the chapter offers invaluable insights into optimizing UHS longevity and efficiency. Chapter 10 intricately navigates microbial considerations essential for UHS success, delineating microbial metabolism's pivotal role in hydrogen storage quality and safety assurance. By meticulously addressing microbial reactions, hydrogen loss phenomena, methane production intricacies, corrosion risks, and robust mitigation strategies, the chapter lays a robust foundation for safeguarding UHS integrity and operational success. Chapter 11 expertly dissects experimental, modeling, and simulation approaches imperative for advancing UHS frontiers. By meticulously detailing core-scale experiments, advanced numerical modeling techniques, and simulation methodologies, the chapter equips stakeholders with invaluable tools to understand and optimize hydrogen behavior within subsurface formations, ensuring optimized UHS outcomes. Chapter 12 provides a detailed examination of government policies, guidelines, initiatives, and supportive frameworks crucial for UHS development and deployment worldwide. By illuminating the pivotal role of governments in promoting hydrogen technology adoption, fostering energy transitions, and creating conducive regulatory environments for UHS proliferation, the chapter offers indispensable insights into navigating complex regulatory landscapes. Chapter 13 captivatingly presents a compelling array of case studies illuminating UHS success stories worldwide. By meticulously showcasing ongoing projects, current operational statuses, challenges, and scientific advancements within diverse storage paradigms, the chapter offers invaluable insights into real-world UHS applications, scientific breakthroughs, and future trajectories. Chapter 14 thoughtfully delineates the promising prospects and formidable challenges characterizing UHS landscapes. By intricately examining geological site selection criteria, operational intricacies, economic viability factors, regulatory challenges, and safety imperatives, the chapter equips stakeholders with a nuanced understanding of UHS intricacies, strategic approaches, and indispensable technological innovations shaping sustainable energy futures.

This chapter presents a comprehensive exploration of hydrogen's importance as a vital energy carrier across various industries, discussing its fundamental properties, characteristics, and significant role in energy production, transportation, and industry applications. It highlights the critical necessity for efficient energy storage solutions, focusing on challenges like high implementation costs, lack of standardization in storage systems, and regulatory limitations. Furthermore, it introduces underground hydrogen storage as a promising solution for large-scale and long-term energy storage, emphasizing geological considerations, storage capacities, and technical requirements. The comparison of subsurface storage options, such as salt caverns, depleted hydrocarbon reservoirs, aquifers, and abandoned mine shafts, provides valuable insights into their usability and scalability. This chapter also addresses key technical and economic challenges associated with underground storage, discussing regulatory compliance, safety protocols, and environmental considerations. It sets the stage for detailed discussions in subsequent chapters, providing a comprehensive overview of essential concepts and frameworks crucial for advancing sustainable energy practices on a global scale.

Acknowledgments

The first author, Achinta Bera, appreciates the support provided by the Department of Petroleum Engineering, School of Energy Technology, Pandit Deendayal Energy University, Gandhinagar-382426, Gujarat, India. Another author, Sunil Kumar, gratefully acknowledges the support of RINOLYST, SciGlyph exploration private limited, India to complete the work.

References

Acar, C., & Dincer, I. (2018). *Hydrogen energy. Comprehensive energy systems* (pp. 568–605). Oxford: Elsevier. Available from https://doi.org/10.1016/B978-0-12-809597-3.00113-9.

Ali, M., Arif, M., Sedev, R., Sánchez-Román, M., Keshavarz, A., & Iglauer, S. (2023). Underground hydrogen storage: The microbiotic influence on rock wettability. *Journal of Energy Storage, 72*. Available from https://doi.org/10.1016/j.est.2023.108405.

Andersson, J., & Grönkvist, S. (2019). Large-scale storage of hydrogen. *International Journal of Hydrogen Energy, 44*(23), 11901–11919. Available from https://doi.org/10.1016/j.ijhydene.2019.03.063, http://www.journals.elsevier.com/international-journal-of-hydrogen-energy/.

Andrews, J., & Shabani, B. (2012). Re-envisioning the role of hydrogen in a sustainable energy economy. *International Journal of Hydrogen Energy, 37*(2), 1184–1203. Available from https://doi.org/10.1016/j.ijhydene.2011.09.137.

Anson, A., Jagiello, J., Parra, J. B., Sanjuan, M. L., Benito, A. M., Maser, W. K., & Martínez, M. T. (2004). Porosity, surface area, surface energy, and hydrogen adsorption in nanostructured carbons. *The Journal of Physical Chemistry B, 108*(40), 15820–15826.

Apostolou, D., & Xydis, G. (2019). A literature review on hydrogen refuelling stations and infrastructure. Current status and future prospects. *Renewable and Sustainable Energy Reviews, 113*. Available from https://doi.org/10.1016/j.rser.2019.109292.

Baufumé, S., Grüger, F., Grube, T., Krieg, D., Linssen, J., Weber, M., Hake, J. F., & Stolten, D. (2013). GIS-based scenario calculations for a nationwide German hydrogen pipeline infrastructure. *International Journal of Hydrogen Energy, 38*(10), 3813–3829. Available from https://doi.org/10.1016/j.ijhydene.2012.12.147.

Bilgen, S., Kaygusuz, K., & Sari, A. (2004). Renewable energy for a clean and sustainable future. *Energy Sources, 26*(12), 1119–1129.

Bosu, S., & Rajamohan, N. (2023). Recent advancements in hydrogen storage-comparative review on methods, operating conditions and challenges. *International Journal of Hydrogen Energy*.

Buttler, A., & Spliethoff, H. (2018). Current status of water electrolysis for energy storage, grid balancing and sector coupling via power-to-gas and power-to-liquids: A review. *Renewable and Sustainable Energy Reviews, 82*, 2440–2454. Available from https://doi.org/10.1016/j.rser.2017.09.003, https://www.journals.elsevier.com/renewable-and-sustainable-energy-reviews.

Bünger, U., Michalski, J., Crotogino, F., & Kruck, O. (2016). *Large-scale underground storage of hydrogen for the grid integration of renewable energy and other applications*, . *Compendium of hydrogen energy* (4). Elsevier BV. Available from http://doi.org/10.1016/b978-1-78242-364-5.00007-5.

Chen, Y., Jin, X., Zeng, L., Zhong, Z., Mehana, M., Xiao, W., Pu, W., Regenauer-Lieb, K., & Xie, Q. (2023). Role of large-scale underground hydrogen storage and its pathways to achieve net-zero in China. *Journal of Energy Storage, 72*. Available from https://doi.org/10.1016/j.est.2023.108448.

Dopffel, N., Jansen, S., & Gerritse, J. (2021). Microbial side effects of underground hydrogen storage–knowledge gaps, risks and opportunities for successful implementation. *International Journal of Hydrogen Energy, 46*(12), 8594–8606.

Dunn, S. (2002). Hydrogen futures: Toward a sustainable energy system. *International Journal of Hydrogen Energy, 27*(3), 235–264. Available from https://doi.org/10.1016/s0360-3199(01)00131-8.

Ebhota, W. S., & Jen, T. C. (2020). Fossil fuels environmental challenges and the role of solar photovoltaic technology advances in fast tracking hybrid renewable energy system. *International Journal of Precision Engineering and Manufacturing - Green Technology, 7*(1), 97–117. Available from https://doi.org/10.1007/s40684-019-00101-9, http://www.springer.com/engineering/production + engineering/journal/40684.

Ehteshami, S. M. M., & Chan, S. H. (2014). The role of hydrogen and fuel cells to store renewable energy in the future energy network - potentials and challenges. *Energy Policy, 73*, 103–109. Available from https://doi.org/10.1016/j.enpol.2014.04.046, http://www.journals.elsevier.com/energy-policy/.

Epelle, E. I., Obande, W., Udourioh, G. A., Afolabi, I. C., Desongu, K. S., Orivri, U., Gunes, B., & Okolie, J. A. (2022). Perspectives and prospects of underground hydrogen storage and natural hydrogen. *Sustainable Energy and Fuels, 6*(14), 3324–3343. Available from https://doi.org/10.1039/d2se00618a, http://www.rsc.org/journals-books-databases/about-journals/sustainable-energy-fuels.

Ershadnia, R., Singh, M., Mahmoodpour, S., Meyal, A., Moeini, F., Hosseini, S. A., Sturmer, D. M., Rasoulzadeh, M., Dai, Z., & Soltanian, M. R. (2023). Impact of geological and operational conditions on underground hydrogen storage. *International Journal of Hydrogen Energy, 48*(4), 1450–1471. Available from https://doi.org/10.1016/j.ijhydene.2022.09.208, http://www.journals.elsevier.com/international-journal-of-hydrogen-energy/.

Le Gallo, Y. (2024). Assessment of hydrogen storage capacity in porous media at the European scale. *International Journal of Hydrogen Energy*, *58*, 400–408.

Germain, J., Hradil, J., Fréchet, J. M. J., & Svec, F. (2006). High surface area nanoporous polymers for reversible hydrogen storage. *Chemistry of Materials*, *18*(18), 4430–4435. Available from https://doi.org/10.1021/cm061186p.

Godula-Jopek, A. (2015). *Hydrogen storage options including constraints and challenges hydrogen production: Electrolysis* (pp. 273–310). Wiley. Available from http://doi.org/10.1002/9783527676507.ch7.

Gondal, I. A., & Sahir, M. H. (2012). Prospects of natural gas pipeline infrastructure in hydrogen transportation. *International Journal of Energy Research*, *36*(15), 1338–1345. Available from https://doi.org/10.1002/er.1915.

Greene, D. L., Ogden, J. M., & Lin, Z. (2020). Challenges in the designing, planning and deployment of hydrogen refueling infrastructure for fuel cell electric vehicles. *ETransportation*, *6*, 100086.

Gür, T. M. (2018). Review of electrical energy storage technologies, materials and systems: challenges and prospects for large-scale grid storage. *Energy and Environmental Science*, *11*(10), 2696–2767.

Haeseldonckx, D., & D'haeseleer, W. (2007). The use of the natural-gas pipeline infrastructure for hydrogen transport in a changing market structure. *International Journal of Hydrogen Energy*, *32*(10-11), 1381–1386. Available from https://doi.org/10.1016/j.ijhydene.2006.10.018.

Hashemi, L., Glerum, W., Farajzadeh, R., & Hajibeygi, H. (2021). Contact angle measurement for hydrogen/brine/sandstone system using captive-bubble method relevant for underground hydrogen storage. *Advances in Water Resources*, *154*. Available from https://doi.org/10.1016/j.advwatres.2021.103964.

Hassan, I. A., Ramadan, H. S., Saleh, M. A., & Hissel, D. (2021). Hydrogen storage technologies for stationary and mobile applications: Review, analysis and perspectives. *Renewable and Sustainable Energy Reviews*, *149*. Available from https://doi.org/10.1016/j.rser.2021.111311, https://www.journals.elsevier.com/renewable-and-sustainable-energy-reviews.

Herzog, A. V., Lipman, T. E., Edwards, J. L., & Kammen, D. M. (2001). Renewable energy: A viable choice. *Environment: Science and Policy for Sustainable Development*, *43*(10), 8–20. Available from https://doi.org/10.1080/00139150109605150.

Hirscher, M., & Panella, B. (2005). Nanostructures with high surface area for hydrogen storage. *Journal of Alloys and Compounds*, *404*, 399–401.

Hosseini, S. E., & Abdul Wahid, M. (2014). The role of renewable and sustainable energy in the energy mix of Malaysia: A review. *International Journal of Energy Research*, *38*(14), 1769–1792. Available from https://doi.org/10.1002/er.3190, http://onlinelibrary.wiley.com/journal/10.1002/(ISSN)1099-114X.

Hosseini, S. E., & Wahid, M. A. (2020). Hydrogen from solar energy, a clean energy carrier from a sustainable source of energy. *International Journal of Energy Research*, *44*(6), 4110–4131.

Hydrogen Council. (2021). Hydrogen for net zero-a critical cost-competitive energy vector.

Ibrahim, M. D., Binofai, F. A., & Mohamad, M. O. (2022). Transition to low-carbon hydrogen energy system in the UAE: Sector efficiency and hydrogen energy production efficiency analysis. *Energies*, *15*(18), 6663.

Idriss, H., Scott, M., & Subramani, V. (2015). *Introduction to hydrogen and its properties. Compendium of Hydrogen Energy* (pp. 3–19). Woodhead Publishing.

Ishimoto, Y., Voldsund, M., Nekså, P., Roussanaly, S., Berstad, D., & Gardarsdottir, S. O. (2020). Large-scale production and transport of hydrogen from Norway to Europe and Japan: Value chain analysis and comparison of liquid hydrogen and ammonia as energy carriers. *International Journal of Hydrogen Energy*, *45*(58), 32865–32883. Available from https://doi.org/10.1016/j.ijhydene.2020.09.017, http://www.journals.elsevier.com/international-journal-of-hydrogen-energy/.

Kanaani, M., & Sedaee, B. (2022). Impact of dilation and irreversible compaction on underground hydrogen storage in depleted hydrocarbon reservoirs. *Energy and Fuels*, *36*(22), 13506–13517. Available from https://doi.org/10.1021/acs.energyfuels.2c02150, http://pubs.acs.org/journal/enfuem.

Kanaani, M., Sedaee, B., & Asadian-Pakfar, M. (2022). Role of cushion gas on underground hydrogen storage in depleted oil reservoirs. *Journal of Energy Storage*, *45*, 103783.

Keçebaş, A., & Kayfeci, M. (2019). *Hydrogen properties. Solar Hydrogen Production* (pp. 3–29). Academic Press.

Kommula, N.D.S. (2021). Analysis of performance and features of hydrogen as an energy carrier. Master's Thesis, Universita Politechnica delle Marche.

Kumari, W. G. P., & Ranjith, P. G. (2023). An overview of underground hydrogen storage with prospects and challenges for the Australian context. *Geoenergy Science and Engineering*, 212354.

Lankof, L., & Tarkowski, R. (2020). Assessment of the potential for underground hydrogen storage in bedded salt formation. *International Journal of Hydrogen Energy, 45*(38), 19479–19492.

Lee, B., Lee, H., Heo, J., Moon, C., Moon, S., & Lim, H. (2019). Stochastic techno-economic analysis of H 2 production from power-to-gas using a high-pressure PEM water electrolyzer for a small-scale H 2 fueling station. *Sustainable Energy and Fuels, 3*(9), 2521–2529. Available from https://doi.org/10.1039/c9se00275h.

Li, B., Chen, M., Ma, Z., He, G., Dai, W., Liu, D., Zhang, C., & Zhong, H. (2021). Modeling integrated power and transportation systems: Impacts of power-to-gas on the deep decarbonization. *IEEE Transactions on Industry Applications, 58*(2), 2677–2693.

Liu, H., Ampah, J. D., Zhao, Y., Sun, X., Xu, L., Jiang, X., & Wang, S. (2022). A perspective on the overarching role of hydrogen, ammonia, and methanol carbon-neutral fuels towards net zero emission in the next three decades. *Energies, 16*(1), 280.

Llera, E., Romeo, L. M., Bailera, M., & Osorio, J. L. (2018). Exploring the integration of the power to gas technologies and the sustainable transport. *International Journal of Energy Production and Management, 3*(1), 1–9. Available from https://doi.org/10.2495/EQ-V3-N1-1-9.

Londe, L. F. (2021). *Four ways to store large quantities of hydrogen. Abu Dhabi international petroleum exhibition and conference*. SPE, D022S189R001.

Ma, N., Zhao, W., Wang, W., Li, X., & Zhou, H. (2023). Large scale of green hydrogen storage: Opportunities and challenges. *International Journal of Hydrogen Energy*. Available from https://doi.org/10.1016/j.ijhydene.2023.09.021, http://www.journals.elsevier.com/international-journal-of-hydrogen-energy/.

Marks-Bielska, R., Bielski, S., Pik, K., & Kurowska, K. (2020). The importance of renewable energy sources in Poland's energy mix. *Energies, 13*(18). Available from https://doi.org/10.3390/en13184624.

Matos, C. R., Carneiro, J. F., & Silva, P. P. (2019). Overview of large-scale underground energy storage technologies for integration of renewable energies and criteria for reservoir identification. *Journal of Energy Storage, 21*, 241–258.

Mazloomi, K., & Gomes, C. (2012). Hydrogen as an energy carrier: Prospects and challenges. *Renewable and Sustainable Energy Reviews, 16*(5), 3024–3033. Available from https://doi.org/10.1016/j.rser.2012.02.028.

Małachowska, A., Łukasik, N., Mioduska, J., & Gębicki, J. (2022). Hydrogen storage in geological formations—the potential of salt caverns. *Energies, 15*(14). Available from https://doi.org/10.3390/en15145038.

Messaoudani, Zl, Rigas, F., Binti Hamid, M. D., & Hassan, C. R. C. (2016). Hazards, safety and knowledge gaps on hydrogen transmission via natural gas grid: A critical review. *International Journal of Hydrogen Energy, 41* (39), 17511–17525. Available from https://doi.org/10.1016/j.ijhydene.2016.07.171, http://www.journals.elsevier.com/international-journal-of-hydrogen-energy/.

Mills, S. (2018). Combining solar power with coal-fired power plants, or cofiring natural gas. *Clean Energy, 2*(1), 1–9. Available from https://doi.org/10.1093/ce/zky004.

Moradi, R., & Groth, K. M. (2019). Hydrogen storage and delivery: Review of the state of the art technologies and risk and reliability analysis. *International Journal of Hydrogen Energy, 44*(23), 12254–12269. Available from https://doi.org/10.1016/j.ijhydene.2019.03.041, http://www.journals.elsevier.com/international-journal-of-hydrogen-energy/.

Muhammed, N. S., Gbadamosi, A. O., Epelle, E. I., Abdulrasheed, A. A., Haq, B., Patil, S., Al-Shehri, D., & Kamal, M. S. (2023). Hydrogen production, transportation, utilization, and storage: Recent advances towards sustainable energy. *Journal of Energy Storage, 73*. Available from https://doi.org/10.1016/j.est.2023.109207, http://www.journals.elsevier.com/journal-of-energy-storage/.

Muhammed, N. S., Haq, B., Al Shehri, D., Al-Ahmed, A., Rahman, M. M., & Zaman, E. (2022). A review on underground hydrogen storage: Insight into geological sites, influencing factors and future outlook. *Energy Reports, 8*, 461–499. Available from https://doi.org/10.1016/j.egyr.2021.12.002, http://www.journals.elsevier.com/energy-reports/.

Navaid, H. B., Emadi, H., & Watson, M. (2023). A comprehensive literature review on the challenges associated with underground hydrogen storage. *International Journal of Hydrogen Energy, 48*(28), 10603–10635.

Okoroafor, E. R., Saltzer, S. D., & Kovscek, A. R. (2022). Toward underground hydrogen storage in porous media: Reservoir engineering insights. *International Journal of Hydrogen Energy, 47*(79), 33781–33802.

Olabi, A. G., bahri, As, Abdelghafar, A. A., Baroutaji, A., Sayed, E. T., Alami, A. H., Rezk, H., & Abdelkareem, M. A. (2021). Large-vscale hydrogen production and storage technologies: Current status and future directions. *Emirates International Journal of Hydrogen Energy, 46*(45), 23498–23528. Available from https://doi.org/10.1016/j.ijhydene.2020.10.110, http://www.journals.elsevier.com/international-journal-of-hydrogen-energy/.

Olaniyi, O., Incer-Valverde, J., Tsatsaronis, G., & Morosuk, T. (2023). Exergetic and economic evaluation of natural gas/hydrogen blends for power generation. *Journal of Energy Resources Technology, 145*(6). Available from https://doi.org/10.1115/1.4056448.

Osman, A. I., Mehta, N., Elgarahy, A. M., Hefny, M., Al-Hinai, A., Al-Muhtaseb, A. H., & Rooney, D. W. (2022). Hydrogen production, storage, utilisation and environmental impacts: A review. *Environmental Chemistry Letters, 20* (1), 153–188. Available from https://doi.org/10.1007/s10311-021-01322-8, http://springerlink.metapress.com/app/home/journal.asp?wasp = d86tgdwvtg0yvw9gvkwp&referrer = parent&backto = browsepublicationsresults,140,541.

Ozturk, M., & Dincer, I. (2021). A comprehensive review on power-to-gas with hydrogen options for cleaner applications. *International Journal of Hydrogen Energy, 46*(62), 31511–31522.

Perera, M. S. A. (2023). A review of underground hydrogen storage in depleted gas reservoirs: Insights into various rock-fluid interaction mechanisms and their impact on the process integrity. *Fuel, 334*. Available from https://doi.org/10.1016/j.fuel.2022.126677.

Preuster, P., Alekseev, A., & Wasserscheid, P. (2017). Hydrogen storage technologies for future energy systems. *Annual Review of Chemical and Biomolecular Engineering, 8*, 445–471.

Quarton, C. J., Tlili, O., Welder, L., Mansilla, C., Blanco, H., Heinrichs, H., Leaver, J., Samsatli, N. J., Lucchese, P., Robinius, M., & Samsatli, S. (2020). The curious case of the conflicting roles of hydrogen in global energy scenarios. *Sustainable Energy and Fuels, 4*(1), 80–95. Available from https://doi.org/10.1039/c9se00833k.

Quintos Fuentes, J. E., & Santos, D. M. F. (2023). Technical and economic viability of underground hydrogen storage. *Hydrogen, 4*(4), 975–1000. Available from https://doi.org/10.3390/hydrogen4040057, https://www.mdpi.com/journal/hydrogen.

Rahman, A., Farrok, O., & Haque, M. M. (2022). Environmental impact of renewable energy source based electrical power plants: Solar, wind, hydroelectric, biomass, geothermal, tidal, ocean, and osmotic. *Renewable and Sustainable Energy Reviews, 161*, 112279.

Rasul, M. G., Hazrat, M. A., Sattar, M. A., Jahirul, M. I., & Shearer, M. J. (2022). The future of hydrogen: Challenges on production, storage and applications. *Energy Conversion and Management, 272*. Available from https://doi.org/10.1016/j.enconman.2022.116326.

Reddy, V. J., Hariram, N. P., Maity, R., Ghazali, M. F., & Kumarasamy, S. (2023). Sustainable E-fuels: Green hydrogen, methanol and ammonia for carbon-neutral transportation. *World Electric Vehicle Journal, 14*(12). Available from https://doi.org/10.3390/wevj14120349, https://www.mdpi.com/journal/wevj.

Salmachi, A., Seyfaee, A., Robert, R. J., Hosseini, T., Nathan, G., Ashman, P., Roberts, A., Jafarian, M., & Simon, C. (2024). Underground hydrogen storage: Integrated surface facilities and fluid flow modelling for depleted gas reservoirs. *International Journal of Hydrogen Energy, 50*, 1055–1069. Available from https://doi.org/10.1016/j.ijhydene.2023.08.335, https://www.sciencedirect.com/science/journal/03603199.

Sambo, C., Dudun, A., Samuel, S. A., Esenenjor, P., Muhammed, N. S., & Haq, B. (2022). A review on worldwide underground hydrogen storage operating and potential fields. *International Journal of Hydrogen Energy, 47*(54), 22840–22880.

Santhanam, K. S., Press, R. J., Miri, M. J., Bailey, A. V., & Takacs, G. A. (2017). *Introduction to hydrogen technology*. John Wiley & Sons.

Sazali, N. (2020). Emerging technologies by hydrogen: A review. *International Journal of Hydrogen Energy, 45*(38), 18753–18771. Available from https://doi.org/10.1016/j.ijhydene.2020.05.021.

Schlichtenmeyer, M., & Klafki, M. (2016). Differences and challenges in salt cavern design for hydrogen, air and natural gas 2016. In *Energy geotechnics: Proceedings of the 1st international conference on energy geotechnics* (pp. 65–69).

Sigfusson, T. I. (2007). Pathways to hydrogen as an energy carrier. *Royal Society, Iceland Philosophical Transactions of the Royal Society A: Mathematical, Physical and Engineering Sciences, 365*(1853), 1025–1042. Available from https://doi.org/10.1098/rsta.2006.1960, http://rsta.royalsocietypublishing.org/.

Simon, J., Férriz, A. M., & Correas, L. C. (2015). HyUnder–hydrogen underground storage at large scale: Case study Spain. *Energy Procedia, 73*, 136–144.

Tarkowski, R. (2019). Underground hydrogen storage: Characteristics and prospects. *Renewable and Sustainable Energy Reviews, 105*, 86–94. Available from https://doi.org/10.1016/j.rser.2019.01.051.

Tarkowski, R., & Uliasz-Misiak, B. (2022). Towards underground hydrogen storage: A review of barriers. *Renewable and Sustainable Energy Reviews, 162*. Available from https://doi.org/10.1016/j.rser.2022.112451.

Zamehrian, M., & Sedaee, B. (2022). Underground hydrogen storage in a partially depleted gas condensate reservoir: Influence of cushion gas. *Journal of Petroleum Science and Engineering, 212*, 110304.

Zgonnik, V. (2020). The occurrence and geoscience of natural hydrogen: A comprehensive review. *Earth-Science Reviews*, *203*, 103140.

Zivar, D., Kumar, S., & Foroozesh, J. (2021). Underground hydrogen storage: A comprehensive review. *International Journal of Hydrogen Energy*, *46*(45), 23436–23462.

Züttel, A. (2004). Hydrogen storage methods. *Naturwissenschaften*, *91*(4), 157–172. Available from https://doi.org/10.1007/s00114-004-0516-x.

Züttel, A., Remhof, A., Borgschulte, A., & Friedrichs, O. (2010). Hydrogen: The future energy carrier. *Philosophical Transactions of the Royal Society A: Mathematical, Physical and Engineering Sciences*, *368*(1923), 3329–3342. Available from https://doi.org/10.1098/rsta.2010.0113, http://rsta.royalsocietypublishing.org/content/368/1923/3329.full.pdf.

Further reading

Glenk, G., & Reichelstein, S. (2019). Economics of converting renewable power to hydrogen. *Nature Energy*, *4*(3), 216–222.

Hinkley, J., Hayward, J., McNaughton, R., Gillespie, R., Matsumoto, A., Watt, M., & Lovegrove, K. (2016). Cost assessment of hydrogen production from PV and electrolysis. In Report to ARENA as part of solar fuels roadmap, project A-3018 (. 1–4).

Hiscock, W. A., & Lindblom, L. (1985). Generic instabilities in first-order dissipative relativistic fluid theories. *Physical Review D*, *31*(4), 725.

International Renewable Energy Agency (IRENA)/Hydrogen Council. (2021). https://commercial.allianz.com/news-and-insights/expert-risk-articles/green-hydrogen.html.

Megia, P. J., Vizcaíno, A. J., Calles, J. A., & Carrero, A. (2021). Hydrogen production technologies: From fossil fuels toward renewable sources. A mini review. *Energy and Fuels*, *35*(20), 16403–16415.

Winter, C. J. (2009). Hydrogen energy—abundant, efficient, clean: A debate over the energy-system-of-change. *International Journal of Hydrogen Energy*, *34*(14), S1–S52.

CHAPTER

2

Hydrogen as an energy carrier

Ramesh Guduru (Kumar)[1], *Robin Singh*[2], *Ricurve Patel*[1] *and Rakesh Kumar Vij*[3]

[1]Department of Mechanical Engineering, School of Technology, Pandit Deendayal Energy University, Gandhinagar, Gujarat, India [2]Department of Chemistry, School of Energy Technology, Pandit Deendayal Energy University, Gandhinagar, Gujarat, India [3]Department of Petroleum Engineering, School of Energy Technology, Pandit Deendayal Energy University, Gandhinagar, Gujarat, India

2.1 Introduction

In the quest for a sustainable and low-carbon energy future, the world is transitioning from conventional fossil fuel–based energy sources to renewable and clean alternatives (Guduru et al., 2024a). Conventional energy sources, such as coal and natural gas, have long been the backbone of global energy production (Lubitz & Tumas, 2007). However, they come with significant environmental drawbacks, including greenhouse gas emissions and air pollution. In response, renewable energy sources like solar, wind, hydropower, and biomass have gained prominence due to their cleaner and abundant nature (International Energy Agency, 2022). Some of the common energy carriers include electricity, natural gas, liquid fuels (gasoline, diesel, and biofuels), batteries, compressed air energy storage (CAES) are shown in Fig. 2.1 (Armaroli & Balzani, 2011).

Among these, electricity is a useful and frequently utilized energy carrier that may be produced from a variety of sources, including fossil fuels, nuclear energy, and renewable sources like sun, wind, and hydroelectric power (Guduru, 2024). Natural gas, which primarily consists of methane, is commonly used as a fossil fuel energy source (Abe et al., 2019). Liquid fuels, including gasoline, diesel, and biofuels, are widely used in transportation, particularly in vehicles with internal combustion engines. Biofuels, such as ethanol and biodiesel, are renewable alternatives to fossil-based liquid fuels (Kothari et al., 2004). Liquid fuels have a high energy density but also produce carbon emissions when burned. Batteries are a form of chemical energy storage and can store electricity. Although batteries have high efficiency, but limited energy density compared to hydrogen which makes them more suitable for short-range applications. Additionally, a huge carbon footprint value is

Subsurface Hydrogen Energy Storage
DOI: https://doi.org/10.1016/B978-0-443-24071-3.00003-0

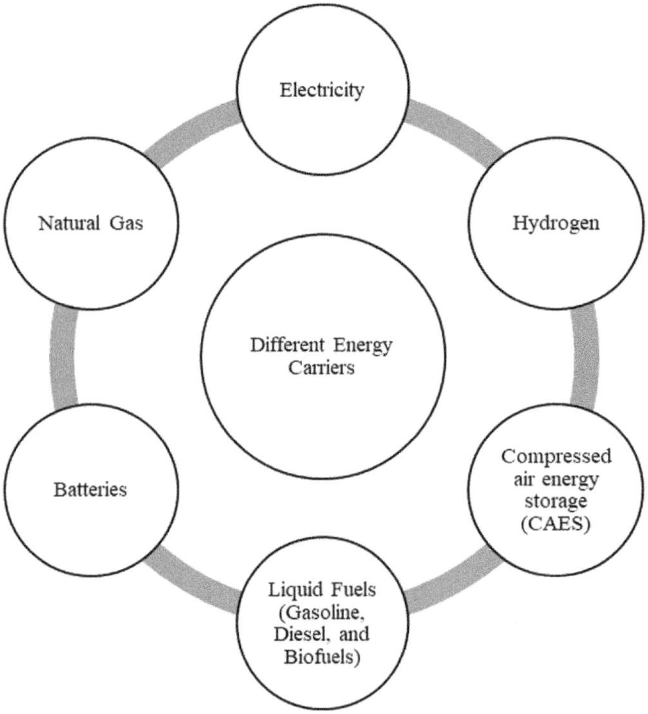

FIGURE 2.1 Different available energy carriers.

associated with the overall process of battery manufacturing which typically involves mining coating, calendaring slitting, electrode making, etc. CAES is a method of storing energy in the form of compressed air and is used in conjunction with other energy sources to generate electricity during periods of high demand (Yue et al., 2021). Hydrogen, which is also termed as Swiss army knife due to its high calorific value ranging between 120 and 140 kJ/kg, has the potential to decarbonize the worldwide economies as shown in Fig. 2.2 (Pistidda, 2021). It can be produced from various primary sources, including fossil fuels (with carbon capture and storage [CCS] for emission reduction) and renewable energy sources (such as solar and wind). Green hydrogen is currently expensive, but it is expected to become cheaper and more competitive compared to brown and gray hydrogen in the future (Jovan & Dolanc, 2020). Brown and gray hydrogen may be more feasible and economical for direct application in certain industries, such as steel and cement, as a greener alternative for now (Guduru et al., 2024b). However, it is important to note that they are not considered completely green and clean technologies (Ajanovic et al., 2022).

Exploring hydrogen's role as an energy carrier reveals its potential to help combat climate change and reduce carbon emissions. Hydrogen is versatile and can be a powerful catalyst for change in the modern energy landscape, offering solutions to complex challenges. As previously discussed, it can be derived from fossil fuels, like natural gas or coal, when integrated with CCS technologies. This approach, known as gray hydrogen production, offers a transitional path to cleaner energy systems while utilizing existing infrastructure. It permits the extraction of energy from fossil fuels with a significantly reduced carbon footprint, providing

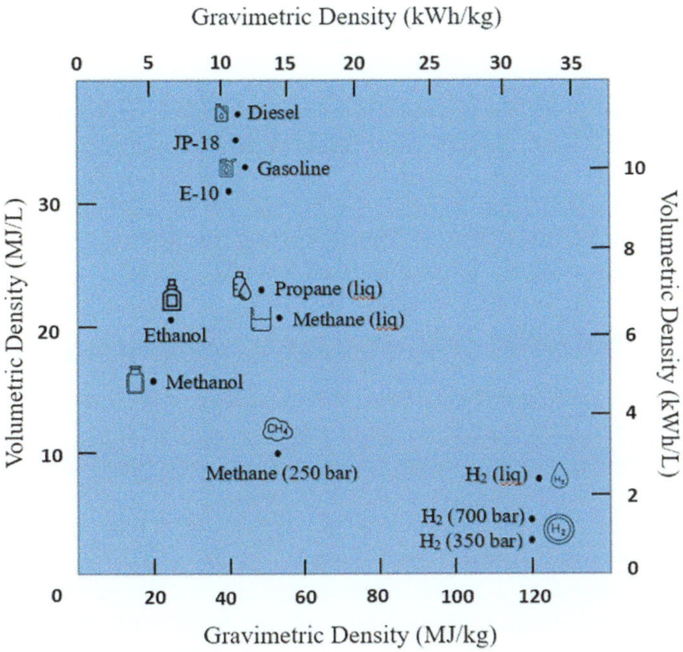

FIGURE 2.2 Comparison of the gravimetric density, commonly known as specific energy (energy per unit mass), and volumetric density, referred to as energy density (energy per unit volume), among various fuels by considering their lower heating values.

a lifeline to industries still seeking fully renewable alternatives (Nicoletti et al., 2015). Conversely, green hydrogen is created using renewable energy sources to split water through electrolysis. This approach can lead to a fully sustainable, emissions-free future that aligns with decarbonization goals. It not only provides an eco-friendly choice but also serves as a means to store and distribute energy when renewable energy is scarce. This helps bridge the gap between intermittent renewable power and the consistent energy supply, improving overall energy system reliabilities (Hassan et al., 2023).

Hydrogen's transformative potential extends beyond production methods. Its adaptability across diverse applications is where it truly shines (Abdin et al., 2020). Hydrogen serves as a clean fuel for various transportation modes, including fuel cell vehicles, promoting sustainability in road, rail, and maritime transport (Singh et al., 2015). It is utilized in industrial processes, such as steel and ammonia production, thereby reducing the carbon intensity of these pivotal sectors (Liu, Zuo et al., 2021). Furthermore, hydrogen plays a central role in energy storage, assisting in managing the intermittency of renewable sources and ensuring a stable energy supply (Ould Amrouche et al., 2016).

Hydrogen plays a vital role in moving toward a cleaner, more sustainable energy future. It acts as a linchpin, connecting different sectors and industries, allowing for a comprehensive approach to reducing carbon emissions. This hydrogen-centered energy system not only reduces greenhouse gas emissions but also opens up possibilities for economic growth, job creation, and technological advancements, by embracing hydrogen as a central component of the energy landscape, a world where economic prosperity and environmental sustainability coexist harmoniously can be achieved, making significant strides in the fight against climate change (Hanley et al., 2018).

2.2 Different energy sources and their limitations

Conventional energy sources have been extensively used for many years and continue to dominate the world's energy supply. They do, however, have limitations in terms of their environmental impact, accessibility, and long-term sustainability. Coal, oil/petroleum, natural gas, nuclear energy, hydropower, biomass, and geothermal energy are among the most widely used conventional energy sources (Dash et al., 2023). Even though coal is a resource that takes millions of years to form and has limited reserves, it is abundant in certain regions. Large quantities of greenhouse gases, especially carbon dioxide (CO_2), are emitted during the combustion of coal, resulting in climate change and air pollution. In addition, coal mining can harm the environment and jeopardize the health of workers (Acar & Dincer, 2019). Crude oil is a versatile energy source that is used in the production of electricity, transportation, and various industrial activities. However, its reserves are limited, and it is a nonrenewable fossil fuel. Oil spills from extraction and transportation can cause significant harm to marine ecosystems. Burning petroleum products results in the production of CO_2 and other pollutants, which have an impact on air quality and contribute to global warming. Compared to coal and oil, natural gas has the cleanest combustion of all fossil fuels, releasing slightly less CO_2 and fewer other pollutants. It is often considered as a fuel for the transition to cleaner energy sources. Natural gas, however, remains a fossil fuel and contributes to greenhouse gas emissions. Water contamination and methane leaks, which are two environmental issues, may be caused by the extraction process, including hydraulic fracturing (fracking). Nuclear energy produces electricity by splitting atomic nuclei through a process called nuclear fission. However, its overall contribution to energy is not equal to that of oil, gas, or petroleum. Nuclear power facilities face challenges with nuclear waste disposal and the potential for disasters, such as the Chernobyl and Fukushima accidents, despite not emitting CO_2 directly during operation. Economic difficulties are also brought on by the significant costs associated with constructing and decommissioning nuclear power plants (Abdalla et al., 2018). Hydropower harnesses the energy of moving water to generate electricity. Despite being a renewable energy source, massive hydroelectric projects can have a negative impact on the environment. Dam construction can interrupt fish migration, alter ecosystems, and displace local inhabitants. Water supply and hydropower efficiency in certain areas may be affected by climate change (Rahman et al., 2022). Organic materials, such as wood, agricultural byproducts, and organic waste, serve as the primary sources of biomass energy. Burning biomass can release CO_2 and other pollutants, even though it is considered a renewable source of energy, which can have a negative impact on air quality. Furthermore, the extensive use of biomass for energy production may lead to deforestation and the loss of biodiversity (Abdalla et al., 2018). Geothermal energy harnesses the heat from deep within the Earth to produce electricity and heat. It is a clean and renewable source of energy. However, it is only applicable in areas where accessible geothermal resources are available. In some instances, drilling and fluid injection used in geothermal operations can also trigger minor earthquakes. The need for a transition to cleaner and more sustainable energy alternatives, such as renewable energy sources like solar, wind, hydro, and geothermal, as well as advancements in energy storage technologies, has emerged. This need exists despite the significant role that conventional energy sources have played in meeting the world's energy demands (Midilli et al., 2021). Although renewable energy source offers numerous advantages over conventional energy sources but they also have certain limitations. Solar and wind energy are both intermittent, meaning they are only available when weather conditions are favorable. Supply and demand must be balanced, which requires energy storage and grid management. Depending on the geographical weather condition including seasonal variation, some renewable energy sources, such as wind and solar, may experience significant fluctuations. Table 2.1 shows the top countries in the wind and solar energy sectors, showcasing their growth and contributions to global renewable energy generation (Energy Institute, 2023). For instance, in areas with heavy cloud cover or during the winter season, solar energy production may be reduced. Renewable energy sources often have lower energy densities than traditional fuels like coal and natural gas, which can affect storage and transportation (Yukesh Kannah et al., 2021). Wind energy is a clean and renewable resource that is abundant in the world's breezy landscapes, providing hope for environmentally friendly power solutions. With turbines towering toward the sky, wind farms represent innovation, transforming nature's gentle whispers into electricity. Despite the challenges, technological advancements are enhancing efficiency and reliability, paving the way for a future where sustainability and power can coexist. As per the data, a minimum wind speed of 2 m/s is required to start rotating most small wind turbines and at 25 m/s maximum, the turbine is stopped or braked (cut-out speed) and wind energy has huge potential to fulfill the energy requirements; however, the seasonal dependency as well their high cost with high maintenance cost is still a barrier.

TABLE 2.1 Top countries in the wind and solar energy sectors.

Country	Wind and solar generation (2015–20)	Total wind and solar generation in 2020	Share of global generation
China	+503 TWh	728 TWh	>20%
EU-27	–	540 TWh	~15%
United States	+503 TWh	469 TWh	~13%
Vietnam	+14 TWh	–	5% (2020)
South Korea	–	21 TWh	<1%
Brazil	+42 TWh	–	–
India	+80 TWh	76 TWh	~2%
Mexico	+21 TWh	–	–
Turkey	+23 TWh	–	–

Note: " + " indicates an increase in wind and solar generation from 2015 to 2020.

2.3 Renewable/green energy versus nonrenewable/fossil-based energy

The transition from renewable/green energy to nonrenewable/fossil-based energy represents a profound shift in the way societies produce and consume power. At the core of this transformation are several critical dimensions that influence economies, the environment, geopolitics, and human well-being (Saravanan et al., 2023). Table 2.2 provides a concise overview of the key dimensions differentiating renewable energy from nonrenewable/fossil-based energy.

2.4 Why hydrogen versus renewable energy sources

Hydrogen is often considered as a potential clean energy carrier and a complement to renewable energy sources. It has some unique characteristics that make it attractive for certain applications, but it also comes with its own set of challenges as mentioned below:

2.4.1 Energy storage

Renewable energy sources like solar and wind generate electricity intermittently, depending on weather conditions and seasonal variations. Effective energy storage systems are crucial to store excess energy during periods of high energy generation and utilize it when the demand is high. Batteries and other storage technologies are only used for this purpose (Hosseini et al., 2015). Hydrogen can be used as an energy storage medium. It can be produced through electrolysis, using excess electricity from renewable sources, and then stored for later use. Hydrogen can act as a flexible and scalable energy carrier, providing long-term storage for periods when renewable energy production is low or in case when energy demand is high (Megia et al., 2021).

TABLE 2.2 Renewable/green energy versus nonrenewable/fossil-based energy.

Dimension	Renewable energy	Fossil fuel energy
Environmental impact	Minimal to no emissions, smaller footprint	Significant emissions, environmental harm during extraction and use
Sustainability	Draws from essentially inexhaustible resources	Faces challenges of resource depletion and environmental damage
Energy security	Enhances security by reducing dependence on imports	Vulnerable to geopolitical tensions and market fluctuations
Economic impact	Fuels economic growth, job creation, and long-term cost reduction	Price volatility, extraction costs, and environmental remediation impact stability
Technological advancements	Benefits from rapid advancements, improving efficiency	Faces ongoing challenges like carbon capture and storage
Global transition driver	Driven by the need to reduce greenhouse gas emissions and combat climate change	Under pressure to adapt to environmental concerns

2.4.2 Emission reduction

Solar, wind, and other renewable sources are clean and produce little to no greenhouse gas emissions during electricity generation. They contribute significantly to reducing carbon emission mitigating climate change (Christopher & Dimitrios, 2012). The production of hydrogen can be emission-free if generated through electrolysis using renewable electricity. When used in fuel cells to produce electricity, hydrogen emits only water vapor, making it a zero-emission energy carrier.

2.4.3 Energy density

Electricity from renewable sources has a lower energy density, which can be a limitation for certain applications that require high energy densities, such as long-haul transportation or heavy industries. Hydrogen has a high energy density, making it a potential candidate for applications where energy density is crucial. It can be compressed or liquefied to enhance storage efficiency (Wang et al., 2019).

2.4.4 Infrastructure and distribution

Renewable energy sources require specific infrastructure for power generation and distribution, such as solar panels, wind turbines, and transmission lines. Grid integration can also present challenges (Tang et al., 2023). Hydrogen requires a dedicated infrastructure for production, distribution, and storage. Existing pipelines and distribution networks need modifications to handle hydrogen safely. The infrastructure development for hydrogen can be costly and time-consuming (Ren et al., 2017).

2.4.5 Efficiency and losses

Direct use of electricity from renewable sources can be more efficient than converting electricity into hydrogen and then back into electricity (via fuel cells) for some applications (Gahleitner, 2013). However, due to the intermittent nature of such energy resources, continuous use of this energy through on-grid is not feasible (Megia et al., 2021). Establishing off-grid arrangement by storing such energy enables continuous utilization of the energy sources (International Energy Agency, 2021). Table 2.3 shows the technologies for hydrogen production. The process of hydrogen production through electrolysis can have energy losses, reducing overall efficiency. Additionally, converting hydrogen back to electricity can result in efficiency losses as well.

In conclusion, both hydrogen and renewable energy sources have their unique strengths and limitations. While hydrogen can be a valuable energy carrier and an option for energy storage, it is essential to remember that the most sustainable approach involves maximizing the direct use of renewable electricity and implementing efficient energy storage solutions like batteries and pumped hydro storage (Moradi & Groth, 2019). Combining the strengths of renewable energy sources with hydrogen as a complementary energy carrier can play a crucial role in building a cleaner and more sustainable energy future.

TABLE 2.3 Technologies for hydrogen production.

Technology	Efficiency (%)	H_2 yield (g/kg feedstock)	Cost (\$/kg of H_2)	Advantages	Drawbacks
Steam reforming	74 − 85	40–130	1.83 − 2.35	Developed technology	Produces carbon coproducts
Gasification	30 − 60	40–190	1.77–2.05	Abundant and cheap feedstocks, carbon-neutral emissions	Requires oxidation agents, hydrogen yield depends on feedstock, tar formation
Electrolysis (water)	55 − 80	111α	4.15 − 10.30	Simplicity, low temperatures, zero carbon emissions	High pressure required, energy storage problems, high capital costs
Thermolysis	20 − 50	111α	7.98 − 8.40	Clean and sustainable, zero carbon emissions	Separation step required, high capital costs
Photo-electrolysis	0.06 − 14	111α	4.98 − 10.36	Contributes to energy supply sustainability, low operating temperature and pressure	Low efficiency, requires significant surface, requires photocatalytic material
Biophotolysis	10 − 15	111α	1.42–2.13	Can produce hydrogen at ambient conditions, consumes CO_2	Requires significant surface area, difficult operations to control bacteria, large reactor volume
Dark fermentation (biomass)	60 − 80	4–44	1.68 − 2.57	Can produce hydrogen at any time, simple reactor design, waste recycling	Hydrogen yields metabolically restricted, high byproduct generation, low efficiency, large reactor volume
Photofermentation (biomass)	0.1 − 12	9–49	2.57–2.83	Hydrogen production from various substrates, waste recycling	Strict control of environmental conditions, requires significant surface area, nitrogenase metabolism restricts economic viability, large reactor volume
Pyrolysis (biomass)	35 − 50	25–65	1.59 − 2.20	Developed technology, abundant and cheap feedstock	Hydrogen yield depends on feedstock, tar formation
Hydrothermal liquefaction	85 − 90	0.3–2	0.54 − 1.26	Abundant and cheap feedstocks, high energy efficiency	Hydrogen yield depends on feedstock, presence of nitrogenated compounds

Note: "α" is calculated from the water-splitting reaction stoichiometry.

2.5 Hydrogen versus fossil fuels

Hydrogen, produced from renewable sources, is a promising competitor to fossil fuels in the fight against greenhouse gas emissions. Green hydrogen, made through electrolysis with renewable electricity, is low in carbon emissions, countering the environmental impact of conventional fossil fuels (Taipabu et al., 2022). However, hydrogen's different storage and transportation requirements present logistical challenges compared to liquid fossil fuels. Hydrogen excels in energy conversion efficiency, especially in fuel cells, while fossil fuel power plants often have lower efficiency due to combustion-related losses. Its versatility allows use in diverse applications. Still, transitioning from fossil fuels to hydrogen demands infrastructure changes, including hydrogen refueling stations and adjustments to industrial processes (Vidas & Castro, 2021). Geographic availability may also vary. The path to hydrogen as a clean energy alternative will require addressing these challenges as the world moves away from fossil fuels toward hydrogen-based energy solutions (Taipabu et al., 2022).

2.6 Hydrogen versus batteries

The global transition to clean and renewable energy sources has sparked a growing interest in effective energy storage solutions. Two prominent contenders in this field are batteries and hydrogen, each offering distinct advantages and challenges (Manisalidis et al., 2020). Hydrogen stands out with its higher energy density, making it efficient for storing substantial energy in a relatively compact space, which is especially advantageous for long-range transportation where quick refueling and extended driving distances are essential (Gibb, 2021). However, cost of production and the development of a robust refueling infrastructure remain significant challenges to hydrogen's widespread adoption.

The choice between hydrogen and battery technology depends on specific usage scenarios and geographic factors. While batteries using readily available materials like sodium, sulfur, and lithium-iron have their practical merits, they face limitations in energy storage capacity, which could result in increased costs due to resource constraints. On the other hand, geological storage methods, such as underground hydrogen storage, offer substantial storage potential without significant material limitations, provided that suitable and environmentally friendly sites can be identified (Hund et al., 2020). In the context of lithium-ion batteries, the global lithium market has experienced significant growth, reaching a value of approximately USD 2.7–4.09 billion in recent years, and it is anticipated to further increase to USD 6.62–8.24 billion by 2027–28, with a compound annual growth rate of 8.1%–19% (Fan et al., 2020). It is worth noting that approximately 96.4% of the world's lithium reserves are concentrated in countries such as Chile, Australia, Argentina, China, the United States, and Zimbabwe. In 2020 Australia, Chile, and China collectively accounted for 86% of global lithium production, which had risen substantially from 28.2 million tonnes in 2010 to an average of around 90 million tonnes from 2018 to 2020 (Chayambuka et al., 2020). The automotive industry, representing over 43% of the market, was a significant driver of demand for both lithium-ion and lithium-ion batteries in 2020. Duration plays a critical role in determining the appropriate storage technology. While lithium-ion batteries excel in short-term applications, the need for extremely long-term storage (lasting multiple days) is anticipated to become crucial only in the later stages of the clean energy transition (Durmus et al., 2020). Hydrogen, with its potential for large-scale and long-duration storage, becomes a more cost-effective option for such scenarios. In the short term, lithium-ion batteries maintain their dominance as the go-to technology for energy storage, particularly for durations of up to 4 hours. Their relatively high energy density and efficiency make them suitable for applications such as grid stabilization and short-duration energy storage. Nevertheless, the limitations of batteries become evident in long-term storage scenarios, both in terms of cost and resource constraints (Eyer & Corey,

2010). India, with its growing renewable energy sector and the imperative to meet climate agreements, may gradually shift its focus toward hydrogen as the ultimate long-term energy storage solution (Maton et al., 2013). Scaling up hydrogen production can significantly reduce capital costs, making it a competitive option, especially when considering the limitations in lithium resources and the need for extensive, prolonged energy storage (Zhang, Lundblad et al., 2016).

In conclusion, in the global pursuit of sustainable energy solutions, both batteries and hydrogen have pivotal roles to play. While batteries continue to dominate short-term storage, hydrogen emerges as a promising candidate for long-duration and large-scale storage needs, offering a viable pathway for countries like India to achieve their clean energy goals effectively (Vidas & Castro, 2021). The key lies in understanding the specific requirements of each application and leveraging the strengths of both technologies to create a resilient and sustainable energy future (Applied Sciences | Free Full-Text | Recent Developments on Hydrogen Production Technologies: State-of-the-Art Review with a Focus on Green-Electrolysis, no date).

A complete comparison of different energy sources is shown in Table 2.4 (Brockway et al., 2019; Do Thi et al., 2021; Elavarasan, 2019; Vate, 1997).

2.7 Hydrogen as a potential energy carrier and its importance

Hydrogen, as an energy carrier, holds significant promise in addressing the intermittent nature of renewables and seasonal supply–demand gaps in the current energy mix. Its high energy density, potential for large-scale storage, and versatility make it a valuable component of the future energy landscape (Abdin et al., 2020). As the world moves toward a sustainable and low-carbon energy future, the importance of hydrogen in providing clean energy solutions cannot be overstated. The upcoming sections will delve into the various aspects of hydrogen as an energy carrier, including its production methods, utilization in different industries, and the challenges and opportunities it presents in transforming the global energy system (Abdin et al., 2020). Amidst the limitations of renewables, hydrogen is gaining attention as a promising energy carrier to address energy storage and seasonal supply–demand gaps. Additionally, hydrogen has one of the highest energy densities by weight, making it suitable for large-scale energy storage and long-distance transportation (Nikolaidis & Poullikkas, 2017). The major advantages of hydrogen as an energy carrier are shown in Fig. 2.3.

2.8 Applications of hydrogen as an energy carrier

Both static and mobile applications for hydrogen exist in a variety of industries. In order to explain the application type and the usage of hydrogen, the terms "static" and "dynamic" are frequently used (Nikolaidis & Poullikkas, 2017). When hydrogen is used as a stationary or nonmoving energy source, this is referred to as a static application. Typically, these applications involve generating heat or electricity at a specific location. In contrast, hydrogen is used in dynamic applications in mobile systems, such as

TABLE 2.4 Comparison for different energy sources.

Energy source	Uses	Limitations	Calorific value (MJ/kg)	Environmental impact	Energy density (MJ/m^3)	Storage potential	Availability/ intermittency	Other parameters
Coal	Electricity generation, industrial processes, heating	High greenhouse gas emissions, air pollution, finite reserves	24–33	High	24–33	Bulk storage	Stable supply	High carbon footprint, resource depletion
Natural gas	Electricity generation, heating, industrial processes, transportation	Greenhouse gas emissions, methane leakage, finite reserves	44–55	Moderate to low	35–45	Onsite and pipeline	Stable supply	Abundant, cleaner than coal
Oil (petroleum)	Transportation, heating, electricity generation, petrochemicals	Greenhouse gas emissions, air pollution, oil spills, finite reserves	42–47	High	34–47	Tank farms	Stable supply	Energy-dense, widely used in transport
Nuclear power	Electricity generation	Radioactive waste disposal, nuclear accidents, high initial costs	N/A	Low	N/A	Onsite	Stable supply	Low carbon emissions, radioactive waste
Solar energy	Electricity generation, heating, water heating, small-scale applications	Intermittent, weather dependent, land use for large-scale installations	N/A	Low	N/A	Batteries, grid	Seasonal variations	Abundant, clean, decentralized energy
Wind energy	Electricity generation, pumping water, small-scale applications	Intermittent, visual and noise impacts, land use for large-scale installations	N/A	Low	N/A	Batteries, grid	Seasonal variations	Abundant, clean, limited footprint
Hydropower	Electricity generation, irrigation, water supply	Environmental impact on ecosystems, limited suitable locations	N/A	Low	N/A	Reservoirs	Seasonal variations	Clean, stable, reservoir-induced seismicity
Biomass energy	Heating, electricity generation, biofuels	Emissions from burning, land use, competition with food crops	14–19	Moderate	15–19	Silos, tanks	Continuous supply	Renewable, carbon-neutral
Geothermal energy	Electricity generation, heating	Limited to geothermal hotspots, resource depletion, hydrogen sulfide emissions	N/A	Low	N/A	Reservoirs	Stable supply	Clean, location-specific resource
Ocean energy	Electricity generation	High capital costs, environmental impact on marine ecosystems	N/A	Low	N/A	Batteries, grid	Seasonal variations	Abundant, diverse technologies available
Hydrogen	Electricity generation, transportation, heating, industrial processes	Hydrogen production methods, storage challenges, high infrastructure costs	120–142	Low	8.5–10.1	High-pressure tanks	Intermittent supply	Clean, versatile, energy storage potential

Note: Calorific values are approximate and can vary based on the type and quality of the energy source. N/A indicates not applicable.

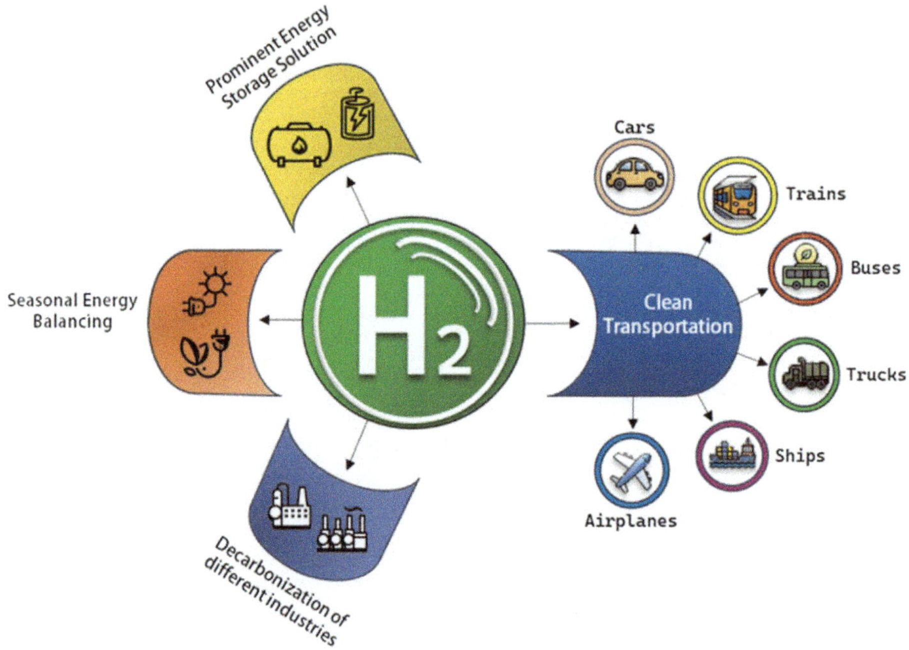

FIGURE 2.3 Applications of hydrogen as an energy carrier.

transportation, where energy is required while the system is in motion (Sharma & Ghoshal, 2015). Fig. 2.4 lists the various static and dynamic applications of hydrogen. The widespread use of hydrogen in the transportation sector faces challenges such as infrastructure development, cost, and hydrogen production methods. These methods include green hydrogen from renewable sources and gray or blue hydrogen from fossil fuels with carbon capture (Salvi & Subramanian, 2015). Despite these obstacles, ongoing research and technological advancements aim to enhance the utilization of hydrogen in stationary and mobile applications, as part of the transition toward a more eco-friendly and low-carbon energy system (Hren et al., 2023).

2.9 Different methods of hydrogen production

To fulfill the rising need for clean energy and solve environmental issues, hydrogen production is crucial. Hydrogen is produced using various techniques, each with its advantages and difficulties. Conventional processes like electrolysis and steam methane reforming (SMR) are commonly employed, while advanced approaches like photoelectrochemical water splitting and biological hydrogen production hold promise for efficient and sustainable hydrogen production (Ahmed et al., 2022). The shift to an energy economy based on hydrogen depends on these techniques' ongoing development and improvement.

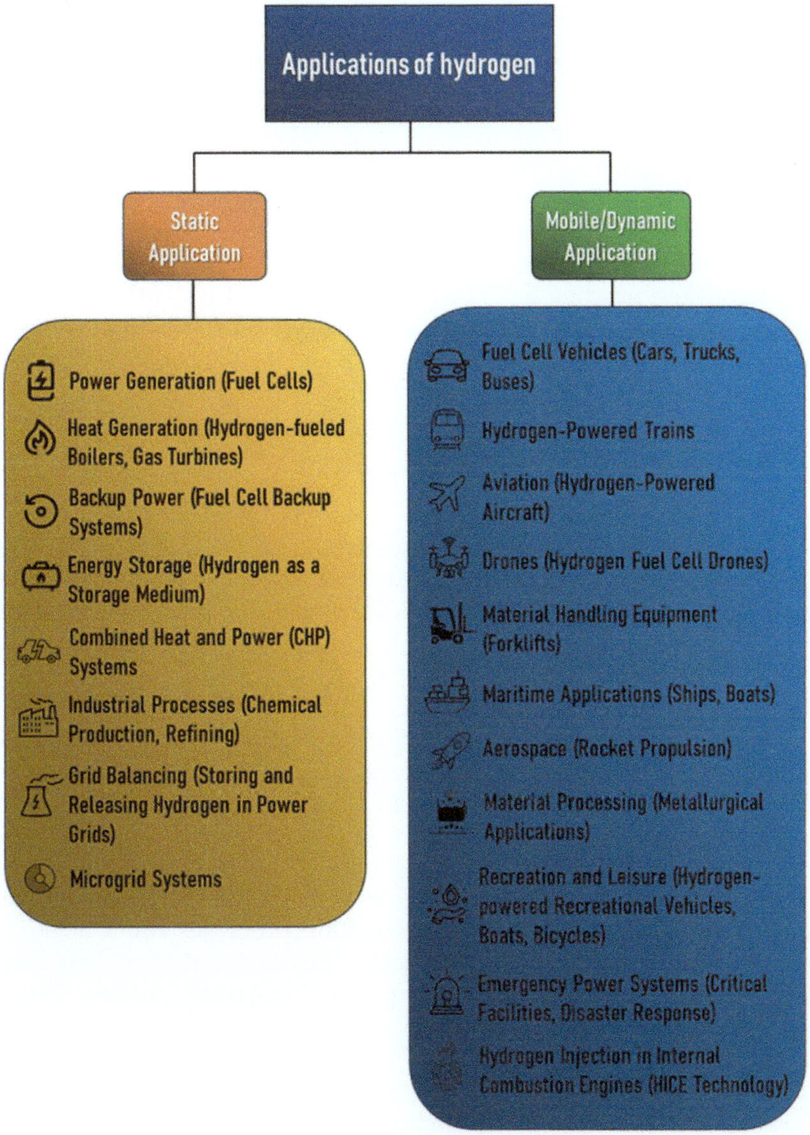

FIGURE 2.4 Applications of hydrogen in different sectors.

2.9.1 Steam methane reforming

SMR is a well-established and extensively utilized technique for hydrogen production, making it a crucial contributor to fulfilling the growing worldwide need for clean energy solutions (Simpson & Lutz, 2007). SMR is a chemical process where methane (CH_4) and steam (H_2O) react in the presence of a catalyst, typically nickel based. The main chemical reaction can be described as:

$$CH_4 + H_2O \rightleftharpoons CO + 3$$

In this energy-absorbing reaction, methane and steam combine to form carbon monoxide (CO) and hydrogen gas (H_2). The resulting hydrogen-rich gas mixture, known as synthesis gas or syngas, undergoes purification to obtain a high-purity hydrogen stream. The pressure swing absorber (PSA), which removes pollutants from the gas, receives the hydrogen-rich gas that exits the reformer. The absorber normally produces 99.9% pure hydrogen. The hydrogen product gas is what is meant by this (Türker Üzden et al., 2023). The SMR process generates over 95% of the hydrogen used worldwide. In this process, CO_2 is also produced. It is known as "gray hydrogen." If CO_2 collection is involved, this is referred to as "blue hydrogen" (Cheng & Ng, 2022).

2.9.2 Water and renewable-powered electrolysis

Electrolysis is the chemical process where water (H_2O) is broken down into its parts, hydrogen (H_2) and oxygen (O_2). Hydrogen gas, which may be utilized as a safe and flexible energy carrier, is created using this procedure. Traditional water electrolysis and electrolysis using renewable energy sources are the two basic forms of electrolysis (Hajjaji et al., 2016; Palmer et al., 2021).

2.9.2.1 Water electrolysis

Traditional water electrolysis involves passing an electrical current through water. Typically, an aqueous solution containing an electrolyte, such as potassium hydroxide or sulfuric acid, is used to do this. The main goal of adding an electrolyte is to increase the solution's total conductivity (Zhou et al., 2021). This electrochemical reaction takes place in an electrolysis cell, a device made up of two different electrodes: an anode and a cathode. The electrolysis process can proceed more quickly since these electrodes are completely immersed in the electrolyte solution. While reduction processes occur at the cathode, oxidation occurs at the anode. The water molecules break down into their component elements, specifically hydrogen, by this meticulously regulated process (Shiva Kumar & Himabindu, 2019).

Anode reaction:

At the anode (positive electrode), water is oxidized, releasing oxygen gas:

$$2H_2O_{(l)} \rightleftharpoons O_{2(g)} + 4H^+{}_{(aq)} + 4e^-$$

Cathode reaction:

At the cathode (negative electrode), hydrogen ions are reduced to form hydrogen gas:

$$4H^+{}_{(aq)} + 4e^- \rightleftharpoons 2H_{2(g)}$$

Overall reaction:

$$2H_2O_{(l)} \rightleftharpoons 2H_{2(g)} + O_{2(g)}$$

The process of water electrolysis can demand significant energy, particularly if powered by nonrenewable sources, and may result in the production of greenhouse gases unless renewable energy options are used.

2.9.2.2 *Renewable-powered electrolysis*

Renewable-powered electrolysis entails using energy generated from environmentally friendly sources such as wind, solar, or hydroelectric power to power the electrolysis process (Dincer, 2012). This approach is usually referred to as "green hydrogen production" since it has the potential to achieve carbon neutrality or even a carbon-free outcome, depending on the renewable energy source used (Kim et al., 2018). In contrast, "gray hydrogen" relies on fossil fuels and emits carbon dioxide during production. Additionally, "yellow hydrogen" is produced through electrolysis, capturing and storing carbon emissions, transitioning toward sustainable hydrogen production. Transitioning from gray to green, with yellow as a middle step, is crucial for reducing the carbon footprint in hydrogen production. Some of the different categories are as follows:

Solar-powered electrolysis: Sunlight is captured by solar panels and transformed into electrical energy, which is then used to power the electrolysis procedure. The creation of hydrogen using solar energy is an environmentally benign process (Kabir et al., 2018).

Wind-powered electrolysis: Wind turbines harness the energy of the wind to produce electricity, and this electricity may be used to speed up the electrolysis procedure. An alternate method for producing ecologically friendly hydrogen is wind-driven electrolysis (Almutairi et al., 2021).

Hydroelectric-powered electrolysis: To produce electricity, hydroelectric power plants use the kinetic energy of moving water. The electrolysis procedure may make use of this sustainable energy source. It is an environmentally friendly way to carry out electrolysis.

By substituting fossil fuels with renewable sources like hydroelectric (82%) and geothermal (18%) for grid energy generation, it is possible to mitigate up to 90% of potential emissions. However, despite its considerable potential, water electrolysis currently contributes to less than 0.1% of the global hydrogen supply (Cho et al., 2023). Renewable-powered electrolysis is recognized as a crucial catalyst in the generation of "green hydrogen," a form of hydrogen production that is entirely devoid of greenhouse gas emissions. Green hydrogen finds diverse utility, serving as a clean transportation fuel, an efficient energy storage solution, and a fundamental resource for a range of industrial applications.

2.9.3 Coal gasification and biomass gasification

Gasification techniques for coal and biomass are separate ways of converting solid carbon-based resources into synthesis gas, often known as syngas (Sarafraz & Christo, 2020). However, there are significant differences in the raw materials employed and the environmental repercussions. Coal gasification is a chemical process that converts coal into syngas, a mixture of hydrogen, carbon monoxide, and other minor gases, by heating coal at high temperatures while controlling the quantities of oxygen and/or steam. This regulated process starts the partial burning of coal, which results in the production of syngas (Messerle et al., 2017). The advantage of this approach is its ability to create a flexible gas that can then be used to make a variety of goods such as fuels, chemicals, and power. As a thermochemical process, biomass gasification is similar to coal gasification, but it concentrates on converting biomass resources such as wood, agricultural waste, and organic waste into syngas (Alamia et al., 2017). To avoid total combustion, this process involves controlled heating of biomass under oxygen-free circumstances (anaerobic settings). This results in syngas generation, which includes hydrogen, carbon monoxide, and other gases, similar to coal gasification. Biomass gasification is distinguished from coal gasification by its environmental friendliness, which is achieved by utilizing renewable feedstocks and contributing to the reduction of greenhouse gas emissions (Mehmeti et al., 2018; Messerle et al., 2017).

In the global trend toward cleaner and more ecologically friendly energy choices, biomass gasification is gaining popularity due to its potential to utilize renewable resources while also helping to reduce carbon emissions.

2.9.4 Metal water reactions

The interaction between particular metals and water is critical in the generation of hydrogen. These interactions set off reactions that result in the release of hydrogen gas (Guduru et al., 2024c). These reactions are divided into two categories: alkali metals and alkaline earth metals (Yang et al., 2015). Alkali metals like sodium and potassium are very reactive in water and produce powerful exothermic reactions (Liu et al., 2014). This produces metal hydroxides and emits hydrogen gas along with heat and occasionally sparks. These responses are reactive, and if not controlled appropriately, they might become dangerous (Liu et al., 2014).

Alkaline earth metals, such as magnesium and calcium, may also react with water to produce hydrogen gas, albeit their reactivity is often lower than that of alkali metals. When compared to alkali metals, these reactions are less strong and explosive. Water interactions with alkaline earth metals produce metal hydroxides and release hydrogen gas (Petrovic & Thomas, 2008).

Transition metals, which include elements such as aluminum, zinc, iron, and titanium, can react with water to produce hydrogen gas under certain conditions. However, as compared to alkali or alkaline earth metals, these reactions are often less strong and exothermic. Water interactions with these transition metals produce metal hydroxides, which release hydrogen gas. The extent and reactivity of the reaction are affected by variables such as metal, temperature, and environmental conditions.

2.10 Advancements in hydrogen production technologies

Hydrogen is developing as a promising sustainable energy carrier due to its ability to decarbonize different sectors, including transportation and industrial activities. As shown in Fig. 2.5, water splitting and biomass technology are two major advances in hydrogen production technologies.

FIGURE 2.5 Advancement in hydrogen production technology.

2.10.1 Water splitting technologies

Thermolysis: Development in thermochemical water-splitting techniques focusing on compounds of iodine and sulfur, along with the utilization of high-temperature gas-cooled reactors (HTGR) and Bunsen reactors, capable of achieving hydrogen production rates of 10 L/h in 8 hours, offering potential for large-scale implementation. However, challenges include the management of formed SO_2 to prevent adverse environmental effects (Noguchi et al., 2017). Significant progress has been made in advancing nuclear hydrogen production using the iodine-sulfur (IS), featuring iodine and sulfur compounds, along with HTGR and Bunsen reactors, capable of achieving hydrogen production rates of 60 NL/h while avoiding CO_2 emissions, showcasing efficient hydrogen generation. Nonetheless, ensuring the consistent and sustainable operation of the closed cycle IS process remains a challenging endeavor (Naterer et al., 2017). Electrolysis: Technoeconomic analysis for the establishment of extensive hydrogen production facilities involving water electrolyzers (with capacities ranging from 4000 to 40,000 kg H_2/d) reveals a carbon-free process that utilizes renewable resources. However, this approach is associated with higher costs (Nguyen et al., 2019). Solid oxide electrolysis cells (SOEC) were utilized to divide untreated seawater, and the electrochemical effectiveness of this process was examined. Seawater electrolyzers demonstrated a flow rate of 183 mL/min, showcasing elevated energy conversion efficiency and reduced electricity demand. However, the presence of sea salt in seawater poses a potential risk of damaging the electrolyzers (Liu, Han et al., 2021). Photo-electrolysis: Investigating the impact of sunlight on hydrogen generation through water electrolysis, this study delves into the utilization of water photovoltaic electrolyzers. With an efficiency of 53%, it reveals that the polarizability and prevalence of sunlight bolster the electric field during water electrolysis, thereby augmenting the water splitting process. However, the interaction of opposing electric fields can lead to a decrease in hydrogen production (Bidin et al., 2017).

2.10.2 Biomass technologies

Thermochemical: Biomass steam gasification is explored to enhance comprehension, specifically assessing the impact on hydrogen yield, as well as thermal and hydrogen efficiency. The study involves wood residue and employs a fluidized bed configuration, resulting in a hydrogen weight percentage of 5.9% and a gasification efficiency of 1.22.

Employing a research-scale fluidized bed gasifier known for its uniform temperature distribution and versatility in handling varied particles, the research yields precise outcomes. However, potential challenges arise from larger bubble sizes that could lead to gas bypassing the bed (Fremaux et al., 2015).

Biological: Bio-photolysis (indirect photolysis), the research explores hydrogen production via immobilized *Synechocystis* sp. cells. This investigation takes place within a photobioreactor and results in a hydrogen yield of 40.6 ± 4.9 mmol H_2 per mg chl[1] (chlorophyll) per hour. The process demonstrates efficient hydrogen production within a short timeframe. Notably, cells immobilized in an alginate-containing culture medium maintained viability for over 30 days (Touloupakis et al., 2016). Challenges include potential cell leakage in cases where immobilization occurs within the culture medium.

Photo-fermentation involves the utilization of molasses for producing hydrogen through photo-fermentative processes in a fed-batch operational mode. This is achieved using *Rhodobacter* capsulates within a tubular photobioreactor, resulting in a hydrogen production rate of 0.31 mol $H_2/m^3/h$ (Kayahan et al., 2017).

2.11 Grades of hydrogen production

Gray hydrogen: Gray hydrogen is commonly generated from hydrocarbons such as natural gas or coal using the SMR process, which is renowned for its cost-effectiveness. However, it is accompanied by substantial CO_2 emissions, making it environmentally unfriendly due to its substantial carbon footprint (Dawood et al., 2020).

Blue hydrogen: Blue hydrogen is derived from hydrocarbons, primarily natural gas, but it incorporates CCS technology to capture and store the CO_2 emissions generated during the production process. This approach significantly reduces CO_2 emissions, making blue hydrogen a more environmentally friendly alternative to gray hydrogen (Dawood et al., 2020).

Turquoise hydrogen: Turquoise hydrogen, a variation of blue hydrogen, distinguishes itself by adopting methane pyrolysis technology for hydrogen production, as opposed to traditional SMR. This innovative approach has the potential to achieve even greater reductions in CO_2 emissions because it prevents the release of CO_2 during the production process, making it an exciting development in environmentally friendly hydrogen production (Hanley et al., 2018).

Green hydrogen: Green hydrogen is generated through electrolysis, a process that utilizes renewable energy sources like wind, solar, or hydropower to split H_2O into H_2 and O_2. This method is regarded as highly sustainable because it produces no direct greenhouse gas emissions. Green hydrogen is of superior environmental quality, making it a favorable choice for eco-friendly hydrogen production (Dawood et al., 2020).

White hydrogen: White hydrogen is produced through the electrolysis of water, with nuclear power as its primary energy source. White hydrogen relies on nuclear energy rather than renewable sources. Notably, white hydrogen is still in its early stages of development and requires ongoing advancements in nuclear technology to enhance its feasibility and become a more widely adopted method of production (Kumar et al., 2023).

Yellow hydrogen: Yellow hydrogen is created through the gasification of biomass or other organic materials. This method involves utilizing agricultural and forestry residues, organic waste, and energy crops as biomass feedstocks. Yellow hydrogen production is regarded as a renewable and low-carbon process, contributing to sustainable hydrogen generation while effectively managing organic waste materials. This eco-friendly approach holds promise for reducing carbon emissions and fostering a more environmentally conscious energy landscape (Kumar et al., 2023).

Brown hydrogen: Brown hydrogen is generated from carbon-intensive fossil fuels like coal or lignite, typically through gasification or similar methods. Unlike gray hydrogen, it does not involve SMR but is still reliant on these fossil fuels. As a result, brown hydrogen is considered an environmentally less friendly option, characterized by higher CO_2 emissions. It is not deemed a sustainable or clean approach to hydrogen production and is associated with a significant environmental footprint (Kumar et al., 2023).

Pink hydrogen: Nuclear power plants offer a promising avenue for the production of "pink hydrogen," presenting various methods that could significantly alleviate greenhouse gas emissions. In contrast to natural gas boilers, nuclear power plants have the potential to generate high-quality steam devoid of pollutant emissions. This steam can be effectively utilized in the SMR process, providing a cleaner alternative. The utilization of high-quality steam in the process can be coupled with innovative technologies such as the solid oxide steam electrolyzer. This technology exhibits high scalability, application flexibility, and the capability to produce pure hydrogen, all while demonstrating enhanced reaction kinetics at elevated temperatures. This makes the solid oxide steam electrolyzer a particularly promising method for hydrogen production. An alternative technique for nuclear hydrogen production involves the thermal decomposition of methane, also known as methane pyrolysis. This process represents another avenue for harnessing the potential of nuclear power plants to produce hydrogen in an environmentally friendly manner (Hosseini, 2022).

2.12 Conversion of hydrogen into electricity or energy

This procedure comprises harnessing the stored energy inside hydrogen molecules and utilizing it to generate electricity via various ways such as fuel cells or combustion mechanisms. By investigating the physics and practical applications of converting hydrogen into useable energy forms, we open the door to novel methods to achieving more environmentally friendly and efficient energy systems.

2.12.1 Fuel cells: electrochemical conversion of hydrogen

Fuel cells are very advanced electrochemical devices that convert hydrogen gas directly into energy. This conversion occurs through a sophisticated mechanism in which hydrogen molecules combine with oxygen to produce energy, heat, and water. Fuel cells' electrochemical properties support their high efficiency, which typically outperforms traditional combustion-based solutions (Peng & Wei, 2020). They provide a clean and efficient source of electricity with little emissions, making them suitable for a wide range of applications such as transportation, stationary power generating, and portable electronic devices.

2.12.2 Combustion: hydrogen in gas turbines and internal combustion engines

The use of hydrogen in gas turbines and internal combustion engines provides an additional method of using its energy capability. The rapid combustion of hydrogen accelerates turbines in gas turbines, creating both mechanical power and electricity. Internal combustion engines, on the other hand, use ignited hydrogen to drive pistons, resulting in mechanical work creation. Notably, hydrogen combustion is nonharmful to the environment, producing water vapor as a byproduct (Zhang et al., 2023). However, it is crucial to note that, while combustion is well-established, it is typically less efficient than the electrochemical processes used by fuel cells.

2.12.2.1 Efficiency and conversion processes

The efficiency of transforming hydrogen into electricity or energy relies on the chosen conversion methods. Hydrogen can be transformed into energy using a range of techniques, each offering distinct levels of efficiency, along with their own sets of benefits and drawbacks. The diversity of methods allows for flexibility in harnessing hydrogen's potential as an energy source. A conversion process, its efficiency, advantages, and disadvantages, is shown in Table 2.5.

The efficiency levels of hydrogen generation processes such as SMR, electrolysis, and methane pyrolysis range from around 60% to more than 80%. Concurrently, storage technologies like as compressed and liquid hydrogen storage, although reaching efficiencies typically in the 90%–95% region, include complexities including issues such as energy losses during compression and expansion procedures (Tashie-Lewis & Nnabuife, 2021).

2.13 Pure hydrogen as an energy carrier

Pure hydrogen, as an energy carrier, can transform global energy use. It stores and delivers energy across sectors, integrating renewables where fossil fuels once dominated. It powers clean industrial processes and fuel cells, driving sustainability for a cleaner future. The term "pure hydrogen" refers to hydrogen gas that contains just hydrogen molecules (H_2) and no additional substances or components (Lubitz & Tumas, 2007). In its pure molecular state, hydrogen, the lightest and simplest element on the periodic table with an atomic number of 1, takes the form of H_2 molecules, in which two hydrogen atoms are tightly bound together.

Producing pure hydrogen offers numerous advantages, mainly tied to its capacity as a clean and adaptable carrier of energy. The following are some of the primary benefits:

- Hydrogen is clean—burning or using it in fuel cells only produces water vapor. This makes it an attractive alternative to polluting fossil fuels (Tarhan & Çil, 2021).
- Using pure hydrogen as fuel can greatly decrease greenhouse gases, helping fight climate change, especially when made from renewable sources like wind, solar, or water power (Amin et al., 2022).
- Extra energy from renewable sources is stored in hydrogen for later use, stabilizing the system by converting it back to electricity or fuel as needed (Zhang, Zhao et al., 2016).
- Fuel cells are powered by pure hydrogen, which uses oxygen to produce electricity while releasing only water and heat as waste. These effective cells have a variety of applications, including stationary electrical and hydrogen transportation (Rahim Malik et al., 2023).
- Hydrogen is a reliable, environmentally friendly energy source that may be added to a sustainable energy mix when produced using clean energy sources (Fischedick et al., 2005).

TABLE 2.5 Conversion process, its efficiency, advantages, and disadvantages.

Method	Process	Efficiency	Advantages	Disadvantages
Fuel cells	Electrochemical devices, fuel cells convert hydrogen and oxygen into energy and water. An electrolyte is sandwiched between an anode (which receives hydrogen) and a cathode (which receives oxygen or air). They generate electricity and create water as the sole residual output by guiding hydrogen ions through the electrolyte (Peters et al., 2024).	Fuel cells have a high efficiency, often ranging between 40% and 60% in most commercial applications.	Emission of water vapor as the sole byproduct of combustion. Utilization of simple infrastructure similar to traditional combustion systems. Capability to launch quickly and operate in a responsive manner.	Elevated material costs. Restricted refueling infrastructure for hydrogen vehicles. Vulnerability to impurities in hydrogen fuel. Issues related to hydrogen storage and transportation.
Combustion	Hydrogen combusts in engines or turbines like fossil fuels release energy for mechanical work (Fan et al., 2018).	Hydrogen combustion is less efficient than fuel cells, with around 25%–35% efficiency in internal combustion engines. Turbines can achieve better efficiency.	Emission limited to water vapor. Utilizes uncomplicated infrastructure akin to current combustion tech. Capable of swift startup and responsive performance.	Lower efficiency than fuel cells. Emissions (nitrogen oxides) during high-temperature, high-pressure combustion. Difficulties in achieving emission-free complete combustion.
Hydrogen direct injection	By using direct injection alongside gasoline or diesel, hydrogen can improve traditional engine performance while also lowering pollutants (Wang et al., 2021).	The efficiency improvements from hydrogen injection vary depending on engine type and hydrogen ratio, improving combustion efficiency and decreasing pollutants.	Enhanced combustion efficiency. Reduced emissions, especially in hybrid systems. Leverages existing internal combustion engine infrastructure.	Modest efficiency gains versus advanced tech like fuel cells. Challenges in precise combustion control for top efficiency.

Efficient and sustainable production of pure hydrogen involves overcoming various challenges. Some of these key obstacles include:

- On Earth, hydrogen is a rare element that is typically found in combination with other elements. Natural gas (by reforming), water (by electrolysis), and biomass are examples of common sources. Each source has difficulties, such as emissions from gas extraction or the electrolysis process' high energy need (Acar & Dincer, 2019).

- The low volume energy density of hydrogen hampers storage and transportation, generating material difficulties and possible leaks. It is critical to develop safe and efficient storage and transportation techniques (Rong et al., 2024).
- The goal of producing hydrogen is to provide a clean energy carrier, but if it generates carbon dioxide (which SMR does), its environmental value decreases. CCS can help, but it is complicated and expensive (Prato-Garcia et al., 2023).

2.14 Blending hydrogen with natural gas

The potential benefits of a blend of H_2 and natural gas, which is mostly made up of methane (CH_4), are being studied for a variety of uses. This mixture has potential as a transitional fuel, especially as we migrate to an energy system that is more ecologically sustainable (Jia et al., 2023). Hydrogen and natural gas are combined when they are blended to create a composite fuel that can easily be used in natural gas appliances and infrastructure that already exist. Beginning with the creation of hydrogen using processes like electrolysis, SMR, or analogous methods, this process involves a number of intricate steps. The produced hydrogen is then carefully blended with natural gas in a variety of ratios, with common blends often containing between 5% and 20% hydrogen, but larger percentages are still technically feasible (Haeseldonckx & D'haeseleer, 2007). Notably, this mixed fuel seamlessly integrates with the current natural gas distribution and pipeline networks, eliminating the need for major adjustments. This mixed gas has several uses, including the production of energy, building space heating, and the facilitation of industrial operations, all without the need for significant equipment and apparatus modification (Haeseldonckx & D'haeseleer, 2007).

This novel strategy has several advantages, with the reduction of carbon emissions, improvement of energy security, and ease of a gradual move away from traditional fossil fuels standing out. This paradigm, however, is not without difficulties; specifically, increased levels of hydrogen incorporation may call for adaptational equipment adjustments, perhaps affecting the dynamics of combustion, and necessitating cautious assurance of secure distribution methods. Notably, this hybrid system's economic viability depends on striking a precise balance between production costs, blending challenges, and the potential need for upgrades to existing infrastructure and technology (Wang et al., 2024). In order to support the overarching goals of reducing carbon emissions and advancing the broader agenda of sustainable energy transformations, hydrogen and natural gas blending has the potential to act as a transitional bridge toward a more hydrogen-oriented energy framework.

2.15 Challenges and opportunities

The key challenges and opportunities associated with hydrogen while used as an energy carrier have been summarized in Table 2.6.

TABLE 2.6 Key challenges and opportunities associated with hydrogen.

Category	Challenges	Opportunities
Infrastructure (Kim et al., 2014)	Establishment of production facilities, storage tanks, and distribution networks. Grid integration into existing energy grids. Global hydrogen supply chain construction (pipelines, ports, transportation systems). Creation of local distribution points.	Job growth and support for local economies. Decentralized production near renewable sources for improved energy efficiency. Green hydrogen hubs centralized around renewable energy for efficient and sustainable production. International collaborations for shared infrastructure development, reducing costs and increasing efficiency.
Technical	Need for efficient and cost-effective production methods (electrolysis). Low volumetric energy density of hydrogen. Safety and energy efficiency in transportation. Advancements in fuel cell technology.	Ongoing research in electrolysis technology for improved energy efficiency. Innovative storage solutions (metal hydrides, liquid organic hydrogen carriers) for increased efficiency and safety. Advancements in fuel cell technology for broader adoption across sectors. Exploration of new materials and processes to enhance the efficiency of hydrogen production and storage.
Commercial (Rasul et al., 2022)	Higher production costs compared to conventional fossil fuels. Regulatory uncertainty impacting private sector investment. Market acceptance challenges requiring a shift in perceptions.	Government incentives, subsidies, and carbon pricing mechanisms to bridge cost gaps and promote competitiveness. Clear and stable government policies to encourage private sector investment. Long-term contracts between producers and consumers for price stability and reduced uncertainty. International hydrogen trade creating new markets and reducing production costs. Public awareness campaigns emphasizing the environmental and economic benefits of hydrogen, fostering market acceptance.
Operational (Yue et al., 2021)	Safety concerns due to flammability in various phases (production, transportation, storage). Managing the complex supply chain. Monitoring and controlling factors (purity, pressure, temperature). Maintaining ultra-purity standards for specific applications.	Development of rigorous safety standards and best practices to mitigate risks. Application of advanced monitoring technologies (sensors, internet of things) for real-time data and control capabilities. Stringent quality control measures for reliability and consistency. Implementation of automation and artificial intelligence-driven systems to enhance operational efficiency and reduce human error. Collaboration with regulatory bodies to establish industry-wide safety standards and practices.

(Continued)

TABLE 2.6 (Continued)

Category	Challenges	Opportunities
Deployment (Maestre et al., 2021)	Limited infrastructure hindering consumer adoption. Competition from established technologies. Resistance in transitioning from existing energy systems.	Pilot projects across industries showcasing feasibility and building confidence. Raising awareness about hydrogen's benefits to drive adoption. International collaboration and knowledge sharing for expedited technology deployment. Investment in building a robust infrastructure to support widespread consumer adoption. Collaboration with existing energy providers to integrate hydrogen solutions into current systems, easing the transition.

2.16 Conclusions

In conclusion, investigating hydrogen's potential as an energy carrier has uncovered a wide range of possibilities that have the ability to completely alter our current energy system. This describes a variety of energy sources, recognizing their shortcomings and the urgent need for sustainable alternatives as the whole world works toward a more ecologically friendly future. Hydrogen stands out as a compelling alternative to conventional renewable energy sources in the complex dance of options that are open to us. Because of its unmatched adaptability and outstanding storage capacities, it can be a key component of a strategy to deal with the intermittent nature of some renewable energy sources. In addition, among the current debate over hydrogen, fossil fuels, and batteries, hydrogen's special qualities stand out as a green, high-energy-density alternative, bringing us one step closer to the vital objective of decarbonization. It is impossible to exaggerate the importance of hydrogen as a potential energy carrier. Because of its flexibility to several industries, including transportation and industrial operations, it offers the potential to drastically reduce carbon emissions. This extensive utility highlights its function as a driving force behind sustainable development, bridging the gap between the generation and use of renewable energy. We have discovered a variety of paths across the complex world of hydrogen generation techniques, each with its own benefits and drawbacks. These techniques give us the instruments we need to successfully exploit hydrogen's potential, effectively charting novel paths for energy production. As this investigation draws to a close, the transformation of hydrogen into useful energy becomes clear as a crucial jigsaw piece. A future when clean energy systems are no longer just ideals but concrete, attainable realities is ushered in by the seamless integration of hydrogen into current energy infrastructure and its conversion to electricity. With careful consideration, the idea of combining natural gas and hydrogen offers a transitional approach that cleverly makes use of existing infrastructures, fusing the past and present for a faster road toward sustainability. As an energy carrier, hydrogen paints a picture of optimism, creativity, and limitless possibilities. It invites us to go beyond the constraints of our present energy system, opening the way for a future that values environmental stewardship and welcomes technological advancement. The potential of hydrogen is within our collective reach, demonstrating our ability to leave behind a lasting legacy of sustainability for future generations.

References

Abdalla, A. M., Hossain, S., Nisfindy, O. B., Azad, A. T., Dawood, M., & Azad, A. K. (2018). Hydrogen production, storage, transportation and key challenges with applications: A review. *Energy Conversion and Management*, *165*, 602–627. Available from https://doi.org/10.1016/j.enconman.2018.03.088.

Abdin, Z., Zafaranloo, A., Rafiee, A., Mérida, W., Lipiński, W., & Khalilpour, K. R. (2020). Hydrogen as an energy vector. *Renewable and Sustainable Energy Reviews*, *120*. Available from https://doi.org/10.1016/j.rser.2019.109620, https://www.journals.elsevier.com/renewable-and-sustainable-energy-reviews.

Abe, J. O., Popoola, A. P. I., Ajenifuja, E., & Popoola, O. M. (2019). Hydrogen energy, economy and storage: Review and recommendation. *International Journal of Hydrogen Energy*, *44*(29), 15072–15086. Available from https://doi.org/10.1016/j.ijhydene.2019.04.068, http://www.journals.elsevier.com/international-journal-of-hydrogen-energy/.

Acar, C., & Dincer, I. (2019). Review and evaluation of hydrogen production options for better environment. *Journal of Cleaner Production*, *218*, 835–849. Available from https://doi.org/10.1016/j.jclepro.2019.02.046, https://www.journals.elsevier.com/journal-of-cleaner-production.

Ahmed, S. F., Mofijur, M., Nuzhat, S., Rafa, N., Musharrat, A., Lam, S. S., & Boretti, A. (2022). Sustainable hydrogen production: Technological advancements and economic analysis. *International Journal of Hydrogen Energy*, *47*(88), 37227–37255. Available from https://doi.org/10.1016/j.ijhydene.2021.12.029, http://www.journals.elsevier.com/international-journal-of-hydrogen-energy/.

Ajanovic, A., Sayer, M., & Haas, R. (2022). The economics and the environmental benignity of different colors of hydrogen. *International Journal of Hydrogen Energy*, *47*(57), 24136–24154. Available from https://doi.org/10.1016/j.ijhydene.2022.02.094, http://www.journals.elsevier.com/international-journal-of-hydrogen-energy/.

Alamia, A., Larsson, A., Breitholtz, C., & Thunman, H. (2017). Performance of large-scale biomass gasifiers in a biorefinery, a state-of-the-art reference. *International Journal of Energy Research*, *41*(14), 2001–2019. Available from https://doi.org/10.1002/er.3758, http://onlinelibrary.wiley.com/journal/10.1002/(ISSN)1099-114X.

Almutairi, K., Hosseini Dehshiri, S. S., Hosseini Dehshiri, S. J., Mostafaeipour, A., Jahangiri, M., & Techato, K. (2021). Technical, economic, carbon footprint assessment, and prioritizing stations for hydrogen production using wind energy: A case study. *Energy Strategy Reviews*, *36*. Available from https://doi.org/10.1016/j.esr.2021.100684, http://www.journals.elsevier.com/energy-strategy-reviews/.

Amin, M., Shah, H. H., Fareed, A. G., Khan, W. U., Chung, E., Zia, A., Rahman Farooqi, Z. U., & Lee, C. (2022). Hydrogen production through renewable and non-renewable energy processes and their impact on climate change. *International Journal of Hydrogen Energy*, *47*(77), 33112–33134. Available from https://doi.org/10.1016/j.ijhydene.2022.07.172, http://www.journals.elsevier.com/international-journal-of-hydrogen-energy/.

Applied Sciences | Free Full-Text | Recent developments on hydrogen production technologies: State-of-the-art review with a focus on green-electrolysis.

Armaroli, N., & Balzani, V. (2011). The hydrogen issue. *ChemSusChem*, *4*(1), 21–36. Available from https://doi.org/10.1002/cssc.201000182, http://www.interscience.wiley.com/jpages/1864-5631.

Bidin, N., Azni, S. R., Abu Bakar, M. A., Johari, A. R., Abdul Munap, D. H. F., Salebi, M. F., Siti, S. N., Sahidan, N. S., & Sulaiman, S. N. A. (2017). The effect of sunlight in hydrogen production from water electrolysis. *International Journal of Hydrogen Energy*, *42*(1), 133–142. Available from https://doi.org/10.1016/j.ijhydene.2016.11.203, http://www.journals.elsevier.com/international-journal-of-hydrogen-energy/.

Brockway, P. E., Owen, A., Brand-Correa, L. I., & Hardt, L. (2019). Estimation of global final-stage energy-return-on-investment for fossil fuels with comparison to renewable energy sources. *Nature Energy*, *4*(7), 612–621. Available from https://doi.org/10.1038/s41560-019-0425-z, http://www.nature.com/nenergy/.

Chayambuka, K., Mulder, G., Danilov, D. L., & Notten, P. H. L. (2020). From Li-Ion batteries toward na-ion chemistries: Challenges and opportunities. *Advanced Energy Materials*, *10*(38). Available from https://doi.org/10.1002/aenm.202001310, http://onlinelibrary.wiley.com/journal/10.1002/(ISSN)1614-6840.

Cheng, C. K., & Ng, K. H. (2022). Editorial SI: Progress in grey, blue and green hydrogen. *Emirates International Journal of Hydrogen Energy*, *47*(98), 41373. Available from https://doi.org/10.1016/j.ijhydene.2022.09.268, http://www.journals.elsevier.com/international-journal-of-hydrogen-energy/.

Cho, H. H., Strezov, V., & Evans, T. J. (2023). A review on global warming potential, challenges and opportunities of renewable hydrogen production technologies. *Sustainable Materials and Technologies*, *35*. Available from https://doi.org/10.1016/j.susmat.2023.e00567, https://www.journals.elsevier.com/sustainable-materials-and-technologies.

Christopher, K., & Dimitrios, R. (2012). A review on exergy comparison of hydrogen production methods from renewable energy sources. *Energy and Environmental Science*, *5*(5), 6640–6651. Available from https://doi.org/10.1039/c2ee01098d.

Dash, S. K., Chakraborty, S., & Elangovan, D. (2023). A brief review of hydrogen production methods and their challenges. *Energies*, *3*(1141). Available from https://doi.org/10.3390/EN16031141.

Dawood, F., Anda, M., & Shafiullah, G. M. (2020). Hydrogen production for energy: An overview. *International Journal of Hydrogen Energy*, *45*(7), 3847–3869. Available from https://doi.org/10.1016/j.ijhydene.2019.12.059, http://www.journals.elsevier.com/international-journal-of-hydrogen-energy/.

Dincer, I. (2012). Green methods for hydrogen production. *International Journal of Hydrogen Energy*, *37*(2), 1954–1971. Available from https://doi.org/10.1016/j.ijhydene.2011.03.173.

Durmus, Y. E., Zhang, H., Baakes, F., Desmaizieres, G., Hayun, H., Yang, L., Kolek, M., Küpers, V., Janek, J., Mandler, D., Passerini, S., & Ein-Eli, Y. (2020). Side by side battery technologies with lithium-ion based batteries. *Advanced Energy Materials*, *10*(24). Available from https://doi.org/10.1002/aenm.202000089, http://onlinelibrary.wiley.com/journal/10.1002/(ISSN)1614-6840.

Elavarasan, R. M. (2019). The motivation for renewable energy and its comparison with other energy sources: A review. *European Journal of Sustainable Development Research*, *3*(1). Available from https://doi.org/10.20897/ejosdr/4005.

Energy Institute. (2023). *Energy Institute Statistical Review of World Energy* (72nd edition). London: EI Statistical Review of World Energy.

Eyer, J., & Corey, G. (2010). *Energy storage for the electricity grid: Benefits and market potential assessment guide: A study for the DOE energy storage systems program. Sandia Report*. Albuquerque, New Mexico: Sandia National Laboratories.

Fan, B., Pan, J., Liu, Y., Zhu, Y., Pan, Z., Chen, W., & Otchere, P. (2018). Effect of hydrogen injection strategies on mixture formation and combustion process in a hydrogen direct injection plus natural gas port injection rotary engine. *Energy Conversion and Management*, *160*, 150–164. Available from https://doi.org/10.1016/j.enconman.2018.01.034.

Fan, E., Li, L., Wang, Z., Lin, J., Huang, Y., Yao, Y., Chen, R., & Wu, F. (2020). Sustainable recycling technology for Li-Ion batteries and beyond: Challenges and future prospects. *Chemical Reviews*, *120*(14), 7020–7063. Available from https://doi.org/10.1021/acs.chemrev.9b00535, http://pubs.acs.org/journal/chreay.

Fischedick, M., Nitsch, J., & Ramesohl, S. (2005). The role of hydrogen for the long term development of sustainable energy systems-a case study for Germany. *Solar Energy*, *78*(5), 678–686. Available from https://doi.org/10.1016/j.solener.2004.09.020.

Fremaux, S., Beheshti, S. M., Ghassemi, H., & Shahsavan-Markadeh, R. (2015). An experimental study on hydrogen-rich gas production via steam gasification of biomass in a research-scale fluidized bed. *Energy Conversion and Management*, *91*, 427–432. Available from https://doi.org/10.1016/j.enconman.2014.12.048.

Gahleitner, G. (2013). Hydrogen from renewable electricity: An international review of power-to-gas pilot plants for stationary applications. *International Journal of Hydrogen Energy*, *38*(5), 2039–2061. Available from https://doi.org/10.1016/j.ijhydene.2012.12.010, http://www.journals.elsevier.com/international-journal-of-hydrogen-energy/.

Gibb, B. C. (2021). The rise and rise of lithium. *Nature Chemistry*, *13*(2), 107–109. Available from https://doi.org/10.1038/s41557-021-00638-w, http://www.nature.com/nchem/index.html.

Guduru, R. K. (2024). *Hydrogen Production From Biomass. From Waste to Wealth* (pp. 327–350). Singapore: Springer. Available from https://doi.org/10.1007/978-981-99-7552-5_15.

Guduru, R. K., Singh, R., & Vij, R. K. (2024a). Desalination of seawater and related technologies: a step towards achieving sustainable development goals. In *Water, The Environment, and the Sustainable Development Goals*, (pp. 305–325). Elsevier. Available from https://doi.org/10.1016/B978-0-443-15354-9.00011-6.

Guduru, R. K., Singh, R., & Vij, R. K. (2024b). Unlocking the Opportunities: Green Hydrogen from Renewable Energy Sources. In *Challenges and Opportunities in Green Hydrogen Production. Energy, Environment, and Sustainability*, (pp. 143–176). Singapore: Springer. Available from https://doi.org/10.1007/978-981-97-1339-4_7.

Guduru, R. K., Singh, R., & Vij, R. K. (2024c). A study on production of hydrogen using Al scrap feedstock. *Environmental Science and Pollution Research*, 1–8. Available from https://doi.org/10.1007/s11356-024-34099-8.

Haeseldonckx, D., & D'haeseleer, W. (2007). The use of the natural-gas pipeline infrastructure for hydrogen transport in a changing market structure. *International Journal of Hydrogen Energy, 32*(10-11), 1381–1386. Available from https://doi.org/10.1016/j.ijhydene.2006.10.018.

Hajjaji, N., Martinez, S., Trably, E., Steyer, J. P., & Helias, A. (2016). Life cycle assessment of hydrogen production from biogas reforming. *International Journal of Hydrogen Energy, 41*(14), 6064–6075. Available from https://doi.org/10.1016/j.ijhydene.2016.03.006, http://www.journals.elsevier.com/international-journal-of-hydrogen-energy/.

Hanley, E. S., Deane, J. P., & Gallachóir, B. P. Ó. (2018). The role of hydrogen in low carbon energy futures–a review of existing perspectives. *Renewable and Sustainable Energy Reviews, 82*, 3027–3045. Available from https://doi.org/10.1016/j.rser.2017.10.034, https://www.journals.elsevier.com/renewable-and-sustainable-energy-reviews.

Hassan, Q., Abdulateef, A. M., Hafedh, S. A., Al-samari, A., Abdulateef, J., Sameen, A. Z., Salman, H. M., Al-Jiboory, A. K., Wieteska, S., & Jaszczur, M. (2023). Renewable energy-to-green hydrogen: A review of main resources routes, processes and evaluation. *International Journal of Hydrogen Energy, 48*(46), 17383–17408. Available from https://doi.org/10.1016/j.ijhydene.2023.01.175, http://www.journals.elsevier.com/international-journal-of-hydrogen-energy/.

Hosseini, S. E. (2022). Retrieving nuclear power plants by producing hydrogen. *Future Energy, 1*(2), 1–2. Available from https://doi.org/10.55670/fpll.fuen.1.2.1.

Hosseini, S. E., Abdul Wahid, M., Jamil, M. M., Azli, A. A. M., & Misbah, M. F. (2015). A review on biomass-based hydrogen production for renewable energy supply. *International Journal of Energy Research, 39*(12), 1597–1615. Available from https://doi.org/10.1002/er.3381, http://onlinelibrary.wiley.com/journal/10.1002/(ISSN)1099-114X.

Hren, R., Vujanović, A., Van Fan, Y., Klemeš, J. J., Krajnc, D., & Čuček, L. (2023). Hydrogen production, storage and transport for renewable energy and chemicals: An environmental footprint assessment. *Renewable and Sustainable Energy Reviews, 173*, 113113. Available from https://doi.org/10.1016/j.rser.2022.113113.

Hund, P., Fabregas, T. L., & Drexhage, J. (2020). *Minerals for climate action: The mineral intensity of the clean energy transition*. Washington, DC: World Bank Publications.

International Energy Agency, I. (2021). Global hydrogen review 2021. http://www.iea.org/t&c.

International Energy Agency, I. (2022). Global hydrogen review 2022.

Jia, G., Lei, M., Li, M., Xu, W., Li, R., Lu, Y., & Cai, M. (2023). Hydrogen embrittlement in hydrogen-blended natural gas transportation systems: A review. *International Journal of Hydrogen Energy, 48*(82), 32137–32157. Available from https://doi.org/10.1016/j.ijhydene.2023.04.266, http://www.journals.elsevier.com/international-journal-of-hydrogen-energy/.

John Petrovic, J., Thomas, G. 2008. Reaction of Aluminum with Water to Produce Hydrogen: A Study of Issues Related to the Use of Aluminum for On-Board Vehicular Hydrogen Storage. pp. 1-26.

Jovan, D. J., & Dolanc, G. (2020). Can green hydrogen production be economically viable under current market conditions. *Slovenia Energies, 13*(24). Available from https://doi.org/10.3390/en13246599, https://www.mdpi.com/1996-1073/13/24/6599/pdf.

Kabir, E., Kumar, P., Kumar, S., Adelodun, A. A., & Kim, K. H. (2018). Solar energy: Potential and future prospects. *Renewable and Sustainable Energy Reviews, 82*, 894–900. Available from https://doi.org/10.1016/j.rser.2017.09.094, https://www.journals.elsevier.com/renewable-and-sustainable-energy-reviews.

Kayahan, E., Eroglu, I., & Koku, H. (2017). A compact tubular photobioreactor for outdoor hydrogen production from molasses. *International Journal of Hydrogen Energy, 42*(4), 2575–2582. Available from https://doi.org/10.1016/j.ijhydene.2016.08.014, http://www.journals.elsevier.com/international-journal-of-hydrogen-energy/.

Kim, J., Jun, A., Gwon, O., Yoo, S., Liu, M., Shin, J., Lim, T. H., & Kim, G. (2018). Hybrid-solid oxide electrolysis cell: A new strategy for efficient hydrogen production. *Nano Energy, 44*, 121–126. Available from https://doi.org/10.1016/j.nanoen.2017.11.074, http://www.journals.elsevier.com/nano-energy/.

Kim, J. W., Boo, K. J., Cho, J. H., & Moon, I. (2014). *Key challenges in the development of an infrastructure for hydrogen production, delivery, storage and use. Advances in hydrogen production, storage and distribution* (pp. 3–31). South Korea: Elsevier Inc. Available from http://www.sciencedirect.com/science/book/9780857097682, 10.1533/9780857097736.1.3.

Kothari, R., Buddhi, D., & Sawhney, R. L. (2004). Sources and technology for hydrogen production: A review. *International Journal of Global Energy Issues, 21*(1-2), 154–178. Available from https://doi.org/10.1504/ijgei.2004.004707, http://www.inderscience.com/ijgei.

Kumar, S., Regassa Jufar, S., Kumar, S., Foroozesh, J., Kumari, S., & Bera, A. (2023). Underground hydrogen storage and its roadmap and feasibility in India toward Net-Zero target for global decarbonization. *Fuel, 350*. Available from https://doi.org/10.1016/j.fuel.2023.128849, http://www.journals.elsevier.com/fuel/.

Liu, W., Zuo, H., Wang, J., Xue, Q., Ren, B., & Yang, F. (2021). The production and application of hydrogen in steel industry. *International Journal of Hydrogen Energy, 46*(17), 10548–10569. Available from https://doi.org/10.1016/j.ijhydene.2020.12.123, http://www.journals.elsevier.com/international-journal-of-hydrogen-energy/.

Liu, Y., Wang, X., Liu, H., Dong, Z., Li, S., Ge, H., & Yan, M. (2014). Improved hydrogen generation from the hydrolysis of aluminum ball milled with hydride. *Energy, 72*, 421–426. Available from https://doi.org/10.1016/j.energy.2014.05.060, http://www.elsevier.com/inca/publications/store/4/8/3/.

Liu, Z., Han, B., Lu, Z., Guan, W., Li, Y., Song, C., Chen, L., & Singhal, S. C. (2021). Efficiency and stability of hydrogen production from seawater using solid oxide electrolysis cells. *Applied Energy, 300*. Available from https://doi.org/10.1016/j.apenergy.2021.117439, https://www.journals.elsevier.com/applied-energy.

Lubitz, W., & Tumas, W. (2007). Hydrogen: An overview. *Chemical Reviews, 107*(10), 3900–3903. Available from https://doi.org/10.1021/cr050200z.

Maestre, V. M., Ortiz, A., & Ortiz, I. (2021). Challenges and prospects of renewable hydrogen-based strategies for full decarbonization of stationary power applications. *Renewable and Sustainable Energy Reviews, 152*. Available from https://doi.org/10.1016/j.rser.2021.111628, https://www.journals.elsevier.com/renewable-and-sustainable-energy-reviews.

Manisalidis, I., Stavropoulou, E., Stavropoulos, A., & Bezirtzoglou, E. (2020). Environmental and health impacts of air pollution: A review. *Frontiers in Public Health, 8*. Available from https://doi.org/10.3389/fpubh.2020.00014, http://journal.frontiersin.org/journal/public-health/section/public-health-education-and-promotion#archive.

Maton, J. P., Zhao, L., & Brouwer, J. (2013). Dynamic modeling of compressed gas energy storage to complement renewable wind power intermittency. *International Journal of Hydrogen Energy, 38*(19), 7867–7880. Available from https://doi.org/10.1016/j.ijhydene.2013.04.030.

Megia, P. J., Vizcaino, A. J., Calles, J. A., & Carrero, A. (2021). Hydrogen production technologies: From fossil fuels toward renewable sources. A mini review. *Energy and Fuels, 35*(20), 16403–16415. Available from https://doi.org/10.1021/acs.energyfuels.1c02501, http://pubs.acs.org/journal/enfuem.

Mehmeti, A., Angelis-Dimakis, A., Arampatzis, G., McPhail, S. J., & Ulgiati, S. (2018). Life cycle assessment and water footprint of hydrogen production methods: From conventional to emerging technologies. *Environments - MDPI, 5*(2), 1–19. Available from https://doi.org/10.3390/environments5020024, https://www.mdpi.com/2076-3298/5/2/24/pdf.

Messerle, V. E., Ustimenko, A. B., & Lavrichshev, O. A. (2017). Plasma coal conversion including mineral mass utilization. *Fuel, 203*, 877–883. Available from https://doi.org/10.1016/j.fuel.2017.05.037, http://www.journals.elsevier.com/fuel/.

Midilli, A., Kucuk, H., Topal, M. E., Akbulut, U., & Dincer, I. (2021). A comprehensive review on hydrogen production from coal gasification: Challenges and opportunities. *International Journal of Hydrogen Energy, 50*, 25385–25412. Available from https://doi.org/10.1016/J.IJHYDENE.2021.05.088.

Moradi, R., & Groth, K. M. (2019). Hydrogen storage and delivery: Review of the state of the art technologies and risk and reliability analysis. *International Journal of Hydrogen Energy, 44*(23), 12254–12269. Available from https://doi.org/10.1016/j.ijhydene.2019.03.041, http://www.journals.elsevier.com/international-journal-of-hydrogen-energy/.

Naterer, G. F., Suppiah, S., Rosen, M. A., Gabriel, K., Dincer, I., Jianu, O. A., Wang, Z., Easton, E. B., Ikeda, B. M., Rizvi, G., Pioro, I., Pope, K., Mostaghimi, J., & Lvov, S. N. (2017). Advances in unit operations and materials for the Cu–Cl cycle of hydrogen production. *International Journal of Hydrogen Energy, 42*(24), 15708–15723. Available from https://doi.org/10.1016/j.ijhydene.2017.03.133, http://www.journals.elsevier.com/international-journal-of-hydrogen-energy/.

Nguyen, T., Abdin, Z., Holm, T., & Mérida, W. (2019). Grid-connected hydrogen production via large-scale water electrolysis. *Energy Conversion and Management, 200*. Available from https://doi.org/10.1016/j.enconman.2019.112108, https://www.journals.elsevier.com/energy-conversion-and-management.

Nicoletti, G., Arcuri, N., Nicoletti, G., & Bruno, R. (2015). A technical and environmental comparison between hydrogen and some fossil fuels. *Energy Conversion and Management*, *89*, 205–213. Available from https://doi.org/10.1016/j.enconman.2014.09.057, https://www.journals.elsevier.com/energy-conversion-and-management.

Nikolaidis, P., & Poullikkas, A. (2017). A comparative overview of hydrogen production processes. *Renewable and Sustainable Energy Reviews*, *67*, 597–611. Available from https://doi.org/10.1016/j.rser.2016.09.044, https://www.journals.elsevier.com/renewable-and-sustainable-energy-reviews.

Noguchi, H., Takegami, H., Kasahara, S., Tanaka, N., Kamiji, Y., Iwatsuki, J., Aita, H., & Kubo, S. (2017). R&D status in thermochemical water-splitting hydrogen production iodine-sulfur process at JAEA. *Energy Procedia*, 113–118. Available from https://doi.org/10.1016/j.egypro.2017.09.459, http://www.sciencedirect.com/science/journal/18766102.

Ould Amrouche, S., Rekioua, D., Rekioua, T., & Bacha, S. (2016). Overview of energy storage in renewable energy systems. *International Journal of Hydrogen Energy*, *41*(45), 20914–20927. Available from https://doi.org/10.1016/j.ijhydene.2016.06.243, http://www.journals.elsevier.com/international-journal-of-hydrogen-energy/.

Palmer, G., Roberts, A., Hoadley, A., Dargaville, R., & Honnery, D. (2021). Life-cycle greenhouse gas emissions and net energy assessment of large-scale hydrogen production via electrolysis and solar PV. *Energy and Environmental Science*, *14*(10), 5113–5131. Available from https://doi.org/10.1039/d1ee01288f, http://pubs.rsc.org/en/journals/journal/ee.

Peng, L., & Wei, Z. (2020). Catalyst engineering for electrochemical energy conversion from water to water: Water electrolysis and the hydrogen fuel cell. *Engineering*, *6*(6), 653–679. Available from https://doi.org/10.1016/j.eng.2019.07.028, http://www.journals.elsevier.com/engineering/.

Peters, R., Pasel, J., & Samsun, R. C. (2024). Technology transfer from fuel processing for fuel cells to fuel synthesis from hydrogen and carbon dioxide. *International Journal of Hydrogen Energy*, *49*, 943–961. Available from https://doi.org/10.1016/j.ijhydene.2023.08.022, https://www.sciencedirect.com/science/journal/03603199.

Pistidda, C. (2021). Solid-state hydrogen storage for a decarbonized society. *Hydrogen*, *2*(4), 428–443. Available from https://doi.org/10.3390/hydrogen2040024.

Prato-Garcia, D., Robayo-Avendaño, A., & Vasquez-Medrano, R. (2023). Hydrogen from natural gas and biogas: Building bridges for a sustainable transition to a green economy. *Gas Science and Engineering*, *111*, 204918. Available from https://doi.org/10.1016/j.jgsce.2023.204918.

Rahim Malik, F., Yuan, H. B., Moran, J. C., & Tippayawong, N. (2023). Overview of hydrogen production technologies for fuel cell utilization. *Engineering Science and Technology, an International Journal*, *43*. Available from https://doi.org/10.1016/j.jestch.2023.101452, http://www.journals.elsevier.com/engineering-science-and-technology-an-international-journal/.

Rahman, A., Farrok, O., & Haque, M. M. (2022). Environmental impact of renewable energy source based electrical power plants: Solar, wind, hydroelectric, biomass, geothermal, tidal, ocean, and osmotic. *Renewable and Sustainable Energy Reviews*, *161*. Available from https://doi.org/10.1016/j.rser.2022.112279, https://www.journals.elsevier.com/renewable-and-sustainable-energy-reviews.

Rasul, M. G., Hazrat, M. A., Sattar, M. A., Jahirul, M. I., & Shearer, M. J. (2022). The future of hydrogen: Challenges on production, storage and applications. *Energy Conversion and Management*, *272*. Available from https://doi.org/10.1016/j.enconman.2022.116326, https://www.journals.elsevier.com/energy-conversion-and-management.

Ren, J., Musyoka, N. M., Langmi, H. W., Mathe, M., & Liao, S. (2017). Current research trends and perspectives on materials-based hydrogen storage solutions: A critical review. *International Journal of Hydrogen Energy*, *42*(1), 289–311. Available from https://doi.org/10.1016/j.ijhydene.2016.11.195, http://www.journals.elsevier.com/international-journal-of-hydrogen-energy/.

Rong, Y., Chen, S., Li, C., Chen, X., Xie, L., Chen, J., & Long, R. (2024). Techno-economic analysis of hydrogen storage and transportation from hydrogen plant to terminal refueling station. *International Journal of Hydrogen Energy*, *52*, 547–558. Available from https://doi.org/10.1016/j.ijhydene.2023.01.187, https://www.sciencedirect.com/science/journal/03603199.

Salvi, B. L., & Subramanian, K. A. (2015). Sustainable development of road transportation sector using hydrogen energy system. *Renewable and Sustainable Energy Reviews*, *51*, 1132–1155. Available from https://doi.org/10.1016/j.rser.2015.07.030, https://www.journals.elsevier.com/renewable-and-sustainable-energy-reviews.

Sarafraz, M. M., & Christo, F. C. (2020). Thermodynamic assessment and techno-economic analysis of a liquid indium-based chemical looping system for biomass gasification. *Energy Conversion and Management, 225*. Available from https://doi.org/10.1016/j.enconman.2020.113428, https://www.journals.elsevier.com/energy-conversion-and-management.

Saravanan, A., Khare, R., Venkatasubramanian, R., Surakasi, R., Boopathi, S., Ray, S., & Sudhakar, M. (2023). *Policy trends of renewable energy and non renewable energy*. Available from https://www.researchgate.net/publication/368227589.

Sharma, S., & Ghoshal, S. K. (2015). Hydrogen the future transportation fuel: From production to applications. *Renewable and Sustainable Energy Reviews, 43*, 1151–1158. Available from https://doi.org/10.1016/j.rser.2014.11.093, https://www.journals.elsevier.com/renewable-and-sustainable-energy-reviews.

Shiva Kumar, S., & Himabindu, V. (2019). Hydrogen production by PEM water electrolysis – a review. *Materials Science for Energy Technologies, 2*(3), 442–454. Available from https://doi.org/10.1016/j.mset.2019.03.002, http://www.keaipublishing.com/en/journals/materials-science-for-energy-technologies/.

Simpson, A. P., & Lutz, A. E. (2007). Exergy analysis of hydrogen production via steam methane reforming. *International Journal of Hydrogen Energy, 32*(18), 4811–4820. Available from https://doi.org/10.1016/j.ijhydene.2007.08.025.

Singh, S., Jain, S., Ps, V., Tiwari, A. K., Nouni, M. R., Pandey, J. K., & Goel, S. (2015). Hydrogen: A sustainable fuel for future of the transport sector. *Renewable and Sustainable Energy Reviews, 51*, 623–633. Available from https://doi.org/10.1016/j.rser.2015.06.040, https://www.journals.elsevier.com/renewable-and-sustainable-energy-reviews.

Taipabu, M. I., Viswanathan, K., Wu, W., Hattu, N., & Atabani, A. E. (2022). A critical review of the hydrogen production from biomass-based feedstocks: Challenge, solution, and future prospect. *Process Safety and Environmental Protection, 164*, 384–407. Available from https://doi.org/10.1016/j.psep.2022.06.006, http://www.elsevier.com/wps/find/journaldescription.cws_home/713889/description#description.

Tang, D., Tan, G. L., Li, G. W., Liang, J. G., Ahmad, S. M., Bahadur, A., Humayun, M., Ullah, H., Khan, A., & Bououdina, M. (2023). State-of-the-art hydrogen generation techniques and storage methods: A critical review. *Journal of Energy Storage, 64*. Available from https://doi.org/10.1016/j.est.2023.107196, http://www.journals.elsevier.com/journal-of-energy-storage/.

Tarhan, C., & Çil, M. A. (2021). A study on hydrogen, the clean energy of the future: Hydrogen storage methods. *Journal of Energy Storage, 40*. Available from https://doi.org/10.1016/j.est.2021.102676, http://www.journals.elsevier.com/journal-of-energy-storage/.

Tashie-Lewis, B. C., & Nnabuife, S. G. (2021). Hydrogen production, distribution, storage and power conversion in a hydrogen economy - a technology review. *Chemical Engineering Journal Advances, 8*. Available from https://doi.org/10.1016/j.ceja.2021.100172, https://www.journals.elsevier.com/chemical-engineering-journal-advances.

Do Thi, H. T., Pasztor, T., Fozer, D., Manenti, F., & Toth, A. J. (2021). Comparison of desalination technologies using renewable energy sources with life cycle, pestle, and multi-criteria decision analyses. *Water, 13*(21). Available from https://doi.org/10.3390/w13213023, https://www.mdpi.com/2073-4441/13/21/3023/pdf.

Touloupakis, E., Rontogiannis, G., Silva Benavides, A. M., Cicchi, B., Ghanotakis, D. F., & Torzillo, G. (2016). Hydrogen production by immobilized Synechocystis sp. PCC 6803. *International Journal of Hydrogen Energy, 41*(34), 15181–15186. Available from https://doi.org/10.1016/j.ijhydene.2016.07.075, http://www.journals.elsevier.com/international-journal-of-hydrogen-energy/.

Türker Üzden, Ş., Seçer, A., Fakı, E., & Hasanoğlu, A. (2023). Utilization of PET (waste) via hydrothermal co–gasification with sorghum for hydrogen rich gas production. *Journal of the Energy Institute, 107*. Available from https://doi.org/10.1016/j.joei.2023.101193, http://www.journals.elsevier.com/journal-of-the-energy-institute.

Vate, J. F. (1997). Comparison of energy sources in terms of their full energy chain emission factors of greenhouse gases. *Energy Policy, 25*.

Vidas, L., & Castro, R. (2021). Recent developments on hydrogen production technologies: State-of-the-art review with a focus on green-electrolysis. *Applied Sciences, 11*(23), 11363. Available from https://doi.org/10.3390/app112311363.

Wang, L., Chen, J., Ma, T., Ma, R., Bao, Y., & Fan, Z. (2024). Numerical study of leakage characteristics of hydrogen-blended natural gas in buried pipelines. *International Journal of Hydrogen Energy, 49*, 1166–1179.

Available from https://doi.org/10.1016/j.ijhydene.2023.07.293, https://www.sciencedirect.com/science/journal/03603199.

Wang, M., Wang, G., Sun, Z., Zhang, Y., & Xu, D. (2019). Review of renewable energy-based hydrogen production processes for sustainable energy innovation. *Global Energy Interconnection*, 2(5), 436–443. Available from https://doi.org/10.1016/j.gloei.2019.11.019, http://www.keaipublishing.com/en/journals/global-energy-interconnection/.

Wang, Y., Zhou, X., & Liu, L. (2021). Study on the mechanism of the ignition process of ammonia/hydrogen mixture under high-pressure direct-injection engine conditions. *International Journal of Hydrogen Energy, 46*(78), 38871–38886. Available from https://doi.org/10.1016/j.ijhydene.2021.09.122, http://www.journals.elsevier.com/international-journal-of-hydrogen-energy/.

Yang, W., Zhang, T., Zhou, J., Shi, W., Liu, J., & Cen, K. (2015). Experimental study on the effect of low melting point metal additives on hydrogen production in the aluminum-water reaction. *Energy, 88*, 537–543. Available from https://doi.org/10.1016/j.energy.2015.05.069, http://www.elsevier.com/inca/publications/store/4/8/3/.

Yue, M., Lambert, H., Pahon, E., Roche, R., Jemei, S., & Hissel, D. (2021). Hydrogen energy systems: A critical review of technologies, applications, trends and challenges. *Renewable and Sustainable Energy Reviews, 146*. Available from https://doi.org/10.1016/j.rser.2021.111180, https://www.journals.elsevier.com/renewable-and-sustainable-energy-reviews.

Yukesh Kannah, R., Kavitha, S., Preethi., Parthiba Karthikeyan, O., Kumar, G., Dai-Viet, N. V., & Banu, J. R. (2021). Techno-economic assessment of various hydrogen production methods – a review. *Bioresource Technology, 319*. Available from https://doi.org/10.1016/j.biortech.2020.124175, http://www.elsevier.com/locate/biortech.

Zhang, F., Zhao, P., Niu, M., & Maddy, J. (2016). The survey of key technologies in hydrogen energy storage. *International Journal of Hydrogen Energy*, 1(33), 14535–14552. Available from https://doi.org/10.1016/j.ijhydene.2016.05.293, http://www.journals.elsevier.com/international-journal-of-hydrogen-energy/.

Zhang, X., Ai, Q., & Wang, W. (2023). Effects of gas models on radiative heat transfer in the combustion chamber of a hydrogen gas turbine. *International Journal of Hydrogen Energy, 48*(77), 30144–30155. Available from https://doi.org/10.1016/j.ijhydene.2023.04.193, http://www.journals.elsevier.com/international-journal-of-hydrogen-energy/.

Zhang, Y., Lundblad, A., Campana, P. E., & Yan, J. (2016). Comparative study of battery storage and hydrogen storage to increase photovoltaic self-sufficiency in a residential building of Sweden. *Energy Procedia, 103*. Available from https://doi.org/10.1016/j.egypro.2016.11.284, http://www.sciencedirect.com/science/journal/18766102.

Zhou, Y., Swidler, D., Searle, S., & Baldino, C. (2021). *White Paper: Life-cycle greenhouse gas emissions of biomethane and hydrogen pathways in the European Union* (pp. 1–40). Washington: International Council on Clean Transportation.

CHAPTER

3

Hydrogen production using different methods

Rahul Saha[1], *Devanshi Raman*[1] *and Ranjan Phukan*[2]

[1]Department of Chemical Engineering, National Institute of Technology Hamirpur, Hamirpur, Himachal Pradesh, India [2]Department of Petroleum Technology, Dibrugarh University, Dibrugarh, Assam, India

3.1 Introduction

The demand in energy has been increasing continuously due to increase in industrialization, population growth, and improved standard of living. As of 2023, around 82% of global major energy needs are entertained by the use of fossil (Balthasar, 1984). Fig. 3.1 shows the drop in fossil fuel consumption from 1965 to 2023 (Rapier, 2023; Global fossil fuel share in energy consumption 1965-2022, 2023).

The supply of fossil fuel is finite and is exhausting at a rapid rate due to its higher global demand. Though, there are several advanced tertiary methods available which focuses on the overall improvement in crude oil production to meet the current energy demands (Phukan & Saha, 2022, Saha et al., 2018a, 2018b, 2018c, Saha et al., 2019, Saha, Ramgopal V.S Uppaluri and Tiwari, 2018; Saha, Ramgopal V.S Uppaluri and Tiwari, 2018; Saha et al., 2019; Saha et al., 2021; Saha et al., 2017; Saha & Phukan, 2023). However, utilization of such nonrenewable energy as a primary source since several decades, has resulted in a tremendous enhancement in the amount of CO_2 level as well as other greenhouse gases. This phenomenon heats up the earth surface and thereby introduces threats in the ecosystem.

The availability of different sources and techniques to reduce the impact of burning fossil fuels are important to minimize/solve issues related to global warming. Moreover, the employment of such source should consider the depletion of fossil fuel resources and the overall rise in energy demand, particularly from the newly emerging economies. This is an important issue in the context of the global energy production landscape. Thus, balancing the world's increasing need for energy with low emissions of harmful gases is likely

Subsurface Hydrogen Energy Storage
DOI: https://doi.org/10.1016/B978-0-443-24071-3.00004-2

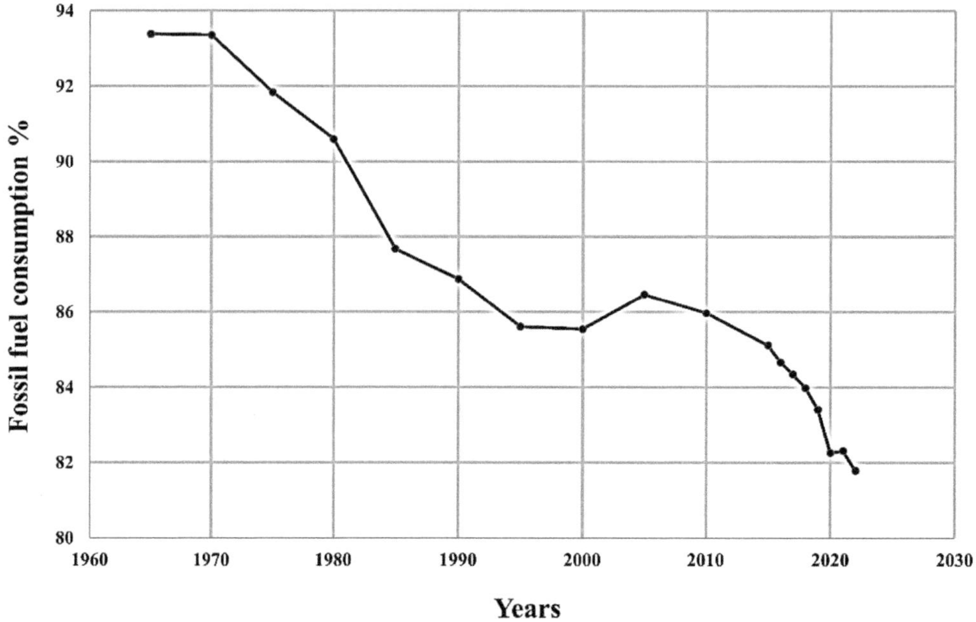

FIGURE 3.1 Fossil fuel energy consumption worldwide from 1965 to 2022 (Rapier, 2023; Global fossil fuel share in energy consumption 1965-2022, 2023).

one of the most significant problems that scientific study will have to solve. The growth of sustainable energy requires us to establish the groundwork in the growing fields of clean and alternative energies (Palma et al., 2016).

Hydrogen, often heralded as the "fuel of the future," has gained prominence in the pursuit of sustainable and clean energy solutions (Dincer & Acar, 2015). It is an essential industrial raw material including refinery. As the world confronts the pressing challenges of climate change and the necessity to shift away from fossil fuels, hydrogen is emerging as a promising alternative to revolutionize the existing energy landscape. Its allure lies in its versatility, high energy content, abundant presence in various forms in the universe and the fact that, when consumed as a fuel, it produces no harmful emissions—only water vapor. Hydrogen energy content is about 122 MJ/kg, which is 2.4 and 2.8 times higher than the methane and gasoline, respectively (Al-Hamamre et al., 2009; Nikolaidis & Poullikkas, 2017). These unique characteristic positions hydrogen as a vital participant in the global effort to combat greenhouse gas emissions and forge a sustainable energy future.

Hydrogen production stands at the forefront of this transformative journey, serving as the critical gateway to harness the advantages of hydrogen as an energy carrier. A myriad of methods and technologies have been developed for the production of hydrogen, each retaining their advantages, limitations, restrictions, and environmental implications (Das & Peu, 2022).

Hydrogen was evolved as distinct element in 1700s and then as hydrogen gas when zinc metal was reacted with hydrochloric acid. With the progress in time around 1800s, the process of electrolysis was evolved in which electric current was passed though water in order to split the water molecules to produce hydrogen. In 1900s, a major breakthrough happened when the idea of replacing the internal combustion for transportation and submarines with hydrogen/hydrogen mixture as introduced by German engineers. Identifying the potential of hydrogen as fuel, the space center in United States then started using hydrogen in liquid form as rocket fuel and in fuel cells. It was in 1990, the first hydrogen plant powered by solar did came into operation. Moreover in 1994, the first electric car (fuel cell vehicle) was introduced in a press conference in Germany. This development resulted in a trust to use hydrogen as future fuel. In fact, energy sector in Unites States decided to invest more than $350-million for hydrogen related research. Germany navy on the other hand has conducted the first trial of submarines powered by fuel cell in deep water (Jonas, 2009). Considering its importance in our daily life, hydrogen has also been deployed for domestic purpose like heating and cooking. Therefore, looking in to the existing scenario, today most of the countries are in a process to develop hydrogen production, developing stations, and transportation system as hydrogen are considered to be clean and reliable fuel for future. The demand for pure hydrogen has been globally enhanced from 18.2 to 73.9 Mt from 1975 to 2018 as per the data reported by Japan in international energy agency 2019 (The Future of Hydrogen, 2019).

Although hydrogen itself is a colorless gas, it is frequently associated with a color to indicate the process of its production as shown in Table 3.1 (Kumar et al., 2023). Though it is extremely rare but hydrogen also occurs naturally in our environment and is found in underground deposits and is sometimes referred to as "white hydrogen." The color designation for hydrogen is derived from the fuel/process used to manufacture it on an industrial basis. However, there is no universal naming convention, thus these color definitions could alter throughout time or even between nations (Lovegrove, 2022; Lubbe et al., 2023).

TABLE 3.1 Color codes for hydrogen.

Color terminology	Energy source and production technology	Hydrogen atom feedstock
Blue	SMR + CCUS Gasification + CCUS	Methane
Turquoise	Pyrolysis	Methane
Gray	SMR	Methane
Brown	Gasification	Brown coal (lignite)
Black	Gasification	Black coal
Golden	Direct solar water splitting	Water
White	Fracking	Naturally occurring
Green	Water electrolysis by means renewable energy	Water or biomass
Yellow	Mixed origin grid powered electrolysis	Water
Red	Thermolysis powered by nuclear or solar heat	Water
Pink	Water electrolysis powered by nuclear energy	Water
Purple	Thermolysis/electrolysis using nuclear heat and electricity	Water

Hydrogen can also be detected in nature considering the geological aspects and has been encountered during mining, underground extraction, etc. Numerous theories have been proposed explaining the existence of natural hydrogen. Those could be due to water–ultrabasic/fresh rock reactions, water–mantle reactions, radiolysis of water, organic matter decomposition, anthropogenic behavior, biological activity, and more. Hence, due to geological heterogeneity, dissolution of hydrogen in water, diffusion, and seepage, the exact mechanism for existence of hydrogen molecules is not clear. Thus, the available hydrogen in geological locations can act as a renewable energy source and therefore can be extracted and utilize accordingly (Zgonnik, 2020).

3.1.1 Hydrogen production: principles and procedures

The increase in demand for hydrogen production due to energy transisiton requires new and more efficient production pathways. It is likely that a combination of different pathways may be used to meet the need for hydrogen in the future.

Major challenges for transition toward hydrogen-based energy systems are unavailability of hydrogen in its free state in nature as well as the requirement for low-cost hydrogen production processes. As illustrated in Fig. 3.2, a wide variety of pathways are available, which may be split into two major types based on feedstock employed namely fossil fuels and renewable resources derived hydrogen (Balthasar, 1984; Ishaq et al., 2022; Kumar & Himabindu, 2019; Kumar et al., 2023; Megía et al., 2021; Vidas & Castro, 2021; Yue et al., 2021). Hydrocarbon reforming, coal gasification, and pyrolysis are three techniques for

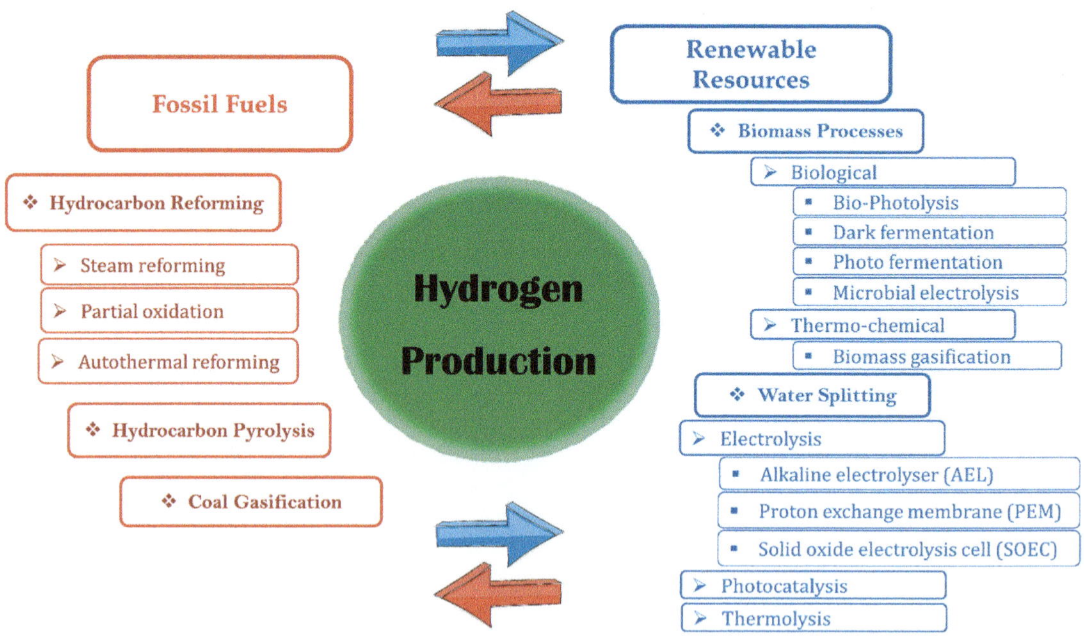

FIGURE 3.2 Methods for production of hydrogen gas.

converting fossil fuels into hydrogen. Hydrogen production from renewable resources comprises of biomass processing and water splitting.

3.2 Hydrogen from fossil fuels

3.2.1 Hydrocarbon reforming

The reforming of hydrocarbon is a process in which specific reforming techniques are used to transform hydrocarbon fuel into hydrogen. Steam and/or oxygen are additional reactants utilized in the reforming process in addition to the hydrocarbon fuel.

Depending on the reactants used during reforming process, hydrocarbons reforming is classified into:

1. Steam reforming (SR)
2. Partial oxidation (POX)
3. Autothermal reforming (ATR)

The above mentioned three methods can produce considerable quantity of carbon monoxide (CO), hence reactors of water−gas shift (WGS) type which are usually at higher followed by lower temperature are deployed. The temperature of the reactors is above 350°C which can induce kinetics at a faster rate but are subjected to thermodynamic limitations considering the quantity of CO which can be shifted. Thus, temperature on the lower side inside the reactor (210°C −330°C) is utilized to reduce the CO levels. WGS reactors of higher temperature uses an iron-based catalyst whereas low temperature reactors frequently uses a copper-based catalyst (Holladay et al., 2009). Table 3.2 compares hydrocarbon reforming technologies and thus summarizes the benefits and drawbacks of each of these reforming methods.

Reforming plant comprises of desulphurization unit, reformer, water−gas shift reactors, gas clean-up section and finally some auxiliary units (pumps, steam generator, heat exchangers, coolers, etc.).

TABLE 3.2 Comparison of hydrocarbons reforming technologies (Megía et al., 2021; Ersöz, 2008).

Technology	Feedstock	Operating temperature	Advantages	Disadvantages
Steam reforming	Light hydrocarbons (natural gas, LPG, light and heavy naptha)	800°C−1000°C	Most extensive commercial experience, oxygen not needed, lowest operating temperature and highest H_2/CO ratio	Air emission is highest
Partial oxidation	Hydrocarbons, heavy fuel oil, and coal	>1000°C	Decreased desulfurization requirement, no catalyst required and low methane slip	Processing temperatures is high, soot formation/handling adds process complexity and air separation unit is required
Autothermal reforming	Light hydrocarbons	>1000°C	Less operating temperature than POX and low methane slip	Commercial experience is limited and air separation unit is also required

3.2.2 Steam reforming

Steam reforming (SR) is a catalytic process which is endothermic in nature and generates hydrogen-rich syngas from light hydrocarbons. The feedstock which are generally used for SR system comprises of natural gas, LPG, and light-heavy naphtha. SR process consists of three steps:

1. Reforming for conversion of hydrocarbon to syngas.
2. WGS reaction to increase hydrogen content of syngas.
3. Gas clean-up system usually pressure swing adsorption (PSA) for separating hydrogen from other gases

To generate hydrogen of desired yield with prohibited coke formation on the reforming catalyst surface, higher temperature with pressure up to 3.5 MPa and steam-to-carbon ratios of 3.5 are specified (Balthasar, 1984; Ersöz, 2008). For SR system, the operating temperature is lower as compared to ATR and POX. However, it can produce syngas with high H_2/CO ratio (3:1) (El-Shafie et al., 2019). However, SR has the highest emissions compared to ATR and POX. SR of methane is the most commonly applied method for hydrogen generation in larger scale, with 74%–85% conversion efficiency (Nikolaidis & Poullikkas, 2017). The chemical reactions in SR are as follows (Holladay et al., 2009):

$$\text{Reformer:}\quad C_nH_m + nH_2O \rightarrow nCO + \left(n + \frac{1}{2}m\right)H_2 \tag{3.1}$$

$$\text{WGS reactor:}\quad CO + H_2O \rightarrow CO_2 + H_2 \tag{3.2}$$

$$\text{Overall methane SR reaction:}\quad CH_4 + 2H_2O \rightarrow CO_2 + 4H_2 \tag{3.3}$$

3.2.3 Partial oxidation

Partial Oxidation (POX) reaction involves the conversion of heavy hydrocarbons like heavy oil or sulfurous organic residues at temperatures above 800°C. However, a challenge arises from solid carbon deposition on the catalyst at such temperature. Thus, the deactivation occurs rapidly thereby hindering gas flow and simultaneously rendering the catalyst ineffectiveness (Wietschel & Ball, 2009). It is an exothermic hydrocarbon conversion which requires oxygen and steam. Because the process is exothermic, no more heat is required.

Compared to syngas generated in SR ($H_2/CO = 3{:}1$), POX generates syngas with more CO ($H_2/CO = 1{:}1$ or 2:1). The POX reactors has thermal efficiency of ranges from 60% to 75% with methane as a fuel (Kalamaras & Efstathiou, 2013). The main drawbacks of POX over SR are the high CO content of the produced gas and the requirement of capital-intensive air separation unit for oxygen production.

The following are the primary chemical reactions that occur in POX:

$$C_nH_m + \frac{1}{2}nO_2 \rightarrow nCO + \frac{1}{2}mH_2 \tag{3.4}$$

It can be classified into:

1. Thermal partial oxidation (TPOX)
2. Catalytic partial oxidation (CPOX)

The TPOX process converts hydrocarbons in the presence of oxygen to syngas at temperatures and pressures ranging from 1300°C–1500°C and 3–8 MPa, respectively (Al-Hamamre et al., 2009). Feedstocks for TPOX mainly includes methane, heavy feedstock like oil residue and coal (Steinberg & Cheng, 1989).

In CPOX process operates at runs at 700°C–1000°C and uses feedstock ranging from natural gas to naphtha. CPOX provides potential solutions to the heat transfer and high external heat energy consumption issues encountered in the SR process.

3.2.4 Autothermal reforming

Autothermal reforming (ATR) process is the combination of exothermic POX that offers required heat and endothermic stream reforming to boost the production of hydrogen. In ATR, steam and oxygen are admitted into the reformer, causing both oxidation and reforming processes to take place simultaneously, and the heat from the exothermic POX reaction balances the heat from endothermic SR reaction.

The following is the primary chemical reactions that occurs in ATR:

$$C_nH_m + \frac{1}{2}nH_2O + \frac{1}{4}nO_2 \rightarrow nCO + \left(\frac{1}{2}n + \frac{1}{2}m\right)H_2 \tag{3.5}$$

It is important to known that for any ATR system to be functional, an appropriate ratio of oxygen and fuel, followed by carbon and steam should be maintained throughout the reaction to control the system temperature, composition of the product and formation factor for coke (Holladay et al., 2009). ATR of natural gas has thermal efficiency of 60%–75%, while optimum operating temperature is at around 700°C.

ATR is advantageous over SR process because it is simpler, less costly, and does not require any external source to supply heat. Another notable benefit of ATR with respect to SR process is the quick response to shut down and start up, additionally generating a greater volume of hydrogen than POX alone (Kalamaras & Efstathiou, 2013).

3.3 Hydrocarbon pyrolysis

Hydrocarbon pyrolysis is another hydrogen production pathway in which hydrocarbon is decomposed in the absence of water or oxygen into hydrogen and carbon. In the hydrocarbon pyrolysis, the only source of hydrogen is hydrocarbon itself, which undergoes thermal breakdown via the following general reaction:

$$C_nH_m \rightarrow nC + \frac{1}{2}mH_2 \tag{3.6}$$

As no oxidant is used in the process, pyrolysis process generates no carbon oxides (CO and CO_2); thus, eliminating the need for secondary reactors (WGS, PrOx, etc.) hence, capital cost of pyrolysis plants is lower than SMR and POX processes resulting in 25%–30% cheaper hydrogen production cost (Muradov, 1993). Consequently, this process has potential for significant emissions reduction. Additionally, for generation of per mole of hydrogen, the energy requirement (37.6 kJ/mol) is much lesser than SMR process (63.3 kJ/mol) and the heat input might be compensated by burning of hydrogen around 15%–20% as produced during the process (Muradov, 2001a, 2001b). At around 980°C and atmospheric pressure, the catalytic decomposition of natural gas is performed (Nikolaidis & Poullikkas, 2017).

3.4 Coal gasification

Coal gasification is a chemical method used to generate fuel gas products. In gasification, carbonaceous feedstocks like coal are partly oxidized in presence of oxygen and steam inside the gasifier at high pressure and temperature to produce syngas.

The pulverized coal combines with oxygen and steam in a specially built gasifier which runs at moderate pressure and higher temperature. This generates a mixture of gas commonly known as syngas. Syngas majorly consists of CO, H_2, with small amount of CO_2, CH_4, and other components. After removal of contaminants from the syngas, the gas stream is admitted to sour WGS rector to convert CO present in the syngas to hydrogen and CO_2. WGS reactor product stream is admitted to gas separation system usually PSA for separation of hydrogen from other gases. Due to the significantly larger carbon content of coal, this method emits substantial CO_2 into the atmosphere compared to other hydrogen generation technologies. Large-scale coal-based hydrogen generation plants are required to be integrated with carbon capture, storage, and utilization methods are used to control highly concentrated CO_2 streams, although this results in increased hydrogen production costs. Although coal gasification is a mature and commercially used technology, hydrogen production through this method has high cost compared to natural gas reforming.

Coal gasification process involves a number of complicated events, including pyrolysis, reforming, POX, water–gas shift, and methanation (Scipioni et al., 2017).

$$C_nH_mO_y \rightarrow \text{tar} + H_2 + CO_2 + CH_4 + C_2 + \ldots \tag{3.7}$$

$$\text{tar} + xH_2O \rightarrow xCO + yH_2 \tag{3.8}$$

$$\text{tar} + CO_2 \rightarrow xCO + yH_2 \tag{3.9}$$

$$C_nH_m + \frac{1}{2}nO_2 \rightarrow nCO + \frac{1}{2}mH_2 \tag{3.10}$$

$$CO + 3H_2 \rightarrow CH_4 + 2H_2O \tag{3.11}$$

3.5 Hydrogen production from renewable resources

3.5.1 Biomass processes

Biomass a renewable energy can be converted to hydrogen by either thermochemical or biological conversion processes. Biomass such as sawdust, wood, corn, pine sawdust, bagasse, cotton stern, waste paper, grain husk, almond shell, straw, animal waste, pinus radiate, crop leftovers, switchgrass, agricultural waste, eucalyptus gobulus, aquatic plants, and sewage sludge has been used for the production of hydrogen (Gao et al., 2023; Ni et al., 2006). Biomass intake CO_2 gas while growing and emits CO_2 when burns, thereby indicating carbon neutral source or zero net balance emission. The various methods available for generation of hydrogen from biomass involve pyrolysis, gasification, biomass to liquid fuels with the application of supercritical extraction, liquification, hydrolysis followed by reforming, and biological conversion (Holladay et al., 2009). The energy required can vary based on the biomass type and specific method used for hydrogen production, and the purity of hydrogen produced is mainly influenced by the specific method adopted.

A sequential procedure is used in the two-step approach to procedure hydrogen from biomass in a way that maximizes the amount of hydrogen gas produced. In the first step, biomass is subjected to various processes that comes under thermochemical conversion which leads to production of syngas (mixture of hydrogen, carbon dioxide, CO, and other gases). It typically comprises of gasification and pyrolysis of biomass. During this step, biomass is heated in the absence of oxygen, resulting in the release of syngas. List of different type of gasification processes that are employed are as follows:

- Catalytic gasification
- Plasma gasification
- Steam gasification
- Hydrothermal gasification
- Oxygen blown gasification
- Sorbent enhanced gasification

In the second step, the syngas is subjected to a reaction, where CO and steam react over a catalyst to produce additional hydrogen gas. List of the many approaches that are employed are listed down below and also briefly discussed throughout this chapter:

- Steam methane reforming
- ATR
- POX
- Water–gas shift reaction
- Membrane separation
- PSA

3.5.2 Biological conversion processes

3.5.2.1 Bio-photolysis

Bio-photolysis makes use of microorganisms (algae and cyanobacteria) which can produce hydrogen by direct photolysis of water with the application of excess solar energy. It is like the photosynthesis of plants and algae. Algae have hydrogen-producing enzymes which can produce hydrogen at definite circumstances. Bio-photolysis may be conducted in direct as well as indirect pathways using a mechanism like microalgae which absorbs energy from solar to generate electrons. The electron then passed to ferredoxin to produce enzyme which is responsible for producing hydrogen from water (Xu et al., 2017).

In direct bio-photolysis, oxygen and hydrogen ions are produced by splitting water molecules via photosynthesis of green algae such as *Chlorococcum littorale, Chlorella fusca, Chlamydomonas reinhardtii, Scenedesmus obliquus,* and *Platymonas subcordiformis* (Cheonh et al., 2022). The hydrogen ions interact with the hydrogenase enzyme which results in the production of hydrogen. However, these enzymes are sensitive to oxygen. Additionally, due to the effect of light saturation, the rate of hydrogen generated is overall poor (Nikolaidis & Poullikkas, 2017). Microalgae mutants with reduced chlorophyll content and enhanced oxygen tolerance exhibited increased hydrogen production (Ni et al., 2006). Green algae coverts water to hydrogen in the following general reaction:

$$2H_2O + \text{lightenergy} \rightarrow 2H_2 + O_2 \tag{3.12}$$

In indirect bio-photolysis, nitrogenase and hydrogenase enzymes both are responsible for production of hydrogen, and the rate of production is similar to that of direct bio-photolysis (Ni et al., 2006). However, indirect bio-photolysis techniques are currently at the theoretical stage. Hydrogen production by this method is cost-effective and sustainable due to the use of solar energy and major greenhouse gas (CO_2). The reaction below illustrate indirect bio-photolysis of water by cyanobacteria, or blue and green algae for production of hydrogen (Nikolaidis & Poullikkas, 2017).

$$12H_2O + 6CO_2 + \text{lightenergy} \rightarrow C_6H_{12}O_6 + 6O_2 \tag{3.13}$$

$$C_6H_{12}O_6 + 12H_2O + \text{lightenergy} \rightarrow 12H_2 + 6CO_2 \tag{3.14}$$

The fundamental feedstock for bio-photolysis technology is selected as the one whose availability and affordability is not restricted such as water. However, to capture sufficient light this method needs surface area which should be larger. During the process, an inappropriate behavior with respect to oxygen production by microbes was observed. These oxygen detection by the organism therefore can restricts hydrogen production. Therefore, to minimize oxygen accumulation, efforts are being made to find/create lower oxygen-sensitive species, separate oxygen and hydrogen cycles, and/or alter the photosynthesis (oxygen generation) to respiration (oxygen consumption) ratio (Holladay et al., 2009).

3.5.3 Dark fermentation

Dark fermentation reactions are carried out without light, where primarily anaerobic bacteria such as *Ruminococcus albus, Enterobacter aerogenes, Enterobacter cloacae, Escherichia coli* and *Citrobacter intermedius* converts substrates which are in rich in carbohydrate (Koutra et al., 2020). It is a technology to produce hydrogen from organic biomass such as agricultural leftovers, agro-industrial wastes, and organic municipal waste. The biomass to be used for fermentation process must retain properties such as biodegradability, large in quantity, inexpensive, and should content high carbohydrate. The availability of pure/simple sugar (glucose and lactose) which are subjected to easy biodegradation are limited and are costly (Holladay et al., 2009). Biomass based feedstocks requires pretreatment for hydrolysis of cellulose and hemicelluloses to carbohydrates (Kapdan & Kargi, 2006). The gas produced in the dark fermentation combines hydrogen, carbon dioxide, methane, CO, and some hydrogen sulfide. Hence, an additional step of separation is necessary for production of hydrogen of high purity. As shown in stoichiometric chemical reactions below, acetate and butyrate fermentation with substrate glucose yields four moles and two moles of hydrogen respectively for per mole of glucose fermentation (Nikolaidis & Poullikkas, 2017).

$$\text{Acetate fermentation:} \quad C_6H_{12}O_6 + 2H_2O \rightarrow 2CH_3COOH + 4H_2 + 2CO_2 \tag{3.15}$$

$$\text{Butyrate fermentation:} \quad C_6H_{12}O_6 + 2H_2O \rightarrow 2CH_3CH_2CH_2COOH + 2H_2 + 2CO_2 \tag{3.16}$$

The raw material glucose used for this process is ideal but are costly and limited in quantity however, it can be derived from agricultural waste. Additionally, starch and cellulose present in plants biomass may be deployed as alternate for this process (Kapdan & Kargi, 2006). The production of hydrogen by this process depends largely on the system pH, and for most effective production, the pH should be between five and six. Moreover, it is observed that increase of system pressure results in decrease in hydrogen production; hence, to increase production, hydrogen must be eliminated as it is produced. Nevertheless, dark fermentation is regarded as a practically effective method due to robustness of the process, higher hydrogen yields, compatibility with a wide range of biomass/organic waste, withstanding at high organic loadings, etc.

3.5.4 Photo fermentation

Photo-fermentative methods, also known as photosynthetic bacterial hydrogen generation, uses the nitrogenase capability of purple non-sulfur bacteria to produce hydrogen. In this mechanism, light harvesting pigments such as chlorophylls, carotenoids, and phycobilin scavenge light energy and transport it to membrane reaction centers. Sunlight turns water into protons, electrons, and oxygen. The nitrogenase catalyst reacts protons and electrons with nitrogen and ATP to produce ammonia, hydrogen, and ADP. As oxygen inhibits the nitrogenase, cyanobacteria isolates nitrogen fixation from oxygen production either spatially or temporally (Holladay et al., 2009).

It converts organic acids (acetic, lactic, and butyric) to hydrogen and carbon dioxide under anerobic conditions. Increase in light intensity results in better hydrogen yields and production rate, but gas has adverse effect on the light conversion efficiency.

The most important benefits of this method include the noninterference of oxygen with the process and the employment of bacteria in different range of condition (i.e., batch processes, continuous cultures, immobilized in carrageenan, activated glass/polyurethane foam, porous glass, agar gel). Whereas, the restriction in organic acids availability, requirement of high energy, hydrogen re-oxidation and slowness in nitrogenase enzymes are the downside (Holladay et al., 2009). The overall reaction for hydrogen production using acetic acid is shown below:

$$CH_3COOH + 2H_2O + \text{lightenergy} \rightarrow 4H_2 + 2CO_2 \tag{3.17}$$

3.5.5 Microbial electrolysis

Microbial aided electrolysis cell (MEC) is a hybrid system combining electrochemistry and biotechnology, where the dark fermentation reaction is enhanced by the small amounts of external energy supplied by renewable sources such as wind and sun. In microbe assisted electrolysis system, the energy required is lesser than electrochemical system and efficiency is higher than biological system due to the enhanced metabolic activities of microbes under small, applied potentials. MEC, also known as bio-electrochemically assisted microbial reactor (BEAMR), uses electrohydrogenesis to directly transform biodegradable materials into hydrogen (Holladay et al., 2009). It uses a special group of bacteria known as electro-active bacteria (EAB) or exoelectrogens having ability to decompose (oxidize) organic matter and transfer electrons to solid electrodes. The breakdown of biomass by enhanced bacterial metabolism releases the desired electrons and protons on one electrode (anode), while they combine at the other electrode (cathode) under mild applied potential, releasing hydrogen.

Hydrogen recovery at the cathode can be utilized at lower redox potential than conventional electrolysis as the requirement is sufficed by the in-situ potential of the system. Only the additional redox potential required to meet the energy required to cross the thermodynamic energy barrier for hydrogen formation is supplied from external renewable sources. Hence, hydrogen can be produced at just -0.414 V (std ref) of cathode potential indicating the combined efforts of EAB and applied potentials.

In microbial electrolysis process, microbes oxidize the substrate at anode is oxidized to produce electrons, protons, and CO_2. The electrons and protons travel to cathode via the external circuit and proton conducting membrane respectively, and form hydrogen by the combined effect of electron and proton. Anode and cathode reactions of MEC are given below:

$$\text{Anode:}\quad CH_3COO^- + 4H_2O \rightarrow 2HCO^{3-} + 9H^+ + 8e^- \tag{3.18}$$

$$\text{Cathode:}\quad 8H^+ + 8e^- \rightarrow 4H_2 \tag{3.19}$$

$$\text{Overall cell reaction:}\quad CH_3COO^- + 4H_2O \rightarrow 2HCO^{3-} + H^+ + 4H_2 \tag{3.20}$$

The key benefit of the MEC process is that it can handle higher organic loadings as well as complex substrates and also expands the theoretical conversion rates of biocatalyst by widening its thermodynamic energy barriers (Muradov, 2001a, 2001b). MEC systems were first designed using components comparable to those as used in PEM fuel cells. Though, designing the electrode into flat form can reduce the surface area for exoelectrogens, while membranes raise the ohmic resistances, prompting the development of alternate designs. The latest design deployed brush of graphite for exoelectrogen substrate (anode) without membrane separation (Holladay et al., 2009).

3.5.6 Thermochemical conversion process

3.5.6.1 *Biomass gasification*

Biomass gasification is regarded as one of the promising routes for low-cost renewable and carbon neutral hydrogen production. It is a thermochemical conversion of biomass into synthesis gas/producer gas. The producer gas is a combination of hydrogen, carbon dioxide, methane, nitrogen, CO, and heavier hydrocarbons (tar). The process is based on the oxidation of components partially and is a variation of pyrolysis. Due to vaporization of moisture that exist in the biomass, the process of gasification is often thermally inefficient. The operation can be executed both in absence and presence of catalyst in a fixed or fluidized bed reactor with latter often providing higher performance. The addition of steam and or oxygen to the gasification process results in SR and produces a syngas stream (H_2:CO ratio of 2:1) (Holladay et al., 2009). Syngas is a very versatile product as it can be used for numerous applications such as production of heat, power, transport fuels, hydrogen, and chemicals. Two well-known processes are—air blown biomass gasification and oxy-steam biomass gasification.

The reactors in gasification process are developed in large/commercial scale and can processed massive quantity of materials in continuous approach. The have the potential to deliver efficiencies in the range of 35%–50% considering lower heating value (Holladay et al., 2009). Some of the restrictions that is usually encountered is that it requires an immense number of resources to transport biomass in large quantity to the central processing facility.

In gasification, plasma method is a prominent technology to produce hydrogen gas. In plasma gasification technology, solid is converted into gaseous state and then into plasma by the addition of enormous amount of heat energy and is also known as the fourth state of material. This fourth state indicates the state of ions, radicals, electrons, atoms as well as neutron particles. The plasma technology is classified into two broad categories as thermal and nonthermal plasma/catalysis technology. In thermal, the heavy molecules in broken down into smaller molecules by chemicals reactions with the application of heat. Whereas in nonthermal plasma technology, the higher temperature is not mandate (temperature are less than 250°C) and is usually carried out at low cost for conversion of hydrogen by the decomposition of ammonia (Moszczyńska et al., 2023) as well as waste plastics (Aminu et al., 2022). The plasma technology however, is an expensive approach and requires additional findings to make it economically convenient.

3.6 Water splitting

It is a process of converting water (H_2O) molecules into hydrogen (H_2) gas and oxygen (O_2) gas as shown below:

$$2H_2O \rightarrow 2H_2 + O_2 \tag{3.21}$$

A minimum of 1.23 eV energy as an input is required for splitting the water. Water dissociation produces hydrogen ions and oxygen anions, which are regulated by the difference in ionization potential and electron affinity. The process of water splitting may be classified into three types:

1. Electrolysis—Involves passing an electrical current through water, which causes the splitting.
2. Thermolysis—Here heat is used for splitting water up. The heat can be provided by solar energy, nuclear energy, or other sources.
3. Photocatalysis—Light energy is used to split water molecules, here light energy is usually provided by sunlight or artificial light source.

This technology is currently at the initial phase, and to develop it in a commercial/larger scale numerous existing barriers should be overcome. A brief description is given below.

3.6.1 Electrolysis

The process of hydrogen production from water by the method of electrolysis was developed since 19th century. Thus, it is an established technology which employs direct current (DC) to produce hydrogen and oxygen gas by splitting water molecules. A typical electrolyzer has a cathode (negatively charged electrode) and an anode (positively charged electrode) immersed in an ionic conductor—the electrolyte as shown in Fig. 3.3 When electrical is current passed by applying potential difference, water splits producing hydrogen at the cathode, while oxygen is evolved at the anode side. This is the only industrial

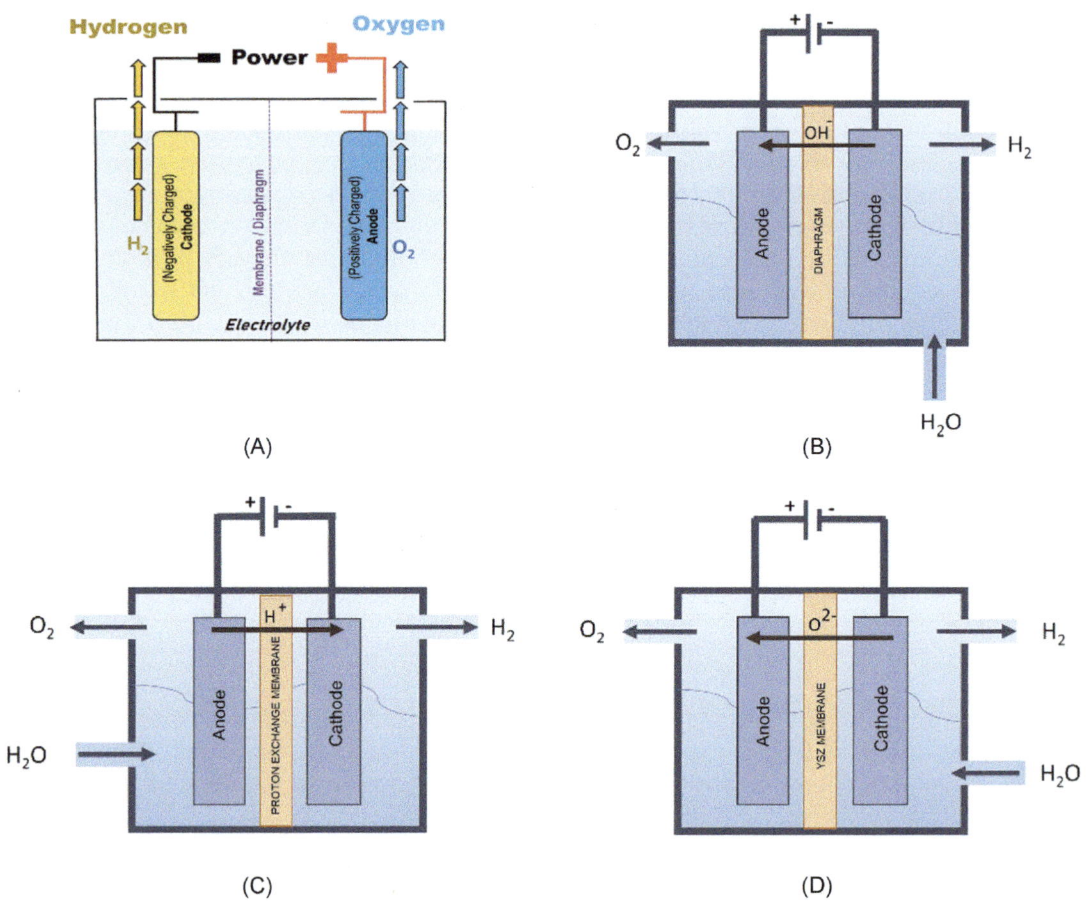

FIGURE 3.3 An overview of electrolysis method for production of hydrogen: (A) Electrolysis, (B) alkaline electrolyzer (AEL), (C) proton exchange membrane (PEM) electrolyzer, and (D) solid oxide electrolysis cell (SOEC) (Le et al., 2023; Shiva Kumar & Lim, 2022).

process to generate nearly 99.999% purity hydrogen with no greenhouse gases emissions by the help of the electricity produced from renewable sources.

Based on the type of electrolyte, type of ionic agent (OH^-, H^+, O^{2-}), and operation temperature; electrolyzers are classified into three parts which are

1. Alkaline electrolyzer (AEL)
2. Proton exchange membrane electrolyzer (PEM)
3. Solid oxide electrolysis cell (SOEC)

The key characteristics comparison are given below in the Table 3.3.

3.6.2 Alkaline electrolyzer

Alkaline water electrolysis is old technology, but this is one of the easiest, simplest, and the most matured technology with dominant market share. This technology is available for larger plant sizes and has an approximate lifespan of 20–30 years (Holladay et al., 2009). AEL systems are easily accessible, effective, can display low capital cost relatively because of the avoidance of noble metals and reasonably mature stack components. The process efficiency lies between 65% and 75% (Vidas & Castro, 2021).

Fig. 3.3B shows the basic set up of alkaline electrolyzer (AEL) highlighting the electrodes (anode and cathode) submerged in electrolyte consisting of aqueous solutions like

TABLE 3.3 Characteristics of different water electrolysis technologies (Burton et al., 2021; Vidas & Castro, 2021).

Characteristic	Alkaline	PEM	SOEC
Technology maturity	Commercial	Near commercial	Demonstration
Temperature (°C)	40–90	20–100	650–1000
Pressure (bar)	<30	<200	<20
Current density (A/cm2)	0.20–0.40	0.60–2.0	0.3–2.0
Voltage (V)	1.8–2.4	1.8–2.2	0.7–1.5
Production rate (m3/h)	<1400	<400	<10
Gas purity (%)	>99.50	>99.99	>99.90
System energy consumption (kWh/Nm^3)	4.5–6.6	4.2–6.6	3.7–3.9
Stack efficiency (% LHV)	63–71	60–68	100
System efficiency (% LHV)	51–60	46–60	76–81
Energy consumption (kWh/m3)	~5.55	~5.40	~3.80
Efficiency (%)	51–60	46–60	76–81
Stack lifetime (kh)	60–120	60–100	8–20
Degradation (%/y)	0.25–1.5	0.5–2.5	3–50
Capital cost (EUR/kWh)	740–1390	1300–2140	>2000
Maintenance cost (% of investment/year)	2–3	3–5	n.a.

sodium hydroxide or potassium hydroxide (NaOH and KOH) with concentration in the range of 20–40 wt.% (Vidas & Castro, 2021). This set up uses nickel/nickel–molybdenum alloy, or nickel catalytic coated with platinum as cathode material and nickel–cobalt alloy or nickel/copper metal covered with metal oxides like manganese, tungsten, or ruthenium as anode material. A diaphragm made of asbestos, polyether sulfone and polyphenylene sulfide is placed between the electrodes thereby separating the anode and cathode region.

Electrical potential is set up between the electrodes to establish DC which can thereby flow through the electrolyte. At high energy, the water molecules dissociate into hydrogen gas and oxygen gas. Water is pumped to the cathode to break into hydrogen and hydroxyl ions (OH^-). The hydroxide ions which are negatively charged are attracted to the anode where they are oxidized to form water, oxygen gas and release electrons. This overall process runs at 40°C–90°C with a pressure of 3 MPa (Burton et al., 2021). The anode and cathode reactions of AEL are shown below:

$$\text{Charge carrier:}\quad OH^- \tag{3.22}$$

$$\text{Cathode reaction:}\quad 2H_2O + 2e^- \rightarrow H_2 + 2OH^- \tag{3.23}$$

$$\text{Anode reaction:}\quad 2OH^- \rightarrow 1/2O_2 + H_2O + 2e^- \tag{3.24}$$

$$\text{Overall cell reaction:}\quad H_2O \rightarrow H_2 + \frac{1}{2}O_2 \tag{3.25}$$

The use of non-noble materials, which results in comparatively cheap material prices is the key benefit while high maintenance costs due to the corrosive nature of electrolyte as well as poor current density are the main primary drawbacks. In traditional design, a gap exists between cathode, diaphragm, and anode. This has a disadvantage which results in high ohmic losses. However, in recent developments, zero-gap electrode design has been introduced to the gap and subsequently reduces the losses.

3.6.3 Proton exchange membrane

For efficient hydrogen, the first water electrolyzer based on a solid polymer electrolyte notion was created by Grubb in the 1950s and general electric further modified it in the 1960s. Proton exchange membrane (PEM) electrolyzer functions similarly as the AEL as shown in Fig. 3.3C. The current efficiency is assumed to be approx. 50%–75% (Burton et al., 2021). However, instead of the liquid electrolyte, PEM electrolysis uses solid sulfonated polystyrene (acid solid polymer) membrane which is named as PEM or polymer electrolyte membrane. This membrane works as both a diaphragm and gas separator. The membrane is backbone of this assembly. Perfluoro sulfonic acid polymers membranes are used such as Nafion, Flemion, and so on (Rashid et al., 2015). These membranes are known for their:

1. High strength and high efficiency.
2. Dimensional stability as temperature changes.
3. Good durability.

The cell is particularly sensitive to water purity, therefore deionized water in pure form without any electrolytic additives are fed in the cell. Hence water must be treated in reverse osmosis followed by demineralization process before admitting. Usually, electrodes are made up of noble metals and their oxides such as iridium, ruthenium, platinum, and rhodium. The operating temperature and pressure ranges from 50°C to 80°C and 1 to 50 bar (Vidas & Castro, 2021).

As the electric potential is induced, water dissociates into oxygen gas, proton/hydrogen ions (H^+), and electrons at the anode side. Oxygen is removed from the system, and the electrons are sent to an external circuit. Protons are sent through the membrane to the cathode, where they combine with electrons and other protons to generate hydrogen gas. The following are anode and cathode reactions of PEM electrolyzer:

$$\text{Charge carrier:} \quad H^+ \tag{26}$$

$$\text{Cathode reaction:} \quad 2H^+ + 2e^- \rightarrow H_2 \tag{27}$$

$$\text{Anode reaction:} \quad H_2O \rightarrow 1/2O_2 + 2\,H^+ + 2e^- \tag{28}$$

$$\text{Overall cell reaction:} \quad H_2O \rightarrow H_2 + \frac{1}{2}O_2 \tag{29}$$

PEM electrolyzers provide several benefits, including small size and mass, lower gas crossover, and power consumption, ability to regulate electrical power variations, and a greater level of safety. Though it also has some disadvantages like high cost of components associated with the required precious component materials, corrosion-acidic environment as the reaction begins at a pH of roughly two and membrane is limited and costly.

3.6.4 Solid oxide electrolysis cell

Solid oxide electrolysis cell (SOEC) has received significant interest because of its efficiency in converting the electrical energy into chemical energy, and generation of pure hydrogen. It runs at high pressures and high temperatures and uses water in the form of steam. It is the least advanced electrolysis technique which are not commercialized. Though the systems are established at laboratory scale, several industrial partners are making every possible effort to bring this technology in the market. It has considerably better efficiency than AEL and PEM of roughly 80%–90% (Burton et al., 2021). Fig. 3.3D illustrates the diagram for SOEC.

High operating temperature is crucial for two reasons. First, it enhances ionic conductivity in the ceramic materials. Second, it reduces the overpotentials at both the anode and cathode, Page | 21 leading to increased efficiency by minimizing energy loss during electrolysis (Rashid et al., 2015). SOEC uses solid ion-conducting ceramics such as yttria stabilized zirconium oxide (YSZ) as the electrolyte, which allows functioning at considerably higher temperatures. Both the electrodes consist of highly dispersed nickel metal mixed with YSZ particles. The operating temperature is very high and it ranges from 650°C to 1000°C and operating pressure is 1–5 bar (Vidas & Castro, 2021).

Vaporized water/steam is used for electrolysis, hence it needs a source of heat, such as nuclear power facilities, solar thermal, or geothermal heat systems, making it more affordable. Vapor-phase electrolysis is also desirable from a thermodynamic perspective, as it allows a greater portion of the requires energy to come from heat instead of electricity (Shaner et al., 2016).

When the electric potential is applied, vaporized water/steam introduced to the cathode side reacts with electrons to dissociate into hydrogen gas and oxide ion (O^{2-}). The negatively charged O^{2-} travels to anode and oxidized to generate oxygen gas while sending out electrons to the external circuit. The following are anode and cathode reactions of SOEC:

$$\text{Charge carrier:}\quad O^{2-} \tag{3.30}$$

$$\text{Cathode reaction:}\quad H_2O + 2e^- \rightarrow H_2 + O^{2-} \tag{3.31}$$

$$\text{Anode reaction:}\quad O^{2-} \rightarrow \frac{1}{2}O_2 + 2e^- \tag{3.32}$$

$$\text{Overall cell reaction:}\quad H_2O \rightarrow H_2 + \frac{1}{2}O_2 \tag{3.33}$$

An SOEC could be used in reverse mode as a fuel cell, producing electricity from hydrogen, allowing it to perform grid balancing services in conjunction with hydrogen storage facilities, boosting the total utilization rate of the equipment. The potential benefits involve low material cost, higher efficiency of electricity, and operate as a fuel cell in reverse mode or in co-electrolysis mode to make syngas from a water stream. On the other hand, material degradation happens because of the high temperatures, resulting in short-term stability.

3.7 Photoelectrochemical water splitting

Photoelectrochemical (PEC) water splitting, also called photocatalysis, is a method that employs sunlight and semiconductor materials to produce hydrogen and oxygen from water. An OEC device can straight away convert solar energy into hydrogen and oxygen using specialized semiconductors called photoelectrochemical materials. PEC use similar semiconductor materials as used in photovoltaic electricity production. Though, in PEC the semiconductor is submerged in a water-based electrolyte for water splitting process which is energized by sunlight (Muradov, 2001a, 2001b). The semiconductor material is exposed to sunlight, it absorbs photons and generates electron-hole pairs. The photoexcited electrons are transferred to the cathode, where they reduce water to generate hydrogen gas, while the holes are transferred to the anode, where they oxidize water to produce oxygen gas.

The effectiveness of light absorption and charge separation depends on the semiconductor material used. Common materials include Metal oxides which are abundant on earth and have relatively large band gaps that results in high photo voltage. All the possible metal oxides that have been widely studied include—TiO_2 (*n*-type, 3.2 eV), Fe_2O_3 (*n*-type, 2.2 eV), ZnO (*n*-type, 3.3–3.4 eV), WO_3 (*n*-type, 2.6–2.8 eV), and Cu_2O (*p*-type, 2.0 eV) (Holladay et al., 2009).

PEC water splitting has received significant attention in recent decades because of the potential of this process to reduce hydrogen production costs. The cost for large-scale application of PEC to produce hydrogen should be lesser that of fossil fuel with additional environmental benefits.

3.8 Thermochemical water splitting

Thermochemical water splitting, also known as thermolysis, employs a series of chemical reactions to decompose water. These reactions involve intermediate steps and chemicals that are continuously recycled, effectively mimicking the direct dissociation of water into hydrogen and oxygen. This method consumes water, acting as a source for hydrogen with other components that be recycled. High temperature from concentrated solar power or waste heat source from nuclear plant are usually considered for thermochemical water splitting (Safari & Dincer, 2020).

In a single process, water may be broken down into hydrogen and oxygen. But because of the unfavorable thermodynamics and extremely high temperature requirement that is of 2500°C for the single step (Holladay et al., 2009), thermochemical cycles are defined as a series of recurring reactions where thermal energy is used to split water at temperatures lower than 900°C, often in two or more parallel steps.

There are around 300 water splitting cycles mentioned in the literature (Holladay et al., 2009). All techniques have greatly decreased the operating temperature from 2500°C, but usually demand greater pressures. However, only a handful of known cycles have been proven to produce considerable amounts of hydrogen. The two famous well studies cycles which are identified as promising processes are:

3.8.1 Sulfur–iodine (S–I) cycle

The sulfur–iodine (S–I) cycle is one of the most promising cycles; since it can couple with high temperature solar and nuclear energy, thus it can generate large amounts of hydrogen. It is a thermochemical cycle with three primary chemical reactions, with oxygen and hydrogen serving as the net products and water serving as the net reactant (Dincer & Acar, 2015).

1. Thermally driven first reaction of S–I cycle, can be expressed as:

$$H_2SO_4(aq) \overset{\text{heat}(300^\circ C-500^\circ C)}{\longrightarrow} H_2O(g) + SO_3(g) \tag{3.34}$$

2. Heats between 800°C and 900°C are used to separate the product gases (H_2O and SO_3). Next, SO_3 gas undergoes thermal breakdown in accordance with the reaction given below:

$$SO_3(g) \overset{\text{heat}(800^\circ C-900^\circ C)}{\longrightarrow} \frac{1}{2}O_2(g) + SO_2(g) \tag{3.35}$$

3. Following its dissociation from O_2, SO_2 spontaneously sets off an exothermic reaction with water and iodine at low temperatures.

$$SO_2(g) + I_2(g) + H_2O(l) \rightarrow 2HI(g) + H_2SO_4(aq) \tag{3.36}$$

4. The HI then undergoes total breakdown to form H_2 using the thermal energy when the temperature of the system is between 425°C and 450°C.

$$2HI(g) \overset{\text{heat}(425^\circ C-450^\circ C)}{\longrightarrow} H_2(g) + I_2(g) \tag{3.37}$$

Energy efficiency of closed loop S–I cycle for hydrogen production depends on operating temperature. It has been reported that integrating an S–I cycle with a high temperature nuclear reactor (~900°C) can achieve an efficiency of 52% (Safari & Dincer, 2020).

Major challenges associated with S–I cycle are:

1. Requirement of higher temperatures (above 800°C).
2. Reactants should not be employed in industrial settings because they are extremely hazardous and corrosive.
3. Separation of HI from H_2SO_4 in an excessive iodine-containing environment when a considerable quantity of iodine is lost.

3.8.2 Copper–chlorine (Cu–Cl) cycle

The copper–chlorine (Cu–Cl) cycle is a viable low temperature cycle for generation of hydrogen. This cycle decomposes the water molecules into oxygen and hydrogen via intermediate copper chloride molecules, in a closed loop that recycles all chemicals constantly. A comparison between Cu–Cl and S–I reveals that for Cu–Cl cycle, the temperature is far lower than (530°C) than that of S–I cycle (850°C) thereby, allowing for simple integration of Cu–Cl cycle with low grade heat source such as nuclear energy (Farsi et al., 2020).

In Cu–Cl cycle, overall process consists of the following four chemical reactions (Rosen et al., 2012).

1. Hydrogen production step: The main or primary step in which hydrochloride (HCl) gas reacts with the Cu solid and produces cupric chloride (CuCl) liquid along with hydrogen gas. This exothermic reaction occurs at temperatures ranging from 430°C to 475°C.

$$2Cu(s) + 2HCl(g) \rightarrow 2CuCl(l) + H_2(g) \tag{3.38}$$

2. Electrolysis step: In this step, electrolysis of cupric chloride (CuCl) liquid is encountered at ambient temperature and voltage required to complete this step ranges between 0.4 and 0.6.

$$2CuCl(l) \rightarrow CuCl_2(aq) + Cu \tag{3.39}$$

3. Hydrolysis step: The step of hydrolysis reaction is endothermic with temperatures between 375°C and 400°C. The reaction occurs between copper chloride ($CuCl_2$) solid and water in gas form. The products are hydrochloric (HCl) gas and Cu_2OCl_2 solid, as shown by the following reaction:

$$2CuCl_2(s) + H_2O(g) \rightarrow Cu_2OCl_2(s) + 2HCl(g) \tag{3.40}$$

4. Oxygen production step: Under heat supply, Cu_2OCl_2 solid decomposes into oxygen gas O_2 and cupric chloride (CuCl) liquid as shown below:

$$2Cu_2OCl_2(s) \rightarrow 4CuCl(l) + O_2(g) \tag{3.41}$$

The step involved in the production of oxygen is most essential because it has the greatest temperature among the different processes that is estimated to be 500°C.

Reported energy efficiency is around 50% respectively. In the above highlighted four-step Cu−Cl cycle, as the temperature of the hydrolysis stage increases with decrease in the reference temperature, the overall energy efficiency of the operation is enhanced (Farsi et al., 2020).

The major challenges that are encountered during the Cu−Cl cycle includes:

1. Limited understanding of corrosion-resistant materials for handling molten Cu−Cl.
2. Solids handling between processes and engineering equipment development for working with corrosive fluids.

TABLE 3.4 Hydrogen production methods, its efficiency and maturity status (Megía et al., 2021).

Technology	Production method	Efficiency (%)	Maturity	References
Fossil fuel based	Steam reforming	70−85	Commercial	Holladay et al. (2009); ; Megía et al. (2021)
	Partial oxidation	60−75	Commercial	Holladay et al. (2009); Megía et al. (2021)
	Autothermal reforming	60−75	Early Commercial	Holladay et al. (2009); ; Megía et al. (2021)
	Pyrolysis	58	Commercial	Megía et al. (2021); Sánchez-Bastardo et al. (2021)
	Coal gasification	60	Commercial	Megía et al. (2021); Sánchez-Bastardo et al. (2021)
Renewable based	Biomass gasification	30−60	Commercial	Dincer and Acar (2015); (Kumar and Himabindu, 2019); Lepage et al. (2021); Megía et al. (2021)
	Pyrolysis	35−50	Commercial	(Kumar and Himabindu, 2019); Lepage et al. (2021); Megía et al. (2021); Nikolaidis and Poullikkas (2017)
	Bio-photolysis	10−15	Research and development	Dincer and Acar (2015); (Kumar and Himabindu, 2019); Megía et al. (2021); Nikolaidis and Poullikkas (2017)
	Dark fermentation	60−80	Research and development	(Kumar and Himabindu, 2019); Lepage et al. (2021); Megía et al. (2021);Nikolaidis and Poullikkas (2017)
	Photo fermentation	0.1−12	Research and development	Dincer and Acar (2015); (Kumar and Himabindu, 2019); Lepage et al. (2021); Megía et al. (2021); Nikolaidis and Poullikkas (2017)
	Microbial electrolysis cells	78	Long term	Call and Logan (2008); Holladay et al. (2009)
	Alkaline electrolysis	65−75	Commercial	(Vidas and Castro, 2021)
	PEM electrolysis	50−75	Commercial	Burton et al. (2021)
	SOEC	80−90	Medium term	Burton et al. (2021)
	Thermochemical	52−60	−	Boretti1 et al. (2021)
	Photoelectrochemical water splitting	12.4	Long term	Holladay et al. (2009); (Turner et al., 2008)

Hydrogen production pathways presented in the earlier sections have different conversion efficiency, production cost and technology maturity level. Comparison of efficiency and technology maturity of different hydrogen production pathways is given in Table 3.4.

3.9 Hydrogen distribution

The distribution of hydrogen in general consists of three major steps which includes pipeline, high-pressure tube trailers and liquefied hydrogen tankers (Ishaq et al., 2022; Energy Technology System Analysis Programme (ETSAP), 2014).

Pipelines—In the pipeline system, hydrogen is transported through pipe connected from source to destination. The process is lower in operating cost as considered to the remaining above two mentioned methods. However, the initial cost is high due to the installation of the pipelines system and the major drawbacks is the distance of the pipelines which limits the capacity of hydrogen transport.

High-pressure tube trailers—In this system, the hydrogen gas is compressed and transported in high-pressure tube trailers by roads using vehicles like trucks, through sea like ships and railways. The process is costly and is limited to short distances transport as compared to pipelines.

Liquefied hydrogen tankers—In this case, the hydrogen gas is liquefied (cryogenic liquefaction) by reducing the temperature of hydrogen. This produces an ease in transport of hydrogen in liquified form over long distances by road, railways, and overseas. The process is expensive and may undergo vaporization of liquid hydrogen which are the limitations of the existing system.

The available literature also reported the supply of hydrogen through four different steps which are onsite, semi centralized, centralized, and intercontinental. The first onsite usually deals with the supply chain of hydrogen from source to fuel station directly. The remaining three categories depicts the supply of hydrogen from source to fuel station through several distances by using tankers (Ishaq et al., 2022).

The current scenario has several challenges for transportation of hydrogen from source to consumers. The nature of hydrogen is the main reason which produces such challenges with respect to its transport and additionally retinas less energy per unit volume with respect to different other fuels. The significant generation of hydrogen at commercial scale is desirable as it reduces the production cost. However, its distribution to the users may be of severe challenges which may require higher capital as well as operating cost. However, hydrogen production near to the user locations may reduce the distribution cost but at the same time enhances the production cost. Therefore, an optimum condition must be achieved with respect to the challenges between production and distribution. Also, there are several research in progress which may provide solutions of the existing challenges in future.

3.10 Current application and cases

Hydrogen can be used as an alternative fuel for heating, electricity generation and energy storage. Hydrogen in industries is used for petroleum refining, producing chemicals, and fertilizers, steel industries, food processing, energy storage, fuel cells for electricity generation, and powering vehicles (Pan et al., 2020).

Hydrogen in oil and gas industries during refining are used in several intermediate processes which includes hydrocracking, hydrogeneration, desulphurization, etc. It is used to produce chemicals like ammonia and methyl alcohol and in metallurgy processes, hydrogen is deployed instead of carbon to reduce the emission of carbon in nature. Hydrogen has also showed greater potential in hybrid system, where energy storage and heat power supply are involved. The perfect example is Yutzla Island in Norway where they deployed this technology and become self-supporting in terms of power generation. In this system, the excess power generated by wind energy were used to generate hydrogen by electrolysis operation of sea water which was stored in a container. When the power generation by wind energy is not sufficient, stored hydrogen is utilized by a fuel cell to generate the electricity. Another scenario of hydrogen application is the joint heat and power supply. In this case, the power or electricity is produced by injection if hydrogen and air in the fuel cells. This results in the formation of DC, which is then converted to alternating current (AC) using invertor. The other part of the joint heat and power system has a provision to collect the heat produced during the power generation in the fuel cells and then it is stored in container. The example of which is demonstrated in Minatojima system in Kobe. The island generated power and heat which is then supplied in sports center and hospitals.

In transportation sector, deploying fuel cells compartment in electric car can be more effective as compared to electric cars. Electric cars usually take long time for charging and at the same time provides lower mileage. Therefore, such power issues can be tackled to some extent by creating a provision of small compartment for fuel cells in the electric cars and in addition maintaining the zero carbon emission. Hydrogen as fuel station is another way of fueling hydrogen/fuel cell cars and promoting its uses toward zero carbon emission. In hydrogen gas station, the gas in converted into dry gas form which is pressurized by compressor inside high-pressure storage containers. The fuel cell cars are then filled accordingly with the usage of hydrogen injector (Pan et al., 2020).

3.11 Challenges

Hydrogen gas is produced by fossil fuel and renewable resources which has their own advantages, disadvantages, and limitations. The primary obstacles are related to the creation and delivery of hydrogen gas from the source to the consumer. The overall cost for production and distribution of hydrogen gas is higher as compared to fuels produced from fossil fuels. Additionally, hydrogen requires a large volume to store because of low density in comparison to other fluids. A hydrogen gas volume of more than 3000 times is required to produce the same energy as produced by other available fuels. Moreover, the storage and transportation of hydrogen could be other possible challenges due to its high flammability nature (Megía et al., 2021). Hydrogen gas must be pressurized or liquified in tube tailers or tankers before it can be stored and transported. Additionally, there are several technical challenges that exist with respect to electrolysis processes for production of hydrogen. Researchers have used impure water instead of pure water to produce hydrogen by electrolysis due to limitation in drinking water. However, the purpose is not solved and additionally resulted in further complications. This includes the designing aspects of electrodes, poisoning effect, identifying inexpensive material, exploring catalyst properties, membrane degradation and durability, membrane efficiency and effectiveness, and influence on electrolysis process due to impurity in water (El-Shafie, 2023). Studies has reported the development of nanostructure noble materials for electrolysis processes but still its commercial scale production is challenging. As discussed above the limitation is due to the stability challenges of the noble material in the reaction medium (Wang et al., 2021). Moreover, methods like plasma techniques for hydrogen production are costly and new findings has to be developed in future to make the process economically viable.

Therefore, today there is a need for development of new techniques to generate hydrogen in eco-friendly nature at commercial scale by consuming minimum energy, thereby making the process economical and substituting fossil fuels as well as hydrogen produced using fossil fuels. Hydrogen production from biomass and water is eco-friendly, however, their efficiency, transportations, and overall cost (economic) are the prime factors which needs to be explored in future research for wide acceptability of hydrogen as fuels in societies.

As hydrogen is flammable and has severe challenges with respect to storage and transport, therefore they can be blended with other gases to store or transport effectively with any danger. Additionally, hydrogen can be directly burned inside combustion engine for transportation purposes, or to generate power to perform work. Also, accurate sensors with fast response time should be developed to detect even minute amount of hydrogen in air so that any kind of accident due to leakage can be prevented.

3.12 Conclusion and future outlooks

The most effective method to produce hydrogen in terms of economy can be achieved by performing SR. Whereas using biomass as a feedstock for hydrogen generation is expensive by almost three times as compared to stream reforming process and is not economical. On the other hand, electrolysis is suitable and effective as it does not involve any by product. However, electrolysis requires large amount of heat which governs the hydrogen production cost. Therefore, an overall balance considering the optimum condition must be achieved or screened looking into the future market demand, supply and eco-friendly environment aspects which need future research and advanced investigations.

The implementation of hydrogen as future fuel replacing fossil fuel will need new infrastructure, social acceptance, latest facilities and new content (Le et al., 2023). Additionally, targeting the CO_2 emission from chemical industries which account for more than 15% of total carbon emission from industries has to be projected for net zero carbon emission in future (Barecka & Ager, 2023).

Therefore, extensive research is necessary to generate hydrogen from renewable resources in an economical way with higher efficiency at commercial scale so that zero or minimum carbon emission can be achieved in future by limiting the usage of fossil fuels. Additionally, the handling, transportation and storage of hydrogen are of huge concerned as hydrogen is flammable. Therefore, the acceptable hydrogen in society majorly depends on the safety issues.

Acknowledgment

The author acknowledges the Science & Engineering Research Board (SERB), Government of India for supporting this research (EEQ/2023/000292) and Department of Chemical Engineering, National Institute of Technology Hamirpur, Himachal Pradesh, India, for providing the necessary resources.

References

Al-Hamamre, Z., Voß, S., & Trimis, D. (2009). Hydrogen production by thermal partial oxidation of hydrocarbon fuels in porous media based reformer. *International Journal of Hydrogen Energy*, *34*(2), 827–832. Available from https://doi.org/10.1016/j.ijhbydene.2008.10.085.

Aminu, I., Nahil, M. A., & Williams, P. T. (2022). Hydrogen production by pyrolysis-nonthermal plasma/catalytic reforming of waste plastic over different catalyst support materials. *Energy and Fuels*, *36*(7), 3788–3801. Available from https://doi.org/10.1021/acs.energyfuels.1c04186, http://pubs.acs.org/journal/enfuem.

Balthasar, W. (1984). Hydrogen production and technology: Today, tomorrow and beyond. *International Journal of Hydrogen Energy*, *9*(8), 649–668. Available from https://doi.org/10.1016/0360-3199(84)90263-5.

Barecka, M. H., & Ager, J. W. (2023). Towards an accelerated decarbonization of the chemical industry by electrolysis. *Energy Advances*, *2*(2), 268–279. Available from https://doi.org/10.1039/d2ya00134a, http://www.rsc.org/journals-books-databases/about-journals/energy-advances/.

Boretti1, A., Nayfeh, J., & Al-Maaitah, A. (2021). Hydrogen production by solar thermochemical water-splitting cycle via a beam down concentrator. *Frontiers in Energy Research*.

Burton, N., Padilla, A., Rose, A., & Habibullah, H. (2021). Increasing the efficiency of hydrogen production from solar powered water electrolysis. *Reneable and Sustainable Energy Reviews*.

Call, D., & Logan, B. E. (2008). Hydrogen production in a single chamber microbial electrolysis cell lacking a membrane. *Environmental Science and Technology*, *42*(9), 3401–3406. Available from https://doi.org/10.1021/es8001822United, http://pubs.acs.org/doi/pdfplus/10.1021/es8001822, States.

Cheonh, P. Y. Y., Kansedo, J., Lau, J. S. Y., & Tan, Y. H. (2022). *Renewable biomass wastes for biohydrogen production* (pp. 273–298). Elsevier BV. Available from 10.1016/b978-0-12-819727-1.00091-1.

Das, A., & Peu, S. D. (2022). A comprehensive review on recent advancements in thermochemical processes for clean hydrogen production to decarbonize the energy sector. *Sustainability* (18).

Dincer, I., & Acar, C. (2015). Review and evaluation of hydrogen production methods for better sustainability. *International Journal of Hydrogen Energy*, *40*(34), 11094–11111. Available from https://doi.org/10.1016/j.ijhydene.2014.12.035.

El-Shafie, M. (2023). Hydrogen production by water electrolysis technologies: A review. *Results in Engineering*, *20*, 101426.

El-Shafie, M., Kambara, S., & Hayakawa, Y. (2019). Hydrogen production technologies overview. *Journal of Power and Energy Engineering*, *7*(1), 107–154. Available from https://doi.org/10.4236/jpee.2019.71007.

Ersöz, A. (2008). Investigation of hydrocarbon reforming processes for micro-cogeneration systems. *International Journal of Hydrogen Energy*, *33*(23), 7084–7094. Available from https://doi.org/10.1016/j.ijhydene.2008.07.062.

Farsi, A., Dincer, I., & Naterer, G. F. (2020). Review and evaluation of clean hydrogen production by the copper–chlorine thermochemical cycle. *Journal of Cleaner Production*, *276*. Available from https://doi.org/10.1016/j.jclepro.2020.123833.

Gao, Y., Wang, M., Raheem, A., Wang, F., Wei, J., Xu, D., Song, X., Bao, W., Huang, A., Zhang, S., & Zhang, H. (2023). Syngas production from biomass gasification: Influences of feedstock properties, reactor type, and reaction parameters. *ACS Omega*, *8*(35), 31620–31631. Available from https://doi.org/10.1021/acsomega.3c03050.

Holladay, J. D., Hu, J., King, D. L., & Wang, Y. (2009). An overview of hydrogen production technologies. *Catalysis Today*, *139*(4), 244–260. Available from https://doi.org/10.1016/j.cattod.2008.08.039.

Ishaq, H., Dincer, I., & Crawford, C. (2022). A review on hydrogen production and utilization: Challenges and opportunities. *International Journal of Hydrogen Energy*, *47*(62), 26238–26264. Available from https://doi.org/10.1016/j.ijhydene.2021.11.149.

Jonas, J. (2009). The history of hydrogen. *Member of International Academy of Science*. Available from https://www.altenergymag.com/article/2009/04/the-history-of-hydrogen/555/.

Kalamaras, C. M., & Efstathiou, A. M. (2013). Hydrogen production technologies: Current state and future developments. *Conference Papers in Energy*, *2013*, 1–9. Available from https://doi.org/10.1155/2013/690627.

Kapdan, I. K., & Kargi, F. (2006). Bio-hydrogen production from waste materials. *Enzyme and Microbial Technology*, *38*(5), 569–582. Available from https://doi.org/10.1016/j.enzmictec.2005.09.015.

Koutra, E., Tsafrakidou, P., Sakarika, M., & Kornaros, M. (2020). *Microalgal biorefinery. Microalgae cultivation for biofuels production* (pp. 163–185). Academic Press Available from https://doi.org/10.1016/B978-0-12-817536-1.00011-4.

Kumar, S., Regassa Jufar, S., Kumar, S., Foroozesh, J., Kumari, S., & Bera, A. (2023). Underground hydrogen storage and its roadmap and feasibility in India toward net-zero target for global decarbonization. *Fuel*, *350*. Available from https://doi.org/10.1016/j.fuel.2023.128849, http://www.journals.elsevier.com/fuel/.

Kumar, S. S., & Himabindu, V. (2019). Hydrogen production by PEM water electrolysis – A review. *Materials Science for Energy Technologies*, *2*(3), 442–454.

Le, P., Trung, V. D., Nguyen, P. L., Phung, T. V. C., Natsuki, J., & Natsuki, T. (2023). The current status of hydrogen energy: an overview. *RSC Advances*, *13*(40), 28262–28287. Available from https://doi.org/10.1039/D3RA05158G.

Lepage, T., Kammoun, M., Schmetz, Q., & Richel, A. (2021). Biomass-to-hydrogen: A review of main routes production, processes evaluation and techno-economical assessment. *Biomass and Bioenergy*, *144*. Available from https://doi.org/10.1016/j.biombioe.2020.105920.

Lovegrove, P. (2022). Hydrogen at risk of being the great missed opportunity of the energy transition. *Hydrogen Forecast to 2050*. Available from https://www.deassociation.ca/newsfeed/hydrogen-at-risk-of-being-the-great-missed-opportunity-of-the-energy-transition.

Lubbe, F., Rongé, J., Bosserez, T., & Martens, J. A. (2023). Golden hydrogen. *Current Opinion in Green and Sustainable Chemistry*, *39*.

Global fossil fuel share in energy consumption 1965-2022. (2023). *Published by Statista Research Department*, October 12

Energy Technology System Analysis Programme (ETSAP), February 2014, Hydrogen Production & Distribution.

Megía, P. J., Vizcaíno, A. J., Calles, J. A., & Carrero, A. (2021). Hydrogen production technologies: From fossil fuels toward renewable sources. A mini review. *Energy & Fuels*, *35*(20), 16403–16415. Available from https://doi.org/10.1021/acs.energyfuels.1c02501.

Moszczyńska, J., Liu, X., & Wiśniewski, M. (2023). Green hydrogen production through ammonia decomposition using non-thermal plasma. *International Journal of Molecular Sciences* (18).

Muradov, N. (1993). How to produce hydrogen from fossil fuels without CO_2 emission. *International Journal of Hydrogen Energy*, *18*(3), 211–215. Available from https://doi.org/10.1016/0360-3199(93)90021-2.

Muradov, N. (2001). Catalysis of methane decomposition over elemental carbon. *Catalysis Communications*, *2*(3–4), 89–94. Available from https://doi.org/10.1016/s1566-7367(01)00013-9.

Muradov, N. (2001). Hydrogen via methane decomposition: An application for decarbonization of fossil fuels. *International Journal of Hydrogen Energy*, *26*(11), 1165–1175. Available from https://doi.org/10.1016/s0360-3199(01)00073-8.

Ni, M., Leung, D. Y. C., Leung, M. K. H., & Sumathy, K. (2006). An overview of hydrogen production from biomass. *Fuel Processing Technology*, *87*(5), 461–472. Available from https://doi.org/10.1016/j.fuproc.2005.11.003.

Nikolaidis, P., & Poullikkas, A. (2017). A comparative overview of hydrogen production processes. *Renewable and Sustainable Energy Reviews*, *67*, 597–611. Available from https://doi.org/10.1016/j.rser.2016.09.044.

Palma, V., Ricca, A., Meloni, E., Martino, M., Miccio, M., & Ciambelli, P. (2016). Experimental and numerical investigations on structured catalysts for methane steam reforming intensification. *Journal of Cleaner Production*, *111*, 217–230. Available from https://doi.org/10.1016/j.jclepro.2015.09.004.

Pan, A., Liu, J., Liu, Z., Yang, Y., Yang, X., & Zhang, M. (2020). Application of hydrogen energy and review of current conditions. *IOP Conference Series: Earth and Environmental Science*, *526*(1). Available from https://doi.org/10.1088/1755-1315/526/1/012124.

Phukan, R., & Saha, R. (2022). Low salinity surfactant alternating gas/CO_2 flooding for enhanced oil recovery in sandstone reservoirs. *Journal of Petroleum Science and Engineering*, *202*, 110253.

Rapier, R. (2023). *Global energy trends: Insights from the 2023 statistical review of world energy, Forbes*. https://www.forbes.com/sites/rrapier/2023/08/06/global-energy-trends-insights-from-the-2023-statistical-review-of-world-energy/.

Rashid, M. M., Mesfer, M. K. A., Naseem, H., & Danish, M. (2015). Hydrogen production by water electrolysis: A review of alkaline water electrolysis, PEM water electrolysis and high temperature water electrolysis. *International Journal of Engineering and Advanced Technology* (4), 80–94.

Rosen, M. A., Naterer, G. F., Chukwu, C. C., Sadhankar, R., & Suppiah, S. (2012). Nuclear-based hydrogen production with a thermochemical copper-chlorine cycle and supercritical water reactor: Equipment scale-up and process simulation. *International Journal of Energy Research*, *36*(4), 456–465. Available from https://doi.org/10.1002/er.1702.

Safari, F., & Dincer, I. (2020). A review and comparative evaluation of thermochemical water splitting cycles for hydrogen production. *Energy Conversion and Management*, *205*. Available from https://doi.org/10.1016/j.enconman.2019.112182.

Saha, R., & Phukan, R. (2023). Retention of surfactant by solid-liquid interaction and the process of improving oil recovery by inducing hydrophilic particles. *Energy and Fuels*, *37*(18), 13734–13750. Available from https://doi.org/10.1021/acs.energyfuels.3c02286, http://pubs.acs.org/journal/enfuem.

Saha, R., Sharma, A., Uppaluri, R. V. S., & Tiwari, P. (2019). Interfacial interaction and emulsification of crude oil to enhance oil recovery. *International Journal of Oil, Gas and Coal Technology*, *22*(1), 1–15. Available from http://www.inderscience.com/ijogct, 10.1504/IJOGCT.2019.102281.

Saha, R., Tiwari, P., & Uppaluri, R. V. S. (2021). *Chemical nanofluids in enhanced oil recovery: Fundamentals and applications*. CRC Press.

Saha, R., Uppaluri, R. V. S., & Tiwari, P. (2017). Effect of mineralogy on the adsorption characteristics of surfactant—reservoir rock system. *Colloids and Surfaces A: Physicochemical and Engineering Aspects*, *531*, 121–132. Available from https://doi.org/10.1016/j.colsurfa.2017.07.039, http://www.elsevier.com/locate/colsurfa.

Saha, R., Uppaluri, R. V. S., & Tiwari, P. (2018a). Influence of emulsification, interfacial tension, wettability alteration and saponification on residual oil recovery by alkali flooding. *Journal of Industrial and Engineering Chemistry*, *59*, 286–296. Available from https://doi.org/10.1016/j.jiec.2017.10.034, http:http://www.sciencedirect.com/science/journal/1226086X.

Saha, R., Uppaluri, R. V. S., & Tiwari, P. (2018b). Silica nanoparticle assisted polymer flooding of heavy crude oil: Emulsification, rheology, and wettability alteration characteristics. *Industrial & Engineering Chemistry Research*, *57*(18), 6364–6376. Available from https://doi.org/10.1021/acs.iecr.8b00540.

Saha, R., Uppaluri, R. V. S., & Tiwari, P. (2018c). Effects of interfacial tension, oil layer break time, emulsification and wettability alteration on oil recovery for carbonate reservoirs. *Colloids and Surfaces A: Physicochemical and Engineering Aspects*, *559*, 92–103. Available from https://doi.org/10.1016/j.colsurfa.2018.09.045.

Saha, R., Uppaluri, R. V. S., & Tiwari, P. (2019). Impact of natural surfactant (reetha), polymer (xanthan gum), and silica nanoparticles to enhance heavy crude oil recovery. *Energy & Fuels*, *33*(5), 4225–4236. Available from https://doi.org/10.1021/acs.energyfuels.9b00790.

Scipioni, A., Manzardo, A., & Ren, J. (2017). *Hydrogen economy: Supply chain, life cycle analysis and energy transition for sustainability* (pp. 1–328). Elsevier Inc. Available from http://www.sciencedirect.com/science/book/9780128111321.

Shaner, M. R., Atwater, H. A., Lewis, N. S., & McFarland, E. W. (2016). A comparative technoeconomic analysis of renewable hydrogen production using solar energy. *Energy and Environmental Science*, *9*(7), 2354–2371. Available from https://doi.org/10.1039/c5ee02573g, http://www.rsc.org/Publishing/Journals/EE/About.asp.

Shiva Kumar, S., & Lim, H. (2022). An overview of water electrolysis technologies for green hydrogen production. *Energy Reports*, *8*, 13793–13813. Available from https://doi.org/10.1016/j.egyr.2022.10.127.

Steinberg, M., & Cheng, H. C. (1989). Modern and prospective technologies for hydrogen production from fossil fuels. *International Journal of Hydrogen Energy*, *14*(11), 797–820. Available from https://doi.org/10.1016/0360-3199(89)90018-9.

Sánchez-Bastardo, N., Schlögl, R., & Ruland, H. (2021). Methane pyrolysis for zero-emission hydrogen production: A potential bridge technology from fossil fuels to a renewable and sustainable hydrogen economy. *Industrial & Engineering Chemistry Research*, *60*(32), 11855–11881. Available from https://doi.org/10.1021/acs.iecr.1c01679.

The future of hydrogen. (2019). Paris: IEA. Licence: CC BY 4.

Turner, J., Sverdrup, G., Mann, M. K., Maness, P. C., Kroposki, B., Ghirardi, M., Evans, R. J., & Blake, D. (2008). Renewable hydrogen production. *International Journal of Energy Research*, *32*(5), 379–407. Available from https://doi.org/10.1002/er.1372.

Vidas, L., & Castro, R. (2021). Recent developments on hydrogen production technologies: State-of-the-art review with a focus on green-electrolysis. *Applied Sciences*, *11*(23), 11363. Available from https://doi.org/10.3390/app112311363.

Wang, S., Lu, A., & Zhong, C. J. (2021). Hydrogen production from water electrolysis: Role of catalysts. *Nano Convergence*, *8*, 4. Available from https://doi.org/10.1186/s40580-021-00254-x.

Wietschel, M., & Ball, M. (2009). *The hydrogen economy: Opportunities and challenges*. Cambridge University Press. Available from http://doi.org/10.1017/CBO9780511635359, 10.1017/CBO9780511635359.

Xu, D., Dong, L., & Ren, J. (2017). *Introduction of hydrogen routines. Hydrogen economy: Supply chain, life cycle analysis and energy transition for sustainability* (pp. 35–54). Elsevier Inc. Available from http://www.sciencedirect.com/science/book/9780128111321, 10.1016/B978-0-12-811132-1.00002-X.

Yue, M., Lambert, H., Pahon, E., Roche, R., Jemei, S., & Hissel, D. (2021). Hydrogen energy systems: A critical review of technologies, applications, trends and challenges. *Renewable and Sustainable Energy Reviews, 146*. Available from https://doi.org/10.1016/j.rser.2021.111180.

Zgonnik, V. (2020). The occurrence and geoscience of natural hydrogen: A comprehensive review. *Earth-Science Reviews, 203*. Available from https://doi.org/10.1016/j.earscirev.2020.103140.

CHAPTER

4

Hydrogen surface storage

Shilpa Singh

Department of Physics and Electronics, St. Xavier's College, Ahmedabad, Ahmedabad, India

4.1 Introduction

The requirement for energy consumption is growing with population expansion and technological improvement. According to one survey, the world's population has expanded by 27% over the last two decades (Ritchie, 2023), while the overall consumption of energy has increased by 37% (Ritchie et al., 2022). The majority of countries import energy from a few countries that are wealthy in fossil fuels to meet their primary energy needs. These sources are limited in quantity, and A significant percentage of fossil fuel reserves has already been depleted. Additionally, the large amount of CO_2 produced by these fuels is a major contributor to global warming. These difficulties drive us to transition to renewable energy sources such as solar, wind, geothermal, hydropower, hydrogen, etc. The utilization of these sources is limited, nevertheless, by technological limitations and geographic disparities. The current technology is not mature enough to make an immediate transition to renewable energy sources. A lot of research is still required to make these sources cost-effective. Due to the intermittent nature of renewable energy sources, the storage of this energy is essential. There is increasing interest in storing it effectively and economically in the form of hydrogen. Since hydrogen is an energy carrier that can be easily stored and transported, it is more valuable for massive industrial uses and long-distance energy transportation than the other sources (Thompson, 2023). Additionally, among all fuels, hydrogen produces the most energy per unit mass and the least amount of CO_2.

During the late 18th century, pioneering experiments led by scientists Henry Cavendish and Antoine Lavoisier were instrumental in the discovery and characterization of hydrogen gas. In the 19th century, hydrogen found widespread use in the production of ammonia and the hydrogenation of oils. By the early 20th century, hydrogen had taken to the skies as a lifting gas for airships, albeit its usage was abandoned following the tragic Hindenburg disaster in 1937. However, in the mid-20th century, the emergence of liquid hydrogen as rocket fuel showcased its potential as a powerful and efficient energy carrier

Subsurface Hydrogen Energy Storage
DOI: https://doi.org/10.1016/B978-0-443-24071-3.00005-4

for space exploration. In more recent years, there has been a significant surge in interest and investment in hydrogen energy worldwide. Governments and research institutions are spearheading ambitious programs to accelerate the transition to cleaner energy sources. Advancements in hydrogen production, storage, and utilization technologies, including the use of renewable sources like solar and wind power, are driving the rapid growth of the hydrogen industry. Hydrogen is increasingly recognized as a critical component in combating climate change, achieving energy independence, and fostering economic growth through innovation and job creation in the hydrogen sector.

However, not all hydrogen is eco-friendly. The hydrogens produced are classified into various categories with different color codes depending on their sources (Hauglustaine, 2024). Gray hydrogen, blue hydrogen, and green hydrogen are the three most common types of hydrogen produced (Sinha, 2023; The hydrogen colour spectrum, 2023). The most common form of hydrogen is gray hydrogen, which is produced from fossil fuels. Greenhouse gasses are released during the production of gray hydrogen. The same procedure is used to make blue hydrogen as gray hydrogen, except greenhouse gasses are captured and stored during the process. Compared to gray hydrogen, blue hydrogen is more environmentally friendly. Green hydrogen is produced through the electrolysis of water, resulting in zero greenhouse gas emissions (Hauglustaine, 2024; Sinha, 2023; The hydrogen colour spectrum, 2023). Green hydrogen could be the key to achieving energy independence and, eventually, net zero emissions.

Only 5% of the newly announced projects for low-emission hydrogen production have solidified investment decisions. In 2022, nearly 400 deals revolving around hydrogen were sealed, with projections indicating a continuous upward trend. By the conclusion of 2030, the global capacity for hydrogen market development is anticipated to exceed 71 million tons per annum. Leading the charge are monumental ventures such as the NEOM Green hydrogen project in Saudi Arabia, Sinopec's Ordos green hydrogen project in China, FFI, and TES green hydrogen project in Germany, Plug Power green hydrogen plants in Finland, and the Western Green Energy Hub in Australia (Largest green hydrogen projects in the world, 2023). All the announced low-emission hydrogen production project worldwide is shown in Fig. 4.1.

While the demand for hydrogen is burgeoning, its utilization remains predominantly rooted in traditional sectors like refining and industrial applications. Surprisingly, only a fraction—less than 0.1%—of hydrogen demand is allocated to long-distance transport and heavy industry. In 2022, global hydrogen demand surged to 95 million tons, marking a 3% increase from 2021. To attain the net zero emission target, it's essential to have policies that prioritize the utilization of low-emission hydrogen. These policies will attract investments to produce more low-emission hydrogen and build the infrastructure needed for it (Global Hydrogen Review, 2023). Fig. 4.1 shows the utilization of hydrogen in different sector and region.

Hydrogen storage is an important part of using hydrogen as a fuel in a variety of applications, including portable power, transportation, and stationary use. The development of safe and effective storage techniques is necessary for the widespread use of hydrogen as a fuel. Energy efficiency, volumetric, and gravimetric capacity, durability, refueling time, cost, technological advancement, life cycle, etc. are important factors for hydrogen storage (Cousins & Zhang, 2019). Targets for the gravimetric density of hydrogen storage systems

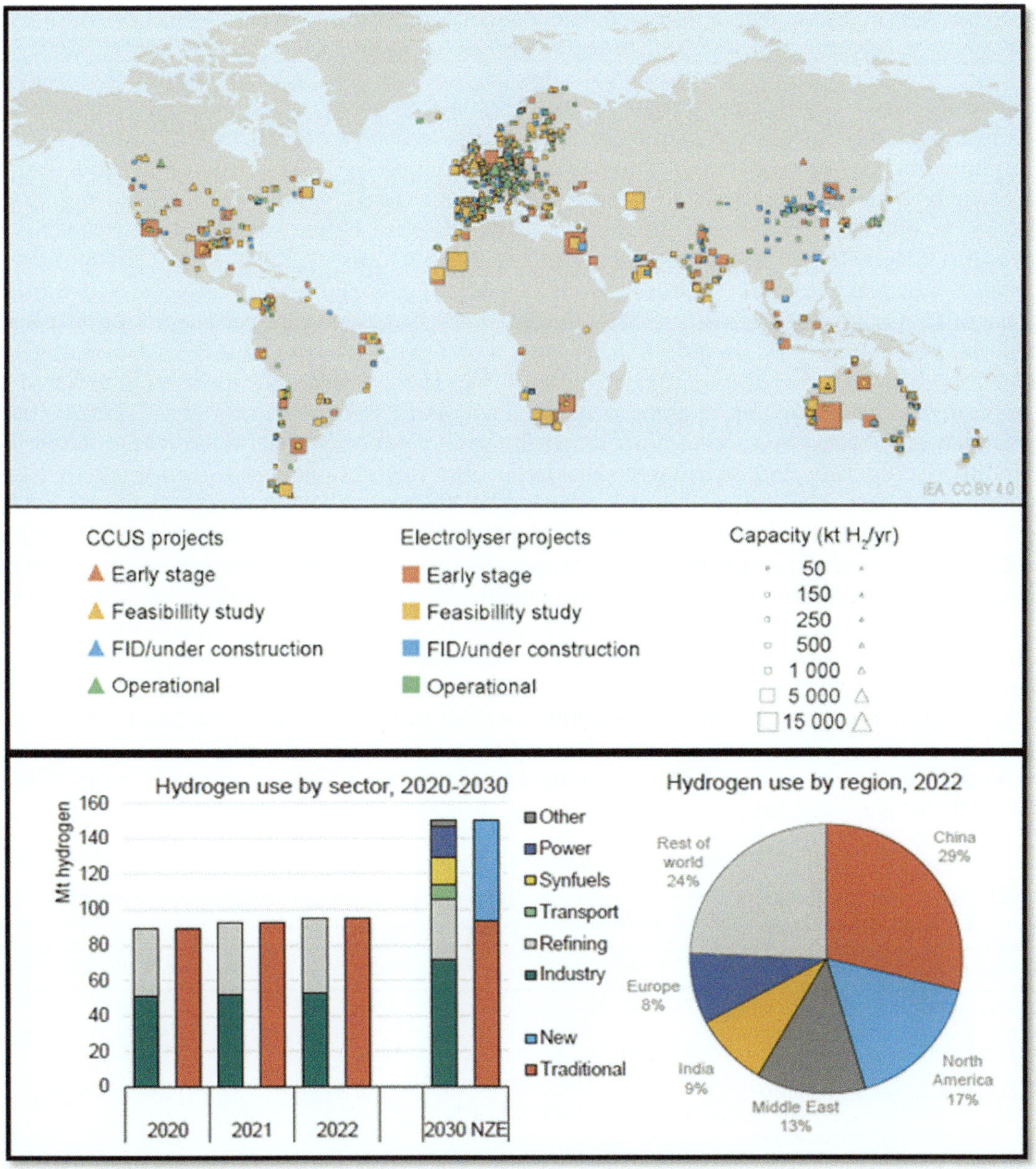

FIGURE 4.1 Low-emission hydrogen production projects in the world and hydrogen consumption in different sectors and region. Map lines delineate study areas and do not necessarily depict accepted national boundaries. Source: *Reproduced from Global Hydrogen Review. (2023). IEA. https://www.iea.org/reports/global-hydrogen-review-2023.*

have been established by the US Department of Energy (DOE). The amount of hydrogen that may be held per unit mass of the storage system is known as its gravimetric density. A target of 6.5 wt.% gravimetric density has been set by the DOE for light-duty vehicles (Materials-Based Hydrogen Storage, n.d.). This indicates that the hydrogen content of the hydrogen storage system should be 6.5% of the system's overall weight, including the container and the storage media. The parameters decided by DOE are given in Table 4.1.

Hydrogen storage methods vary based on specific goals, safety considerations, energy efficiency, and cost. Hydrogen can be stored in pressure tanks as either compressed gas or liquid (Barthélémy, 2012; Barthélémy et al., 2017; Mannan, 2012), or within materials through physisorption or chemisorption, constituting surface-based storage techniques (Jena, 2011; Langmi, 2022; Zhou, 2005). Lately, there has been a global surge in interest in material-based storage techniques, particularly with regard to nanomaterials. Nanomaterials offer a high surface-to-volume ratio, facilitating increased active sites for hydrogen adsorption and thereby enhancing storage capacity. Their reduced dimensions enable accelerated hydrogen adsorption and desorption kinetics. Certain nanomaterials demonstrate superior stability during repeated hydrogen absorption and desorption cycles compared to their bulk counterparts, thus extending the operational lifespan of the material. These materials can be precisely engineered at the atomic or molecular level to possess tailored structures and properties optimized for hydrogen storage. Examples include carbon nanotubes (CNT), metal-organic frameworks (MOFs), and nanostructured metal hydrides, which can be specifically designed to maximize hydrogen uptake while minimizing weight and volume. Furthermore, nanomaterials offer versatility in their composition and structure, enabling the development of a wide range of storage solutions suitable for diverse applications and operating conditions. Alternatively, for extended storage durations, hydrogen can be stored in aquifers or caverns, representing subsurface hydrogen storage techniques.

Initially, hydrogen was predominantly stored as compressed gas in metal cylinders. While this method was straightforward and efficient, it suffered from low storage capacity and

TABLE 4.1 The technical target decided by department of energy for the year 2025 and the long-term future (DOE Technical Targets for Onboard Hydrogen Storage for Light-Duty Vehicles, n.d.).

Storage parameter	Units	2025	Ultimate
System gravimetric capacity	kg H_2/kg system	0.055	0.065
System volumetric capacity	kg H_2/L system	0.040	0.050
Operating ambient temperature	°C	−40/60 (sun)	−40/60 (sun)
Min/Max delivery temperature	°C	−40/85	−40/85
Operational cycle life	Cycles	1500	1500
System fill time (5 kg H_2)	Min	3–5	3–5
Fuel purity	%H_2	99.7% (dry basis)	99.7% (dry basis)
Loss of usable H_2	(g/h) kg H_2 stored	0.05	0.05

Reproduced from DOE Technical Targets for Onboard Hydrogen Storage for Light-Duty Vehicles. (n.d.). Hydrogen and fuel cell technologies office. https://www.energy.gov/eere/fuelcells/doe-technical-targets-onboard-hydrogen-storage-light-duty-vehicles.

safety concerns. In the early 20th century, metal hydrides emerged as a promising alternative for hydrogen storage. These compounds could store hydrogen at relatively high densities and provided the potential for safe and reversible storage. Extensive research in this field led to the discovery and development of various metal hydride systems. Another significant development occurred in the mid-20th century with the adoption of liquid hydrogen as a storage method, driven by advancements in cryogenic technology. Later in the 20th century, carbon-based materials like activated carbon and CNT garnered attention for their ability to adsorb hydrogen molecules onto their surfaces. This method, known as physisorption, offered room temperature storage with potential for high capacities. In recent decades, there has been a substantial focus on advanced materials and nanotechnology for hydrogen storage. This has led to the development of novel materials such as MOFs, porous organic polymers (POPs), and graphene-based materials. These materials hold promise for achieving high hydrogen storage capacities, customizable properties, and enhanced kinetics. MOFs, with their high surface area and customizable pore size, show promise for hydrogen storage. Researchers have been designing tailored architectures for MOFs to improve both hydrogen adsorption capacities and kinetics. Carbon-based nanomaterials like CNTs, graphene, and activated carbon offer prime hydrogen adsorption sites, with methods like functionalization enhancing storage capacity and reversibility. Nanoporous materials such as zeolites, MOPs (metal organic polyhedra), and COFs (covalent organic frameworks), known for well-defined nanopores, are optimized for high-density hydrogen storage. Nanocomposites blending metal hydrides with nanoscale additives synergize storage properties, while nanocatalysts, like metal nanoparticles, enhance hydrogen sorption kinetics on substrates like carbon nanomaterials or metal oxides.

Hydrogen storage technologies are still the subject of extensive research. The goal of ongoing research and development is to increase the economic feasibility, safety, and efficiency of various storage techniques in order to make hydrogen a more viable and widely used energy source.

4.2 Properties of hydrogen

Hydrogen is the most abundant element in the universe and the lightest element in the periodic table. Hydrogen constitutes nearly 75% of all matter. On Earth, hydrogen is primarily found as water; the atmospheric concentration of hydrogen gas is only 0.53 parts per million. Hydrogen is the most fundamental element, with only one electron and one proton in its nucleus. Deuterium and tritium, two others naturally occurring isotopes of it, have one and two neutrons in their respective nuclei (Fig. 4.2). Among these three isotopes, tritium is radioactive, whereas hydrogen and deuterium are stable, with natural abundances of 99.9844% and 0.0156%, respectively (Krishnamurthy, 1998). The amount of energy released as a fuel burn is referred to as the enthalpy of combustion. The enthalpy of combustion is classified into two types: lower heating value (LHV) and higher heating value (HHV). The difference between LHV and HHV is the latent heat of vaporization. Hydrogen has the highest value of HHV and LHV, that is, it produces highest amount of energy per unit mass. The LHV and HHV for different fuels are reported in Table 4.2. A chemical reaction takes place when oxygen and hydrogen gas interact. A small amount

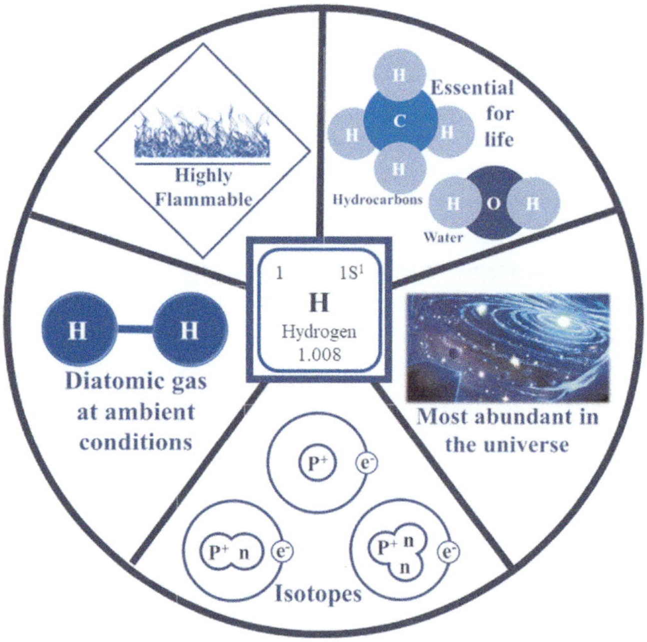

FIGURE 4.2 Basic properties of hydrogen element.

TABLE 4.2 The enthalpy of combustion of different fuels at 25°C and 1 atm.

Enthalpy of combustion	Hydrogen	Methane	Gasoline	Diesel	Methanol
LHV (kJ/g)	119.93	50.02	44.50	42.50	18.05
HHV (kJ/g)	141.86	55.53	47.50	44.80	19.96

of activation energy (~3 eV at STP in absence of any catalyst) is needed to start this reaction. Once this energy is supplied, a chemical reaction begins, and the energy that is produced serves as an activation energy for further gas molecules. As a result, using hydrogen fuels, a small quantity of input energy can produce a huge amount of heat and water as an end product, and the generation of greenhouse gases can be reduced significantly. Since less heat is generated than required to generate hydrogen, hydrogen cannot be regarded as a source of energy. It may, however, be transformed into many forms of energy, making it a useful energy carrier (Panfilov, 2016).

4.3 Hydrogen economy

The hydrogen economy refers to the use of hydrogen as a primary source of energy consumption across many sectors. The main objective of the hydrogen economy is to minimize CO_2 emissions and reduce dependency on fossil fuels by replacing traditional fuels with hydrogen energy. However, there are still a number of obstacles to overcome before we can switch to using hydrogen as main fuel. Energy demand, energy supply, logistics and infrastructure, and transition are the key issues that any new fuel source encounters (Bossel & Eliasson, 2003; Committee on Alternatives and Strategies for Future Hydrogen Production and Use, n.d.).

Energy demand is determined by market preferences as well as energy competition. Hydrogen energy must successfully compete with the current fuels in order to be economically viable. The physical characteristics or quantity of primary energy consumed is determined by the technologies that convert energy into services or products, as well as the kind of the services that the market requires. For instance, whereas electronic gadgets use electricity and can be powered by less energy-dense solid or gaseous fuels, transportation requires fuels with high energy density and quick refueling possibilities (Bossel & Eliasson, 2003; Committee on Alternatives and Strategies for Future Hydrogen Production and Use, n.d.).

Energy supply implies that commercial hydrogen generation should have mature and cost-effective technology. Nuclear power, fossil fuels, and renewable energy sources can all be used to make hydrogen. These production technologies can be used at various scales to generate hydrogen as needed. Hydrogen is categorized as gray hydrogen, blue hydrogen, or green hydrogen depending on the method it is produced. Among these, green hydrogen is the only clean fuel because it emits no greenhouse gases as a by-product (Bossel & Eliasson, 2003; Committee on Alternatives and Strategies for Future Hydrogen Production and Use, n.d.).

The four logistics procedures that occur between hydrogen generation and consumption at any level are packaging, distribution, dispensing, and storage. The schematics of these logistics are shown in Fig. 4.3. To package hydrogen simply means to have it in a form that is appropriate for shipping, such as compressed gas, liquid, chemical compound, etc. The term "distribution" refers to the delivery of hydrogen to the point of consumption. There are numerous ways to transfer hydrogen, including via pipelines, roadways, and railroads. The term transfer simply refers to the transfer of hydrogen fuel from retailers

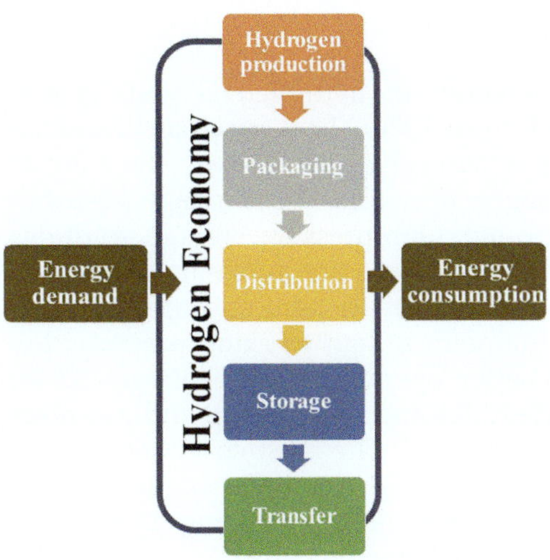

FIGURE 4.3 The schematic of the processes to determine hydrogen economy.

to consumers. In the period between generation and consumption, hydrogen must be securely stored. It could be stored in a pressurized or cryogenic container, solid-state materials, or an underground cavern. Currently, all of these processes are open to research and development (Bossel & Eliasson, 2003; Committee on Alternatives and Strategies for Future Hydrogen Production and Use, n.d.).

The transition to a hydrogen economy won't be smooth or linear. To switch to hydrogen energy, we'll need new technologies and infrastructures. Infrastructure and technology required during a transition period can be entirely different from those required when a hydrogen economy is fully developed.

In order to realize the hydrogen economy idea, a lot of research is being done. Currently, the focus of research is on commercial hydrogen production and storage (Bossel & Eliasson, 2003; Committee on Alternatives and Strategies for Future Hydrogen Production and Use, n.d.).

4.4 Hydrogen storage methods

A key aspect of the hydrogen economy is the large-scale storage of hydrogen. The procedure for storing and transporting hydrogen is still in the development stage and is equally crucial to the hydrogen economy as hydrogen production. The cost-effective hydrogen storage method has to be efficient and secure, and it should offer a way to utilize it whenever and wherever required. Hydrogen storage can be broadly categorized into two main types: subsurface based hydrogen storage and surface-based hydrogen storage (Fig. 4.4). Subsurface storage is particularly advantageous for large-scale, long-term storage of hydrogen (Epelle, 2022). On the other hand, surface-based storage methods are well-developed, established, and economically feasible (Epelle, 2022).

4.4.1 Subsurface hydrogen storage

The cheapest approach to store a significant amount of hydrogen is underground storage, which is similar to natural gas storage (Taylor, 1986). The principal distinction between hydrogen and natural gas is in their physiochemical characteristics. There are no environmental issues associated with storing hydrogen in naturally occurring subsurface structures (Walters, 1976). Compared to above-ground high-pressure storage solutions, UHS offers a higher capacity (Schaber et al., 2004). Geological structures are a desirable option for large-scale hydrogen storage due to the ability to create large subsurface storage cavities or access to natural reservoirs, ability to maintain a high working pressure, low specific cost, small surface facility footprint, and better operating safety (Bünger, 2016). Geological sites can be classified as porous media (aquifer and depleted oil and gas reservoirs) and cavern storage (salt caverns) as shown in Fig. 4.5. The most cost-effective geological formation is discovered to be salt caverns. Only 1% per cycle of the injected hydrogen is lost throughout operation, with the first cycle's injection losing 0.4% due to the development of hydrogen solution in brine (Carden & Paterson, 1979).

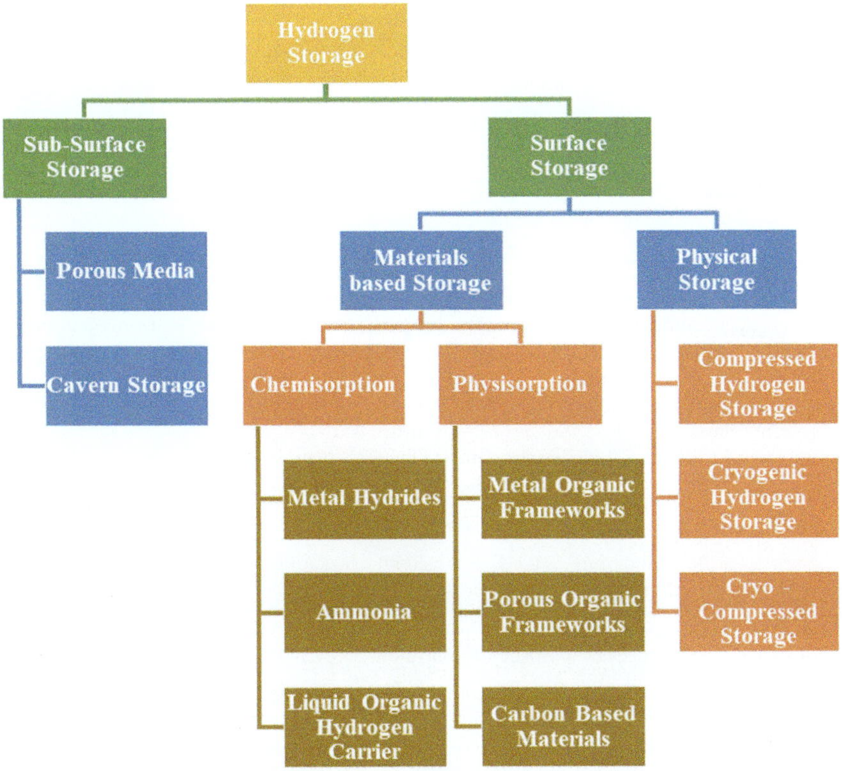

FIGURE 4.4 Classification of hydrogen storage medium.

However, the topic of subsurface hydrogen storage is not the focus of this chapter and will be explored in detail in the Chapter 6.

4.4.2 Surface hydrogen storage methods

Hydrogen storage finds application in three primary areas: stationary, transportation, and portable uses. Each application demands specific criteria for an appropriate storage method. In transportation, the primary focus lies on achieving high storage capacity, along with moderate working pressure and temperature, rapid kinetics, and cost-effectiveness in terms of recharging and recycling. Conversely, for stationary applications, the key concern revolves around material selection rather than solely focusing on capacity or working conditions. In surface-based hydrogen storage methods, hydrogen is primarily stored either in gaseous or liquid states under high pressure, or it's adsorbed/absorbed into materials. Based on the storage process, surface hydrogen storage techniques can be categorized into two main types: material-based hydrogen storage and physical hydrogen storage. Currently, the most prevalent surface hydrogen storage method is high-pressure storage.

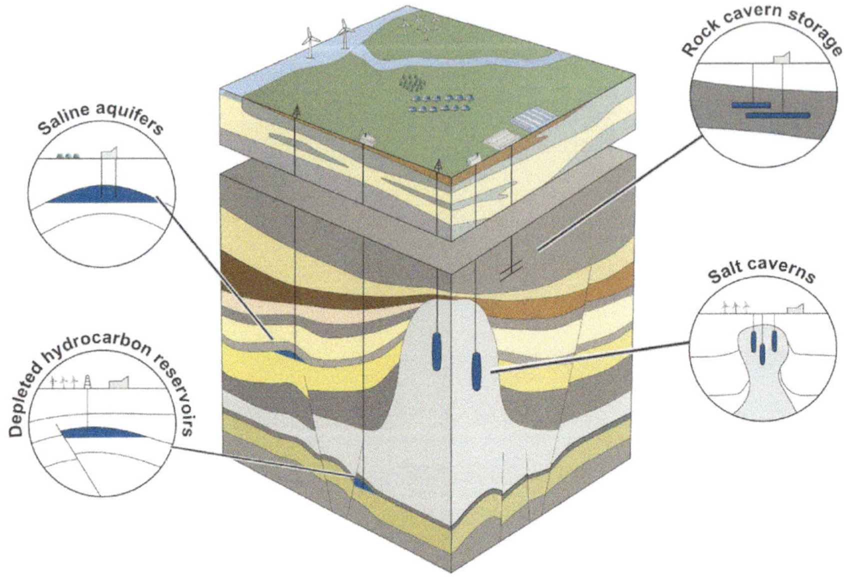

FIGURE 4.5 Illustration of underground hydrogen storage in different geological structures (Miocic, 2023). Source: *Reproduced from Miocic, J., Heinemann, N., Edlmann, K., Scafidi, J., Molaei, F., & Alcalde, J. (2023). Underground hydrogen storage: A review.* Geological Society Special Publication, 528, *73–86. https://doi.org/10.1144/SP528-2022-88.*

4.4.3 Physical storage

The most popular way to store hydrogen is through physical storage. In a physical manner, hydrogen is either held as a gas or a liquid in its pure state, free from any interaction with other materials in a high pressure vessel. Pressure vessels are subject to specific design criteria, including considerations such as material selection, stress assessment, maximum operating pressure, maximum allowable working pressure, and burst pressure. Additionally, these vessels are susceptible to various forms of damage overtime, including aging, abrasion, impact, scratches, and hydrogen embrittlement. Physical hydrogen storage is further categorized as compressed hydrogen storage, cryogenic hydrogen storage, and cryo-compressed hydrogen storage. The gravimetric density of hydrogen storage increases as pressure increases and temperature decreases (Klier, 2012). The Fig. 4.6 illustrates the temperature and pressure influence on hydrogen storage in physical storage methods.

4.4.4 Compressed hydrogen storage

In this procedure, hydrogen is compressed and kept in high-pressure containers. A schematic of such containers is given in Fig. 4.7. The material used to store hydrogen at high pressure should have a low density, high tensile strength, and be inert and impermeable to hydrogen (Mannan, 2012). There are four types of vessels available for storing hydrogen, and depending on the use and cost of storage, any of the four vessels can be employed (Barthélémy, 2012; Barthélémy et al., 2017). Although these vessels are typically cylindrical in shape, some composite vessels can also be polymorph or toroidal.

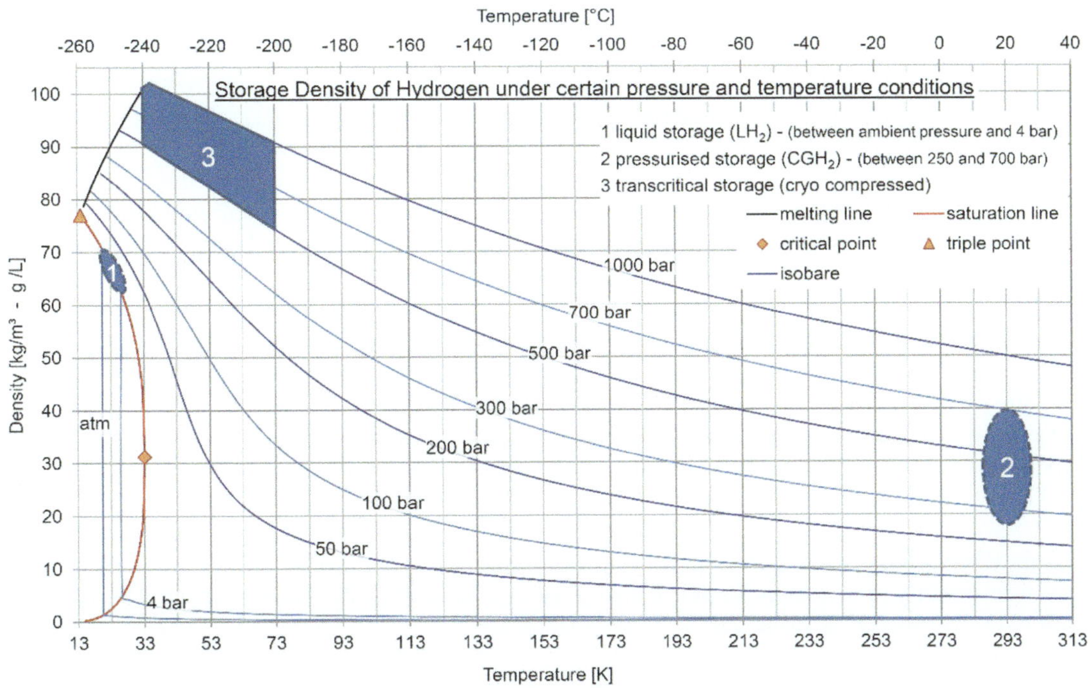

FIGURE 4.6 The variation in storage density of hydrogen with temperature at different pressure (Hydrogen Storage, n.d.). Source: *Reproduced from Hydrogen storage. (n.d.). Retrieved from https://en.wikipedia.org/wiki/Hydrogen_storage#cite_note-target-7.*

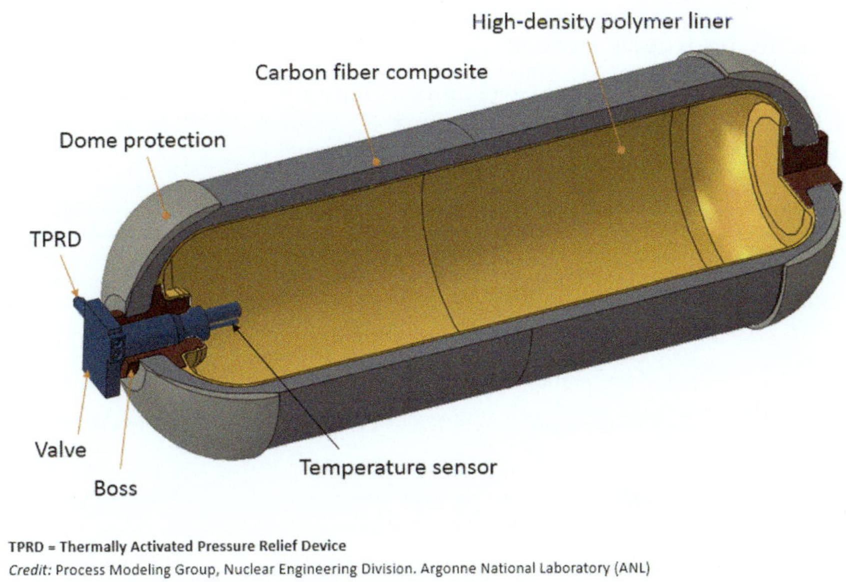

FIGURE 4.7 High pressure storage vessel for compressed hydrogen (Physical Hydrogen Storage, n.d.). Source: *Reproduced from Physical Hydrogen Storage. (2024). Original work published 2024.*

The earliest vessels are type 1 vessels, which were introduced in the 19th century. These vessels are built of steel and operate at pressures ranging from 150 to 300 bar. Type 1 vessels have a very low gravimetric density of 1%. These are the least expensive vessels and are mostly utilized in industrial applications (Barthélémy, 2012; Barthélémy et al., 2017; Hassan, 2021).

Type 2 vessels have a thick metallic liner. A hoop constructed of firer–resin composite is used to encircle the cylindrical portion. Although they are also employed in stationary applications, these vessels have a higher operating pressure than type 1 vessels (Barthélémy, 2012; Barthélémy et al., 2017; Hassan, 2021).

Type 3 vessels are constructed of a metallic liner entirely wrapped in carbon fiber inserted in a polymer matrix. The metal lining provides more than 5% of the mechanical resistance. These containers have a working pressure of up to 450 bar and are utilized for portable applications (Barthélémy, 2012; Barthélémy et al., 2017; Hassan, 2021).

Type 4 pressure vessels are constructed with a polymeric liner or an extremely thin metal liner that is entirely encased in a fiber–resin composite. Pressures of up to 1100 bar can be maintained in these vessels. These vessels are also employed in mobile applications. A gravimetric density of about 4.5 wt.% has been attained using type 4 vessels (Barthélémy, 2012; Barthélémy et al., 2017; Hassan, 2021). Types 3 and 4 vessels cost significantly more than Types 1 and 2 vessels (Hassan, 2021). A comparison of these vessels are reported in Table 4.3.

This method's major drawbacks include low storage density, high storage costs, and expensive compression. The safety concern brought on by high pressure storage is another significant drawback of this approach.

4.4.5 Cryogenic hydrogen storage

Another commercially used storage technique is liquid hydrogen. In this procedure, hydrogen is kept in cryogenic tanks at a temperature of around 20.3K at atmospheric pressure.

TABLE 4.3 Comparison between different types of vessels for compressed hydrogen storage.

Generation	Type 1	Type 2	Type 3	Type 4
Construction	Metallic liner	Metallic liner with fiber–resin composite hoop encircling cylindrical portion	Metallic liner completely wrapped by carbon fiber inserted in polymer matrix	Extremely thin metal liner or polymeric liner entirely encased in a fiber–resin composite
Pressure range	150–300 bar	No pressure limit	Up to 450 bar	Up to 1100 bar
Cost performance	Lowest cost	Low cost	High cost	High cost
Weight performance	Very low gravimetric capacity	Low gravimetric capacity	High gravimetric capacity	High gravimetric capacity

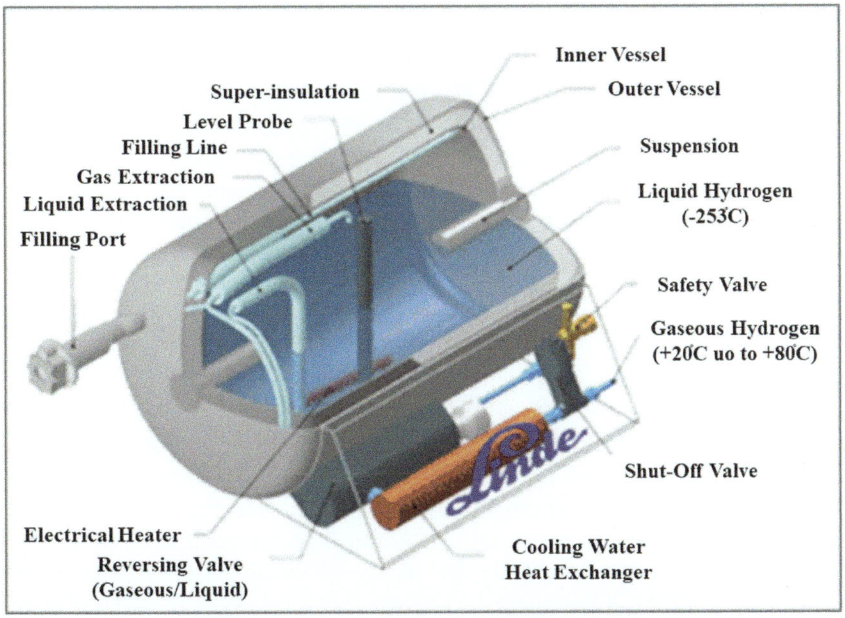

FIGURE 4.8 Schematic of tank to store liquid hydrogen (Krishna, 2012). Source: *Reproduced from Krishna, R., Titus, E., Salimian, M., Okhay, O., Rajendran, S., Rajkumar, A., Sousa, J. M. G., Ferreira, A. L. C., Campos Gil, & Gracio, J. (2012).* Hydrogen storage for energy application. *InTech. https://doi.org/10.5772/51238.*

A schematic of liquid hydrogen storage tank is shown in Fig. 4.8. Compared to compressed gas systems, liquid hydrogen storage offers higher gravimetric densities.

Hydrogen is liquefied using the Linde-cycle, which works on the principle of Joule–Thompson effects. In a heat exchanger, the hydrogen gas is cooled after being compressed to a high pressure. It then passes through a throttle valve, where Joule–Thompson expansion occurs and liquid forms. The cooled gas is then separated from the liquid and returned to the compressor. Since hydrogen's inversion temperature is approximately 200K, it must be cooled below that temperature before expanding (Mannan, 2012; Zhang, 2005). Three phases of heat transfer are involved in liquifying hydrogen. The temperature is initially reduced (from 300K to 20.3K) by extracting heat, after which, hydrogen gas is condensed (at 20.3K) by extracting latent heat. Finally, heat is removed to be used in ortho-para conversion (Zhang, 2005). The total heat removed in this procedure is 5153 KJ/kg.

Efficiency of the procedure and boil-off of liquid are two significant difficulties with this approach (Eberle et al., 2009; Mannan, 2012). The energy used to store hydrogen in this approach is 35% of the hydrogen's stored energy value (Hassan, 2021). The cryogenic tanks used in this procedure are extremely insulating, and multilayer insulation is utilized to prevent the transfer of radiation heat. The shape and size of the tanks have an impact on the boil-off rate as well. The amount of heat lost during this process depends on the vessel's surface-to-volume ratio (Mannan, 2012; Rivard et al., 2019). These shortcomings have restricted the use of liquid hydrogen in flight and space applications (Hassan, 2021; Zhou, 2005).

4.4.6 Cryo-compressed hydrogen storage

In this technique, hydrogen is kept in a high-pressure cryo-compressed vessel at a cryogenic temperature (Ahluwalia et al., 2016; Rivard et al., 2019). Liquid hydrogen, supercritical cryogenic hydrogen, or hydrogen in saturated liquid–vapor phase can all be stored in these tanks. These tanks utilize cold temperatures and medium pressure, providing a hybrid solution that combines elements of both liquid hydrogen storage and compressed gas hydrogen storage. This approach achieves significantly higher storage densities compared to the methods mentioned above. Cryo-compressed tanks are classified into three generations: Gens 1, 2, and 3 (Table 4.4).

The Gen 1 vessels had an interior volume of 135 L and had a liquid hydrogen storage capacity of 9.6 kg. These vessels were made to function at a maximum pressure of 245 atm (Ahluwalia et al., 2016).

In comparison to the first generation, the second-generation vessel had a larger volume. It had an internal volume of 151 L, could function at a greater pressure of 340 atm, and was built to hold 10.7 kg of liquid hydrogen (Ahluwalia et al., 2016; Technical Assessment: Cryo-Compressed Hydrogen Storage for Vehicular Applications, 2006).

The third-generation vessel is an enhanced version of the second-generation vessel. The interior volume and hydrogen capacity of these tanks are identical (Ahluwalia, 2010). However, the modifications were done to reduce the total weight and volume of the system by 23%. The price of these vessels is decreased since cryo-compressed tanks require 85% less carbon fiber than compressed hydrogen storage (Ahluwalia et al., 2016). Due to the high permissible pressure, increased heat sensitivity, higher thermal capacity of the vessel, and heat absorbed during the para to ortho hydro conversion, the dormancy period is also prolonged over liquid hydrogen storage (Ahluwalia et al., 2016).

4.4.7 Material-based storage

Material-based storage serves as a long-term solution for both stationary and portable applications, while also aiming to resolve the current challenges associated with onboard automobile storage. The current target set by the DOE requires a gravimetric storage of 5.5 wt.% hydrogen, with an ultimate target of 6.5 wt.%. The objective of investigating material-based hydrogen storage is to enhance storage capacity, both gravimetrically and volumetrically, and to decrease the necessary storage pressure.

An ideal material for hydrogen storage must possess high storage capacity, rapid kinetics, robust mechanical strength, ample pore volume, and surface area, as well as cost-effectiveness (Deng, 2006). The storage capacity of a material depends on factors such as

TABLE 4.4 Comparison of different generation of cryo-compressed vessels for hydrogen storage.

Generation	Volume	Hydrogen capacity	Pressure
Gen1	135	9.6	245
Gen2	151	10.7	340
Gen3	151	10.7	1

its porosity, surface area, and chemical properties. Higher surface area and pore volume generally lead to increased storage capacity. The speed at which hydrogen is adsorbed or desorbed from the material is also crucial, as slow kinetics can result in longer refueling times for hydrogen fuel cell vehicles. The energy dynamics of hydrogen sorption reactions impact the feasibility of storage. Enhancing the thermodynamics of storage is essential for efficient operation at ambient conditions. Material storage performance typically varies with temperature and pressure. Ideally, materials should maintain stable storage properties across a range of conditions.

Material-based storage is divided into adsorption storage and chemical hydrogen storage, depending on how hydrogen interacts with materials (Deng, 2006). Hydrogen can be stored in materials through two primary processes: physisorption, characterized by weak bonding between hydrogen and materials, and chemisorption, which involves stronger bonding. Physisorption and chemisorption represent two distinct adsorption processes characterized by the interaction between adsorbate and adsorbent. They differ primarily in their bonding nature, adsorption energy, and reversibility. Physisorption relies on weak van der Waals interactions between the adsorbate and adsorbent, resulting in lower adsorption energy and reversible behavior. Conversely, chemisorption entails the formation of chemical bonds between the adsorbate and adsorbent, leading to higher adsorption energy and often irreversible binding, unless significant energy input is applied to break the formed bonds. Fig. 4.9 illustrates these contrasting processes.

Nanomaterials have sparked significant interest in the field of materials-based storage. Nanomaterials offer a versatile platform for hydrogen storage, allowing for structural engineering to enhance storage capacity and regulate storage kinetics. Nanoparticles exhibit stronger binding with hydrogen compared to bulk materials, and in certain cases, the electric field of metal ions can polarize H_2 molecules, further augmenting binding strength. Table 4.5 shows the hydrogen storage capacity of different materials.

4.4.8 Adsorption storage

The mechanism behind hydrogen storage in adsorption is physisorption. Physisorption is caused by a weak van der Waal interaction between the adsorbent surface and the adsorbate.

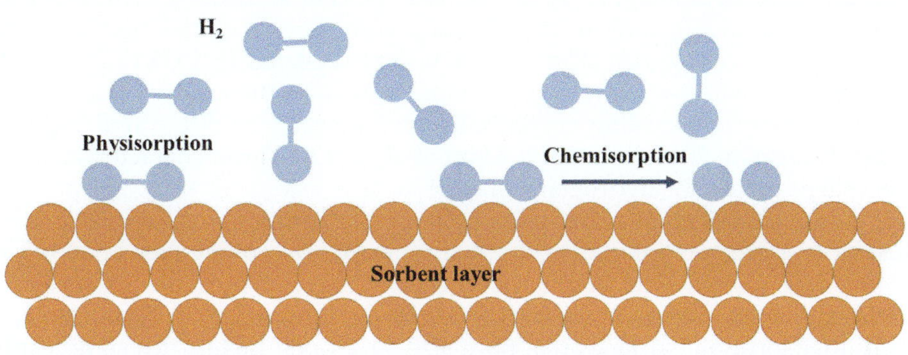

FIGURE 4.9 A representation of physisorption and chemisorption of H_2 gas on sorbent material.

TABLE 4.5 Hydrogen storage capacity of different materials.

Material	Storage capacity
Single walled CNT	4.2 wt.% at 300K/100 bar (Liu, 1999)
Multiwalled CNT	3.6 wt.% at 298K/70 bar (Lueking & Yang, 2003)
BN nanotube	4.2 wt.% at 298K/108 bar (Tang, 2002)
TiS_2 nanotube	2.5 wt.% at 298K/40 bar (Chen, 2003)
MOF-5	4.5 wt.% at 77K/0.75 bar (Rowsell, 2004)
Mg_2NiH_4	3.6 wt.% at ~500K/1 bar (Sandrock, 1999)
$NaAlH_4$	8.0 wt.% at 403K/90 bar (Zaluska et al., 2001)
CPP-g-PEI copolymers	11.26 wt.% at 77K/50 bar (Chen, 2017)

In this instance, hydrogen gas serves as the adsorbate while the adsorbent surface refers to the material's surface. Any material's total storage capacity is the sum of its absolute and excess adsorption. The interaction between the adsorbent surface and the adsorbate results in absolute adsorption. It can be described as the quantity of gas that has been solely retained by adsorption on the surface. The amount that is trapped in the pores in the absence of physisorption is known as excess adsorption. As a result, porous materials may hold a higher concentration of H_2 gas.

Adsorption mechanisms vary depending on adsorbent geometry and adsorption temperature. Multilayer adsorption would take place on an open surface. on the contrary, in a pore narrower than 2 nm, volume filling will occur. in a pore sized between 2 and 50 nm, capillary condensation could occur. But a pore that is bigger than 50 nm behaves like an open surface. At temperatures higher than critical-temperature, none of the aforementioned mechanisms could take place. The physisorption is further divided into three categories based on the adsorbent type: metal-organic frameworks (MOFs), POPs, and carbon-based nanomaterials.

MOFs are built up of metal ions or clusters, referred to as secondary building units (SBUs), that are coupled to organic ligands (Fig. 4.10), referred to as linkers (Bhattacharyya & Maji, 2022; Tranchemontagne, 2009). Alfred Werner's discovery of the connection between metal ions and ligands in coordination compounds in 1893 led to a significant interest in MOFs (Diercks, 2018). The first MOF was reported in 1995 (Yaghi et al., 1995). Ever since, MOFs have drawn a lot of attention because of their substantial high stability, specific surface area, outstanding crystallinity and number of SBUs and organic linkers (James, 2003; Kitagawa, 2014; Suh, 2012; Zhou et al., 2012). These characteristics also make them an excellent option for hydrogen storage.

In 2003, the first application of MOF for hydrogen adsorption was reported. (Rosi, 2003). Metal clusters and organic linkers are the two types of adsorption sites observed in MOF. Organic linkers are identified as secondary adsorption sites, with metal clusters acting as the primary adsorption site (Sagara et al., 2004; Yildirim & Hartman, 2005; Zhang et al., 2007).

The heat of adsorption, surface area, pore size, and open metal sites significantly affect the amount of H_2 that may be adsorbed by MOFs. H_2 adsorption primarily depend on

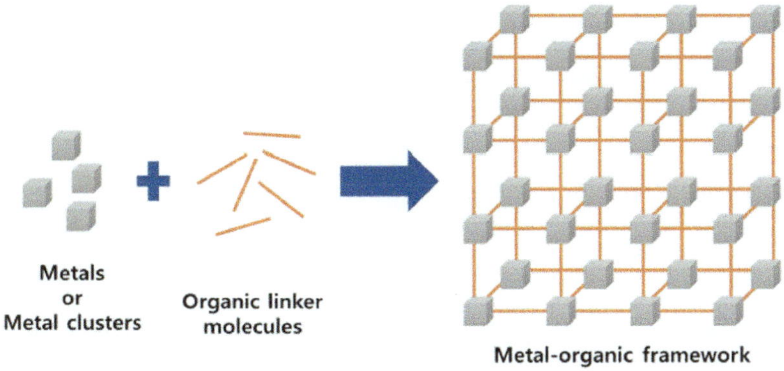

FIGURE 4.10 Simplified representation of metal-organic frameworks (Heo, 2020). Source: *Reproduced from Heo, D. Y., Do, H. H., Ahn, S. H., & Kim, S. Y. (2020). Metal-organic framework materials for perovskite solar cells.* Polymers, 12(9). *https://doi.org/10.3390/POLYM12092061.*

TABLE 4.6 Hydrogen storage capacity of different MOFs.

MOFs	Total gravimetric capacity (wt.%)
MOF-5	1.32 at 78K/1 bar (Rowsell, 2004; Yang, 2013)
MOF-177	7.5 at 77K/70 bar (Furukawa et al., 2007)
MIL-53-Cr	3.04 at 77K/150 bar (Rowsell, 2004)
MIL-101	6.1 at 77K/80 bar (Latroche, 2006)
MOF-210	17.6 at 77K/80 bar (Furukawa, 2010)
Mg-MOF-74	3.2 at 77K/15 bar (Pham, 2015)
NU-100	9.95 at 77K/70 bar (Ryan et al., 2008)

MOF, Metal-organic frameworks.

heat of adsorption, surface area and free volume at cryogenic temperature and low pressure, intermittent pressure and high pressure respectively (Frost et al., 2006).

Due to the low binding energy of MOF and H_2 at room temperature, MOFs have a very low gravimetric capacity. Storage of hydrogen requires an adsorption heat of 15.1 KJ/mol at ambient temperature (Bhatia & Myers, 2006). However, the reported value for MOFs ranges from 4 to 7 KJ/mol (Dinca, 2006; Latroche, 2006). Optimizing the MOF structure with the proper metal cation could increase the binding energy (Dinca, 2006; Latroche, 2006). According to studies, the binding energy of MOFs can reach from 10 to 50 KJ/mol when Mn, Sc, Ti, and V are used (Sun et al., 2007). By doping alkali metals or catalysts, binding energy can also be enhanced (Han & Goddard, 2007; Li & Yang, 2006; Sabo, 2007). H_2 intake can be increased further by increasing pore size, catenation, ligand functionalization, and introducing open metal sites (Bao, 2023; Murray et al., 2009; Suh, 2012). Since metal clusters are the main binding sites, creating active metal sites can also increase the binding energy of H_2. While mixed ligand systems can enhance pore volume, elongation of ligands may increase surface area. Both of these processes can enhance the intake of hydrogen. The storage capacity of different MOFs have been reported in Table 4.6.

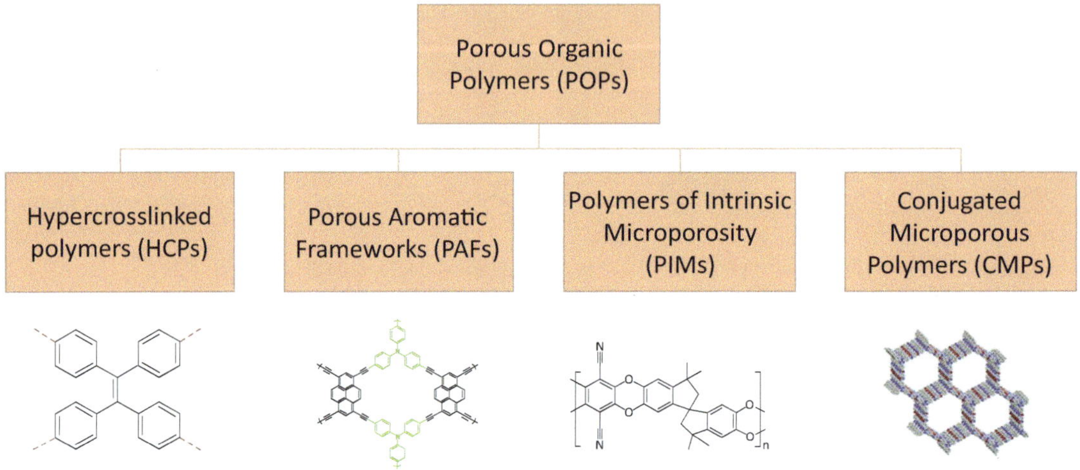

FIGURE 4.11 Different types of porous organic polymers (Loh, 2023; Mohamed, 2022). Source: *Reproduced with permission from Loh, C. Y. et al. (2023). Toward sustainable synthesis: A life cycle assessment of polymer of intrinsic microporosity (PIM-1) by green mechanosynthesis.* RSC *Sustainability, 1(9), 2287–2295. https://doi.org/10.1039/d3su00340j; Mohamed, M. G., El-Mahdy, A. F. M., Kotp, M. G., & Kuo, S. W. (2022). Advances in porous organic polymers: Syntheses, structures, and diverse applications.* Materials Advances, *3(2), 707–733. https://doi.org/10.1039/d1ma00771h.*

POPs are highly porous and stable materials which are result of polymerization of different structured organic units (Wang, 2023). The large surface to mass ratio, flexible pore size and structure, and high porosity (Wang, 2023) makes them an attractive candidate for hydrogen storage. Four types of POPs (Fig. 4.11) have been investigated widely for application in hydrogen storage, (1) hypercrosslinked polymers (HCPs), (2) polymer of intrinsic microporosity (PIM), (3) conjugated microporous polymers (CMPs), and (4) porous aromatic frameworks (PAFs) (Cousins & Zhang, 2019).

HCPs are fabricated through Friedel-crafts method (Veverka & Jeřábek, 1999). These structures can have a hydrogen storage capacity of 5 wt.% at 77K and 80 bar. But at ambient temperature, that is, 298K and 90 bar pressure, storage is only as high as 0.2 wt.% (Germain et al., 2009). HCPs of chlorinated polypropylene (CPP) grafted with polyethylenimine (PEI), CPRg-PEI showed a hydrogen storage of 11.26 wt.% at 77K and 50 bar. At 300K and 50 bar pressure, this HCP showed a storage of 2.47 wt.% (Chen, 2017).

PIM was first synthesized by McKeown and Budd (2006) by using planar phthalocyanine macrocycles. The adsorption storage of original PIM, that is, PIM-1 is nearly 1.7 wt.% at 77K and 10 bar. PIM-1 is highly stable and it only loses hydrogen from 2.60 to 1.90 wt.% over 400 days (Rochat, 2019). Triptycene based microporous polymers TMP-3, has shown highest hydrogen storage capacity of 4.42 wt.% at 77K and 1 bar (Bera et al., 2018).

CMPs were discovered by Cooper et al. (Cooper, 2009; Jiang et al., 2008) and are 3D network of amorphous polymers connected via π-conjugated bonds. CMPs synthesized with phenyl monomers can reach up to 3.07 wt.% at 77K and 60 bar (Yuan, 2010).

PAFs are derived from aromatic building units of carbon (Tian & Zhu, 2020). The PAF-1 derived from tetrakis(4-bromophenyl)methane have an absolute H_2 storage of 10 wt.% and excess storage of 7 wt.% at 77K and 48 bar (Ben, 2009). PAFs derived from tetrahedral precursors X(C6H4Br), where X = C, Ge and Si can have excess hydrogen adsorption up to 9.1 wt.% at 77K and 55 bar (Yuan, 2011).

Carbon-based materials carbon porous nanostructures are also a promising option for hydrogen storage and have received a lot of attention. Examples of porous carbon structures include graphene, fullerene, CNT, graphene nanofibers (GNF), activated carbon, and template carbon (Mohan, 2019; Sdanghi, 2020).

Graphene is a monolayer of carbon atoms organized in a hexagonal honeycomb arrangement. Graphene has a low hydrogen storage capacity despite its excellent polarizability. The interlayer spacing between layers can be increased by adding alkali metals (Selig & Ebert, 1980), which can boost the hydrogen storage capacity. The theoretical results show a storage capacity of 7.42 wt.% of pristine graphene at cryogenic temperature but experimental studies found much lower storage capacity.

GNFs are also an attractive option for hydrogen storage. GNF are made up of graphene platelets inclined at different angle with respect to fiber axis (Mohan, 2019; Ströbel, 2006). One study reported hydrogen adsorption of 67.5 wt.% at 298K and 11.3 MPa (Chambers, 1998). However, this reported values were not reproduced by any other group. Incorporation of transition metals such as Ni can increase the H_2 intake in GNFs (2 wt.% storage at 298K and 10 MPa) (Kim & Park, 2006; Kim et al., 2008).

Fullerene, another porous allotrope of carbon, is closed or partially closed spherical, ellipsoid and tubular structure. It consists of rings of five–seven carbon atoms. Fullerenes with incorporated transition metals can store up to 9 wt.% of hydrogen (Yildirim et al., 2005; Zhao, 2005).

The cylindrical fullerenes are also called CNT. These are basically cylindrically folded structure of graphene. The storage capacity of CNT vary with its geometry, pretreatments, structural defects, applied pressure, and temperature etc. (Mohan, 2019). the hydrogen can be stored on the outer and inner surface of CNT as well as between tubes in case of multi wall (MW) CNTs. A storage capacity of 20 wt.% was observed in alkali metal doped CNT (Chen, 1999). However, later studies confirmed that moisture has affected this storage capacity and in dry hydrogen only 2.5 wt.% storage was observed (Yang, 2000). MWCNTs can store up to 6.3 wt.% at 14.8 MPa (Hou, 2002). At ambient temperature and 4 MPa pressure, a storage of 1.97 wt.% was observed (Lee, 2002). The hydrogen storage capacity of different carbon structure is given in Table 4.7.

Various nanostructures, akin to carbon-based materials have been extensively investigated for hydrogen storage purposes. Nanosheets, nanotubes, and nanoparticles composed of lighter elements have shown promise in this regard. Remarkably, a storage capacity as high as 14.2 wt.% has been reported for lithium-functionalized boron graphdiyne sheets using density functional theory (Hussain, 2019).

4.4.9 Chemical hydrogen storage

In chemical hydrogen storage method, hydrogen is stored via chemisorption process. In chemisorption, the adsorbate forms chemical bonds with the adsorbent material (Jena, 2011). However, in chemisorption the adsorbate–adsorbent bonds are weaker than chemical compounds (Králik, 2014). In this process, H_2 molecule dissociates into hydrogen

TABLE 4.7 The hydrogen storage capacity of different carbon allotropes.

Carbon allotrope	Storage capacity (wt.%)
Activated carbon	1.4 at 77K/0.1 MPa (Wang, 2011)
Activated carbon (KOH treated)	6.6 at 77K/4 MPa (Zhao, 2012)
Ni/porous graphite	4.48 at 298K/10 MPa (Ahluwalia et al., 2015)
Aligned SWCNT	4 at 298K/11 MPa (Tozzini & Pellegrini, 2013)
Well-aligned MWCNT	3 at 290K/10 MPa (Broom, 2016)
Li-doped MWCNT	20 at 200–400K/0.1 MPa (Dillon, 1997)
Carbon nanofiber	12.38 at 77K/12 MPa (Ariharan et al., 2018)
Graphite nanofiber	10 at 300K/12 MPa (Tozzini & Pellegrini, 2013)

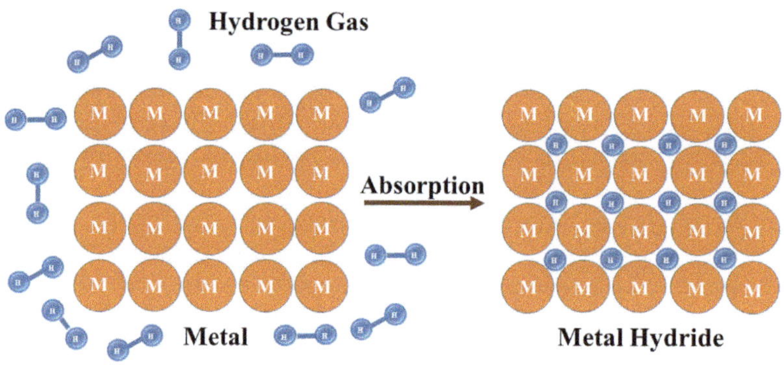

FIGURE 4.12 A schematic of hydrogen storage in metal hydrides (HYSTORSYS, n.d.). Source: *Reproduced from Hystorsys. (n.d.). Retrieved from https://www.hystorsys.no/our-story/.*

atoms and forms a chemical bond with adsorbent material. This method can further be divided into three categories, metal hydrides, liquid organic hydrogen carriers (LOHCs), and ammonia.

In metal hydrides, metals are bonded with hydrogen. Depending on type of binding, metal hydrides are classified as interstitial hydrides and non-interstitial hydrides (Young, 2018).

In interstitial hydrides hydrogen forms metallic bond with transition metals (Liu et al., 2021). Hydrogen atom mostly occupies tetrahedral or octahedral sites of interstitial hydrides as shown in Fig. 4.12. In these hydrides, hydrogen decompose and recombine on the surface of structure while hydrogen diffusion occurs in the bulk structure (Chen, 2021). Pure metals form a stable hydride and therefore, they release hydrogen at higher temperature while the transition metals that doesn't form hydride can release hydrogen at ambient temperature (Henderson, 2011). Several types of alloys like solid solution alloys, AB_5 type, AB_3 type, A_2B_7 type, A_2B type, and AB_2 type has been investigated for hydrogen storage application, where A is hydride forming metal and B is no-hydride forming metal. It has been found that V rich BCC alloy interstitial hydride can reach a hydrogen storage of 4 wt.% (Balcerzak, 2020; Kondo & Sakurai, 2006).

In non-interstitial hydrides, hydrogen atom bonds with neighboring elements ionically or covalently. These hydrides are mainly made up of lighter elements and can reach very high storage capacity. In non-interstitial hydrides, desorption of hydrogen occurs via decomposition of hydrides while hydrogen absorption occurs via recombination of decomposed product which results in diffusion of hydrogen as well as other constituent elements (Chen, 2021). These hydrides can store hydrogen up to 5 wt.% and can even reach up to approximately 20 wt.%. However, some hydrides are less stable and cannot absorb hydrogen after dehydrogenation (Gutowska, 2005; Saitoh et al., 2014).

The major challenge in these materials is hydrogenation and dehydrogenation process. Not only these process are complex but also have a low reversibility (Babu, 2014; Lai, 2015). The dehydrogenation process in these hydrides is less efficient too. The storage capacity of some hydrides are reported in Table 4.8.

The LOHC system is a pair of organic compounds in which one compound is hydrogen-lean (LOHC −) and other hydrogen compound is hydrogen-rich (LOHC +). In this system, hydrogen is released by catalytic reaction of dehydrogenation in which LOHC − converts into LOHC + and vice versa (Teichmann, 2011; Teichmann et al., 2012).

The process of hydrogenation and dehydrogenation is reversible and hydrogen can be released without breaking LOHC compound (Fig. 4.13) (Preuster et al., 2017; Stark, 2015; Xu, 2015). LOHC + compound can store hydrogen for a very long time (Geburtig, 2016). LOHC includes all hydrogen storage systems with high hydrogen concentration in liquid form.

A suitable LOHC compound for hydrogen storage should have a low melting point and a high boiling point (Cooper, 2005). The heat of desorption of hydrogen in these compounds should be low so that dehydrogenation can occur at ambient temperature. These materials should also have high hydrogen content, should be stable against hydrogenation, and dehydrogenation of the compound (Abdin, 2021).

The most investigated LOHC till now are benzene and cyclohexane, toluene and methylcyclohexane, naphtlene, and decalin etc. (Ali & Rippin, 1995; Ferreira-Aparicio et al., 2002;

TABLE 4.8 Hydrogen storage capacity of different interstitial and non-interstitial metal hydrides.

Type	Hydride	Storage capacity (wt.%)
AB_5	$LaNi_5H_6$	1.40 (Sandrock & Thomas, 2001)
AB_3	$LaMg_2Ni_9H_9$	1.3 (Denys & Yartys, 2011)
AB_2	$TiCr_2H_4$	2.57 (Chen et al., 2005)
AB	$TiFeH_2$	1.89 (Young, 2018)
BCC	$V_{22}Ti_{33}Cr_{43}$	2.5 54 (Mazzolai et al., 2008)
Alanate	$3NaAlH_4 = Na_3AlH_6 + 2Al + 3H_2$	3.7 (Bhakta et al., 2012)
Amide	$LiNH_2 + LiH = Li_2NH + H_2$	6.5 (Song & Guo, 2006)
Borohydride	$LiBH_4 = LiH + B + 3/2H_2$	13.9 (Mauron et al., 2008)
Mg-based Hydride	$MgH_2 = Mg + H_2$	7.6 (Zhang et al., 2017)

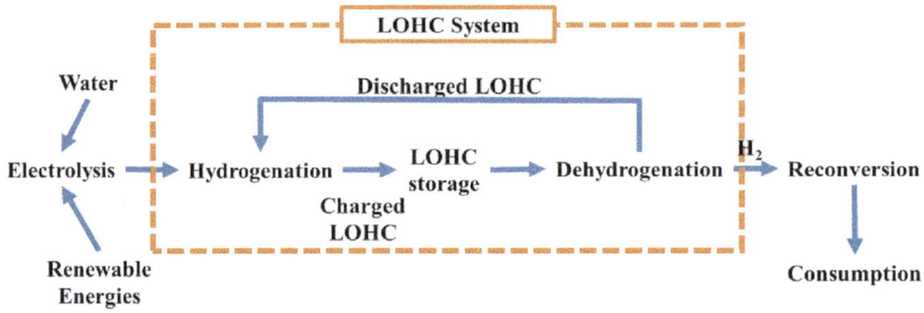

FIGURE 4.13 The process of hydrogen storage in liquid organic hydrogen carriers systems. Source: *Reproduced from Modisha, P. M., Ouma, C. N. M., Garidzirai, R., Wasserscheid, P., & Bessarabov, D. (2019). The prospect of hydrogen storage using liquid organic hydrogen carriers.* Energy and Fuels, *33(4), 2778–2796. https://doi.org/10.1021/acs.energyfuels.9b00296.*

Hodoshima et al., 2003; Jiang, 2012; Kou, 2013; Koutsonikolas, 2012; Oda, 2010). The benzene and cyclohexane system have a hydrogen storage capacity of 7.19 wt.% (Kariya et al., 2002; Lee, 2005). but low boiling point restricts its application in H_2 storage. Toluene and methylcyclohexane system have lower hydrogen storage capacity (6.2 wt.%) than benzene/cyclohexane system (Modisha, 2019). However, they are less toxic and can be used over a wide range of temperature. The storage capacity of hydrogen is approximately 7.3 wt.% (Hodoshima, 2005) in naphthalene/decalin system but the melting point of naphthalene is high. The formation of large amount of by-products and high temperature for dehydrogenation are some major challenges for these compounds (Abdin, 2021; Modisha, 2019).

Recently, formic acid has attracted interest as a hydrogen carrier. At low temperatures and high pressures, the hydrogen in formic acid can be liberated (Müller et al., 2017). Formic acid can be decomposed catalytically in two different ways. The first one is the production of hydrogen from the dehydrogenation of formic acid. The second approach involves producing CO through the dehydration of formic acid.

Catalysts are needed in order to produce hydrogen exclusively and prevent the production of unwanted CO. In 2008, two research described the generation of hydrogen from formic acid without any CO contamination (Fellay et al., 2008; Loges, 2008). Ru served as a catalyst for the production of hydrogen in these researches. Since then, numerous studies have been conducted on the production of hydrogen from formic acid utilizing both precious and non-precious metals as catalysts. The most energy-effective method for storing hydrogen based on formic acid has been discovered to be the biocarbonate/formate system (Müller et al., 2017). The main challenge with this approach is the choice of catalyst. The catalysts that can effectively create hydrogen without CO contamination are primarily precious metals and are expensive.

Methanol is also a viable LOHC for hydrogen storage. For the evolution of hydrogen from methanol, both homogeneous and heterogeneous catalysts have been reported. High temperatures are necessary for these reactions. However, at mild conditions, H_2 is produced via methanol dehydrogenation using a Ru catalyst (Nielsen, 2013).

However, the breakdown of methanol produces CO_2. The separation of produced CO_2 is an energy consuming process. Furthermore, CO_2 production is unsustainable.

4.4.9.1 Ammonia

Another excellent possibility for storing hydrogen is ammonia. It can be produced from both fossil fuels and renewable energy sources. The Haber–Bosch process is the conventional method for producing ammonia (Vojvodic, 2014). Other method to produce ammonia involves electrochemical processing (Juangsa et al., 2021; Wang, 2022) and thermochemical cycle (Aziz et al., 2020; Juangsa et al., 2021). Further, ammonia can also be produced using surplus electricity via hydrogen conversion (Ajiwibowo et al., 2019). Ammonia has a very high hydrogen capacity of 17.8 wt.%. Due to its excellent stability, it can be used for both long-term storage and transportation (Fig. 4.14) (Ikäheimo, 2018). Ammonia already has numerous applications in industries like agriculture, refrigerant gas, explosives, etc., therefore it already have an established infrastructure for hydrogen storage (Fecke et al., 2016). The volume of hydrogen stored in liquid ammonia is 1.7 times higher than liquid hydrogen. The hydrogen can be released through decomposition of ammonia. The decomposition occurs through a thermo-catalytic reaction. Ammonia breaks down through a sequential dehydrogenation reaction in the presence of metal catalysts (Mukherjee, 2018). Ammonia adsorption on a metal catalyst initiates the process, which is then proceeds via ammonia dehydrogenation, and concludes with nitrogen and hydrogen recombinative desorption. The degree of decomposition and rate-limiting step is determined by the catalyst utilized. Particle size, type of support material, catalytic metal, surface area, etc., all play a significant role in the catalytic activity of catalysts. (Makepeace, 2019). The rate-determining step for precious metals (such as Ru, Rh, Ir, etc.) is NH cleavage; for non-precious metals (such as Fe, Co, Ni, etc.), the rate-determining step is N2 desorption. A number of different catalysts are being investigated for ammonia decomposition. It is discovered that TM(N)-alkali metal amide/imide composite catalysts are a potential candidate for higher conversion efficiency at low temperatures.

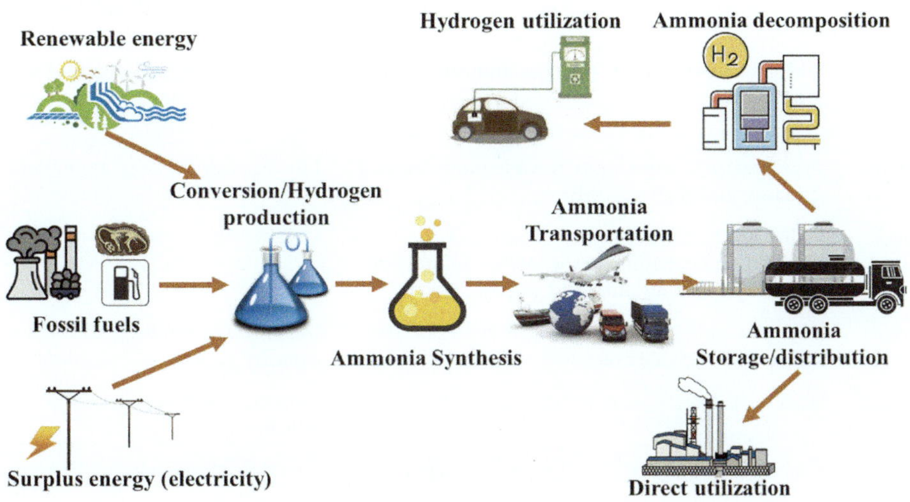

FIGURE 4.14 The process of production and consumption of ammonia in different sectors. Source: *Reproduced from Aziz, M., Wijayanta, A. T., & Nandiyanto, A. B. D. (2020). Ammonia as effective hydrogen storage: A review on production, storage and utilization.* Energies, 13(12). *https://doi.org/10.3390/en13123062.*

The rate-determining step varies with temperature as well. The partial pressures of hydrogen and ammonia determine the reaction kinetics at low temperatures, whereas only the partial pressure of ammonia is important at high temperatures (Mukherjee, 2018).

The efficiency of ammonia conversion at low temperatures is influenced by fast kinetics and limitations in heat transfer. Thus, even in the presence of various catalysts, high temperatures are required for the complete breakdown of ammonia. Additionally, it is still difficult to discover a catalyst that is both affordable for large-scale applications and has a high conversion efficiency at low temperatures (Mukherjee, 2018). Furthermore, the separation and purifying procedure is energy intensive (Aziz et al., 2020).

4.5 Challenges and future scope

The major challenge with surface hydrogen storage is achieving high storage capacity under ambient conditions. Table 4.9 provides a comparison of all the surface-based hydrogen storage method. Currently, the predominant approach for storing hydrogen is through high-pressure compression. However, physical hydrogen storage methods still face several challenges. One major drawback is their lower energy density compared to other hydrocarbon

TABLE 4.9 Comparison of different storage methods.

Storage method	Advantages	Disadvantages
Compressed gas storage	Fast charging and discharging, low cost and mature technology	Low storage capacity, safety issues, and high energy consumption
Cryogenic storage	High storage capacity and bulk density	High requirement for conversion technology and storage material and higher cost
Cryo-compressed storage	Bridge between compressed gas and cryogenic storage	Complex design of tank and high cost of maintaining extremely low temperatures, safety issues
Porous organic polymers	large surface to mass ratio, flexible pore size and structure and high porosity	Low storage capacity at ambient condition
Metal-organic frameworks	high stability, high specific surface area, outstanding crystallinity, and higher number of secondary building units and organic linkers	Low storage capacity at ambient condition
Carbon-based materials	High storage density, lightweight, high porosity, and specific surface area	Immature technology, low storage capacity at ambient condition
Metal hydrides	High bulk density, easy to transport	Immature technology and low quality efficiency
Liquid organic hydrogen carrier	High storage density, recycling of hydrogen storage materials, low cost	Low dehydrogenation efficiency, ample of bi-products
Ammonia	can be utilize directly as fuel, mild storage condition	Stronger corrosiveness, low conversion efficiency at ambient temperature

fuels, especially concerning hydrogen fuel cell vehicles. Due to hydrogen's low energy density, a significant storage volume is necessary to contain an adequate amount of hydrogen. While accommodating large-volume hydrogen storage poses less of a challenge for larger vehicles, it presents a major issue for light-duty vehicles. To fulfill the minimum requirement of a 300-mile range for light vehicles, onboard storage of at least 5–13 kg of hydrogen would be necessary. Increasing storage pressure can potentially boost storage capacity, but this necessitates stronger, heavier tanks. Given the space constraints in vehicles, reducing tank size and weight while ensuring safety and storage capacity remains a significant hurdle. Lightweight yet durable tank materials are crucial to meet these demands. Additionally, compressed hydrogen poses risks like leaks, ruptures, and accidents. Preventing such occurrences involves managing hydrogen embrittlement, ensuring tank integrity under extreme conditions, and implementing effective leak detection and mitigation strategies. Fast refueling poses another challenge for hydrogen fuel cell vehicles. Optimizing refueling infrastructure, developing efficient compression and dispensing technologies, ensuring safety during refueling, and establishing a widespread network of hydrogen refueling stations are key areas to address. Making these advancements economically viable is also critical.

While material-based hydrogen storage is envisioned as a solution for hydrogen fuels, it encounters its own set of challenges. One significant challenge it faces is storage capacity. While some materials exhibit exceptional gravimetric storage capacity at cryogenic temperatures and high pressure, most fail to meet the DOE's target of 6 wt.% at ambient conditions. Enhancing storage capacity under ambient conditions is currently a major focus. Kinetics of hydrogen adsorption and desorption are crucial for effective storage materials; slow kinetics can lead to impractically long refueling times for hydrogen-powered vehicles. Optimizing the thermodynamics of hydrogen sorption processes to operate at ambient conditions is highly desirable. Addressing material degradation, fatigue, and loss of storage capacity over extended use is imperative. Finally, finding cost-effective materials and scalable production methods is essential to make material-based hydrogen storage economically viable.

To tackle these challenges, we need both short-term and long-term solutions. In the short-term solution, the focus lies on enhancing vessel design to boost hydrogen capacity by maintaining high pressure, reducing vessel weight, and cutting down system costs. The long-term solution involves both cryo-compressed and material-based hydrogen storage methods, both of which are intended to ultimately meet the DOE's target for hydrogen storage.

4.6 Conclusions

Hydrogen storage stands as a pivotal component in the advancement of the hydrogen economy and the promotion of clean, sustainable energy solutions. Depending on the intended application—be it transportation, stationary, or portable—a tailored storage method becomes imperative. Each application dictates its crucial factors for storage. For instance, transportation necessitates high storage capacity and optimal working conditions, whereas stationary applications prioritize the selection of storage materials.

In this chapter, various storage options for hydrogen have been explored. Presently, the most prevalent method involves physically storing hydrogen within high-pressure vessels.

However, these vessels undergo rigorous testing to ensure their integrity under stress evaluation, operating temperature and pressure, and to mitigate hydrogen embrittlement risks. Material-based storage emerges as a promising avenue for hydrogen storage in the future. Nonetheless, none of the materials have yet met the gravimetric or volumetric density targets set by the DOE for hydrogen storage under ambient conditions. Both physical and material-based storage methods remain subjects of ongoing investigation aimed at achieving economic viability for hydrogen storage in the future.

Despite successful surface-based storage solutions, underground storage, similar to natural gas storage, emerges as the most cost-effective option without environmental concerns within the hydrogen industry. Geological formations are favored for large-scale storage due to their ability to create expansive subsurface storage, maintain high pressure, minimize costs and space requirements, and ensure operational safety.

Acknowledgment

The author would like to express sincere gratitude to St. Xavier's College and Gujarat University for generously providing the resources necessary for the research endeavors.

AI disclosure

During the preparation of this work the author(s) used ChatGpt in order to improve the quality of English, remove grammatical error and a bit of literature details. After using this tool/service, the author(s) reviewed and edited the content as needed and take(s) full responsibility for the content of the publication.

References

Abdin, Z., Tang, C., Liu, Y., & Catchpole, K. (2021). Large-scale stationary hydrogen storage via liquid organic hydrogen carriers. *Iscience, 24*(9).

Ahluwalia, R. K., Hua, T. Q., Peng, J.-K., Lasher, S., McKenney, K., Sinha, J., & Gardiner, M. (2010). Technical assessment of cryo-compressed hydrogen storage tank systems for automotive applications. *International Journal of Hydrogen Energy, 35*(9), 4171–4184.

Ahluwalia, R. K., Peng, J.-K., & Hua, T. Q. (2015). Sorbent material property requirements for on-board hydrogen storage for automotive fuel cell systems. *International Journal of Hydrogen Energy, 40*(19), 6373–6390.

Ahluwalia, R. K., Peng, J.-K., & Hua, T. Q. (2016). 5 – Cryo-compressed hydrogen storage. In R. B. Gupta, A. Basile, & T. N. B. T.-C. ofH.E. Veziroğlu (Eds.), *Woodhead publishing series in energy* (pp. 119–145). Woodhead Publishing. Available from https://doi.org/10.1016/B978-1-78242-362-1.00005-5.

Ajiwibowo, M. W., Darmawan, A., & Aziz, M. (2019). A conceptual chemical looping combustion power system design in a power-to-gas energy storage scenario. *International Journal of Hydrogen Energy, 44*(19), 9636–9642.

Ali, J. K., & Rippin, D. W. T. (1995). Comparing mono-and bimetallic noble-metal catalysts in a catalytic membrane reactor for methylcyclohexane dehydrogenation. *Industrial & Engineering Chemistry Research, 34*(3), 722–729.

Ariharan, A., Viswanathan, B., & Nandhakumar, V. (2018). Nitrogen-incorporated carbon nanotube derived from polystyrene and polypyrrole as hydrogen storage material. *International Journal of Hydrogen Energy, 43*(10), 5077–5088.

Aziz, M., Wijayanta, A. T., & Nandiyanto, A. B. D. (2020). Ammonia as effective hydrogen storage: A review on production, storage and utilization. *Energies, 13*(12), 3062.

Babu, A. R. V., Devunuri, N., Manisha, D. R., Prashanthi, Y., Merugu, R., & Teja, A. J. R. R. (2014). Magnesium hydrides for hydrogen storage: A mini review. *International Journal of ChemTech Research, 6*(7), 3451–3455.

Balcerzak, M. (2020). Hydrogenation properties of nanocrystalline TiVMn body-centered-cubic alloys. *International Journal of Hydrogen Energy*, *45*(31), 15521–15529.

Bao, W., Yu, J., Chen, F., Du, H., Zhang, W., Yan, S., Lin, T., Li, J., Zhao, X., & Zhu, B. (2023). Controllability construction and structural regulation of metal-organic frameworks for hydrogen storage at ambient condition: A review. *International Journal of Hydrogen Energy*, [Preprint].

Barthélémy, H. (2012). Hydrogen storage – Industrial prospectives. *International Journal of Hydrogen Energy*, *37* (22), 17364–17372.

Barthélémy, H., Weber, M., & Barbier, F. (2017). Hydrogen storage: Recent improvements and industrial perspectives. *International Journal of Hydrogen Energy*, *42*(11), 7254–7262.

Ben, T., Ren, H., Ma, S., Cao, D., Lan, J., Jing, X., Wang, W., Xu, J., Deng, F., & Simmons, J. M. (2009). Targeted synthesis of a porous aromatic framework with high stability and exceptionally high surface area. *Angewandte Chemie International Edition*, *48*(50), 9457–9460.

Bera, R., Mondal, S., & Das, N. (2018). Triptycene based microporous polymers (TMPs): Efficient small gas (H_2 and CO_2) storage and high CO_2/N_2 selectivity. *Microporous and Mesoporous Materials*, *257*, 253–261.

Bhakta, R. K., et al. (2012). Thermodynamics and kinetics of $NaAlH_4$ nanocluster decomposition. *Physical Chemistry Chemical Physics*, *14*(22), 8160–8169.

Bhatia, S. K., & Myers, A. L. (2006). Optimum conditions for adsorptive storage. *Langmuir: The ACS Journal of Surfaces and Colloids*, *22*(4), 1688–1700.

Bhattacharyya, S., & Maji, T. K. (2022). Multi-dimensional metal-organic frameworks based on mixed linkers: Interplay between structural flexibility and functionality. *Coordination Chemistry Reviews*, *469*, 214645.

Bossel, U.,& Eliasson, B. (2003). Hydrogen economy, alternative fuels data center (AFDC) Report USA [Preprint].

Broom, D. P., et al. (2016). Outlook and challenges for hydrogen storage in nanoporous materials. *Applied Physics A*, *122*, 1–21.

Bünger, U., Michalski, J., Crotogino, F., & Kruck, O. (2016). Large-scale underground storage of hydrogen for the grid integration of renewable energy and other applications. *Compendium of hydrogen energy* (pp. 133–163). Elsevier.

Carden, P. O., & Paterson, L. (1979). Physical, chemical and energy aspects of underground hydrogen storage. *International Journal of Hydrogen Energy*, *4*(6), 559–569.

Chambers, A., Park, C., Baker, R. T. K., & Rodriguez, N. M. (1998). Hydrogen storage in graphite nanofibers. *The Journal of Physical Chemistry. B*, *102*(22), 4253–4256.

Chen, J., Li, S.-L., Tao, Z.-L., Shen, Y.-T., & Cui, C.-X. (2003). Titanium disulfide nanotubes as hydrogen-storage materials. *Journal of the American Chemical Society*, *125*(18), 5284–5285.

Chen, P., Wu, X., Lin, J., & Tan, K. L. (1999). High H2 uptake by alkali-doped carbon nanotubes under ambient pressure and moderate temperatures. *Science (New York, N.Y.)*, *285*(5424), 91–93.

Chen, S., Liu, J., Li, Z., Wang, H., Wang, X., & Xu, Y. (2017). Hydrogen storage properties of highly cross-linked polymers derived from chlorinated polypropylene and polyethylenimine. *International Journal of Hydrogen Energy*, *42*(36), 23028–23034.

Chen, X.-Q., et al. (2005). Ab initio study of ground-state properties of the Laves phase compounds $TiCr_2$, $ZrCr_2$, and $HfCr_2$. *Physical Review B*, *71*(17), 174101. Available from https://doi.org/10.1103/PhysRevB.71.174101.

Chen, Z., Ma, Z., Zheng, J., Li, X., Akiba, E., & Li, H.-W. (2021). Perspectives and challenges of hydrogen storage in solid-state hydrides. *Chinese Journal of Chemical Engineering*, *29*, 1–12.

Cooper, A. C., Fowler, D. E., Scott, A. R., Abdourazak, A. H., Cheng, H., Wilhelm, F. C., Toseland, B. A., Campbell, K. M., & Pez, G. P. (2005). Hydrogen storage and delivery by reversible hydrogenation of liquid-phase hydrogen carriers. *Papers of the American Chemical Society*, *50*, 271.

Cooper, A. I. (2009). Conjugated microporous polymers. *Advanced Materials*, *21*(12), 1291–1295.

Committee on Alternatives and Strategies for Future Hydrogen Production and Use. (n.d.). *The hydrogen economy: Opportunities, costs, barriers, and R&D needs.* https://doi.org/10.2172/882095.

Cousins, K., & Zhang, R. (2019). Highly porous organic polymers for hydrogen fuel storage. *Polymers*, *11*(4), 690.

Deng, S. (2006). Sorbent technology. *Encyclopedia of Chemical Processing*, 2825–2845.

Denys, R. V., & Yartys, V. A. (2011). Effect of magnesium on the crystal structure and thermodynamics of the La3 – xMgxNi9 hydrides. *Journal of Alloys and Compounds*, *509*, S540–S548.

Diercks, C. S., Kalmutzki, M. J., Diercks, N. J., & Yaghi, O. M. (2018). Conceptual advances from Werner complexes to metal–organic frameworks. *ACS Central Science*, *4*(11), 1457–1464.

Dillon, A., et al. (1997). Storage of hydrogen in single-walled carbon nanotubes. *Nature*, *386*(6623), 377–379.

Dinca, M., Dailly, A., Liu, Y., Brown, C. M., Neumann, D. A., & Long, J. R. (2006). Hydrogen storage in a microporous metal – organic framework with exposed Mn^{2+} coordination sites. *Journal of the American Chemical Society, 128*(51), 16876–16883.

DOE Technical Targets for Onboard Hydrogen Storage for Light-Duty Vehicles. (n.d.). Hydrogen and fuel cell technologies office. https://www.energy.gov/eere/fuelcells/doe-technical-targets-onboard-hydrogen-storage-light-duty-vehicles.

Eberle, U., Felderhoff, M., & Schueth, F. (2009). Chemical and physical solutions for hydrogen storage. *Angewandte Chemie International Edition, 48*(36), 6608–6630.

Epelle, E. I., Obande, W., Udourioh, G. A., Afolabi, I. C., Desongu, K. S., Orivri, U., Gunes, B., & Okolie, J. A. (2022) Perspectives and prospects of underground hydrogen storage and natural hydrogen, Sustainable Energy & Fuels [Preprint].

Fecke, M., Garner, S. and Cox, B. (2016) Review of global regulations for anhydrous ammonia production, use, and storage, Institution of Chemical Engineers Symposium Series, 1–11.

Fellay, C., Dyson, P. J., & Laurenczy, G. (2008). A viable hydrogen-storage system based on selective formic acid decomposition with a ruthenium catalyst. *Angewandte Chemie International Edition, 47*(21), 3966–3968.

Ferreira-Aparicio, P., Rodriguez-Ramos, I., & Guerrero-Ruiz, A. (2002). On the performance of porous Vycor membranes for conversion enhancement in the dehydrogenation of methylcyclohexane to toluene. *Journal of Catalysis, 212*(2), 182–192.

Frost, H., Düren, T., & Snurr, R. Q. (2006). Effects of surface area, free volume, and heat of adsorption on hydrogen uptake in metal – organic frameworks. *The Journal of Physical Chemistry. B, 110*(19), 9565–9570.

Furukawa, H., et al. (2010). Ultrahigh porosity in metal-organic frameworks. *Science (New York, N.Y.), 329*(5990), 424–428.

Furukawa, H., Miller, M. A., & Yaghi, O. M. (2007). Independent verification of the saturation hydrogen uptake in MOF-177 and establishment of a benchmark for hydrogen adsorption in metal–organic frameworks. *Journal of Materials Chemistry, 17*(30), 3197–3204.

Gardiner, G. (2023). Cryo-compressed hydrogen, the best solution for storage and refueling stations? Available at: https://www.compositesworld.com/articles/cryo-compressed-hydrogen-the-best-solution-for-storage-and-refueling-stations.

Geburtig, D., Preuster, P., Bösmann, A., Müller, K., & Wasserscheid, P. (2016). Chemical utilization of hydrogen from fluctuating energy sources–Catalytic transfer hydrogenation from charged liquid organic hydrogen carrier systems. *International Journal of Hydrogen Energy, 41*(2), 1010–1017.

Germain, J., Fréchet, J. M. J., & Svec, F. (2009). Nanoporous polymers for hydrogen storage. *Small (Weinheim an der Bergstrasse, Germany), 5*(10), 1098–1111.

Global Hydrogen Review. (2023). IEA. https://www.iea.org/reports/global-hydrogen-review-2023.

Gutowska, A., Li, L., Shin, Y., Wang, C. M., Li, X. S., Linehan, J. C., Smith, R. S., Kay, B. D., Schmid, B., & Shaw, W. (2005). Nanoscaffold mediates hydrogen release and the reactivity of ammonia borane. *Angewandte Chemie, 117*(23), 3644–3648.

Han, S. S., & Goddard, W. A. (2007). Lithium-doped metal-organic frameworks for reversible H2 storage at ambient temperature. *Journal of the American Chemical Society, 129*(27), 8422–8423.

Hassan, I. A., Ramadan, H. S., Saleh, M. A., & Hissel, D. (2021). Hydrogen storage technologies for stationary and mobile applications: Review, analysis and perspectives. *Renewable and Sustainable Energy Reviews, 149*, 111311.

Hauglustaine, D. (2024). The climate impact of a future hydrogen economy. https://doi.org/10.5194/egusphere-egu24-11016.

Henderson, N. (2011). Rare earths: Crucial elements of advanced technologies. *Material Matters, 6*(2).

Heo, D. Y., Do, H. H., Ahn, S. H., & Kim, S. Y. (2020). Metal-organic framework materials for perovskite solar cells. *Polymers*. Available from https://doi.org/10.3390/polym12092061.

Hodoshima, S., Takaiwa, S., Shono, A., Satoh, K., & Saito, Y. (2005). Hydrogen storage by decalin/naphthalene pair and hydrogen supply to fuel cells by use of superheated liquid-film-type catalysis. *Applied Catalysis A: General, 283*(1–2), 235–242.

Hodoshima, S., Arai, H., & Saito, Y. (2003). Liquid-film-type catalytic decalin dehydrogeno-aromatization for long-term storage and long-distance transportation of hydrogen. *International Journal of Hydrogen Energy, 28*(2), 197–204.

Hou, P., Yang, Q., Bai, S., Xu, S., Liu, M., & Cheng, H. (2002). Bulk storage capacity of hydrogen in purified multiwalled carbon nanotubes. *The Journal of Physical Chemistry. B, 106*(5), 963–966.

Hussain, T., Mortazavi, B., Bae, H., Rabczuk, T., Lee, H., & Karton, A. (2019). Enhancement in hydrogen storage capacities of light metal functionalized boron–graphdiyne nanosheets. *Carbon, 147*, 199–205.

Hydrogen storage. (n.d.). Wikipedia. https://en.wikipedia.org/wiki/Hydrogen_storage#cite_note-target-7.

Hystorsys. (n.d.). https://www.hystorsys.no/our-story/.

Ikäheimo, J., Kiviluoma, J., Weiss, R., & Holttinen, H. (2018). Power-to-ammonia in future North European 100% renewable power and heat system. *International Journal of Hydrogen Energy, 43*(36), 17295–17308.

James, S. L. (2003). Metal-organic frameworks. *Chemical Society Reviews, 32*(5), 276–288.

Jena, P. (2011). Materials for hydrogen storage: Past, present, and future. *The Journal of Physical Chemistry Letters, 2*(3), 206–211.

Jiang, J.-X., Su, F., Niu, H., Wood, C. D., Campbell, N. L., Khimyak, Y. Z., & Cooper, A. I. (2008). Conjugated microporous poly (phenylene butadiynylene)s. *Chemical Communications, 4*, 486–488.

Jiang, N., Rao, K. S. R., Jin, M.-J., & Park, S.-E. (2012). Effect of hydrogen spillover in decalin dehydrogenation over supported Pt catalysts. *Applied Catalysis A: General, 425*, 62–67.

Juangsa, F. B., Irhamna, A. R., & Aziz, M. (2021). Production of ammonia as potential hydrogen carrier: Review on thermochemical and electrochemical processes. *International Journal of Hydrogen Energy, 46*(27), 14455–14477.

Kariya, N., Fukuoka, A., & Ichikawa, M. (2002). Efficient evolution of hydrogen from liquid cycloalkanes over Pt-containing catalysts supported on active carbons under "wet–dry multiphase conditions,". *Applied Catalysis A: General, 233*(1–2), 91–102.

Kim, B.-J., Lee, Y.-S., & Park, S.-J. (2008). A study on the hydrogen storage capacity of Ni-plated porous carbon nanofibers. *International Journal of Hydrogen Energy, 33*(15), 4112–4115.

Kim, B.-J., & Park, S.-J. (2006). Preparation of nanoporous carbons from graphite nanofibres. *Nanotechnology, 17*(17), 4395.

Kitagawa, S. (2014). Metal–organic frameworks (MOFs). *Chemical Society Reviews, 43*(16), 5415–5418.

Klier, J. et al. (2012). A new cryogenic high-pressure H_2 test area: First results, in Proceedings of the 12th cryogenics 2012 IIR international conference, Dresden, Germany, 11–14.

Kondo, T., & Sakurai, Y. (2006). Hydrogen absorption–desorption properties of Mg–Ca–V BCC alloy prepared by mechanical alloying. *Journal of Alloys and Compounds, 417*(1–2), 164–168.

Kou, Z., Zhi, Z., Xu, G., An, Y., & He, C. (2013). Investigation of the performance and deactivation behavior of Raney-Ni catalyst in continuous dehydrogenation of cyclohexane under multiphase reaction conditions. *Applied Catalysis A: General, 467*, 196–201.

Koutsonikolas, D., Kaldis, S., Zaspalis, V. T., & Sakellaropoulos, G. P. (2012). Potential application of a microporous silica membrane reactor for cyclohexane dehydrogenation. *International Journal of Hydrogen Energy, 37*(21), 16302–16307.

Krishna, R., Titus, E., Salimian, M., Okhay, O., Rajendran, S., Rajkumar, A., Sousa, J. M. G., Ferreira, A. L. C., Gil, J. C., & Gracio, J. (2012). *Hydrogen storage for energy application. Hydrogen storage*. IntechOpen.

Krishnamurthy, R. V. (1998). *Hydrogen isotopes BT – Geochemistry* (pp. 326–330). Springer. Available from https://doi.org/10.1007/1-4020-4496-8_160.

Králik, M. (2014). Adsorption, chemisorption, and catalysis. *Chemical Papers, 68*(12), 1625–1638.

Lai, Q., Paskevicius, M., Sheppard, D. A., Buckley, C. E., Thornton, A. W., Hill, M. R., Gu, Q., Mao, J., Huang, Z., & Liu, H. K. (2015). Hydrogen storage materials for mobile and stationary applications: Current state of the art. *ChemSusChem, 8*(17), 2789–2825.

Langmi, H. W., Engelbrecht, N., Modisha, P. M., & Bessarabov, D. (2022). Chapter 13 – Hydrogen storage. In T. Smolinka, & J. B. T.-E. P. S. F. Garche (Eds.), *Systems and applications* (pp. 455–486). Elsevier. Available from https://doi.org/10.1016/B978-0-12-819424-9.00006-9.

Largest green hydrogen projects in the world. (2023). Future fuels. https://www.linkedin.com/pulse/largest-green-hydrogen-projects-world-futurefuels/.

Latroche, M., Surblé, S., Serre, C., Mellot-Draznieks, C., Llewellyn, P. L., Lee, J., Chang, J., Jhung, S. H., & Férey, G. (2006). Hydrogen storage in the giant-pore metal–organic frameworks MIL-100 and MIL-101. *Angewandte Chemie, 118*(48), 8407–8411.

Lee, H., Kang, Y.-S., Kim, S.-H., & Lee, J.-Y. (2002). Hydrogen desorption properties of multiwall carbon nanotubes with closed and open structures. *Applied Physics Letters, 80*(4), 577–579.

Lee, H., Lee, J., Kim, D. Y., Park, J., Seo, Y.-T., Zeng, H., Moudrakovski, I. L., Ratcliffe, C. I., & Ripmeester, J. A. (2005). Tuning clathrate hydrates for hydrogen storage. *Nature, 434*(7034), 743–746.

Li, Y., & Yang, R. T. (2006). Significantly enhanced hydrogen storage in metal – organic frameworks via spillover. *Journal of the American Chemical Society, 128*(3), 726–727.

Liu, C., Fan, Y. Y., Liu, M., Cong, H. T., Cheng, H. M., & Dresselhaus, M. S. (1999). Hydrogen storage in single-walled carbon nanotubes at room temperature. *Science (New York, N.Y.)*, *286*(5442), 1127–1129.

Liu, Y., Chabane, D., & Elkedim, O. (2021). Intermetallic compounds synthesized by mechanical alloying for solid-state hydrogen storage: A review. *Energies*, *14*(18), 5758.

Loges, B., Boddien, A., Junge, H., & Beller, M. (2008). Controlled generation of hydrogen from formic acid amine adducts at room temperature and application in H_2/O_2 fuel cells. *Angewandte Chemie International Edition*, *47* (21), 3962–3965.

Loh, C. Y., Huang, R., Bell, R., & Xie, M. (2023). Towards sustainable synthesis: A life cycle assessment of polymer of intrinsic microporosity (PIM-1) by green mechanosynthesis††Electronic supplementary information (ESI) available. *RSC Sustainability*, *1*(9), 2287–2295. Available from https://doi.org/10.1039/d3su00340j.

Lueking, A., & Yang, R. T. (2003). Hydrogen storage in carbon nanotubes: Residual metal content and pretreatment temperature. *AIChE Journal*, *49*(6), 1556–1568.

Makepeace, J. W., He, T., Weidenthaler, C., Jensen, T. R., Chang, F., Vegge, T., Ngene, P., Kojima, Y., de Jongh, P. E., & Chen, P. (2019). Reversible ammonia-based and liquid organic hydrogen carriers for high-density hydrogen storage: Recent progress. *International Journal of Hydrogen Energy*, *44*(15), 7746–7767.

Mannan, S.B.T.-L.L.P. (2012). Chapter 22 – Storage, in The P.I. (4th ed.). Butterworth-Heinemann, 1889–1985. Available from https://doi.org/10.1016/B978-0-12-397189-0.00022-7.

Materials-Based Hydrogen Storage. (n.d.). https://www.energy.gov/eere/fuelcells/materials-based-hydrogen-storage.

Mauron, P., et al. (2008). Stability and reversibility of LiBH4. *The Journal of Physical Chemistry. B*, *112*(3), 906–910. Available from https://doi.org/10.1021/jp077572r.

Mazzolai, G., et al. (2008). Hydrogen-storage capacities and H diffusion in bcc TiVCr alloys. *Journal of Alloys and Compounds*, *466*(1), 133–139. Available from https://doi.org/10.1016/j.jallcom.2007.11.040.

McKeown, N. B., & Budd, P. M. (2006). Polymers of intrinsic microporosity (PIMs): Organic materials for membrane separations, heterogeneous catalysis and hydrogen storage. *Chemical Society Reviews*, *35*(8), 675–683.

Miocic, J., Heinemann, N., Edlmann, K., Scafidi, J., Molaei, F., & Alcalde, J. (2023). Underground hydrogen storage: A review.

Modisha, P. M., Ouma, C. N. M., Garidzirai, R., Wasserscheid, P., & Bessarabov, D. (2019). The prospect of hydrogen storage using liquid organic hydrogen carriers. *Energy & Fuels*, *33*(4), 2778–2796.

Mohamed, M. G., EL-Mahdy, A. F. M., Kotp, M. G., & Kuo, S.-W. (2022). Advances in porous organic polymers: Syntheses, structures, and diverse applications. *Materials Advances*, *3*(2), 707–733. Available from https://doi.org/10.1039/d1ma00771h.

Mohan, M., Sharma, V. K., Kumar, E. A., & Gayathri, V. (2019). Hydrogen storage in carbon materials—A review. *Energy Storage*, *1*(2), e35.

Mukherjee, S., Devaguptapu, S. V., Sviripa, A., Lund, C. R. F., & Wu, G. (2018). Low-temperature ammonia decomposition catalysts for hydrogen generation. *Applied Catalysis B: Environmental*, *226*, 162–181.

Murray, L. J., Dincă, M., & Long, J. R. (2009). Hydrogen storage in metal–organic frameworks. *Chemical Society Reviews*, *38*(5), 1294–1314.

Müller, K., Brooks, K., & Autrey, T. (2017). Hydrogen storage in formic acid: A comparison of process options. *Energy & Fuels*, *31*(11), 12603–12611.

Nielsen, M., Alberico, E., Baumann, W., Drexler, H.-J., Junge, H., Gladiali, S., & Beller, M. (2013). Low-temperature aqueous-phase methanol dehydrogenation to hydrogen and carbon dioxide. *Nature*, *495*(7439), 85–89.

Oda, K., Akamatsu, K., Sugawara, T., Kikuchi, R., Segawa, A., & Nakao, S. (2010). Dehydrogenation of methylcyclohexane to produce high-purity hydrogen using membrane reactors with amorphous silica membranes. *Industrial & Engineering Chemistry Research*, *49*(22), 11287–11293.

Panfilov, M. (2016). *Underground and pipeline hydrogen storage. Compendium of hydrogen energy* (pp. 91–115). Elsevier.

Pham, T., et al. (2015). Understanding the H2 sorption trends in the M-MOF-74 series (M = Mg, Ni, Co, Zn). *The Journal of Physical Chemistry C*, *119*(2), 1078–1090.

Physical Hydrogen Storage. (n.d.). https://www.energy.gov/eere/fuelcells/physical-hydrogen-storage.

Preuster, P., Papp, C., & Wasserscheid, P. (2017). Liquid organic hydrogen carriers (LOHCs): Toward a hydrogen-free hydrogen economy. *Accounts of Chemical Research*, *50*(1), 74–85.

Ritchie, H., Rodés-Guirao, L., Mathieu, E., Gerber, M., Ortiz-Ospina, E., Hasell, J., Roser, M. (2023) Population growth, our world in data [Preprint].

Ritchie, H., Roser, M. and Rosado, P. (2022) Energy, our world in data [Preprint].

Rivard, E., Trudeau, M., & Zaghib, K. (2019). Hydrogen storage for mobility: A review. *Materials, 12*(12), 1973.

Rochat, S., Polak-Kraśna, K., Tian, M., Mays, T. J., Bowen, C. R., & Burrows, A. D. (2019). Assessment of the long-term stability of the polymer of intrinsic microporosity PIM-1 for hydrogen storage applications. *International Journal of Hydrogen Energy, 44*(1), 332–337.

Rosi, N. L., Eckert, J., Eddaoudi, M., Vodak, D. T., Kim, J., O'Keeffe, M., & Yaghi, O. M. (2003). Hydrogen storage in microporous metal-organic frameworks. *Science (New York, N.Y.), 300*(5622), 1127–1129.

Rowsell, J. L. C., Millward, A. R., Park, K. S., & Yaghi, O. M. (2004). Hydrogen sorption in functionalized metal – organic frameworks. *Journal of the American Chemical Society, 126*(18), 5666–5667.

Ryan, P., Broadbelt, L. J., & Snurr, R. Q. (2008). Is catenation beneficial for hydrogen storage in metal–organic frameworks? *Chemical Communications, 35*, 4132–4134.

Sabo, M., Henschel, A., Fröde, H., Klemm, E., & Kaskel, S. (2007). Solution infiltration of palladium into MOF-5: Synthesis, physisorption and catalytic properties. *Journal of Materials Chemistry, 17*(36), 3827–3832.

Sagara, T., Klassen, J., & Ganz, E. (2004). Computational study of hydrogen binding by metal-organic framework-5. *The Journal of Chemical Physics, 121*(24), 12543–12547.

Saitoh, H., Kato, S., & Katagiri, M. (2014). Hydrogenation of anodized aluminum and crystal growth of formed hydride at high pressure and high temperature. *Materials Transactions, 55*(8), 1114–1116.

Sandrock, G. (1999). A panoramic overview of hydrogen storage alloys from a gas reaction point of view. *Journal of Alloys and Compounds, 293*, 877–888.

Sandrock, G., & Thomas, G. (2001). The IEA/DOE/SNL on-line hydride databases. *Applied Physics A, 72*, 153–155.

Schaber, C., Mazza, P., & Hammerschlag, R. (2004). Utility-scale storage of renewable energy. *The Electricity Journal, 17*(6), 21–29.

Sdanghi, G., Canevesi, R. L. S., Celzard, A., Thommes, M., & Fierro, V. (2020). Characterization of carbon materials for hydrogen storage and compression. *Journal of Carbon Research, 6*(3), 46.

Selig, H., & Ebert, L. B. (1980). *Graphite intercalation compounds, in. Advances in inorganic chemistry and radiochemistry* (pp. 281–327). Elsevier.

Sinha, H. (2023). Understanding the different types of hydrogen: Grey, blue, and green. https://www.linkedin.com/pulse/understanding-different-types-hydrogen-grey-blue-green-himanshu-sinha/.

Song, Y., & Guo, Z. X. (2006). Electronic structure, stability and bonding of the Li-N-H hydrogen storage system. *Physical Review B, 74*(19), 195120. Available from https://doi.org/10.1103/PhysRevB.74.195120.

Stark, K., Emel'yanenko, V. N., Zhabina, A. A., Varfolomeev, M. A., Verevkin, S. P., Müller, K., & Arlt, W. (2015). Liquid organic hydrogen carriers: thermophysical and thermochemical studies of carbazole partly and fully hydrogenated derivatives. *Industrial & Engineering Chemistry Research, 54*(32), 7953–7966.

Ströbel, R., Garche, J., Moseley, P. T., Jörissen, L., & Wolf, G. (2006). Hydrogen storage by carbon materials. *Journal of Power Sources, 159*(2), 781–801.

Suh, M. P., Park, H. J., Prasad, T. K., & Lim, D.-W. (2012). Hydrogen storage in metal–organic frameworks. *Chemical Reviews, 112*(2), 782–835.

Sun, Y. Y., Kim, Y.-H., & Zhang, S. B. (2007). Effect of spin state on the dihydrogen binding strength to transition metal centers in metal-organic frameworks. *Journal of the American Chemical Society, 129*(42), 12606–12607.

Tang, C., Bando, Y., Ding, X., Qi, S., & Golberg, D. (2002). Catalyzed collapse and enhanced hydrogen storage of BN nanotubes. *Journal of the American Chemical Society, 124*(49), 14550–14551.

Taylor, J. B., Alderson, J. E. A., Kalyanam, K. M., Lyle, A. B., & Phillips, L. A. (1986). Technical and economic assessment of methods for the storage of large quantities of hydrogen. *International Journal of Hydrogen Energy, 11*(1), 5–22.

Technical Assessment: Cryo-Compressed Hydrogen Storage for Vehicular Applications (2006). https://www.energy.gov/eere/fuelcells/articles/technical-assessment-cryo-compressed-hydrogen-storage-vehicular#:~:text=In/this/concept/2C hydrogen is, compressed hydrogen at low temperatures.

Teichmann, D., Arlt, W., Wasserscheid, P., & Freymann, R. (2011). A future energy supply based on liquid organic hydrogen carriers (LOHC). *Energy & Environmental Science, 4*(8), 2767–2773.

Teichmann, D., Arlt, W., & Wasserscheid, P. (2012). Liquid organic hydrogen carriers as an efficient vector for the transport and storage of renewable energy. *International Journal of Hydrogen Energy, 37*(23), 18118–18132.

The hydrogen colour spectrum. (2023). https://www.nationalgrid.com/stories/energy-explained/hydrogen-colour-spectrum.

Thompson, J. (2023) Hydrogen vs. solar and wind: Comparing renewable energy sources. https://www.linkedin.com/pulse/hydrogen-vs-solar-wind-comparing-renewable-energy-sources-thompson/.

Tian, Y., & Zhu, G. (2020). Porous aromatic frameworks (PAFs). *Chemical Reviews, 120*(16), 8934–8986. Available from https://doi.org/10.1021/acs.chemrev.9b00687.

Tozzini, V., & Pellegrini, V. (2013). Prospects for hydrogen storage in graphene. *Physical Chemistry Chemical Physics, 15*(1), 80–89.

Tranchemontagne, D. J., Mendoza-Cortés, J. L., O'keeffe, M., & Yaghi, O. M. (2009). Secondary building units, nets and bonding in the chemistry of metal–organic frameworks. *Chemical Society Reviews, 38*(5), 1257–1283.

Veverka, P., & Jeřábek, K. (1999). Mechanism of hypercrosslinking of chloromethylated styrene–divinylbenzene copolymers. *Reactive and Functional Polymers, 41*(1–3), 21–25.

Vojvodic, A., Medford, A. J., Studt, F., Abild-Pedersen, F., Khan, T. S., Bligaard, T., & Nørskov, J. K. (2014). Exploring the limits: A low-pressure, low-temperature Haber–Bosch process. *Chemical Physics Letters, 598*, 108–112.

Walters, A.B. (1976). Technical and environmental aspects of underground hydrogen storage, in 1st World Hydrogen Energy Conference, Volume 2, 2B_65-2B_79.

Wang, B., Li, T., Gong, F., Othman, M. H. D., & Xiao, R. (2022). Ammonia as a green energy carrier: Electrochemical synthesis and direct ammonia fuel cell – A comprehensive review. *Fuel Processing Technology, 235*, 107380. Available from https://doi.org/10.1016/j.fuproc.2022.107380.

Wang, Y., et al. (2011). Surface functionalization-enhanced spillover effect on hydrogen storage of Ni–B nanoalloy-doped activated carbon. *International Journal of Hydrogen Energy, 36*(21), 13663–13668.

Wang, Y., Cui, X., Zhang, P., Wang, Y., & Lu, W. (2023). Synthesis of porphyrin porous organic polymers and their application of water pollution treatment: A review. *Environmental Technology & Innovation, 29*, 102972. Available from https://doi.org/10.1016/j.eti.2022.102972.

Xu, R., Chakraborty, S., Yuan, H., & Jones, W. D. (2015). Acceptorless, reversible dehydrogenation and hydrogenation of N-heterocycles with a cobalt pincer catalyst. *ACS Catalysis, 5*(11), 6350–6354.

Yaghi, O. M., Li, G., & Li, H. (1995). Selective binding and removal of guests in a microporous metal–organic framework. *Nature, 378*(6558), 703–706.

Yang, J., et al. (2013). The hydrogen storage capacity of mono-substituted MOF-5 derivatives: An experimental and computational approach. *Microporous and Mesoporous Materials, 171*, 65–71.

Yang, R. T. (2000). Hydrogen storage by alkali-doped carbon nanotubes–revisited. *Carbon, 38*(4), 623–626.

Yildirim, T., & Hartman, M. R. (2005). Direct observation of hydrogen adsorption sites and nanocage formation in metal-organic frameworks. *Physical Review Letters, 95*(21), 215504.

Yildirim, T., Íñiguez, J., & Ciraci, S. (2005). Molecular and dissociative adsorption of multiple hydrogen molecules on transition metal decorated C 60. *Physical Review B, 72*(15), 153403.

Young, K. (2018). *Metal hydrides☆. Reference module in chemistry molecular sciences and chemical engineering*. Elsevier. Available from 10.1016/B978-0-12-409547-2.05894-7.

Yuan, D., Lu, W., Zhao, D., & Zhou, H. (2011). Highly stable porous polymer networks with exceptionally high gas-uptake capacities. *Advanced Materials, 23*(32), 3723–3725.

Yuan, S., Dorney, B., White, D., Kirklin, S., Zapol, P., Yu, L., & Liu, D.-J. (2010). Microporous polyphenylenes with tunable pore size for hydrogen storage. *Chemical Communications, 46*(25), 4547–4549.

Zaluska, A., Zaluski, L., & Ström-Olsen, J. O. (2001). Structure, catalysis and atomic reactions on the nano-scale: A systematic approach to metal hydrides for hydrogen storage. *Applied Physics A, 72*, 157–165.

Zhang, J., Fisher, T. S., Ramachandran, P. V., Gore, J. P., Mudawar, I. (2005). A review of heat transfer issues in hydrogen storage technologies.

Zhang, J., et al. (2017). Controllable fabrication of Ni-based catalysts and their enhancement on desorption properties of MgH_2. *Journal of Alloys and Compounds, 715*, 329–336. Available from https://doi.org/10.1016/j.jallcom.2017.05.011.

Zhang, L., Wang, Q., & Liu, Y.-C. (2007). Design for hydrogen storage materials via observation of adsorption sites by computer tomography. *The Journal of Physical Chemistry. B, 111*(17), 4291–4295.

Zhao, W., et al. (2012). Impact of synthesis conditions of KOH activated carbons on their hydrogen storage capacities. *International Journal of Hydrogen Energy, 37*(19), 14278–14284.

Zhao, Y., et al. (2005). Hydrogen storage in novel organometallic buckyballs. *Physical Review Letters, 94*(15), 155504.

Zhou, H.-C., Long, J. R., & Yaghi, O. M. (2012). *Introduction to metal–organic frameworks. Chemical Reviews* (pp. 673–674). ACS Publications.

Zhou, L. (2005). Progress and problems in hydrogen storage methods. *Renewable and Sustainable Energy Reviews, 9* (4), 395–408.

CHAPTER

5

Navigating hydrogen storage and transport networks: grid connectivity and storage site logistics

Gourav Kumar Rath[1], Gaurav Pandey[1,4], Ajay Mittal[2], Sakshi Singh[1], Swagata Sharma[1], Dharmendra Kumar Gupta[3], Sanket Joshi[5] and Nadezhda Molokitina[4]

[1]Department of Petroleum Engineering and Earth Science, Energy Cluster, School of Advanced Engineering, UPES, Dehradun, Uttarakhand, India [2]Department of Chemical Engineering, Energy Cluster, School of Advanced Engineering, UPES, Dehradun, Uttarakhand, India [3]Department of Sustainability, School of Advanced Engineering, UPES, Dehradun, Uttarakhand, India [4]Earth Cryosphere Institute, Tyumen Scientific Center SB, RAS, Tyumen, Russia [5]Amity Institute of Microbial Technology, Amity University, Jaipur, Rajasthan, India

5.1 Introduction

Hydrogen, the most abundant element in the universe, has been widely acknowledged for its immense energy potential over the years. Hydrogen is a versatile fuel that can be effectively utilized across various sectors, such as power generation, transportation, heating purposes, and industrial applications. One of the primary benefits of hydrogen is its remarkable capacity to generate energy without releasing greenhouse gases or other detrimental pollutants into the environment, primarily when used in fuel cells or combusted in engines. The environmental ramifications of hydrogen utilization represent a pivotal factor that must be carefully considered. Although hydrogen stands out as a clean and eco-friendly fuel source, the techniques employed for its production can exhibit varying degrees of environmental consequences. To comprehensively assess the overall environmental efficiency of hydrogen systems, life cycle assessments (LCAs) are employed.

Subsurface Hydrogen Energy Storage
DOI: https://doi.org/10.1016/B978-0-443-24071-3.00006-6

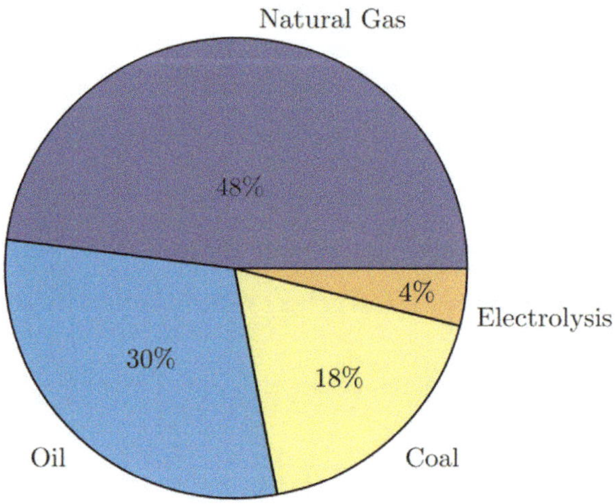

FIGURE 5.1 Major Hydrogen production souces (Pashchenko, 2024).

These assessments are crucial as they consider the complete supply chain of hydrogen, ranging from its production phase to its ultimate utilization stage (Hassan et al., 2024). Fig. 5.1 illustrates the contribution percentage of hydrogen produced from different sources, which essentially helps in understanding the environmental impact of current H_2 producing methods (Mittal & Mahajan, 2023). The worldwide need for hydrogen surged to 255.3 **billion cubic meters in** 2013, and forecasts suggest a rise to 150 **million metric tons** by 2030. As indicated in the analysis conducted by Frank Umbach, who serves as the head of research at EUCERS/CASSIS, University of Bonn, it was highlighted that in the year 2021, the global demand for hydrogen stood at 94 million tons, which is approximately 2.5% of the total energy consumption across the globe. This data underscores the growing significance of hydrogen as a critical component in meeting energy demands worldwide (IEA, 2023; Sikiru et al., 2024).

The utilization of seawater, with neutral pH levels, is considered a promising approach to achieving energy sustainability, especially when combined with renewable electricity to produce hydrogen. It is crucial to recognize seawater's immense availability and cost-effectiveness in this context. However, it is imperative to conduct pretreatment procedures to eliminate solid particles, organic substances, and dissolved salts from seawater. This pretreatment is essential to prevent various issues such as corrosion, fouling, undesired chemical reactions, electrode and membrane damage, and the accumulation of salt within the electrolyzer (Bolar et al., 2021; Gao et al., 2022; Mohammed-Ibrahim & Moussab, 2020; Sun et al., 2021).

The generation of hydrogen in remote offshore areas has the potential to facilitate the production of cost-effective green hydrogen by leveraging economies of scale and higher capacity utilization rates. The design of such systems is intricate and multifaceted, requiring careful consideration of factors including energy expenditures, marine-related challenges, safety protocols, tough operational environments, and transportation costs. The development of advanced technologies for electrolysis processes and the maintenance of offshore platforms in challenging marine settings represent significant hurdles that must be addressed in this context. The impediment posed by the substantial costs associated with the deployment and operation of offshore platforms represents a formidable challenge that requires strategic solutions. One possible approach to overcoming this challenge is to focus on achieving economies of scale, which involves maximizing efficiency and productivity to reduce overall costs. In addition, developing innovative funding models can also help address the financial barriers to offshore platform projects, potentially opening up new opportunities for investment and development in this sector. Furthermore, efforts to streamline expenses can be made by adopting modular designs and utilizing off-the-shelf components, which can help lower costs without compromising quality or safety standards. When it comes to addressing environmental concerns related to offshore platforms, applying impact assessments and implementing mitigation measures are crucial steps in minimizing the ecological footprint of these structures. Moreover, compliance with relevant national and international regulations is essential, necessitating ongoing communication with regulatory agencies to ensure adherence to all applicable laws and regulations and staying abreast of any changes in the regulatory landscape to maintain operational compliance (Ramakrishnan et al., 2024).

The use of offshore wind electricity to produce H_2 from electrolysis of seawater has grabbed the attention of researchers and industries throughout the world (Wu et al., 2022). Offshore wind to H_2 (OWH) systems produces hydrogen using renewable energy sources as an alternative of natural gas (NG) or coal contributing toward reducing greenhouse gas emission. Compared to earlier offshore wind power initiatives, OWH systems reduce instability caused by green energy grid connections and provide an efficient way to use abundant offshore wind energy. This energy management system was developed based on project data and verified via computational simulations. OWH systems can use current NG pipelines to transfer power, reducing the need for new electrical distribution facilities and overall investment costs. Hydrogen pipelines have lower estimated capital and specific expenses for energy delivery compared to high-voltage DC lines in remote offshore areas (McDonagh et al., 2020; Patonia et al., 2023; Wu et al., 2019).

There are various ways for hydrogen storage, including liquefied hydrogen (LH_2), compressed hydrogen (CH_2), liquid organic hydrogen carriers (LOHCs), solid-state hydrogen storage, and underground storage. Historically, it was stored as either a liquid or compressed gas to enhance storage density, often reaching pressures of up to 700 bar (Chu et al., 2023; Hurskainen, 2019). LOHCs are considered a highly efficient method for hydrogen storage under ambient conditions without any loss. The storage process begins with the exothermic hydrogenation of a hydrogen-deficient molecule, which typically occurs at the hydrogen production site, transforming it into a hydrogen-rich molecule (Lin & Bagnato, 2024). Hydrate application for hydrogen gas storage has been proposed since 2002 by Mao et al. (2002) followed by continuous research in the field since (Mao et al., 2002). Many researchers are interested in storage of hydrogen due to low environmental damages, potential for significant storage space, and need of reusable hydrate forming components like water. The high-pressure and low-temperature conditions for hydrate formation pose a hurdle to hydrogen storage. Researchers proposed use of different additives as promoters for hydrogen hydrates. These additives shift the hydrate formation conditions closer to ambient temperature and pressure conditions to address this issue (Saikia et al., 2023). This chapter discusses different technologies of hydrogen production, transportation and storage is detail. The issues associated with these technologies and the available solutions are also conferred in the chapter.

5.2 Hydrogen as fuel

The worldwide need for hydrogen as a fuel is estimated to reach 180 million tons by 2030. Hydrogen generates more energy per mass than traditional energy sources and is regarded a clean fuel. At 298K, hydrogen has a very high heating value that ranges from 120 to 141.8 MJ/kg. Processes like thermochemistry, sun splitting, water electrolysis, and biomass conversion are used to manufacture hydrogen from fossil fuels, biomass, and electrolysis. Coal gasification, steam methane reforming, and biomass gasification are processes that produce hydrogen. Nearly 70% of hydrogen generation systems use NG steam reforming, with fossil fuel-based steam conversion making up nearly 80% of these systems (Singh et al., 2023).

5.3 Hydrogen production

The selection of H_2 production methods relies on energy sources, feedstock availability, and required purity levels. Integrating purification into production pathways is vital for seamless operation and energy utilization. Hydrogen's versatility allows for diverse production from various resources and technological approaches, including fossil fuels and renewables. Production methods vary based on factors like resource availability, environmental impact, and technological progress. These methods are crucial for future energy systems and transition. Techniques used in hydrogen generation determine cleanliness, cost-effectiveness, and viability of hydrogen production pathways (HPPs). HPP choice depends on the hydrogen source, energy, and catalyst materials involved, along with factors like feedstock type and production scale (Dawood et al., 2020).

Green energy resources like solar, biomass, hydro, geothermal, wind, and ocean thermal energy conversion are identified for H_2 production. However, the irregular nature of wind and solar energy poses challenges for consistent energy generation. Hydrogen's versatility as a fuel, storage medium, and energy carrier makes it a favorable choice. With its benefits and potential for carbon-free energy, hydrogen is gaining popularity worldwide as a competitive fuel substitute and innovative energy mobility solution (Ishaq et al., 2022).

5.3.1 Methods of H_2 production

Several technologies exist for hydrogen production from fossil fuels, with hydrocarbon reforming and pyrolysis being prominent. These methods, extensively developed and widely used, meet almost all hydrogen demand. Historically, hydrogen generation heavily depended on fossil fuels: with a minimum of 18% from coal, followed by 30% from heavy oils and naphtha, and a maximum of 48% from NG (Balat & Kırtay, 2010; Dincer & Acar, 2015; Kothari et al., 2008). Fossil fuels are indispensable for global hydrogen demand due to cost-effectiveness linked to fuel prices, emphasizing the necessity for sustainable technologies in transportation, industry, and energy sectors. Ongoing research is critical for reducing fossil fuel reliance, enhancing production efficiency, and promoting environmental sustainability, inevitably shifting toward renewable technologies in the long run (Nikolaidis & Poullikkas, 2017; Singh et al., 2023). Fig. 5.2 depicts different H_2 producing methods, depending on the source of the hydrogen.

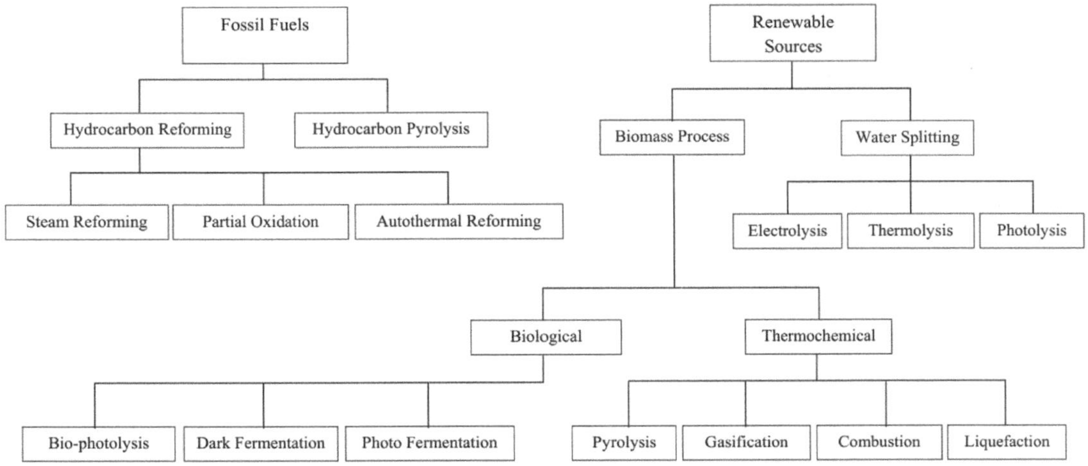

FIGURE 5.2 Various methods of producing hydrogen (Nikolaidis and Poullikkas, 2017).

Coal gasification, an ancient method, produces gas containing approximately 60% hydrogen and substantial carbon monoxide. This process contributes 18% to global hydrogen output. Coal heated to 900°C becomes gaseous, combined with steam and oxygen, then catalyzed at 857°C, often with nickel or $FeO-CrO_2-ThO_2$. Despite its age, this traditional method remains a significant contributor to hydrogen production, showcasing the effectiveness and longevity of such processes in hydrogen production technologies (Kayfeci et al., 2019).

$$2C + O_2 + H_2O \rightarrow H_2 + CO_2 + CO \tag{5.1}$$

Gasification of coal or other carbonaceous materials has been a well-established method for nearly two hundred years to produce synthesis gas (syngas) and hydrogen. While not the most preferred due to high temperatures and resulting pollution, it remains effective. The process involves reacting coal with steam and oxygen at elevated temperatures to obtain H_2, CO, and CO_2 (Kayfeci et al., 2019; National Energy Technology Laboratory, 2018). Steam reforming, the main hydrogen production method, extracts hydrogen and carbon oxides by reacting steam and hydrocarbons with a catalyst, typically nickel-based, at high temperatures (750°C–1000°C). External heat input is essential due to the reaction's endothermic nature, ensuring efficient hydrogen yield (Casper, 1978; National Energy Technology Laboratory, 2018).

Biomass stands as a major renewable hydrogen source, boasting abundant supply and sustainable production potential. In the United States alone, annual lignocellulosic biomass resources range from 400,000 to 1.3 billion dry tons, with costs under $50 per dry ton, making it commercially viable. Recent advancements in crop science and harvesting techniques further enhance its feasibility. Conservatively, biomass could annually yield 40 million tons of hydrogen, powering over 150 million fuel cell automobiles, highlighting its substantial role in renewable energy (U.S. Department of Energy, 2006, n.d.). Fig. 5.3 shows the various biomass treatment processes used for producing H_2 from bio-wastes. Gasification systems hold vast potential for large-scale operations, especially in regions abundant in feedstock. They easily integrate with pipelines for hydrogen transport and

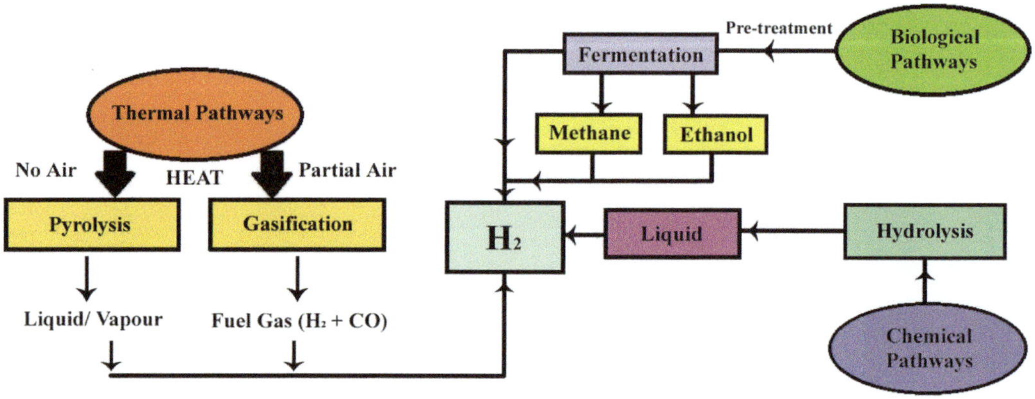

FIGURE 5.3 Pathways of producing hydrogen from biomass (Turner et al., 2008).

offer cogeneration opportunities with fuels like ethanol in fully integrated biomass refineries. Steam gasification processes yield up to 17% hydrogen from lignocellulosic biomass, indicating a stoichiometric reaction converting all carbon to CO_2 and a significant portion of hydrogen sourced from water molecules. (Turner et al., 2008).

Among various crop residues studied, such as rice straw, soybean straw, husk, and sugarcane tops, dark fermentation of rice straw shows the most significant potential for hydrogen production, capable of generating an impressive 58,002 mm^3 of biohydrogen annually. Additionally, incorporating pretreatment and anaerobic digestion stages could enhance both economic viability and overall hydrogen production volume and energy output (Assis & Gonçalves, 2022; Bundhoo, 2019; Cremonez et al., 2021).

Hydrogen production via metals can be categorized into two groups. The first group includes metals like lithium, sodium, potassium, rubidium, cesium, and francium, which inherently react with water, even under standard conditions. The second group comprises metals like magnesium and aluminum, which require external stimuli to provide activation energy for the metal–water reaction to occur. These external inputs trigger the chemical reaction between the metal and water (Abdelkareem et al., 2024). Fig. 5.4 elucidates the pathways used for H_2 production using waste metals in the form of a flowchart, further depicting the purification, storage and applications of the fuel.

Solid oxide electrolysis cell (SOEC) is an innovative technology for high-temperature hydrogen production. Unlike traditional electrolysis methods, SOEC operates at elevated temperatures, improving thermodynamic conditions and reaction kinetics, lowering power requirements (Guo et al., 2024). Proton exchange membrane (PEM) water electrolysis offers inexpensive, high-purity hydrogen production from renewable sources, with only oxygen as a by-product. Despite its efficiency, economic challenges limit its industrial use, though there's renewed interest in green hydrogen production with PEM electrolysis (Farhana et al., 2024). OWH projects reduce carbon emissions and enhance environmental cleanliness by tapping into abundant offshore wind resources for cost-effective green hydrogen production (Ramakrishnan et al., 2024). Fig. 5.5 shows the process flow diagram for the production of H_2 from spent metal items like lithium, aluminium, and sodium.

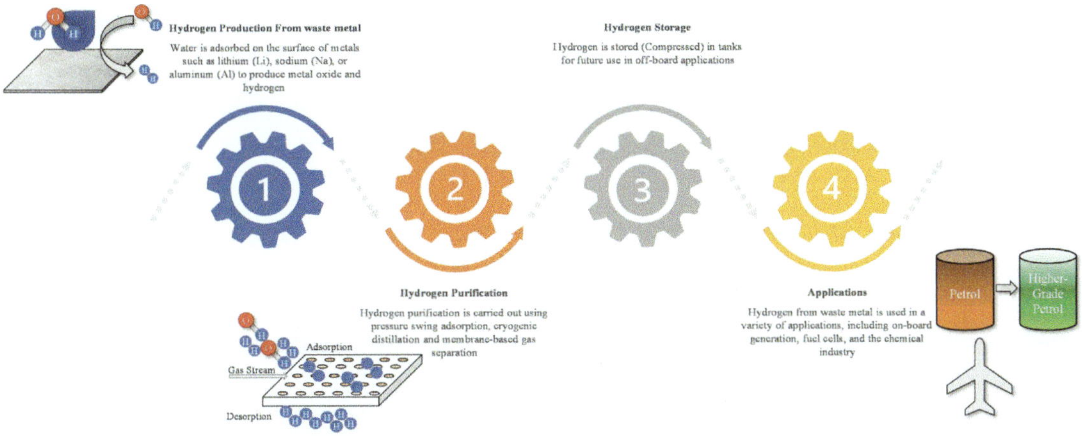

FIGURE 5.4 Flowchart depicting the production of hydrogen using waste metal (Abdelkareem et al., 2024).

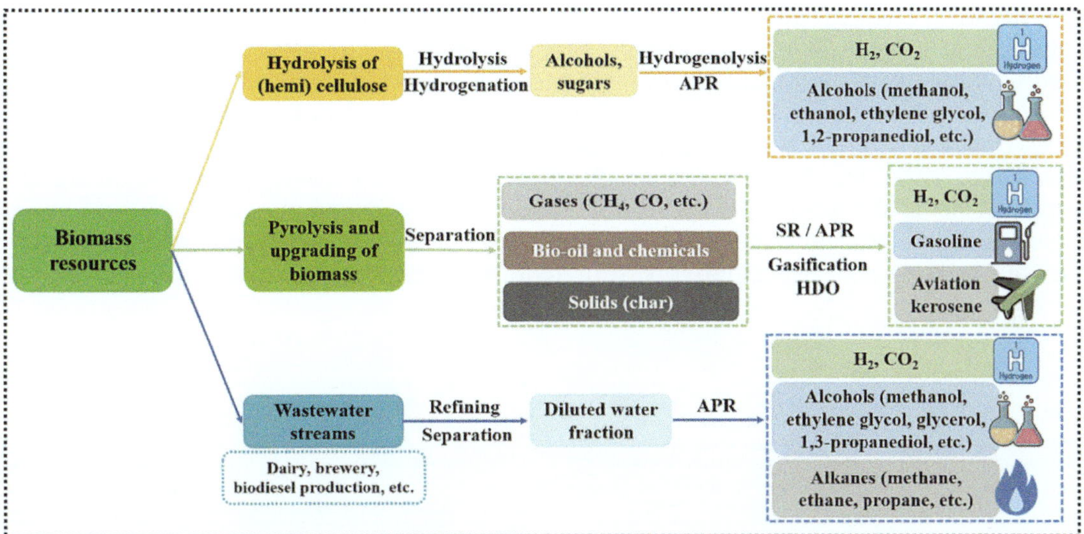

FIGURE 5.5 Process flow diagram showing how various biomass sources are converted into hydrogen and other products (Tian et al., 2024).

Carbon nanomaterials work as catalysts and enhance the rate of biological processes, increasing active sites and catalytic efficiency (Kumar et al., 2024). Material advancements in hydrogen production are vital for effectiveness and affordability across various processes like water electrolysis, coal gasification, and steam methane reforming. Wind power aids in emission reduction, while geothermal energy, with stable output, enhances hydrogen production via PEM electrolyzers (Alamiery, 2023; M & G, 2023).

The projected investment costs for alkaline water electrolysis (AWE) and PEM systems are, respectively, 850–1590 \$/kW and 1485–2230 \$/kW. Other studies, however, show that a lower capital cost range—500–1000 \$ kW − 1 for AWE and 700–1400 \$ kW − 1 for PEMWE—can be attained by streamlining manufacturing, boosting productivity, and enhancing durability (Niblett et al., 2024).

It may be more affordable and practical to produce hydrogen through water vapor plasmolysis. A sustainable way to meet the increasing demand for energy on a worldwide scale may be found in hydrogen's role as a secondary energy carrier (Younas et al., 2022).

5.4 Transportation of hydrogen

The availability of hydrogen for commercial application chiefly depends on the distribution chain of the gas. The price and delivery systems of hydrogen are heavily influenced by where it's produced and distributed. For instance, a large-scale, cost-effective production might incur high transportation costs if the production and consumption sites are far apart. Conversely, production charges may be significant, but delivery costs are typically moderate (Riera et al., 2023). Hydrogen transportation has been a concern for industries over decades. The extremely small size of H_2 makes it difficult to store and transfer H_2 gas without any leakage or damage concerns. However, there are several methods currently used to transport H_2 with minimum loses possible. The chief methods used for hydrogen transportation are transportation through tankers (trucks and rail), transportation through tubes (rail and trailer), ships and solidified hydrogen hydrates (Borah et al., 2024; Out-Law News, 2022). Fig. 5.6 depicts a process flow diagram for Bio-H_2 production using different methods, along with the final products and their applications. Gas compression and liquefaction are widely used methods for hydrogen storage but come with drawbacks such as leakage concerns, high pressure requirements, energy-intensive processes and

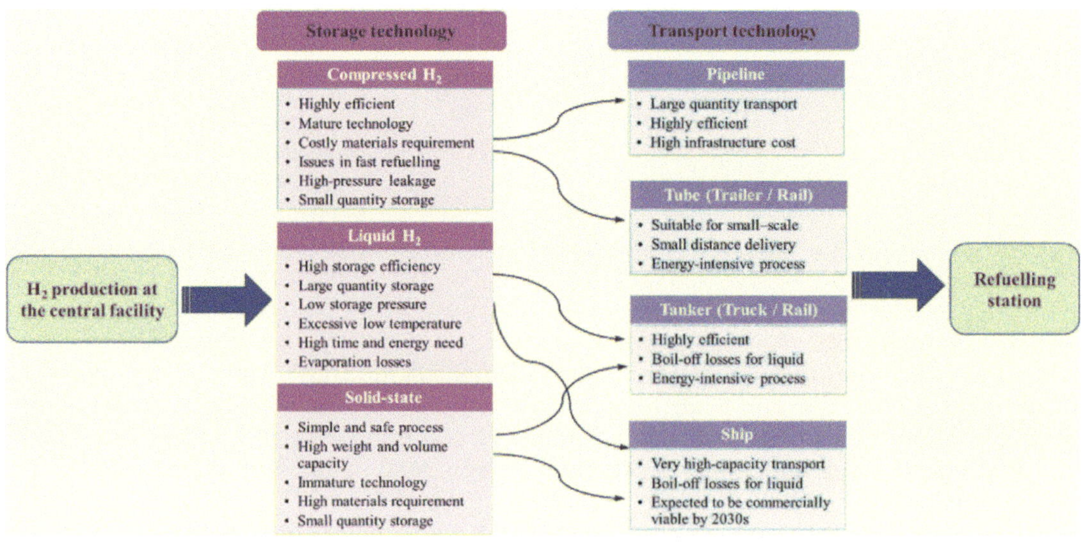

FIGURE 5.6 Different techniques for hydrogen storage and transportation technologies (Halder et al., 2024).

boil-off losses. To overcome these challenges, Lawrence Livermore National Laboratory developed the cryo-compressed H_2 storing device. The device can hold H_2 as liquid hydrogen (lH_2), cold compressed H_2, and/or hydrogen in both vapor and liquid phases at cryogenic pressures and temperatures between 250 and 350 atm (Aceves et al., 2010; Ahluwalia et al., 2010; Kumar et al., 2022; Yue et al., 2021). Transportation and storage of hydrogen is convenient when the H_2 gas is in liquid state. However, it is essential to maintain a temperature of $-253°C$ is required to store H_2 in liquid form, which requires a strong refrigerating unit and consumption of high energy as compare to the compressed gas systems (Cheng et al., 2024; Fuel Cell Electric Buses Knowledge Base, 2023). The transportation of hydrogen in its gaseous state commonly relies on pipes and trucks. Given that gaseous hydrogen is typically generated at low pressures ranging from 20 to 30 bar, compression is necessary before transportation. The compression process elevates the pressure to approximately 180 bars, or roughly 2600 psig, enabling it to be loaded into specialized tubes and trailers. However, this pressure level may pose challenges when transporting long cylinders stacked on the trailer of the towing truck (Baldwin, 2017).

Hydrogen gas fuel is transported from site of production to storage sites in two states, gaseous H_2 and liquid H_2 forms. In the gaseous form H_2 is compressed at high pressures and further transported in specialized containers, confirming safety and regulatory compliance. This approach is frequently adopted due to low cost and utilization of existing setup. Gas compression is achieved through mechanical methods such as centrifugal compressors and piston and diaphragm compression. PEM electrolysis principles are employed for designing advanced techniques involving ionic liquid piston and electrochemical density. Metal hydride hydrogen compressors are gaining popularity due to their advantages, including environmental friendliness, enhanced safety, negligible noise and vibration, efficient hydrogen purification, active sealing, low repair costs and, friction-free operation (Muhammed et al., 2023; Noh et al., 2023; Sdanghi et al., 2019; Zhang et al., 2024). Gaseous H_2 is mainly transported using pipeline network and compressed cylinders. Pipelines offer advantages like large-scale supply for various users, constant sources, environment friendly and cost-effectiveness for power plants and industrial applications. Reports suggest employing the existing NG pipelines for H_2 gas transport, to reduce the initial cost for pipeline installations (Makaryan et al., 2022; Pellegrini et al., 2020).

Compressed gas cylinders offer a versatile option for delivering hydrogen, especially for smaller-scale uses. These cylinders typically made of steel or composites, store hydrogen at pressures from 200 to 700 bars. They enable hydrogen transportation to areas lacking pipelines, suiting temporary or mobile applications like fueling stations. However, their limited storage capacity requires multiple cylinders for large quantities, adding logistical complexity. Weight and size challenges arise, particularly for long-distance transport. Proper handling is crucial to manage high-pressure risks and potential leaks. Access to refueling or exchange facilities is vital for a reliable supply chain, necessitating infrastructure establishment and regulatory compliance (Massarweh et al., 2023; Muhammed et al., 2023; Faye et al., 2022).

Transporting hydrogen in liquid form, typically at ultra-low temperatures, offers advantages over gaseous methods by enabling efficient long-distance movement. Various liquefaction technologies increase hydrogen's transport density. Cryogenic tanks maintain the hydrogen in its liquid state during transportation, mainly by road but potentially by rail or ship. Upon arrival, it is transferred to storage tanks or processing

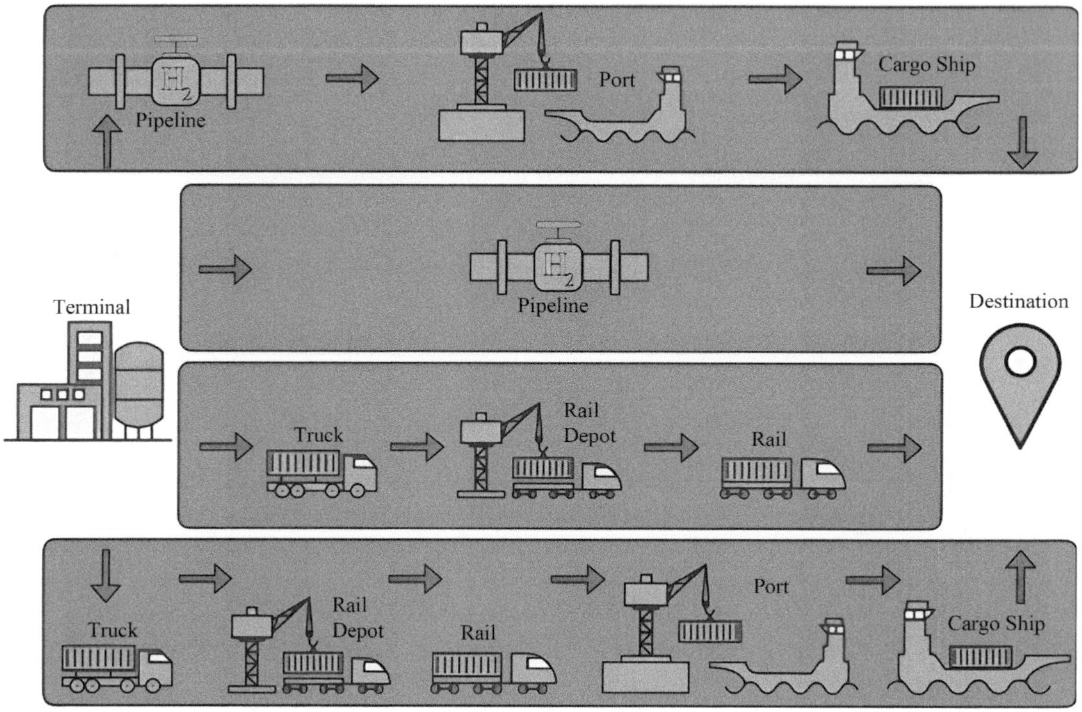

FIGURE 5.7 Schematic diagram of possible hydrogen transportation (combination of road, rail, over sea and pipeline transportation used in the industry) (Borsboom-Hanson et al., 2022).

units, often requiring vaporization before use. Ongoing research aims to improve efficiency, cost-effectiveness, and safety, including advancements in tanker insulation and pressure relief systems. Challenges such as high costs, energy intensity, and safety concerns persist, demanding continuous innovation and regulatory compliance for sustainable lH_2 transport (Faye et al., 2022; Klebanoff et al., 2017; Makaryan et al., 2022; Massarweh et al., 2023; Pellegrini et al., 2020; Zhang et al., 2023). Fig. 5.7 represents the various available H_2 transport methods, it depicts the process of H_2 transport as combination of different methods.

5.4.1 Transportation through tankers

Hydrogen is generally transferred and stored in different types of tanks. The tankers used for H_2 transfer and storage are generally made of metals, composites or a mixture of both. There are five types of tankers used for H_2 storage, distinguished by the manufacturing material (Barthelemy et al., 2017). A detailed discussion about the tankers is done below:

1. Type I: These tanks are made up of metal base only, such as steel and aluminum alloys. They are the most convenient and cost effective to manufacture. Keeping in mind the quality of material, these tanks can be designed to store large capacity of gas at high

working pressure of up to 200 bars. However, due to high density material, these tanks are heavy and possess inconvenience in moving over long distances, thus limiting their application to stationary gas storage and/or movement over shorter distances (Barthélémy, 2012; Barthelemy et al., 2017; Cheng et al., 2024; Kumar, 2024; MPA, 2023).

2. Type II: Type II hydrogen tanks are akin to Type I, comprising all-metal cylinders with the distinction of having carbon fiber or glass fiber filament wound around their straight body section. These tanks find applications in both stationary and mobile settings, with a working pressure typically around 300 bar, surpassing that of Type I tanks due to the supplementary high-strength fiber covering (Barthelemy et al., 2017; Cheng et al., 2024).
3. Type III: The third types of tanks are made with composite material of fiberglass or carbon fiber. The type III tanks are wrapped with aluminum liners to increase the tensile strength of the structure further reducing the weight. These tanks are generally used in automotive applications, both for storing gaseous fuel and transportation of gases for industrial applications (Barthelemy et al., 2017; Cheng et al., 2024; Zhang et al., 2015). The use of a composites and aluminum wrap makes the tank a great competition of types I and II tanks. Type III hydrogen tanks utilize AA6061 or AA7000 metals for their liner, chosen for their favorable mechanical strength and capability for heat treatment to enhance its performance. Type III tanks can endure a maximum pressure ranging from 305 (for glass/Al tanks) to 700 bars (for carbon/Al tanks) depending on the material used for construction (Barral & Barthélémy, 2006; Cheng et al., 2024).
4. Type IV: Same as type III, type IV tanks are also made of composite base material like carbon fiber. But types III and IV differ from each other on the basis of the external coating on these tanks. Unlike the former, type IV tanks have composite wrapping around the tank body making it further light weight and easy to use for H_2 transportation. These tanks generally use carbon fiber T700S as the coating material, whereas the body is made of T700S or T800S, same as type III. Type IV tanks have less than half the thickness of a type I tank, to endure the same amount of pressure, due to the high tensile strength of tank body (Barral & Barthélémy, 2006; Barthélémy, 2012; Cheng et al., 2024; Zhang et al., 2015). The type IV tanker have a maximum pressure control of 350 bars to 700 bars, and is generally used as vehicular fuel tanks and H_2 transport (IEA, 2019).
5. Type V: The type V tankers are a new tank type, currently under development, which is approximately 20% lighter as compare to type IV. Type V tanks are made of 100% composite material which signifies a high cost of the final product. Type V tankers have ability to withstand pressure as high as 1000 bar, depending on the material used in the composite material (Hoon, 2021; Su et al., 2021).

5.4.2 Transportation through pipeline

Hydrogen pipelines are primarily designed to transmit hydrogen to major users such as chemical factories and refineries. They provide a convenient means for high-volume delivery over long distances (Baldwin, 2017). Hydrogen can utilize existing NG pipelines for transportation. America and Europe are known to have approximately 1000 and 2000 km of NG pipelines dedicated to hydrogen transport, respectively. Additional networks are present in South Korea and Thailand as well. Table 5.1 lists the total length of NG pipeline network in different

TABLE 5.1 Length of pipeline networks in different countries globally (Gondal, 2016).

Rank	Country	Gas pipeline length (in km)
1	The United States	548,665
2	Russia	158,767
3	Canada	74,980
4	China	28,132
5	Ukraine	33,327
6	Argentina	28,657
7	Australia	26,719
8	Germany	25,094
9	Mexico	22,705
10	Iran	19,161

countries of the world. The pipeline connections cater to the safety requirements for H_2 transport, which makes them a suitable solution for the transportation problem (Dryer et al., 2007; Shinnar, 2003). At defined pressure, gaseous H_2 has a 36% higher density than NG. For equivalent energy to be carried, the H_2 flux should be 280% (2.8 times) bigger when equated to the NG flux. Nonetheless, H_2 viscosity (8.92×10^{-6} Pa/s) is considerably lower than NG (11.2×10^{-6} Pa/s) which ensures a quicker and less obstructed movement of gas particles. The following formula provides the least power P needed to pump a gas via a pipe:

$$P = 8 \times \pi \times l \times v^2 \times \eta \tag{5.2}$$

where, l is the length of the pipe, v is the gas velocity, and η is the gas's dynamic viscosity. Given that the velocity of H_2 and NG are kept same, for a defined length of pipe, the power required to transfer a unit H_2 gas will be lower than the power needed for unit NG transfer. Additionally, pipelines can be utilized to manage supply and demand variation and provide temporary storage solutions for the gaseous fuel (Damen et al., 2007; Ipsakis et al., 2009). However, using existing NG networks for hydrogen transport raises safety concerns due to the differing characteristics of materials in the pipeline system like hydrogen Embrittlement, gas pressure reduction over long distance, leakage, and formation of hydrogen hydrides in the pipeline (Bockris, 2002; Verykios, 2003). Fig. 5.8 illustrates a gas pipelines used for long distance transportation of NG and H_2.

5.4.3 Hydrogen transfer using tube trailers

Tube trailers are long cylindrical metal structures loaded on long trucks and trains for transportation. Highly CH_2 (180 bar ~ 2600 psig) gas in filled in the containers and further carried to the destination. Fig. 5.9 shows a tube trailer truck designed to carry high-pressure hydrogen. Tube trailers transport hydrogen using steel or composite pressure vessels. Choice between steel and composite depends on cost, weight restrictions, and

FIGURE 5.8 Connection of pipelines for H_2 and NG transfer (HFCTO, 2021).

FIGURE 5.9 Tube trailer carried by trucks for compressed gas transportation (HFCTO, 2022).

payload needs. Steel tube trailers are frequently utilized and typically hold around 380 kg on board, with their load capacity controlled by their individual weights. In recent times, composite storing containers have emerged, boasting capacities ranging from 560 to 900 kg of H_2 gas per trailer. Tube trailers are presently deployed for delivering compressed NG in various other nations (HFCTO, 2022).

Composite tube trailers employ Type III (carbon-fiber compound with aluminum liner) or Type IV (carbon-fiber compound with polyethylene liner) pressure vessels. These vessels, lighter and capable of withstanding pressures up to 500 bar, allow for transporting a larger H_2 consignment, up to 1000 kg per trip. This technological leap offers enhanced efficiency and capacity compared to traditional steel tube trailers, enabling more flexible and substantial hydrogen distribution across diverse applications (Elgowainy et al., 2014;

Federal size regulations for commercial motor vehicles, 2004; Su et al., 2021). Evolving hydrogen transportation infrastructure may overcome weight limitations, fostering broader hydrogen usage.

The tube-trailer system has two different delivery pathways, depending on the accessibility of the hydrogen production facility. Firstly, for remotely situated H_2 producing industries, the hydrogen made is initially transported to a gas distribution terminal through a pipeline. The gas is then compressed at high pressure and filled in the pressure vessel mounted on a truck (thereafter known as a tube trailer) and moved to fueling stations. But, for H_2 production facilities at accessible locations, the produced gas is directly loaded onto the tube trailer and moved to distribution points. The second scenario results in a faster delivery service with less fuel loss, expenditure and much less chaos (Bolat & Thiel, 2014; Elgowainy et al., 2014).

5.4.4 Hydrogen transportation through ships

Long-distance H_2 transportation that cannot be done with the help of pipelines, are done through the offshore routes using ships. The first ever H_2 transportation using ships was undertaken from Australia to Japan in 2020, to collect information about existing problems in this process (Collins, 2022). Hydrogen is transported through three different forms in ships overseas, that is, lH_2, LOHCs, and ammonia. These techniques are discussed below for further understanding:

1 lH_2: The gaseous H_2 is converted in liquefied state by freezing the gas to $-253°C$ at the loading sites before filling it in the highly insulated tanks. The liquefaction process consumes high energy and also utilizes $25\% - 35\%$ of the initial H_2 quantity. Currently only one ship, the Suiso Frontier, built by Kawasaki, has been tested successfully to have carried lH_2 safely (Harding, 2019).

2 LOHCs: The tendency of organic compounds to successfully dissolve H_2 in its bulk is identified as a prominent method to both store and transport H_2 using ships. LOHC is low energy process as compared to lH_2 as it does not require cooling. As the process of LOHC involves organic liquids as a medium of transport, this process can be done with the help of the existing oil and gas infrastructure (Niermann et al., 2019).

3 Ammonia: Hydrogen can undergo a transformation into ammonia through a reaction with N_2 from the air, using only air, electricity and water. NH_3 possesses a significantly higher energy density compared to H_2, enabling the trading of a larger energy volume. Leveraging the well-established international trade in ammonia is viable. It currently serves as an industrial raw material, notably in fertilizer manufacturing units. Additionally, it holds promise as a decarbonization fuel in sectors such as power generation and shipping. However, the toxicity of NH_3 is a concern globally as it can contribute to NO_2 emissions (IRENA, 2022).

5.5 Storage of hydrogen

The most popular containers for compressing hydrogen storage are carbon composite or pressurized steel cylinders. Hydrogen's low volumetric density has the benefit of being compressed into higher densities, which means less storage space is needed. Thus, the application of H_2 liquefaction and the investigation of supplementary chemical transporters, including ammonia, have grown in favor of H_2 transportation and storage. The different methods of H_2 storage are discussed below.

5.5.1 Compression

Compression is used inside pressurized tanks with mechanical pressure control mechanisms to stock H_2 in its gaseous form. While composite tanks can store hydrogen at pressures of up to 800 bar, steel tanks can usually only hold it at 200 bar. Although this method of storing hydrogen is well-established, the low volumetric densities of the gas may make it unfeasible to use on a wide scale. The practice of storing compressed gas inside a network of pipelines by adjusting the pipeline pressure is known as line packing. Large amounts of hydrogen gas can be kept inside pipelines for days at a time and then released when needed, like during times of high demand. Injection of hydrogen into salt rock is another method of compressing hydrogen into subterranean salt caverns, usually for long-term storage. But the availability of salt caverns limits the use of this storage technique (Bruce et al., 2018).

5.5.2 Liquefaction

A multistage compression and chilling procedure can be used to liquefy hydrogen, which is subsequently kept in cryogenic tanks at $-253°C$ (Bruce et al., 2018). Due to its significantly higher volumetric density, LH_2 offers more advantages in terms of cost when it comes to storage capacity. It should be mentioned, nevertheless, that about 30% of the energy in the hydrogen is usually consumed during the liquefaction process. Moreover, evaporation, or "boil-off," results in hydrogen losses, and the procedure itself is costly (Pascal, 2022).

Currently, cryogenic tanker trucks with enhanced insulation are used to transport hydrogen over longer distances in liquid form. After liquefaction, delivery vehicles transport the lH_2 to supply facilities, where it is vaporized to produce a high-pressure gaseous produce that can be distributed. Since a liquid tanker truck can carry a significantly larger mass of hydrogen than a gaseous tube trailer, carrying lH_2 over long distances is more economical than transporting gaseous hydrogen. One problem with liquid transportation is that it may boil off in the delivery process (de Moura Altea & Yanagihara, 2024; Dincer & Acar, 2015; DOE, n.d.a, n.d.b).

5.5.3 Material based storage

The adsorption or absorption of hydrogen on suitable metal surfaces and porous material is an easy method of capturing and transporting hydrogen as compare to conventional H_2 storage solutions. This technique is divided into physical and chemical absorption solutions which make use of metals and nonmetal compounds like alloys and ammonia (DOE, n.d.a, n.d.b).

The Haber–Bosch process can be used to combine hydrogen and nitrogen to create ammonia. Green ammonia, produced by using green hydrogen, has multiple uses such as fuel, mineral fertilizer, and climate-friendly energy carriers. Ammonia is easier to store and transport since it requires a lot less energy to liquefy than hydrogen. Ammonia is seen as an alternative fuel for long-haul ships because it burns with no carbon emissions. Currently, 2% of greenhouse gas emissions worldwide are attributed to shipping, with long-haul shipping making up 80% of these emissions. Thus, using green ammonia as a marine fuel can help us get closer to achieving net-zero emissions. Furthermore, frequently utilized as fertilizer is ammonia (McMahon, 2020; Statkraft, 2021).

Other substances like palladium, aluminum, or steels and several other metals and alloys are also used for adsorption-based storage of hydrogen (Hirscher et al., 2020; Jadhav, 2020).

5.5.4 Geological hydrogen storage

Hydrogen can also be stored in geological formation for a longer period of time with comparatively lower Operating and maintenance cost as compare to the existing storage solutions. Generally, depleted oil and gas reserves and hydrate beds are used for long time storage of gases like hydrogen and carbon dioxide in such situations.

Subsurface hydrogen storage reservoirs promise safe and efficient storage of large quantities of hydrogen. Interdisciplinary research spanning geochemistry, microbiology, hydrodynamics, geo-mechanics, and reservoir engineering enhances our understanding of underground storage. Geochemical studies ensure reservoir stability, while microbiological investigations focus on storage integrity. Hydro-dynamical and geo-mechanical studies evaluate hydrogen behavior and reservoir performance. Reservoir-scale simulations optimize operations. Despite progress, challenges such as potential changes in reservoir characteristics and optimizing subsurface hydrogen production persist. Continued research is crucial for enhancing storage capacity, efficiency, and safety, advancing subsurface hydrogen storage for climate change mitigation (Ma et al., 2024).

Gas hydrates offer a unique solution for hydrogen storage, composed mainly of water molecules. This technique releases only water vapor when used, aligning with eco-friendly goals. However, research is needed to improve gas hydrate formation efficiency in diverse conditions. This article reviews research on the hydrogen storage capacity of gas hydrates. While early studies focused on pure hydrogen storage, recent work explores modifying production rates and conditions with various promoters, impacting kinetics and thermodynamics. Innovative approaches aim to increase storage capacity by trapping more hydrogen molecules in a single hydrate cage. Additionally, research on materials like amino acids, surfactants, and porous materials aims to enhance water-gas interactions and accelerate formation. Developing these techniques could make gas hydrates a more viable hydrogen storage option, crucial for reducing greenhouse gas emissions and environmental impact (Borah et al., 2024; Kim & Ahn, 2024; Saikia et al., 2023). Gas hydrates have been investigated for long for gas storage applications. Storage of gases like CO_2, CH_4, and H_2 under different conditions has been studied significantly by researchers (Pandey & Sangwai, 2021, 2020; Pandey et al., 2022,

TABLE 5.2 Global geological hydrogen storage capacity (Kruck et al., 2013; Liebscher et al., 2016; Tarkowski, 2019).

Field and project name	Type of storage	H_2 (%)	Working pressure	Depth (m)	Volume (m^3)	Status
United Kingdom (Teeside)	Bedded-salt	95	45 bar	365[a]	210,000	Operating
United States (Clemens)	Salt-dome	95	70–137 bar	1000[a]	580,000	Operating
United States (Moss Bluff)	Salt-dome		55–152 bar	1200[a]	566,000	Operating
United States (Spindletop)	Salt-dome	95	68–202 bar	1340[a]	906,000	Operating
Germany (Kiel)	Salt-caverne	60	80–100 bar		3200	Losed
Germany (Ketzin)	Aquifer	62	Not reported	200–250	Not reported	Operating with NG
France (Beynes)	Aquifer	50	Not reported	430	3.3×10^8	Operating with NG
Czech Republic (Lobodice)	Aquifer	50	90 bar/34°C	430	Not reported	Operating
Argentina (Diadema)	Depleted gas reservoir	10	10 bar/50°C	600	Not reported	No report
Austria (Underground sun storage)	Depleted gas reservoir	10	78 bar/40°C	1000	Not reported	Operating

[a]*Represent the depth registered for salt caverns.*
NG, Natural gas.

2019, 2017a; Rath et al., 2023a; Poothia, 2022; Poothia, 2023). Gas hydrate systems' stability and gas uptake are crucial for storage performance. Researchers study various additives to control and enhance stability and gas volume. Studies examine synthetic and bio-based additives in different gas environments (Kibkalo et al., 2023; Pandey et al., 2018, 2017b; Rath et al., 2023b). Table 5.2 enlists the geological hydrogen storage capacity in different countries of the world, as estimated till 2019.

5.6 Challenges in hydrogen used as a fuel

The application of hydrogen as a source of clean energy shows great promises, irrespective of its source. But smooth implementation of these techniques is challenged by several significant obstacles. These obstacles encompass issues related to storage, transportation, and safety concerns, which must be addressed for safe commercial application. The methods employed for storing hydrogen differ depending on the intended use, like either gas or liquid storage, or for both gas and liquid storage simultaneously. One of the key challenges associated with using hydrogen as a fuel in compression-ignition (CI) engines is its notably high auto-ignition temperature (576°C). Nevertheless, this particular aspect presents an area that necessitates further exploration and study, given that H_2 has an impressive octane number of 130, the fuel allows attainment of a higher compression ratio (CR) giving more power as output. This concept identifies the potential to enhance the overall performance of the engine through the use of H_2, highlighting the importance of investigating hydrogen as a fuel option in greater depths (El-Adawy et al., 2024). H_2 gas storing presents a substantial challenge because of low-density gas nature, with a density of 40.8 g/m^3. This characteristic makes the storage of hydrogen a complex task that requires innovative solutions. Additionally, H_2 is considered a secondary energy source, produced from a primary source through different methodologies. However, the production of hydrogen involves various conversion processes that result in energy losses, leading to higher production costs compared to the energy input required. Despite these challenges, hydrogen holds great promise in decarbonizing the maritime industry. One potential solution is to utilize off-grid offshore wind energy for the production of blue hydrogen, which can be kept in the form of NH_3. This ammonia can then be used for national and international export/import purposes, offering a sustainable energy solution for the maritime sector. Furthermore, NH_3 can be simply split to produce H_2 on-site, showcasing the versatility and potential of hydrogen as an environmentally friendly energy source for various applications (Ishaq et al., 2022).

The current situation regarding the capital cost of the system and the cost of H_2 manufacture is uncompetitive due to extensive application of H_2 in industrialized surroundings. The use of H_2O and rare elements has constrained sustainability efforts in the development of hydrogen technologies. Additionally, the effectiveness and longevity of electrolyzer and fuel cell systems have not met expectations, resulting in increased operational and maintenance expenses. Through a thorough examination of existing literature, it becomes evident that various advancements are necessary in the foreseeable future. It is imperative to focus on research and development initiatives aimed at reducing costs while simultaneously enhancing the efficiency and durability of these systems. Moreover, policymakers need to implement strategies that facilitate the addition of H_2 into current markets and foster the growth of hydrogen-based energy systems (Yue et al., 2021).

5.7 Environmental impact of H_2 production, transportation, and storage methods

The environmental consequences of water electrolysis are contingent upon the primary energy origin responsible for supplying the electricity, a crucial determinant in assessing its sustainability. Presently, this technology does not qualify as environmentally benign within the existing electricity framework, raising concerns about its ecological footprint. However, electrolysis systems powered by renewable sources such as wind, solar, nuclear, and biomass offer a promising solution to this environmental challenge. Such an approach holds significant promise in the ongoing transition toward a more sustainable power infrastructure. Particularly, the application of thermochemical cycles relying on nuclear energy stands out for its cost-effectiveness and minimal environmental impact compared to other alternatives. Moreover, electrolysis and thermolysis processes utilizing nuclear energy exhibit a higher level of cost competitiveness when contrasted with those relying on different energy sources. Given these advantages, it is imperative to thoroughly assess the technological maturity and safety aspects of nuclear

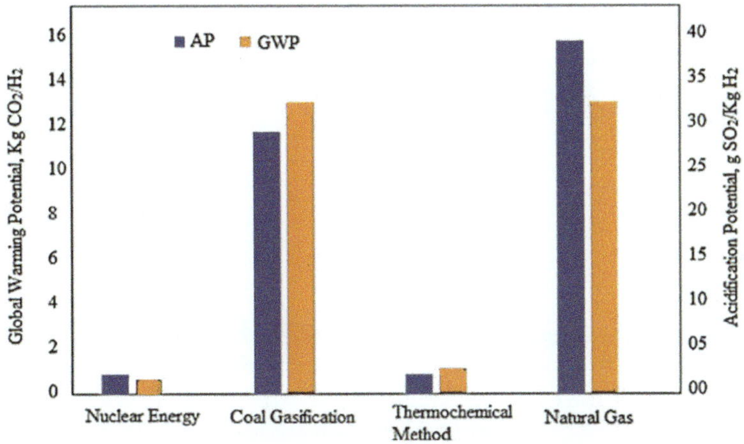

FIGURE 5.10 Schematic diagram of possible hydrogen transportation (combination of road, rail, over sea and pipeline transportation used in the industry) (Borsboom-Hanson et al., 2022).

energy, with the anticipation that it will emerge as a key catalyst for hydrogen production in the foreseeable future (Ji & Wang, 2021).

Hydrogen possesses the potential to significantly decrease CO_2 emissions when derived from sustainable, renewable, and environmentally friendly energy sources. It is imperative to conduct a comprehensive LCA study to fully grasp and evaluate the carbon dioxide emissions associated with a particular process. The evaluation of the environmental impact of different H_2 gas preparation technologies typically involves the use of indicators such as the global warming potential (GWP) and the acidification potential (AP). The measurement of CO_2 emissions is typically quantified using the GWP articulated in weight of equivalent CO_2, while AP is grams of sulfur dioxide (SO_2) required for changing the soil and water acidity. Hydrogen production from nonrenewable energy sources like coal gasification and NG stands out as the most environmentally impactful methods, characterized by significantly high AP and GWP. Conversely, the production of H_2 from nuclear energy and thermochemical processes demonstrates a lower ecological footprint when compared to other hydrogen production technologies reliant on nonrenewable energy resources. The assessment of the environmental implications of hydrogen production is crucial for making informed decisions regarding the utilization of different production methods. Fig. 5.10 illustrates the level of GWP and acidification potential of different H_2 production methods. It is observed that H_2 production from NG has the highest AP and GWP ratings as compared to other methods. Stakeholders can connect their climate change mitigation aims with more sustainable choices by having a better awareness of the environmental effects of various hydrogen generation systems. Identification of more ecologically friendly options is made possible by considering the global warming and acidification potential, which offer important insights into the environmental effects of H_2 generation processes. To create a cleaner and more sustainable energy future, assessing the environmental effects of H_2 generation requires a comprehensive approach (Amin et al., 2022).

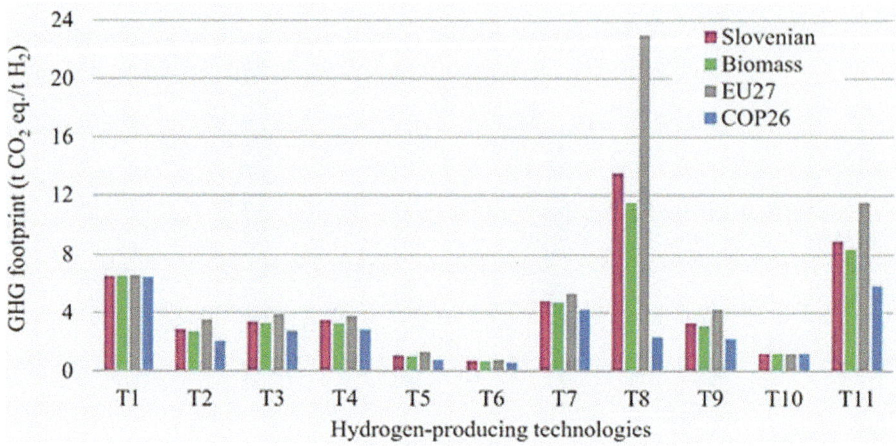

FIGURE 5.11 Various hydrogen-producing technologies and their respective energy footprints (Hren et al., 2023).

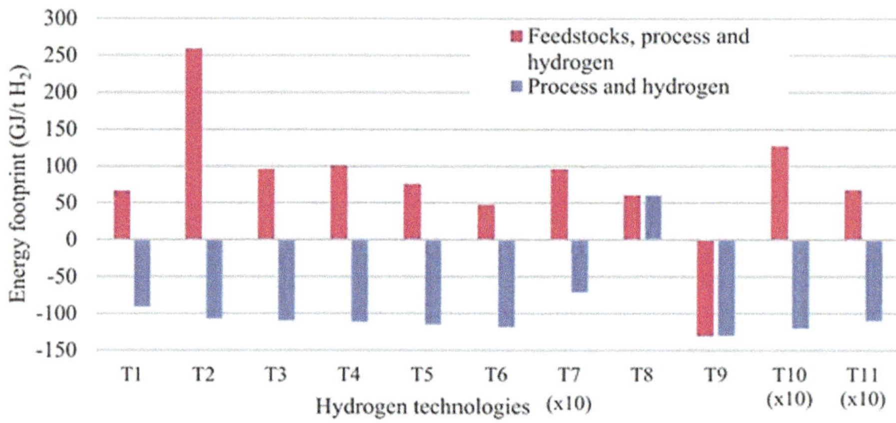

FIGURE 5.12 Connection of pipelines for H_2 and natural gas transfer (HFCTO, 2021).

Research revealed that, as reported during the COP26 conference, the technologies with the smallest greenhouse gas (GHG) trails are water electrolysis (T8), with ~2330 kg CO_2 eq/t H_2, and alcohol and glycerol waste steam reforming, with values of 586.7 and 768.1 kg CO_2 equivalent per ton of hydrogen (H_2) produced respectively. On the other hand, when using the electrical mix of the EU27 region, water electrolysis was shown to have the greatest GHG footprint, measuring 22,964.4 kg CO_2 eq/t H_2. Notably, water electrolysis sticks out as a process that uses more energy than it produces, generating 180.3 GJ/ton of H_2, compared to 120.0 GJ/ton of H_2, according to analysis of the energy footprint data. On the other hand, aluminum combustion is emphasized for its ability to produce more energy than it consumes, especially when secondary aluminum manufacturing is involved (Hren et al., 2023). Fig. 5.11 and 5.12, elucidate the GHG footprints and energy footprints of different H_2 producing technologies as studied in literature.

Photocatalysis is considered to have significant potential when viewed from an overall environmental performance perspective. According to existing literature, photonic catalysis, which is based on solar energy, is regarded as the most optimal choice in terms of environmental performance, standing shoulder to shoulder with biological, thermal, and electrical-based hydrogen technologies. Consequently, it occupies the lowest economic position due to its status as a nascent technology, with ongoing efforts focusing on mitigating the initial costs associated with its implementation. The advancement of a hydrogen economy hinges greatly on the development of hydrogen production processes, which are pivotal in generating hydrogen at costs that are comparable to those of conventional methods while also being environmentally sustainable in contrast to the utilization of fossil fuels (Younas et al., 2022). Strict regulations on heavy metal emissions due to global warming necessitate hydrogen for oil hydro-processing, including crude, vegetable, and nonedible oils. Hydrogen fuel cells, with quick refueling and lighter weight compared to batteries, offer eco-friendliness and affordability in vehicle refueling infrastructure. The demand for hydrogen grows due to its role in meeting environmental regulations and addressing climate change, pivotal in transitioning to sustainable energy solutions (Muhammed et al., 2023). In assessing environmental impacts, pumped hydro storage (PHS) demonstrates its sustainability. It boasts a significantly lower exergy cost of total resources at 84% and notably reduces CO_2 emissions to just 23% compared to the H_2 alternative. This highlights PHS's superior eco-friendly profile in both resource utilization and greenhouse gas emissions (de Moura Altea & Yanagihara, 2024).

5.8 Conclusion

Hydrogen is emerging as a versatile and promising solution to meet various energy needs across sectors while reducing greenhouse gas emissions and pollutants. H_2's environment friendly nature, makes it an option for achieving sustainable energy transition. However, significant challenges remain in its production, storage, and transportation, hindering widespread adoption. Efforts to address these challenges include advancements in electrolysis technologies, such as offshore hydrogen production using wind energy (OWH), and innovative storage methods like LOHCs. Additionally, the shift toward renewable energy sources for hydrogen production, such as biomass gasification and electrolysis powered by solar and wind, signals a growing momentum toward sustainable hydrogen solutions. Despite the dominance of fossil fuel-based hydrogen production methods historically, the increasing emphasis on renewable technologies reflects a global push toward environmentally sustainable energy solutions. However, to realize hydrogen's full potential, ongoing research, policy support, and investment are crucial to overcome technical and economic barriers. In inference H_2 holds vast potential as a green, clean and adaptable fuel, offering significant potential for decarbonizing many industries ensuring a cleaner and sustainable future. Continued efforts toward advancing hydrogen technologies and promoting its widespread adoption are essential for realizing its full benefits and addressing global energy and environmental challenges.

Acknowledgment

We would like to acknowledge this work under National project (124022900010-9) from TSC SB-RAS at Earth cryosphere institute, Tyumen, Russia.

References

Abdelkareem, M. A., Ayoub, M., Al Najada, R. I., Alami, A. H., & Olabi, A. G. (2024). Hydrogen from waste metals: Recent progress, production techniques, purification, challenges, and applications. *Sustainable Horizons, 9*. Available from https://doi.org/10.1016/j.horiz.2023.100079, https://www.sciencedirect.com/science/journal/27727378.

Aceves, S. M., Espinosa-Loza, F., Ledesma-Orozco, E., Ross, T. O., Weisberg, A. H., Brunner, T. C., & Kircher, O. (2010). High-density automotive hydrogen storage with cryogenic capable pressure vessels. *International Journal of Hydrogen Energy, 35*(3), 1219–1226. Available from https://doi.org/10.1016/j.ijhydene.2009.11.069.

Ahluwalia, R. K., Hua, T. Q., Peng, J.-K., Lasher, S., McKenney, K., Sinha, J., & Gardiner, M. (2010). Technical assessment of cryo-compressed hydrogen storage tank systems for automotive applications. *International Journal of Hydrogen Energy, 35*(9), 4171–4184. Available from https://doi.org/10.1016/j.ijhydene.2010.02.074.

Alamiery, A. (2023). Advancements in materials for hydrogen production: A review of cutting-edge technologies. *ChemPhysMater., 3*(1). Available from https://doi.org/10.1016/j.chphma.2023.09.002.

Amin, M., Shah, H. H., Fareed, A. G., Khan, W. U., Chung, E., Zia, A., Rahman Farooqi, Z. U., & Lee, C. (2022). Hydrogen production through renewable and non-renewable energy processes and their impact on climate change. *International Journal of Hydrogen Energy, 47*(77), 33112–33134. Available from https://doi.org/10.1016/j.ijhydene.2022.07.172, http://www.journals.elsevier.com/international-journal-of-hydrogen-energy/.

Assis, T. I., & Gonçalves, R. F. (2022). Valorization of food waste by anaerobic digestion: A bibliometric and systematic review focusing on optimization. *Journal of Environmental Management, 320*. Available from https://doi.org/10.1016/j.jenvman.2022.115763, https://www.sciencedirect.com/science/journal/03014797.

Balat, H., & Kırtay, E. (2010). Hydrogen from biomass – Present scenario and future prospects. *International Journal of Hydrogen Energy, 35*(14), 7416–7426. Available from https://doi.org/10.1016/j.ijhydene.2010.04.137.

Baldwin, D. (2017). FINAL REPORT – Development of high pressure hydrogen storage tank for storage and gaseous truck delivery. Golden, CO (United States). Available from https://doi.org/10.2172/1373926.

Barral, K., & Barthélémy, H. (2006). Hydrogen high pressure tanks storages: Overview and new trends due to H_2 energy specifications and constraints. 16th World Hydrogen Energy Conference 2006, WHEC 2006 1515-1524 France 2.

Barthelemy, H., Weber, M., & Barbier, F. (2017). Hydrogen storage: Recent improvements and industrial perspectives. *International Journal of Hydrogen Energy, 42*, 7254–7262. Available from https://doi.org/10.1016/j.ijhydene.2016.03.178.

Barthélémy, H. (2012). Hydrogen storage – Industrial prospectives. *International Journal of Hydrogen Energy, 37*(22), 17364–17372. Available from https://doi.org/10.1016/j.ijhydene.2012.04.121.

Bockris, J. (2002). The origin of ideas on a hydrogen economy and its solution to the decay of the environment. *International Journal of Hydrogen Energy, 27*(7-8), 731–740. Available from https://doi.org/10.1016/s0360-3199(01)00154-9.

Bolar, S., Shit, S., Murmu, N. C., & Kuila, T. (2021). Progress in theoretical and experimental investigation on seawater electrolysis: Opportunities and challenges. *Sustainable Energy & Fuels, 5*(23), 5915–5945. Available from https://doi.org/10.1039/d1se01347e.

Bolat, P., & Thiel, C. (2014). Hydrogen supply chain architecture for bottom-up energy systems models. Part 1: Developing pathways. *International Journal of Hydrogen Energy, 39*(17), 8881–8897. Available from https://doi.org/10.1016/j.ijhydene.2014.03.176.

Borah, U. B., Pandey, G., Sarma, S., Molokitina, N., & Chauhan, G. (2024). Affordable and clean energy. *Clean and renewable energy production* (pp. 87–121). Wiley. Available from https://doi.org/10.1002/9781394174805.ch4.

Borsboom-Hanson, T., Patlolla, S. R., Herrera, O. E., & Mérida, W. (2022). Point-to-point transportation: The economics of hydrogen export. *International Journal of Hydrogen Energy, 47*(74), 31541–31550. Available from https://doi.org/10.1016/j.ijhydene.2022.07.093, http://www.journals.elsevier.com/international-journal-of-hydrogen-energy/.

Bruce, S., Temminghoff, M., Hayward, J., Schmidt, E., Munnings, C., & Palfreyman, D. H. P. (2018). Australia National Hydrogen Roadmap. - 92.

Bundhoo, Z. M. A. (2019). Potential of bio-hydrogen production from dark fermentation of crop residues: A review. *International Journal of Hydrogen Energy, 44*(32), 17346–17362. Available from https://doi.org/10.1016/j.ijhydene.2018.11.098, http://www.journals.elsevier.com/international-journal-of-hydrogen-energy/.

Casper, M. S. (1978). Hydrogen manufacture by electrolysis, thermal decomposition and unusual techniques.

Cheng, Q., Zhang, R., Shi, Z., & Lin, J. (2024). Review of common hydrogen storage tanks and current manufacturing methods for aluminium alloy tank liners. *International Journal of Lightweight Materials and Manufacture, 7*(2), 269–284. Available from https://doi.org/10.1016/j.ijlmm.2023.08.002.

Chu, C., Wu, K., Luo, B., Cao, Q., & Zhang, H. (2023). Hydrogen storage by liquid organic hydrogen carriers: Catalyst, renewable carrier, and technology – A review. *Carbon Resources Conversion, 6*(4), 334–351. Available from https://doi.org/10.1016/j.crcon.2023.03.007.

Collins, L. (2022). Why shipping pure hydrogen around the world might already be dead in the water.

Cremonez, P. A., Teleken, J. G., Weiser Meier, T. R., & Alves, H. J. (2021). Two-Stage anaerobic digestion in agroindustrial waste treatment: A review. *Journal of Environmental Management, 281*. Available from https://doi.org/10.1016/j.jenvman.2020.111854, https://www.sciencedirect.com/journal/journal-of-environmental-management.

Damen, K., Troost, M. van, Faaij, A., & Turkenburg, W. (2007). A comparison of electricity and hydrogen production systems with CO_2 capture and storage—Part B: Chain analysis of promising CCS options. *Progress in Energy and Combustion Science, 33*(6), 580–609. Available from https://doi.org/10.1016/j.pecs.2007.02.002.

Dawood, F., Anda, M., & Shafiullah, G. M. (2020). Hydrogen production for energy: An overview. *International Journal of Hydrogen Energy, 45*, 3847–3869. Available from https://doi.org/10.1016/j.ijhydene.2019.12.059.

de Moura Altea, C., & Yanagihara, J. I. (2024). Energy, exergy and environmental impacts analyses of pumped hydro storage (PHS) and hydrogen (H_2) energy storage processes. *Journal of Energy Storage, 76*. Available from https://doi.org/10.1016/j.est.2023.109713.

Dincer, I., & Acar, C. (2015). Review and evaluation of hydrogen production methods for better sustainability. *International Journal of Hydrogen Energy, 40*(34), 11094–11111. Available from https://doi.org/10.1016/j.ijhydene.2014.12.035.

DOE. (n.d.a). Liquid hydrogen delivery [WWW document]. Energy Efficiency Renewable Energy. https://www.energy.gov/eere/fuelcells/liquid-hydrogen-delivery.

DOE. (n.d.b). Hydrogen storage [WWW document]. Energy Efficiency Renewable Energy. https://www.energy.gov/eere/fuelcells/hydrogen-storage.

Dryer, F. L., Chaos, M., Zhao, Z., Stein, J. N., Alpert, J. Y., & Homer, C. J. (2007). Spontaneous ignition of pressurized releases of hydrogen and natural gas into air. *Combustion Science and Technology, 179*(4), 663–694. Available from https://doi.org/10.1080/00102200600713583.

El-Adawy, M., Nemitallah, M. A., & Abdelhafez, A. (2024). Towards sustainable hydrogen and ammonia internal combustion engines: Challenges and opportunities. *Fuel, 364*, 131090. Available from https://doi.org/10.1016/j.fuel.2024.131090.

Elgowainy, A., Reddi, K., Sutherland, E., & Joseck, F. (2014). Tube-trailer consolidation strategy for reducing hydrogen refueling station costs. *International Journal of Hydrogen Energy, 39*(35), 20197–20206. Available from https://doi.org/10.1016/j.ijhydene.2014.10.030.

Farhana, K., Shadate Faisal Mahamude, A., & Kadirgama, K. (2024). Comparing hydrogen fuel cost of production from various sources – A competitive analysis. *Energy Conversion and Management, 302*, 118088. Available from https://doi.org/10.1016/j.enconman.2024.118088.

Faye, O., Szpunar, J., & Eduok, U. (2022). A critical review on the current technologies for the generation, storage, and transportation of hydrogen. *International Journal of Hydrogen Energy, 47*, 13771–13802. Available from https://doi.org/10.1016/j.ijhydene.2022.02.112.

Federal size regulations for commercial motor vehicles. (2004). *US Department of Transportation, Federal Highway Adminsitration.*

Fuel Cell Electric Buses Knowledge Base. (2023). Liquefied hydrogen. https://fuelcellbuses.eu/wiki/concept-hydrogen-refueling-station/tank-trailer-liquid-hydrogen.

Gao, Y., Xue, Y., He, F., & Li, Y. (2022). Controlled growth of a high selectivity interface for seawater electrolysis. *Proceedings of the National Academy of Sciences, 119*(36). Available from https://doi.org/10.1073/pnas.2206946119.

Gondal, I. A. (2016). Hydrogen transportation by pipelines. *Compendium of hydrogen energy* (pp. 301–322). Elsevier. Available from https://doi.org/10.1016/B978-1-78242-362-1.00012-2.

Guo, X., Zhu, H., & Zhang, S. (2024). Overview of electrolyser and hydrogen production power supply from industrial perspective. *International Journal of Hydrogen Energy, 49*, 1048–1059. Available from https://doi.org/10.1016/j.ijhydene.2023.10.325.

Halder, P., Babaie, M., Salek, F., Haque, N., Savage, R., Stevanovic, S., Bodisco, T. A., & Zare, A. (2024). Advancements in hydrogen production, storage, distribution and refuelling for a sustainable transport sector: Hydrogen fuel cell vehicles. *International Journal of Hydrogen Energy, 52*, 973–1004. Available from https://doi.org/10.1016/j.ijhydene.2023.07.204, https://www.sciencedirect.com/science/journal/03603199.

Harding, R. (2019). Japan launches first liquid hydrogen carrier ship. Financial Times.

Hassan, Q., Algburi, S., Jaszczur, M., Al-Jiboory, A. K., Al Musawi, T. J., Ali, B. M., Viktor, P., Fodor, M., Ahsan, M., Salman, H. M., & Sameen, A. Z. (2024). Hydrogen role in energy transition: A comparative review. *Process Safety and Environmental Protection., 184*, 1069–1093. Available from https://doi.org/10.1016/j.psep.2024.02.030, https://www.sciencedirect.com/science/journal/09575820.

HFCTO. (2021). Hydrogen pipelines. DOE. https://www.energy.gov/eere/fuelcells/hydrogen-pipelines.

HFCTO. (2022). Hydrogen tube trailer. DOE.

Hirscher, M., Yartys, V. A., Baricco, M., Bellosta von Colbe, J., Blanchard, D., Bowman, R. C., Broom, D. P., Buckley, C. E., Chang, F., Chen, P., Cho, Y. W., Crivello, J. C., Cuevas, F., David, W. I. F., de Jongh, P. E., Denys, R. V., Dornheim, M., Felderhoff, M., Filinchuk, Y., ... Zlotea, C. (2020). Materials for hydrogen-based energy storage – Past, recent progress and future outlook. *Journal of Alloys and Compounds, 827*. Available from https://doi.org/10.1016/j.jallcom.2019.153548, https://www.journals.elsevier.com/journal-of-alloys-and-compounds.

Hoon, Y. T. (2021). Hanwha solutions developing type 5 hydrogen tank.

Hren, R., Vujanović, A., Van Fan, Y., Klemeš, J. J., Krajnc, D., & Čuček, L. (2023). Hydrogen production, storage and transport for renewable energy and chemicals: An environmental footprint assessment. *Renewable and Sustainable Energy Reviews, 173*. Available from https://doi.org/10.1016/j.rser.2022.113113.

Hurskainen, M. (2019). Liquid organic hydrogen carriers (LOHC) – Concept evaluation and techno-economics.

IEA. (2019). Hydrogen as a fuel for fuel cell electric vehicles. Advanced Fuel Cells. https://www.ieafuelcell.com/index.php?id = 33.

IEA. (2023). Global hydrogen review 2023. *Global Hydrogen Review, 2023*. Available from https://doi.org/10.1787/cb2635f6-en.

Ipsakis, D., Voutetakis, S., Seferlis, P., Stergiopoulos, F., & Elmasides, C. (2009). Power management strategies for a stand-alone power system using renewable energy sources and hydrogen storage. *International Journal of Hydrogen Energy, 34*(16), 7081–7095. Available from https://doi.org/10.1016/j.ijhydene.2008.06.051.

IRENA. (2022). Geopolitics of the energy transformation: The hydrogen factor. Abu Dhabi.

Ishaq, H., Dincer, I., & Crawford, C. (2022). A review on hydrogen production and utilization: Challenges and opportunities. *International Journal of Hydrogen Energy, 47*, 26238–26264. Available from https://doi.org/10.1016/j.ijhydene.2021.11.149.

Jadhav, R. (2020). Effect of nanoparticles on the viscosity alteration of vacuum residue. *Proceedings of the Fifth International Conference in Ocean Engineering (ICOE2019)*. Available from https://doi.org/10.1007/978-981-15-8506-7_36.

Ji, M., & Wang, J. (2021). Review and comparison of various hydrogen production methods based on costs and life cycle impact assessment indicators. *International Journal of Hydrogen Energy, 46*(78), 38612–38635. Available from https://doi.org/10.1016/j.ijhydene.2021.09.142.

Kayfeci, M., Keçebaş, A., & Bayat, M. (2019). Hydrogen production. *Solar hydrogen production* (pp. 45–83). Elsevier. Available from https://doi.org/10.1016/B978-0-12-814853-2.00003-5.

Kibkalo, A. A., Pandey, G., Pletneva, K. A., Molokitina, N. S., Kumar, A., Zhingel, P., & Grigoriev, B. V. (2023). Enhanced methane hydrate formation kinetics in frozen particles of biopolymer solutions: Applicable to methane storage. *Energy and Fuels, 37*(18), 13928–13936. Available from https://doi.org/10.1021/acs.energyfuels.3c02344, http://pubs.acs.org/journal/enfuem.

Kim, M. K., & Ahn, Y. H. (2024). Gas hydrates for hydrogen storage: A comprehensive review and future prospects. *Korean Journal of Chemical Engineering, 41*(1), 73–94. Available from https://doi.org/10.1007/s11814-024-00025-4, https://www.springer.com/journal/11814.

Klebanoff, L. E., Pratt, J. W., & LaFleur, C. B. (2017). Comparison of the safety-related physical and combustion properties of liquid hydrogen and liquid natural gas in the context of the SF-BREEZE high-speed fuel-cell ferry. *International Journal of Hydrogen Energy, 42,* 757–774. Available from https://doi.org/10.1016/j.ijhydene.2016.11.024.

Kothari, R., Buddhi, D., & Sawhney, R. L. (2008). Comparison of environmental and economic aspects of various hydrogen production methods. *Renewable and Sustainable Energy Reviews, 12,* 553–563. Available from https://doi.org/10.1016/j.rser.2006.07.012.

Kruck, O., Crotogino, F., Prelicz, R., Rudolph, T. (2013). D3.1. Overview on all known underground storage technologies for hydrogen (H2020 Project HyUnder – Grant Agreement).

Kumar, A., Muthukumar, P., Sharma, P., & Kumar, E. A. (2022). Absorption based solid state hydrogen storage system: A review. *Sustainable Energy Technologies and Assessments, 52.* Available from https://doi.org/10.1016/j.seta.2022.102204.

Kumar, A. (2024). Fast switching hydrogen gas leakage identification using FPGA. *International Journal of Hydrogen Energy, 69,* 1157–1165. Available from https://doi.org/10.1016/j.ijhydene.2024.05.061.

Kumar, D., Abraham, J. E., Varghese, M., George, J., Balachandran, M., & Cherusseri, J. (2024). Nanocarbon assisted green hydrogen production: Development and recent trends. *International Journal of Hydrogen Energy, 50,* 118–141. Available from https://doi.org/10.1016/j.ijhydene.2023.07.257, https://www.sciencedirect.com/science/journal/03603199.

Liebscher, A., Wackerl, J., & Streibel, M. (2016). Geologic storage of hydrogen – Fundamentals, processing, and projects. *Hydrogen science and engineering: Materials, processes, systems and technology* (pp. 629–658). Wiley. Available from https://doi.org/10.1002/9783527674268.ch26.

Lin, A., & Bagnato, G. (2024). Revolutionising energy storage: The latest breakthrough in liquid organic hydrogen carriers. *International Journal of Hydrogen Energy, 63,* 315–329. Available from https://doi.org/10.1016/j.ijhydene.2024.03.146.

M, A., & G, P. K. (2023). Hydrogen towards sustainable transition: A review of production, economic, environmental impact and scaling factors. *Results in Engineering, 20.* Available from https://doi.org/10.1016/j.rineng.2023.101456, https://www.journals.elsevier.com/results-in-engineering.

Ma, N., Zhao, W., Wang, W., Li, X., & Zhou, H. (2024). Large scale of green hydrogen storage: Opportunities and challenges. *International Journal of Hydrogen Energy, 50,* 379–396. Available from https://doi.org/10.1016/j.ijhydene.2023.09.021.

Makaryan, I. A., Sedov, I. V., Salgansky, E. A., Arutyunov, A. V., & Arutyunov, V. S. (2022). A Comprehensive review on the prospects of using hydrogen–methane blends: Challenges and opportunities. *Federation Energies, 15*(6). Available from https://doi.org/10.3390/en15062265, https://www.mdpi.com/1996-1073/15/6/2265/pdf.

Mao, W. L., Mao, H. K., Goncharov, A. F., Struzhkin, V. V., Guo, Q., Hu, J., Hu, J., Hemley, R. J., Somayazulu, M., & Zhao, Y. (2002). Hydrogen clusters in clathrate hydrate. *Science (New York, N.Y.), 297*(5590), 2247–2249. Available from https://doi.org/10.1126/science.1075394.

Massarweh, O., Al-khuzaei, M., Al-Shafi, M., Bicer, Y., & Abushaikha, A. S. (2023). Blue hydrogen production from natural gas reservoirs: A review of application and feasibility. *Journal of CO_2 Utilization., 70.* Available from https://doi.org/10.1016/j.jcou.2023.102438.

McDonagh, S., Ahmed, S., Desmond, C., & Murphy, J. D. (2020). Hydrogen from offshore wind: Investor perspective on the profitability of a hybrid system including for curtailment. *Applied Energy., 265.* Available from https://doi.org/10.1016/j.apenergy.2020.114732.

McMahon, M. (2020). New technology seamlessly converts ammonia to green hydrogen.

Mittal, A., & Mahajan, R. (2023). Generation of green hydrogen using semiconductor-based nanomaterials. *Springer Nature.* Available from https://doi.org/10.1007/978-981-99-2870-5_18.

Mohammed-Ibrahim, J., & Moussab, H. (2020). Recent advances on hydrogen production through seawater electrolysis. *Materials Science for Energy Technologies, 3,* 780–807. Available from https://doi.org/10.1016/j.mset.2020.09.005.

MPA. (2023). Singapore hosted the world's first bulk liquefied hydrogen carrier, suiso frontier, to the port of Singapore. MPA Singapore, A Singapore Gov. Agency Website. https://www.mpa.gov.sg/media-centre/details/singapore-hosted-the-world-s-first-bulk-liquefied-hydrogen-carrier-suiso-frontier-to-the-port-of-singapore.

Muhammed, N. S., Gbadamosi, A. O., Epelle, E. I., Abdulrasheed, A. A., Haq, B., Patil, S., Al-Shehri, D., & Kamal, M. S. (2023). Hydrogen production, transportation, utilization, and storage: Recent advances towards sustainable energy. *Journal of Energy Storage, 73.* Available from https://doi.org/10.1016/j.est.2023.109207, http://www.journals.elsevier.com/journal-of-energy-storage/.

National Energy Technology Laboratory. (2018). Coal gasification for hydrogen production [WWW Document]. https://netl.doe.gov/carbon-management/gasification.

Niblett, D., Delpisheh, M., Ramakrishnan, S., & Mamlouk, M. (2024). Review of next generation hydrogen production from offshore wind using water electrolysis. *Journal of Power Sources, 592*. Available from https://doi.org/10.1016/j.jpowsour.2023.233904.

Niermann, M., Drünert, S., Kaltschmitt, M., & Bonhoff, K. (2019). Liquid organic hydrogen carriers (LOHCs) – Techno-economic analysis of LOHCs in a defined process chain. *Energy & Environmental Science, 12*(1), 290–307. Available from https://doi.org/10.1039/c8ee02700e.

Nikolaidis, P., & Poullikkas, A. (2017). A comparative overview of hydrogen production processes. *Renewable and Sustainable Energy Reviews, 67*, 597–611. Available from https://doi.org/10.1016/j.rser.2016.09.044.

Noh, H., Kang, K., & Seo, Y. (2023). Environmental and energy efficiency assessments of offshore hydrogen supply chains utilizing compressed gaseous hydrogen, liquefied hydrogen, liquid organic hydrogen carriers and ammonia. *International Journal of Hydrogen Energy, 48*, 7515–7532. Available from https://doi.org/10.1016/j.ijhydene.2022.11.085.

Out-Law News. (2022). World's first liquefied hydrogen ship completes voyage in Japan, Pinsent Masons.

Pandey, G., Bhattacharjee, G., Veluswamy, H. P., Kumar, R., Sangwai, J. S., & Linga, P. (2018). Alleviation of foam formation in a surfactant driven gas hydrate system: Insights via a detailed morphological study. *ACS Applied Energy Materials, 1*(12), 6899–6911. Available from https://doi.org/10.1021/acsaem.8b01307, pubs.acs.org/journal/aaemcq.

Pandey, G., Kumar, A., Veluswamy, H. P., Sangwai, J., & Linga, P. (2017a). Morphological studies of mixed methane tetrahydrofuran hydrates in saline water for energy storage application. *Energy Procedia, 143*, 786–791. Available from https://doi.org/10.1016/j.egypro.2017.12.763.

Pandey, G., Linga, P., & Sangwai, J. S. (2017b). High pressure rheology of gas hydrate formed from multiphase systems using modified Couette rheometer. *The Review of Scientific Instruments, 88*. Available from https://doi.org/10.1063/1.4974750.

Pandey, G., Poothia, T., & Kumar, A. (2022). Hydrate based carbon capture and sequestration (HBCCS): An innovative approach towards decarbonization. *Applied Energy, 326*, 119900. Available from https://doi.org/10.1016/j.apenergy.2022.119900.

Pandey, G., & Sangwai, J. S. (2020). High pressure rheological studies of methane hydrate slurries formed from water-hexane, water-heptane, and water-decane multiphase systems. *Journal of Natural Gas Science and Engineering, 81*, 103365. Available from https://doi.org/10.1016/j.jngse.2020.103365.

Pandey, G., & Sangwai, J. S. (2021). High pressure rheology of gas hydrate in multiphase flow systems. *Lecture Notes in Civil Engineering, 106*, 321–327. Available from https://doi.org/10.1007/978-981-15-8506-7_26, http://www.springer.com/series/15087.

Pandey, G., Veluswamy, H. P., Sangwai, J., & Linga, P. (2019). Morphology study of mixed methane–tetrahydrofuran hydrates with and without the presence of salt. *Energy & Fuels, 33*, 4865–4876. Available from https://doi.org/10.1021/acs.energyfuels.9b00490.

Pascal. (2022). Different methods of storing, transporting, and distributing Hydrogen.

Pashchenko, D. (2024). Green hydrogen as a power plant fuel: What is energy efficiency from production to utilization? *Renewable Energy, 223*. Available from https://doi.org/10.1016/j.renene.2024.120033.

Patonia, A., Lenivova, V., Poudineh, R., & Nolden, C. (2023). Hydrogen pipelines vs. HVDC lines: Should we transfer green molecules or electrons? 35.

Pellegrini, M., Guzzini, A., & Saccani, C. (2020). A preliminary assessment of the potential of low percentage green hydrogen blending in the Italian natural gas network. *Energies, 13*, 5570. Available from https://doi.org/10.3390/en13215570.

Poothia, T (2022). Role of CO_2 capture and sequestration (CCS) using gas hydrate based technology for sustainable energy. *Energy Proceedings, 20*. Available from https://www.energy-proceedings.org/wp-content/uploads/icae2021/1642960161.pdf.

Poothia, T. (2023). A comprehensive review on CO_2 capture and storage (CCS) technology in global scenario. *AIP Conference Proceedings*. Available from https://doi.org/10.1063/5.0113040.

Ramakrishnan, S., Delpisheh, M., Convery, C., Niblett, D., Vinothkannan, M., & Mamlouk, M. (2024). Offshore green hydrogen production from wind energy: Critical review and perspective. *Renewable and Sustainable Energy Reviews, 195*. Available from https://doi.org/10.1016/j.rser.2024.114320.

Rath, G. K., Pandey, G., Singh, S., Molokitina, N., Kumar, A., Joshi, S., & Chauhan, G. (2023a). Carbon dioxide separation technologies: Applicable to net zero. *Energies, 16*, 4100. Available from https://doi.org/10.3390/en16104100.

Rath, G.K., Singh, S., Poothia, T., Pandey, G., Molokitina, N. (2023b). Role of additives in phase equilibrium and kinetic studies of gas hydrates. p. 040008. https://doi.org/10.1063/5.0168135.

Riera, J. A., Lima, R. M., & Knio, O. M. (2023). A review of hydrogen production and supply chain modeling and optimization. *International Journal of Hydrogen Energy, 48*, 13731–13755. Available from https://doi.org/10.1016/j.ijhydene.2022.12.242.

Saikia, T., Patil, S., & Sultan, A. (2023). Hydrogen hydrate promoters for gas storage—A review. *Energies, 16*, 2667. Available from https://doi.org/10.3390/en16062667.

Sdanghi, G., Maranzana, G., Celzard, A., & Fierro, V. (2019). Review of the current technologies and performances of hydrogen compression for stationary and automotive applications. *Renewable and Sustainable Energy Reviews, 102*, 150–170. Available from https://doi.org/10.1016/j.rser.2018.11.028.

Shinnar, R. (2003). The hydrogen economy, fuel cells, and electric cars. *Technology in Society, 25*(4), 455–476. Available from https://doi.org/10.1016/j.techsoc.2003.09.024.

Sikiru, S., Oladosu, T. L., Amosa, T. I., Olutoki, J. O., Ansari, M. N. M., Abioye, K. J., Rehman, Z. U., & Soleimani, H. (2024). Hydrogen-powered horizons: Transformative technologies in clean energy generation, distribution, and storage for sustainable innovation. *International Journal of Hydrogen Energy, 56*, 1152–1182. Available from https://doi.org/10.1016/j.ijhydene.2023.12.186, https://www.sciencedirect.com/science/journal/03603199.

Singh, S., Pandey, G., Rath, G. K., Veluswamy, H. P., & Molokitina, N. (2023). Life cycle assessment of biomass-based hydrogen production technologies: A review. *International Journal of Green Energy*. Available from https://doi.org/10.1080/15435075.2023.2245453, http://www.tandf.co.uk/journals/titles/15435075.asp.

Statkraft (2021). Green ammonia: Clime friendly fuel for long distances and heavy tasks.

Su, Y., Lv, H., Zhou, W., & Zhang, C. (2021). Review of the hydrogen permeability of the liner material of type IV on-board hydrogen storage tank. *World Electric Vehicle Journal, 12*(3). Available from https://doi.org/10.3390/wevj12030130.

Sun, F., Qin, J., Wang, Z., Yu, M., Wu, X., Sun, X., & Qiu, J. (2021). Energy-saving hydrogen production by chlorine-free hybrid seawater splitting coupling hydrazine degradation. *Nature Communications, 12*(1). Available from https://doi.org/10.1038/s41467-021-24529-3.

Tarkowski, R. (2019). Underground hydrogen storage: Characteristics and prospects. *Renewable and Sustainable Energy Reviews, 105*, 86–94. Available from https://doi.org/10.1016/j.rser.2019.01.051.

Tian, Z., Lu, Y., Wang, J., Shu, R., Wang, C., & Chen, Y. (2024). Advances in hydrogen production by aqueous phase reforming of biomass oxygenated derivatives. *Fuel, 357*. Available from https://doi.org/10.1016/j.fuel.2023.129691.

Turner, J., Sverdrup, G., Mann, M. K., Maness, P. C., Kroposki, B., Ghirardi, M., Evans, R. J., & Blake, D. (2008). Renewable hydrogen production. *International Journal of Energy Research, 32*(5), 379–407. Available from https://doi.org/10.1002/er.1372.

U.S. Department of Energy. (2006). Hydrogen, fuel cells and infrastructure technologies program, multi-year research, development and demonstration plan. U.S. Department of Energy, Office of Energy Efficiency & Renewable Energy Washington, DC.

U.S. Department of Energy. (n.d.). Biomass feedstocks. DOE. https://www.energy.gov/eere/bioenergy/bioenergy-technologies-office.

Verykios, X. (2003). Catalytic dry reforming of natural gas for the production of chemicals and hydrogen. *International Journal of Hydrogen Energy, 28*(10). Available from https://doi.org/10.1016/s0360-3199(02)00215-x.

Wu, Y., Liu, F., Wu, J., He, J., Xu, M., & Zhou, J. (2022). Barrier identification and analysis framework to the development of offshore wind-to-hydrogen projects. *Energy, 239*. Available from https://doi.org/10.1016/j.energy.2021.122077.

Wu, Y., Xu, C., Zhang, B., Tao, Y., Li, X., Chu, H., & Liu, F. (2019). Sustainability performance assessment of wind power coupling hydrogen storage projects using a hybrid evaluation technique based on interval type-2 fuzzy set. *Energy, 179*, 1176–1190. Available from https://doi.org/10.1016/j.energy.2019.04.195.

Younas, M., Shafique, S., Hafeez, A., Javed, F., & Rehman, F. (2022). An overview of hydrogen production: Current status, potential, and challenges. *Fuel, 316*. Available from https://doi.org/10.1016/j.fuel.2022.123317.

Yue, M., Lambert, H., Pahon, E., Roche, R., Jemei, S., & Hissel, D. (2021). Hydrogen energy systems: A critical review of technologies, applications, trends and challenges. *Renewable and Sustainable Energy Reviews, 146*. Available from https://doi.org/10.1016/j.rser.2021.111180.

Zhang, L., Jia, C., Bai, F., Wang, W., An, S., Zhao, K., Li, Z., Li, J., & Sun, H. (2024). A comprehensive review of the promising clean energy carrier: Hydrogen production, transportation, storage, and utilization (HPTSU) technologies. *Fuel, 355*. Available from https://doi.org/10.1016/j.fuel.2023.129455.

Zhang, T., Uratani, J., Huang, Y., Xu, L., Griffiths, S., & Ding, Y. (2023). Hydrogen liquefaction and storage: Recent progress and perspectives. *Renewable and Sustainable Energy Reviews, 176*. Available from https://doi.org/10.1016/j.rser.2023.113204.

Zhang, Y., Jia, Z., Yuan, Z., Yang, T., Qi, Y., & Zhao, D. (2015). Development and application of hydrogen storage. *Journal of Iron and Steel Research International, 22*(9), 757–770. Available from https://doi.org/10.1016/s1006-706x(15)30069-8.

CHAPTER

6

Subsurface underground hydrogen storage

Jagruti Thakur[1] *and Ahmed Elberry*[2,3]

[1]Department of Energy Technology, KTH Royal Institute of Technology, Stockholm, Sweden [2]Faculty of Science (HIMS), University of Amsterdam, Amsterdam, The Netherlands [3]TNO Energy and Materials Transition, Amsterdam, The Netherlands

6.1 Introduction

In the drive toward a sustainable, low-carbon future, hydrogen emerges as a potent clean energy carrier, poised to play a fundamental role in global energy systems A pivotal aspect of this role centers on its storage, particularly due to the inherent intermittency of renewable power sources like wind and solar energy (Seo et al., 2020). With production peaks and troughs varying throughout the day and across seasons, a mechanism to store excess power for later use becomes indispensable (Elberry, Thakur, & Veysey, 2021). In this context, hydrogen, effectively stored and managed, emerges as a key element of a reliable and resilient renewable energy system (Berna-Escriche et al., 2022; Preuster et al., 2017).

The challenges associated with hydrogen storage are multifaceted and rooted in the physical and chemical properties of the element (Armaroli & Balzani, 2011). As the smallest and lightest of gases, hydrogen poses unique containment difficulties (Zohuri, 2018). Its low volumetric energy density means that even when stored at high pressures or in liquefied form, vast volumes or specialized infrastructures are necessary to store appreciable amounts (Krishna et al., 2012; Møller et al., 2017; Ni, 2006). For prolonged storage durations, issues such as hydrogen embrittlement, permeation, and potential degradation of storage materials become prominent (Laadel et al., 2022; Meda et al., 2023). These factors are particularly challenging in surface-based hydrogen storage systems, where the exposure to varying environmental conditions exacerbates these issues. Together, they create a demanding environment for hydrogen storage, necessitating innovation and research to overcome these barriers and unlock hydrogen's potential as a versatile energy carrier. As part of this quest, subsurface hydrogen storage stands out as a highly

Subsurface Hydrogen Energy Storage
DOI: https://doi.org/10.1016/B978-0-443-24071-3.00007-8

promising solution. This storage approach draws its foundation from conventional natural gas storage practices, repurposed and tailored for hydrogen (Liu et al., 2021).

The attributes of underground storage are multifold, the high storage capacity and ability to store hydrogen over longer periods stand as one of the most appealing (Elberry, Thakur, Santasalo-Aarnio, & Larmi, 2021). This type of storage can contain substantially larger volumes of hydrogen compared to surface-level storage facilities such as tanks (Elberry, Thakur, Santasalo-Aarnio, & Larmi, 2021). The safety factor associated with underground storage is also significant, as it addresses some of the key challenges posed by hydrogen's properties (e.g., high flammability) (Najjar, 2013); by placing it beneath the surface, the risks associated with accidental ignition are substantially reduced (Epelle et al., 2022). Furthermore, the relative stability of subterranean conditions, in terms of temperature and pressure, provides stable environments in terms of temperature and pressure, contributes to more consistent storage conditions (Thiyagarajan et al., 2022; Zivar et al., 2021). This constancy is something surface storage struggles to maintain. Another compelling advantage of UHS is the potential re-utilization of existing infrastructure, such as depleted oil or gas reservoirs. This not only offers practical solutions but also makes the approach economically attractive, as it can substantially minimize both costs and environmental footprint tied to the creation of new facilities (Okoroafor et al., 2022). However, it is crucial to recognize that due to the different physiochemical properties of hydrogen compared to oil or natural gas, some modifications are necessary for both surface and subsurface facilities and equipment.

The field of subsurface storage is diverse and complex, encompassing a range of geological formations such as caverns, aquifers, and depleted hydrocarbon reservoirs. The intricacy of this topic expands further when delving into the types and properties of porous geological formations for determining their suitability for hydrogen storage. Factors such as rock, mineral, petrophysical, physiochemical, and mechanical properties, all play a significant role in determining the potential success of hydrogen storage (Muhammed, Haq, Al Shehri, Al-Ahmed, et al., 2023; Raza et al., 2022). Alongside this, operational factors, such as hydrogen purity, injection and withdrawal cycles, and the choice and role of cushion gas are crucial due to their impact on storage efficiency and safety (Muhammed, Haq, Al Shehri, 2023). Furthermore, potential mechanical challenges, such as hysteresis affecting wells, tubing, and pipes need careful evaluation. This is essential to reduce risks and enhance the overall productivity of the storage system (Otto et al., 2022; Veziroglu et al., 2005).

Underground hydrogen storage (UHS) is being more recognized as a critical tool for helping nations make the shift to a low-carbon economy. Compared to other hydrogen storage techniques, UHS has a number of benefits, such as a large storage capacity, extended storage life, and safety (Jafari Raad et al., 2022). Governments are therefore putting several measures into place to aid in the creation and application of UHS technologies. Through its hydrogen and fuel cell technologies office, the US Department of Energy (DOE) funds research and development of UHS technologies. Additionally, the DOE is creating guidelines and criteria for UHS institutions. Pipeline transportation of hydrogen is now governed by the US Department of Transportation (DOT). Regulations for the secure subterranean pipeline transportation of hydrogen are being developed by the DOT. By 2030, the European Commission (EC) wants to produce 40 GW of hydrogen using renewable energy sources. Through its Horizon Europe program, the EC is also providing funds for UHS technology research and development. In 2019, the European Parliament passed a regulation requiring Europe to establish a hydrogen infrastructure. The rule contains guidelines for subterranean hydrogen storage. By 2030, the Japanese government wants to produce 30 GW of hydrogen using renewable energy sources. Through its New Energy and Industrial Technology Development Organization, the government also funds UHS technology research and development. Japan has a long history of storing hydrogen underground in salt caverns, and the government is thinking of adding hydrogen storage to this purpose. Natural gas resources are plentiful in Australia, and they could be exploited to create blue hydrogen. Under its national hydrogen strategy, the Australian government is supporting UHS technology research and development. Australia is also investigating the possibility of storing hydrogen in depleted gas and oil reservoirs.

The advancement of UHS is being aided not just by national policy but also by several international efforts. For instance, a roadmap outlining the essential steps required to introduce this technology to the market has been developed by the International Energy Agency (IEA) for UHS. In order to exchange best practices and encourage cooperation on UHS projects, the IEA is actively collaborating with other international organizations.

Considering the above, it is evident that the deployment of UHS strategies necessitates an intricate balance between capitalizing on benefits and addressing potential complexities and risks. However, the overarching potential of UHS as a tool for large-scale, long-term energy storage remains compelling. This chapter provides an integrated overview of these multifaceted aspects, offering a deep and critical examination of the science, engineering, and operational logistics involved in subsurface hydrogen storage. The chapter is divided into four main sections which explore the types of UHS and their respective advantages and disadvantages, focusing on porous geological formations like caverns, aquifers, and depleted hydrocarbon reservoirs. It further delves into the properties of these porous geological formations that render them more efficient and economical than other options, and the role of cushion gas, injection and withdrawal cycles, challenges related to hydrogen purity. The chapter also analyses the types and requirements of caprock, storage efficiency, and potential alternative porous geological formations for hydrogen storage. Finally, a comprehensive and systematic outline of activities pertinent to the subsurface project identification, operational planning, and well engineering in relation to hydrogen storage in both aquifers and depleted reservoirs is presented. It also offers a brief glimpse into the relevant ongoing UHS projects around the world. Finally, this chapter provides concluding remarks and future outlooks in the field of UHS.

6.1.1 Advantages and disadvantages of subsurface storage

As compared to aboveground storages, UHS offers several advantages (See Table 6.1) as described below:

- **Capacity**—Due to subsurface formations, UHS offers large volumes of storage capacity of hydrogen in salt caverns and porous reservoirs.
- **Safety and durability**—UHS is usually not affected by external weather conditions or climate change and hence offers long-term storage stability. In addition, it is not

TABLE 6.1 Advantages and disadvantages of underground hydrogen storage.

Characteristics	Advantages	Disadvantages
Storage capacity	Large storage capacity	Dependent on-site selection and availability
Safety	Natural barrier to leaks due to isolation from external factors	Prone to mechanical stress on well and tubing due to injection and withdrawal cycles
Durability	Unaffected by external weather conditions or climate change	Prone to leakage if there is a breach in caprock
Land use efficiency	UHS aids in reduction of land footprint	Depending on the current status of reservoir, high initial investments may be needed

UHS, Underground hydrogen storage.

affected by external human intervention and is able to give leakage protection thereby ensuring increased safety.
- **Efficiency**—Due to subsurface formations, UHS also leads to lower land footprint needed for storage leading to land use efficiency.
- In addition, they can also be integrated smoothly with urban planning with some modifications and in many cases are more economic than other alternatives.

6.2 Types of underground hydrogen storage and their efficacy in porous geological formations

In principle, underground storage can be divided in five types of geological storage as shown in Fig. 6.1.

6.2.1 Salt caverns

Salt caverns are artificially created by dissolution of rock salts through injection of water while mining. Due to injection of water, a brine solution is created which is then extracted leading to formation on underground cavern. Due to peculiar characteristics of impermeability which can avoid leakage, these salt caverns are usually used as storage. The common use of such salt caverns is to store natural gas. In case of low, excess gas can be stored in such caverns. In some places, these caverns are used for storing crude oil as a strategic move to provide oil in case of emergencies or in case of disruption in the supply of oil. In addition to this, these salt caverns can be used for storing other petroleum products like liquefied petroleum gas, ethylene or propylene (PHMSA, 2021).

Exploring the possibility of utilizing salt caverns for hydrogen storage is becoming more popular given the global movement to switch to greener energy sources. Long-term storage can be possible in geologically stable settings seen in salt deposits. As salt forms are impermeable, there is less chance of leakage, which makes them appealing for storing hydrogen, which can be difficult to store because of its small molecular size. Salt caverns can provide substantial amounts of storage space. They can be rather enormous in size, which

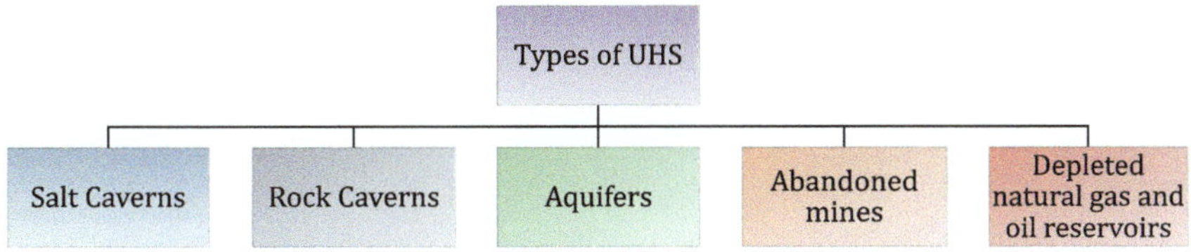

FIGURE 6.1 Types of underground hydrogen storage.

makes it possible to store a lot of hydrogen. They also provide operational flexibility to accommodate changing demand and supply requirements since they are relatively easy to access and allow rapid injection and withdrawal rates. (Linde, 2024; Panfilov, 2015).

As hydrogen can cause metals to embrittle, the infrastructure that stores and retrieves hydrogen from salt caverns has difficulties. Care must be taken while choosing building materials and methods to prevent hydrogen-related problems. To stop leaks and keep hydrogen contained inside the caverns, it is essential to maintain the integrity of the seals and wellbore casings. To find any potential leakages or integrity problems, methods for continuous monitoring must be in place. Hydrogen is flammable and reactive, thus safety precautions need to be taken. Salt cavern development for hydrogen storage is economically feasible and sometimes do not rely on a number of elements, including building costs, operating costs, and regulatory constraints (Dias et al., 2023; Engie, 2021).

Salt caverns are a popular choice for storing hydrogen because of their favorable geological conditions and features. The presence of halite, a kind of rock salt largely constituted of sodium chloride (NaCl), is the primary factor influencing the mineralogical composition of salt caverns. Salt caverns usually originate where halite formations like sedimentary rocks are left behind by the evaporation of ancient seas. Crystals of sodium chloride predominate in the mineralogical composition in this case. Solution mining can be used to create caves in halite formations because these rocks typically have high porosity and permeability. Solution mining leads to dissolution of water by injecting water which leads to formation of a cavern. After that, the water-salt solution is removed, leaving a space that can be utilized to store hydrogen. Salt dissolves to provide subterranean spaces that can be used to store gasses, such as hydrogen. Anhydrite or other nonporous rock formations are frequently from the caprock, the impermeable layer that seals the salt cavern. This caprock stops gas leaks and guarantees the cavern's structural integrity.

Research is being carried out to create materials that are resistant to hydrogen embrittlement. Cutting-edge coatings and alloy compositions could lessen this problem. To improve safety and guarantee the integrity of storage activities, advanced monitoring technologies can be implemented, such as sensors and real-time monitoring systems. A strong regulatory framework must be established in order to handle operational, environmental, and safety concerns. In addition, standards and recommendations for salt cavern hydrogen storage can be developed by regulatory bodies in collaboration with industry players (Ineris, 2022; Portarapillo & Di Benedetto, 2021).

Thereby despite the fact that salt caverns have the ability to store hydrogen due to their geological features and storage capacity, it is imperative that technical, safety, and financial issues be resolved. To fully utilize the potential of salt caverns for hydrogen storage, industry, government, and academic institutions must continue their research, innovate, and work together.

6.2.2 Rock caverns

Compared to salt caverns, rock caverns offer different advantages and challenges as a possible subterranean hydrogen storage solution. Certain rock formations can offer geologically stable conditions ideal for long-term hydrogen storage, much as salt caverns. The possibility of hydrogen leaking can be reduced in part by the structural soundness and impermeability of appropriate rock formations. The location of storage facilities near demand centers or renewable hydrogen production sites can be more flexible when it comes to rock formations appropriate for cavern construction, as they may be more widely distributed geographically than suitable salt formations. Potential integration of rock caverns with already-existing subterranean infrastructures or facilities, like mines or depleted gas reservoirs, could result in cost and synergistic benefits (Muhammed et al., 2022).

Like salt caverns, rock cavern construction and lining materials need to be compatible with hydrogen to avoid embrittlement and other possible problems and therefore choosing the right building supplies and methods is essential. In order to maintain containment and prevent possible leaks in rock caverns, maintaining the integrity of the seals and barriers is essential. To overcome possible permeability difficulties, strong sealing materials and methods are needed. Also, to determine whether a rock formation is suitable for storing hydrogen, its geological features—such as its porosity, permeability, and mechanical qualities must be thoroughly examined. Thorough geological surveys and evaluations are necessary to explore the best possible locations and maximize the utilization of cavern design. Another crucial aspect to consider is to identify and reduce any possible problems related to storing hydrogen in rock caverns and hence it is essential to design thorough safety precautions and monitoring systems to guarantee secure and dependable operations (IEA, 2021; Portarapillo & Di Benedetto, 2021).

To create cutting-edge building materials, coatings, and construction methods specifically suited for hydrogen storage in rock caverns, research and development activities are still underway. It is important to conduct comprehensive geological, geotechnical, and environmental investigations to identify acceptable rock formations and maximize cavern design. Also to realize, the potential of the caverns, sophisticated modeling and simulation tools can be used to assess the storage capacity, performance, and safety factors. In addition, the standards related to permit, environmental, safety, and social factors must be addressed early in the planning and development phase by interacting with regulatory bodies, local communities, and stakeholders (Londe, 2021; Patanwar et al., 2024).

Geological formations other than salt are frequently utilized in rock caverns intended for storing hydrogen. Depending on the type of rock and the geological circumstances, rock caverns can have different mineralogical compositions. Several rock caverns that are used to store hydrogen were formerly depleted gas or oil sources. The type of hydrocarbon in the rock determines the mineralogical makeup of these caverns. Shale, limestone, and sandstone are examples of common rock types. Hydrogen can be stored in porous rock formations like limestone aquifers or sandstone. Depending on the particular type of rock, the mineralogical composition usually consists of minerals like quartz, feldspar, and calcite. The use of metamorphic rocks, which change in texture and mineral composition as a result of heat and pressure, is being investigated in certain hydrogen storage projects. Marble, schist, and gneiss are a few examples of metamorphic rocks wherein each has unique mineralogical properties. Sandstone, limestone, and dolomite are examples of sedimentary rocks that are frequently used for storing hydrogen. Minerals including quartz, dolomite, calcite, and clay minerals may be present in these rocks. Igneous rocks, such as granite or basalt, may be used to store hydrogen in specific situations. For example, basalt's porosity and propensity for mineral carbonation reactions are well-known. Granite, on the other hand, might still be useful for storing while being less permeable.

Overall, rock caverns present a viable substitute as UHS, enhancing the capabilities of aboveground facilities and salt caverns. Reaching the full potential of rock caverns for hydrogen storage and facilitating the shift to sustainable energy requires tackling technical, safety, environmental, and regulatory issues via innovation, cooperation, and research.

6.2.3 Aquifers

Another strategy that has drawn attention is using aquifers to store hydrogen. This is due to the fact that there is a growing need for large-scale energy storage options to support efforts to integrate renewable energy sources and reduce carbon emissions. Under certain specific circumstances, aquifers which are natural subterranean formations may be useful for storing hydrogen (Zeng et al., 2023).

Depending on the size and geological features of the aquifer, aquifers can have considerably larger storage capabilities than rock or salt caverns. Geographically, suitable aquifers may be more widely distributed, giving storage facilities greater flexibility in their location relative to renewable hydrogen production locations or demand centers (Haddad et al., 2022). Impermeable rock layers that naturally surround aquifers act as a built-in containment system. The integrity and security of the hydrogen that has been stored can be preserved with the aid of the aquifer's geological features.

To determine the aquifer's suitability for storing hydrogen, geological features such as porosity, permeability, depth, and structural integrity needs to be evaluated. Finding appropriate aquifers and streamlining storage operations can be aided by carrying out thorough geological surveys, evaluations, and modeling investigations (Jafari Raad et al., 2022). One important factor to take into account while using aquifers for storage is preventing hydrogen migration and leakage. To maintain containment and reduce dangers, it is essential to protect the integrity of the aquifer's natural containment system and installation of additional barriers, such as engineered seals and casings. It is also essential to create thorough safety procedures, risk evaluations, and emergency action plans in order to reduce any possible risks related to storing hydrogen in aquifers. Using recent technology for mitigation, control, and monitoring can contribute to dependable and secure storage operations (Haddad et al., 2022; Zeng et al., 2023).

As aquifers are permeable and porous underground geological formations with the ability to store and transfer water, they can be used to store hydrogen. Aquifers' mineralogical makeup varies according to the kind of rock that makes up the aquifer. Several rock types, such as limestone, sandstone, and other sedimentary rocks, can make up aquifers. These rocks' porosity and permeability are influenced by their mineralogical makeup, which qualifies them as hydrogen storage materials. The sand-sized grains that make up sandstone aquifers are mostly made of minerals like feldspar, quartz, and other minerals. Hydrogen may be stored in sandstone due to its high porosity. Carbonate rocks, which are abundant in minerals like calcite and dolomite, make up limestone and dolomite aquifers. These rocks may have increased porosity due to the breakdown of carbonate minerals, which allows hydrogen to be stored there. Aquifer systems may have aquitards, which are impermeable layers that limit the flow of water. Clay minerals like kaolinite, illite, or smectite are frequently found in the mineralogical composition of aquitards.

Some of the potential solutions to maximize the potential usage of aquifers include technical assessment, innovation in technologies and design of standards. Evaluating aquifer suitability, storage capacity, performance, and safety factors requires comprehensive site characterization, geological modeling, and risk assessments. Seismic imaging, well logging, and reservoir simulation are examples of technologies that can be used to reduce risks and optimize storage operations (Ajayi et al., 2019). Maintaining the integrity, effectiveness, and dependability of storage operations can be facilitated by creating strong engineering designs, construction methods, and operational protocols specifically designed for the aquifer storage (Jafari Raad et al., 2022). To address environmental, safety, social, and economic concerns, a thorough regulatory framework, permitting procedure, and stakeholder engagement plan must be established. It is possible to make project approval, implementation, and continuous management easier by collaborating closely with stakeholders, local communities, and regulatory bodies (Thiyagarajan et al., 2022).

Aquifers present viable chances for large-scale hydrogen storage, enhancing existing storage choices and aiding in the shift to sustainable energy sources. Realizing the full potential of aquifers for hydrogen storage requires addressing geological, technical, safety, environmental, regulatory, and stakeholder factors through thorough planning, assessment, cooperation, and innovation.

6.2.4 Abandoned mines

Utilizing closed mines to store hydrogen offers a unique opportunity to repurpose current subterranean infrastructure for long-term, sustainable energy storage (Muhammed et al., 2022). Both potential benefits and difficulties for hydrogen storage can be found in abandoned mines that are no longer in operation for mining purposes (Osman et al., 2022).

Repurposing abandoned mines for hydrogen storage could save money, time, and environmental damage compared to building new storage facilities because abandoned mines already have subterranean chambers and equipment in place. Certain closed mines may provide geologically stable settings appropriate for long-term hydrogen storage, depending on the particulars of the mine, including depth, structural integrity, and geological conditions (Miocic et al., 2022). Repurposing abandoned mines for hydrogen storage can improve integration and efficiency by enabling localized storage solutions closer to demand centers, renewable hydrogen production locations, or current energy infrastructures.

Determining the suitability of abandoned mines for hydrogen storage requires conducting thorough geological investigations. To minimize risks and maximize storage operations, factors like water content, rock stability, structural integrity, and possible geological dangers must be carefully considered. Repurposing abandoned mines for hydrogen storage necessitates addressing potential safety issues such as mine collapses, gas accumulations, or environmental contamination. Ensuring the safety of employees, the general public, and the environment requires the implementation of strict safety regulations, monitoring systems, and risk mitigation measures. When using abandoned mines for hydrogen storage, it is critical to evaluate and mitigate any potential environmental implications, such as groundwater contamination, subsidence, or disturbances to ecosystems. To address environmental considerations and ensure regulatory compliance, it is imperative to implement thorough environmental monitoring programs (Miocic et al., 2022; Peppas et al., 2023).

The suitability of rock formations in abandoned mines for storing hydrogen is largely dependent on their mineralogical composition. The geological features of the area influence the mineralogical composition of closed mines. Sedimentary rocks (like sandstone, limestone), metamorphic rocks (like gneiss, schist), and igneous rocks (like granite, basalt) are frequently found in abandoned mines. Sand-sized grains mostly made of minerals like feldspar and quartz can be found in sandstone mining. Sandstone's high permeability and porosity make it an excellent choice for hydrogen storage. The carbonate rocks found in limestone mines are abundant in minerals like calcite. The rock's porosity can be increased by the breakdown of carbonate minerals, which frees up hydrogen storage space. Common igneous rocks discovered in abandoned mines are basalt and granite. These rocks may have the ability to store hydrogen due to their joints and fissures, while generally having less porosity than sedimentary rocks. Depending on the particular geological history, mines in metamorphic rock formations, such as gneiss or schist, may show different mineralogical compositions. There may be folds and cracks in these rocks that are connected, which increases storage capacity.

To determine whether an abandoned mine is suitable for storing hydrogen, a comprehensive site evaluation involving geological, geotechnical, and environmental assessments must be completed. Mitigating hazards and optimizing storage operations can be achieved by the implementation of remediation methods, such as stabilizing structures, managing water intrusion, or resolving pollution issues (Alms et al., 2023). Ensuring the integrity, efficiency, and dependability of storage operations can be achieved by creating strong engineering designs, building methods, and operational processes specifically designed for the storage of abandoned mines. Innovation along with the regulatory compliance can all be facilitated by working together with engineering firms, regulatory bodies, and mining professionals. To address legal, environmental, safety, social, and economic factors, a thorough regulatory framework, permitting procedure, and stakeholder engagement plan must be established. It is possible to streamline project approval, implementation, ongoing management, and decommissioning by collaborating closely with regulatory bodies, local communities, stakeholders, and mine owners (Colas et al., 2023).

In conclusion, using already-existing subterranean infrastructures for sustainable energy storage solutions may be possible through the conversion of abandoned mines into hydrogen storage facilities. To fully utilize abandoned mines for hydrogen storage and aid in the transition to a sustainable energy future, careful planning, assessment, cooperation, innovation, and remediation are necessary to address geological, technical, safety, environmental, regulatory, and stakeholder considerations.

6.2.5 Depleted natural gas and oil reservoirs

Hydrogen storage in exhausted oil and natural gas reservoirs is becoming more and more popular as a practical way to reuse subterranean infrastructure for clean energy applications. Hydrogen storage poses unique benefits and concerns due to depleted reservoirs (Zeng et al., 2023). The following summarizes the factors to be taken into account while utilizing depleted natural gas and oil reservoirs for this purpose.

Existing underground infrastructures, such as wells, pipelines, and storage facilities, are available from depleted natural gas and oil reserves and can be adapted or converted for the storage of hydrogen. This may lessen the time, money, and environmental damage that come with developing new storage facilities. Previous exploration and production of depleted reservoirs has yielded comprehensive geological data and knowledge regarding the reservoir's properties, including porosity, permeability, pressure, and structural integrity. Site selection, reservoir characterization, and storage operations optimization can all be made easier with the use of this existing information (Camargo et al., 2024). By repurposing exhausted reservoirs for hydrogen storage, integration, efficiency, and system reliability can be improved by localizing storage options closer to demand centers, renewable hydrogen production locations, or existing energy infrastructures (Sekar et al., 2023).

In order to fully realize the underlying potential, it is crucial to assess the depleted reservoir's geological, physical, and chemical suitability for storing hydrogen. To guarantee secure, effective, and long-lasting storage operations, a number of factors need to be carefully considered, including rock characteristics, fluid dynamics, reservoir conditions, and possible interactions between hydrogen and reservoir materials. Preventing hydrogen leakage, migration, or contamination requires maintaining the reservoir's integrity and containment, as well as the wellbores and surface facilities. Throughout the storage life cycle, containment must be maintained and possible integrity issues must be addressed by putting strong well construction, completion, monitoring, and maintenance procedures into place. It is crucial to have thorough safety procedures, risk analyses, emergency action plans, and monitoring systems in order to reduce any possible risks related to storing hydrogen in depleted reservoirs. For storage operations to be safe and dependable, concerns such pressure control, gas migration, seismicity, and environmental effects must be addressed (Delshad et al., 2022; Perera, 2023).

Depleted natural gas and oil reservoirs are viable options for storing hydrogen as they have underground cavities and infrastructure built for hydrocarbon extraction. These reservoirs' suitability for storing hydrogen is influenced by their mineralogical compositions. Sedimentary rock formations, such as sandstone, limestone, shale, or a combination of these, are usually home to depleted reservoirs. Sand-sized grains made of minerals including feldspar, quartz, and other such grains are frequently found in sandstone reservoirs. Sandstone's high permeability and porosity make it an excellent choice for hydrogen storage. Also, carbonate rocks that contain minerals like calcite and dolomite can be found in reservoirs within limestone and dolomite formations. Carbonate mineral dissolution can increase porosity, which gives hydrogen storage area. Fine-grained sedimentary rocks that are abundant in clay minerals, quartz, and other minerals form shale reservoirs. Despite having less porosity than sandstone, shale formations' natural and fractured cracks may increase their ability to store hydrogen.

To determine depleted reservoir suitability, storage capacity, injectivity, and containment factors, comprehensive reservoir characterization, modeling, simulation, and performance assessments are necessary (Quintos Fuentes & Santos, 2023). System performance may be improved, risks can be reduced, and storage operations can be optimized with the use of technological innovation, data analytics, and modeling tools. Integrity, efficiency, dependability, and safety can be improved by creating strong engineering designs, building methods, operational protocols, and monitoring systems specifically for depleted reservoir storage. Working together with specialists in the fields of reservoir engineering, geosciences, and business can promote innovation, the creation of best practices, and regulatory compliance. To address legal, environmental, safety, social, and economic factors, a thorough regulatory framework, permitting procedure, and stakeholder engagement plan must be established (Geologic CO_2 Sequestration Technology and Cost Analysis—Technical Support Document, 2008). The approval, implementation, continuing management, and decommissioning of projects can be facilitated by collaborating closely with regulatory bodies, industry stakeholders, local communities, and partners (Delshad et al., 2022).

Hence, using already-existing subsurface infrastructures for sustainable energy storage solutions may be possible through the conversion of exhausted natural gas and oil reserves into hydrogen storage facilities. To realize the full potential of depleted reservoirs for hydrogen storage and support the transition to a sustainable energy future, comprehensive planning, assessment, collaboration, innovation, and management are necessary to address geological, technical, safety, environmental, regulatory, and stakeholder considerations.

6.2.6 Porous geological formations versus caverns for hydrogen storage

One of the most important components of the shift to cleaner energy is hydrogen storage. Aquifers and reservoirs are two examples of the several geological formations that may be used as storage locations. It is interesting to compare the porous formations with caverns also considering the economic aspects. For the storage of hydrogen, aquifers' inherent structure and composition can offer improved containment and safety. Conversely, caverns may not provide the same degree of natural containment and usually need human intervention. Due to their large volume and area, porous formations frequently offer higher storage capacities. Large volumes of hydrogen may be accommodated with this scalability, especially as demand rises as the hydrogen economy expands. In addition, appropriate permeability features in porous formations, such as aquifers and reservoirs, enable effective hydrogen injection, storage, and extraction. Furthermore, the caprocks or natural seals connected to these formations can aid in containing hydrogen, reducing the possibility of leaks. Large-scale porous formations can result in economies of scale, which lowers the cost of infrastructure and operations for hydrogen storage per unit.

6.2.7 Hydrocarbon reservoirs and their suitability

Using hydrocarbon reservoirs to store hydrogen is a very cost-effective solution. Significant infrastructure, such as wells, pipelines, and facilities, already exists in these reservoirs and can be converted or repurposed for the storage of hydrogen. Utilizing current infrastructure can speed up rollout and significantly save capital costs. Hydrocarbon reservoirs have a wealth of historical information, including information on temperature, pressure, fluid composition, and production history. Engineers and geoscientists can evaluate reservoir suitability, create the best storage plans, and successfully manage hazards thanks to this abundance of information. In addition, the benefits for storing hydrogen are offered by the hydrocarbon reservoirs connected to developed and planned fields. Established infrastructure, continuous technological investments, and regulatory compliance are advantageous for these areas. Project timescales can be accelerated by repurposing these fields for hydrogen storage, which can make use of already-existing infrastructure. Also, a number of nations have put in place frameworks governing the production, storage, and exploration of hydrocarbons. The legal frameworks already in place can be advantageous for the repurposing of hydrocarbon reservoirs for the storage of hydrogen, since they guarantee environmental preservation, safety, and adherence to industry norms.

In conclusion, hydrocarbon reservoirs are one of the best choices for storing hydrogen because of their developed infrastructure, regulatory framework, economic viability, and availability of reservoir history. However, porous geological formations such as aquifers and reservoirs also offer inherent advantages over caverns. Hydrocarbon reservoirs can be used for hydrogen storage (Fig. 6.2), which can speed up deployment and ease the shift to cleaner energy sources.

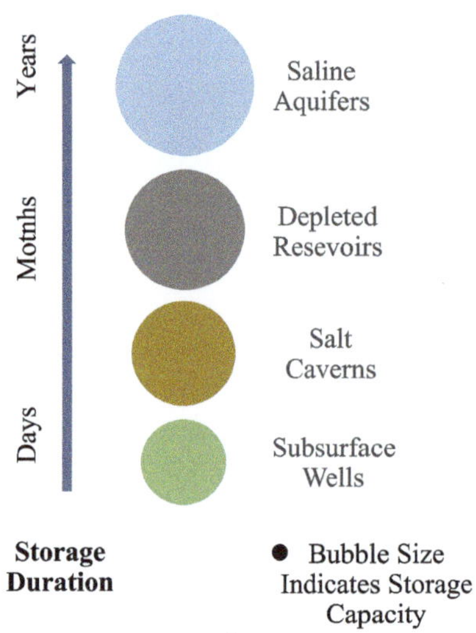

FIGURE 6.2 The storage duration and capacity of different geological formulations (Chen et al., 2023).

6.3 Caprock analysis: examining caprocks in depleted oil and gas reservoirs and aquifers

Before delving into the specifics of caprocks, let us first revisit the basic structure (Fig. 6.3) and components of an oil and reservoirs and aquifers. Both are composed of several layers, each having distinct properties and roles in the accumulation and trapping of hydrocarbons or other fluids (Crotogino, 2016). The primary components of these geological formations structure include:

Source rock: This is the birthplace of oil and gas. Here, organic matter decomposes under the influence of heat and pressure, eventually forming hydrocarbons. Typically, sedimentary rock is the primary source rock for hydrocarbons (Curiale & Curtis, 2016; Gautier, 2004). For aquifers, the term "source rock" is not relevant. However, in the context of a broader geological sequence or sedimentary basin, an aquifer could indeed be closely associated with or adjacent to source rocks. But the aquifer's primary function is to store and transmit water, not to generate hydrocarbons (Badawy & Ganat, 2021).

Reservoir rock: This is the porous and permeable rock layer where oil, gas, or other fluids such as water in the case of aquifers, accumulate and are stored (Alamooti & Malekabadi, 2018; Magoon, 2004). The porosity and permeability are key petrophysical properties of rocks:

- **Porosity:** When we talk about porosity, we mean the amount of empty space or pores within the rock. It is these pores that store the oil and gas. A rock with high porosity has more spaces, allowing it to store more hydrocarbons (Wheaton, 2016).
- **Permeability:** This refers to the ability of a rock to let fluids pass through its pores. High permeability means that fluids can move more easily through the rock, which is

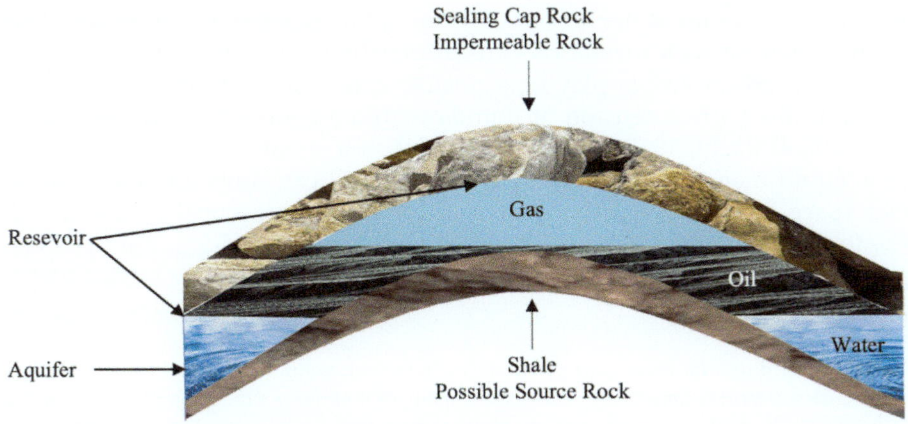

FIGURE 6.3 A reservoir structure (Badawy & Ganat, 2021).

crucial during the extraction process (Tiab & Donaldson, 2016). Permeability in reservoir rocks may range from 0.1 mD to more than 10 D.[1]

Sandstones and carbonates are common for both reservoir rocks and aquifer formations (Alva et al., 2018; Klemme, 1977). They are notable for their substantial porosity and permeability, for example, sandstones typically displaying porosity levels between 10% and 40%, and carbonates ranging from 5% to 25%. These characteristics allow them to store and transmit significant volumes of fluids, whether hydrocarbons or water (Zivar et al., 2021).

Grain size and sorting can significantly affect the rock's porosity and permeability. Smaller grains create smaller pores and pore throats, acting like miniature roadblocks for fluids trying to pass through. Conversely, coarse-grained rocks offer larger, more open pathways. Besides that, the distribution of the grains is also a crucial factor (Selley, 1998). A wide distribution of large and small grains where the smaller particles fill in the spaces between larger ones, ultimately decreasing the permeability. Another important factor is the presence of certain minerals (e.g., clays), which can also reduce porosity by filling pore spaces. Henkel et al. (2013), studied reservoir rocks in three different locations in Germany that varied among other things in grain size and the presence of clay. Fig. 6.4 shows their results where they found variations in the porosity and permeability of rocks even within the same location.

Caprock: Acting as a barrier, this impermeable (meaning fluids cannot pass through it) rock layer lies above the reservoir rock and prevents the fluids from escaping. It's like a lid that keeps fluids contained within the reservoir rock below. Without it, fluids would continue migrating upwards and might eventually reach the surface, which would render the accumulation nonviable. Various materials, like clays, shales, carbonates, and evaporites, can serve as caprocks in both reservoirs and aquifers (Lyons & Plisga, 2004).

6.3.1 Caprock properties

Some caprocks are quite selective in what they allow to pass through. For instance, there are rocks that permit water to traverse but act as barriers to oil or gas. These selective interactions within rock formations are primarily governed by their physicochemical properties, with capillary forces playing a notable role. Capillary forces refer to the forces that arise due to the surface tension of liquids within a confined space, such as the pores within a rock (Golf-Racht, 1982). When a liquid is in a small tube or pore, the interaction between the liquid molecules and the tube or pore wall can cause the liquid to rise or fall within that space, with no regard to external forces (e.g., gravity) (Cai et al., 2022). As such, these forces play a critical role in moving and trapping fluids within porous rocks in oil and gas reservoirs. Several factors such as fluid properties and rock attributes influence the capillary forces:

[1] The Darcy, the standard unit for measuring permeability in the petroleum industry, is frequently used in millidarcies (1 mD = 0.001 D). One Darcy is the permeability that allows a fluid with a viscosity of 1 centipoise (a measure of the fluid's thickness or resistance to flow) to flow through a material at a rate of 0.01 m/s when there is a pressure difference of 0.0001 atm/m^2 across the material.

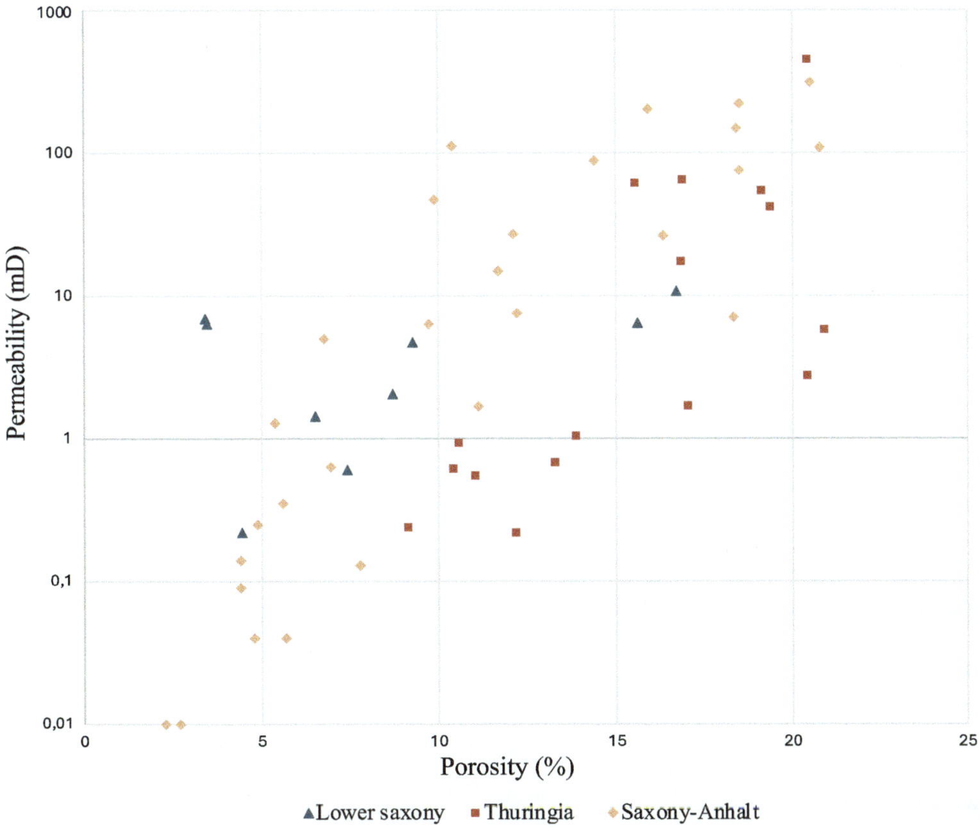

FIGURE 6.4 Porosity–permeability variations in the reservoir rocks of studied areas (Henkel et al., 2013).

Fluid properties

Fluids with different densities and viscosities will interact with the caprock in varying ways. Fluids with higher densities, such as oil, tend to move slower through porous media compared to fluids with lower densities, like gas or water. The denser a fluid is, the more force it requires to move through the rock's pore spaces. On the other hand, high-viscosity fluids move more slowly through the porous medium due to greater internal resistance, which affect the rate of capillary rise; it will be slower for a high-viscosity fluid as it struggles against its own resistance to flow through narrow spaces (Civan, 2007; McPhee et al., 2015).

Rock attributes

The structure and wettability of the rock also have a substantial impact. These attributes determine how the rock interacts with different fluids and whether it will allow them to pass through or not (Puntervold et al., 2022).

Pore size

Capillary forces are especially influenced by the size of the pores in the rock (Fanchi, 2002). These forces are significant in smaller pores while in larger pores, the influence of capillary forces is almost negligible, allowing fluids to pass more freely (Kadkhodaie & Kadkhodaie, n.d.).

It should be noted that subsurface conditions such as temperature, pressure, and geochemistry drastically affect both the rock and fluid characteristics. Thus, what may serve as a seal for one fluid might not serve as a seal for another. The sealing capabilities of a rock are not uniform and can significantly vary depending on the fluid it interacts with, allowing a rock to switch roles between being a caprock and a reservoir rock based on the fluids present.

6.3.2 Types of caprocks in reservoirs and their effectiveness in storing hydrogen

Caprocks can be broadly classified into the following primary types (Hyne, 2001), each with its unique properties:

Shale caprocks

Predominantly composed of clay (fine-grained minerals), which can impede or allow fluid flow, depending on their composition, structure, and the presence of other minerals. When compacted, they form shale layers, that can trap hydrocarbons due to their fine-grained, layered structures, and low permeability (Tao et al., 2023).

Anhydrite caprocks

Anhydrite is usually found in evaporite deposits where it forms due to the evaporation of seawater or saline lakes. It can also be found in hydrothermal veins and metamorphic rocks (Webb & White, 2013). This type of caprocks exhibit low permeability and are usually rigid and compact, providing an effective seal of hydrocarbons. The sealing efficacy can, however, be influenced by factors like temperature, pressure, and the presence of other minerals (Hangx et al., 2010).

Salt caprocks

Primarily composed of sodium chloride (rock salt) but can also contain other evaporite minerals. Exceptionally impermeable and plastic, meaning they can deform and flow over geological timescales to seal any potential leaks (Lyu et al., 2023). As such, they are recognized as one of the most effective barriers for hydrocarbon migration.

When considering the storage of hydrogen, the effectiveness of each type of caprock can vary. Given hydrogen's tiny molecular size and its ability to escape through even minuscule pathways, making the selection of an appropriate caprock crucial to prevent leakage. Table 6.2 provides a comparative overview of the effectiveness of the different caprock types in preventing hydrogen leakage and the associated challenges (Box 6.1).

While each caprock type exhibits inherent barrier properties essential for hydrogen storage, salt caprocks are generally deemed the most effective, followed by anhydrite and

TABLE 6.2 Comparative effectiveness of caprock types in hydrogen storage.

Type of caprocks	Effectiveness	Challenges[a]
Shale caprocks	Despite their low permeability and finely stratified structure, are generally considered the least effective among the three, for hydrogen storage	Variations in mineral content and structural anomalies (e.g., microfractures or high porosity) can compromise the permeability of shale caprocks, potentially allowing the escape of hydrogen molecules (Ilgen et al., 2018; van Gessel, 2023)
Anhydrite caprocks	With their inherent rigidity and compact structure, also serve as competent barriers against hydrogen migration, though their effectiveness is somewhat inferior to that of salt caprocks	The sealing efficacy of anhydrite is contingent upon prevailing environmental conditions such as temperature and pressure, and any structural compromise or alteration in these conditions can potentially precipitate leakage (Aghaei et al., 2023; van Gessel, 2023).
Salt caprocks	Universally recognized as the most effective barriers for hydrogen storage, attributed to their exceptionally low permeability and unique self-sealing properties that ensure a robust and hermetic seal, essential for preventing hydrogen escape	Despite their efficacy, careful monitoring is required as their structural integrity is prone to compromises due to mechanical stresses or dissolution phenomena, which can affect their sealing capabilities (Nuclear Energy Agency, 2014)

[a]*See Box 6.1.*

then shale caprocks. Each type, however, presents its unique set of challenges, requiring specialized considerations and management practices to ensure they effectively seal the reservoirs and maintain their sealing properties over time. Regular assessments and monitoring are therefore universally crucial across all caprock types to detect and address any issues promptly and maintain the viability of hydrogen storage in depleted oil and gas reservoirs or aquifers.

Operating conditions

The underground gas that is stored is of two types one is working gas and another is cushion gas. Cushion gas, which can be defined as mass, volume, or energy delivered out of the reservoir per unit of time, is the base gas of the underground storage reservoir that is intended to maintain gas deliverability at a sufficient level of pressure (Mazarei et al., 2019; Panfilov, 2015). Cushion gas, which can be thought of as the minimum charge level of storage similar to batteries, is therefore meant to remain permanently in the reservoir and, since it is mostly trapped inside the geological structure, it also becomes nonrecoverable (Feldmann et al., 2016; van der Meer & Obdam, 2008). The real storage capacity, on the other hand, that can be charged and discharged again is known as working gas (Flanigan, 1995; Uniper, 2022). It is worth noting that certain technologies may make it possible to extract a portion of the cushion gas from the underground storage; however, this partial recovery of the gas would be largely restricted to the extent of potential damage to the underground formation that could arise from the extraction process (Speight, 2019).

BOX 6.1

Clarifications on specific terms related to caprock challenges in hydrogen storage.

Shale caprocks (Challenge: variations in mineral content and structural anomalies, such as microfractures or high porosity.)

Explanation:

Microfractures: these are tiny cracks in the rock structure. Hydrogen, due to its small molecular size, can easily exploit these microfractures, leading to potential leakage. As microfractures widen or multiply, the risk of leakage intensifies.

High porosity: this refers to the volume of open spaces within a rock. While porosity can enable storage, high porosity in shale can indicate inconsistent and irregular pore distribution, which might not always be effectively sealed, allowing hydrogen to find escape routes.

Anhydrite caprocks *(Challenge: the sealing efficacy is influenced by environmental conditions such as temperature and pressure.)*

Explanation:

Temperature and pressure sensitivity: anhydrite caprocks can undergo thermal expansion or contraction and may become brittle or undergo phase changes at certain temperatures or pressures. These alterations might introduce fractures or pores, potentially allowing hydrogen, which is sensitive to minute escape paths due to its small size, to leak.

Salt caprocks *(Challenge: their structural integrity is susceptible to compromises due to mechanical stresses or dissolution phenomena.)*

Explanation:

Mechanical stresses: over time and under certain conditions, salt can deform. While this plasticity can often seal minor escape routes, significant mechanical stress, such as from tectonic forces, can create gaps or fractures. Given hydrogen's propensity to migrate through the smallest of openings, this poses a risk.

Dissolution phenomena: salt can dissolve in the presence of water or certain other liquids. If such liquids are present or introduced, the salt caprock could thin out or develop holes. This could create pathways for the stored hydrogen to escape.

In depleted natural gas and oil reservoirs, injection wells are used to inject gas below the reservoir's surface (Amid et al., 2016). The caprocks that cover the reservoir's top are the main component of the confining layer. These are naturally occurring rocks with special qualities, the most significant of which is their impermeability, which guarantees the tightness of the gas (Chen et al., 2014). Caprocks, which were originally naturally occurring in the depleted reservoirs and helped to keep gas from escaping, are frequently salt or shale rocks (D3.1_Overview-of-all-known-underground-storage-technologies, n.d.). A minimal budget (1.23 USD/kg) is probably needed for the conversion of depleted hydrogen storage reservoirs to underground storage (Tarkowski, 2019). This is mostly because the reservoir is already there and has all the facilities and equipment required. Furthermore, as geological tests and studies already exist, none are required to examine storage behavior (Reitenbach et al., 2015; Tarkowski, 2019). The fact that there is still gas left over after depletion, which can serve as cushion gas, is another benefit. Aquifers do not contain hydrocarbon leftovers, although there are worries about possible hydrogen loss and contamination from reactions with nearby minerals and microbes (Crotogino, 2016). Aquifers also have the disadvantage of having high cushion gas needs, meaning that a significant amount of gas will be permanently trapped in the aquifer with little chance of recovery in the future (Tarkowski, 2019). Concerns about possible land subsidence and contaminated subsurface water are among the major environmental issues (De & Prabu, 2017; Kapusta et al., 2013). The latter issue affects all underground gas storage types since one of the deep layers that helps seal these storages overall may be underground water.

Based on the properties of each storage medium, there are specific aspects for the injection and withdrawal cycles for hydrogen storage in different geological formations. The type of storage reservoir, operational efficiency, and energy demand are some of the variables that determine the optimal cycles for injection-storage and production-withdrawal in underground hydrogen storage. The number of injection and withdrawal cycles per year is estimated to be between 10 and 12 based on research and industry practices (Ruiz Maraggi & Moscardelli, 2023). Specific operational cycles, however, can differ. According to some sources, a once-yearly cycle that consists of 6 months of injection and 6 months of production is usual (Tutorial 5 – Hydrogen Storage, 2024). The balance of supply and demand as well as storage efficiency should be taken into account while choosing the best cycles for UHS. Furthermore, in order to maximize storage operations and reduce hydrogen loss during the course of storage operations, the characterization of hydrogen storage potential and life cycle assessments are ongoing (Lackey et al., 2023).

6.3.3 Salt caverns

The optimal injection-storage and production-withdrawal cycles for subterranean hydrogen storage in salt caverns depend on a number of variables, including the kind of storage reservoir, operating effectiveness, and energy consumption. The frequency of injection and withdrawal cycles usually falls between 10 and 12 cycles annually, according to research and industry norms. Specific operational cycles, however, can differ. Some sources indicate a once-a-year cycle, which consists of 6 months of injection and 6 months of production. Choosing the optimal cycles for UHS in salt caverns requires consideration of supply and demand as well as storage efficiency (Linde, 2024).

Injection (storage) phase

- Surplus energy identification: find times when renewable energy is abundant enough to electrolyze hydrogen.
- Production of hydrogen: generate hydrogen when there is an excess of energy.
- Compression and injection: reduce the hydrogen's pressure to its storage level. Fill the salt cavern with compressed hydrogen by using designated injection wells.
- Monitoring and control: to guarantee safe and effective hydrogen injection, put monitoring and control systems in place. Monitor the temperature, pressure, and other pertinent factors in the cavern.
- Storage time: when there is extra output or little demand, store hydrogen in the salt cavern.

Withdrawal (production) phase

- Determining energy demand: determine whether there is a large demand for energy or when there is not enough supply of renewable energy.
- Hydrogen recovery: utilizing certain production wells, start removing hydrogen from the salt cavern.
- Distribution and decompression: distribute the extracted hydrogen to the appropriate pressure for distribution. Deliver hydrogen to businesses or final consumers.
- Monitoring and control: design and implement procedures for efficient and safe hydrogen removal. While withdrawal, monitor the hydrogen quality and cavern conditions.
- Utilization: use extracted hydrogen for industrial processes, electricity generation, or other purposes as needed.
- Reinjection (optional): if demand is minimal, surplus hydrogen generated during times of high renewable energy may occasionally be reinjected into the salt cavern.

6.3.4 Rock caverns

The optimal injection-storage and production-withdrawal cycles for UHS in rock caverns depend on a number of variables, including the kind of storage reservoir, operating effectiveness, and energy consumption. The frequency of injection and withdrawal cycles usually falls between 10 and 12 cycles annually, according to research and industry norms. Specific operational cycles, however, can differ. Some sources indicate a once-a-year cycle, which consists of 6 months of injection and 6 months of production. The balance of supply and demand as well as storage efficiency are critical factors to take into account while choosing the best cycles for subterranean hydrogen storage in rock caverns (Miocic et al., 2022; Smart Energy International, 2022). Similar to salt caverns, rock caverns use the rock medium for safe hydrogen storage and adhere to injection-production cycles.

6.3.5 Aquifers

Aquifer type, operating efficiency, energy demand, and other variables determine the optimal cycles for injection, storage, and production-withdrawal for hydrogen storage. The frequency of injection and withdrawal cycles usually falls between 10 and 12 cycles annually, according to research and industry norms. Specific operational cycles, however, can differ. Some sources indicate a once-a-year cycle, which consists of 6 months of injection and 6 months of production. The balance of supply and demand as well as storage efficiency must be taken into account while choosing the best cycles for subterranean hydrogen storage in aquifers (Alms et al., 2023; Li et al., 2022).

Injection (storage) phase

- Surplus energy identification: calculate the periods of excess renewable energy.
- Production of hydrogen: produce hydrogen when there is an excess of energy.
- Injection and compression: reduce hydrogen pressure to a storage level. Utilizing designated wells, inject compressed hydrogen into the aquifer.

- Monitoring and control: implement monitoring systems to ensure the safe injection of hydrogen. Monitor the reservoir's temperature, pressure, and other elements.
- Storage: during times of surplus production or low demand, store hydrogen in the aquifer.

Withdrawal (production) phase

- Determining energy demand: determine when there is a high demand for energy.
- Recovery of hydrogen: extract hydrogen out of the aquifer with production wells designated for this purpose.
- Distribution and decompression: extracted hydrogen should be decompressed to the appropriate distribution pressure. Distribute hydrogen for final usage.
- Monitoring and control: implement monitoring systems to ensure safe hydrogen extraction. While withdrawal, monitor the quality of the hydrogen and the aquifer.
- Application: utilize extracted hydrogen for industrial operations, power generation, or other purposes.

6.3.6 Depleted reservoirs

Depending on particular operational and reservoir parameters, different production-withdrawal and injection-storage cycles may be more or less suitable for hydrogen storage in abandoned mines. Research and industry practices indicate that the frequency of injection and withdrawal cycles usually falls between 10 and 12 cycles annually, while there is no universal standard. Specific operational cycles, however, can differ. Some sources indicate a once-a-year cycle, which consists of 6 months of injection and 6 months of production. When choosing the best cycles for subterranean hydrogen storage in abandoned mines, it is critical to take supply and demand balance and storage efficiency into account (Delshad et al., 2022). Depleted reservoirs use underground space for hydrogen storage and adhere to injection-production cycles, much like salt and rock caves.

6.3.7 Abandoned mines

Based on unique operational and reservoir features, different production-withdrawal and injection-storage cycles may be more or less suitable for storing hydrogen in depleted oil and natural gas reservoirs. The frequency of injection and withdrawal cycles usually falls between 10 and 12 cycles annually, according to research and industry norms. Specific operational cycles, however, can differ. Some sources indicate a once-a-year cycle, which consists of 6 months of injection and 6 months of production (Tutorial 5 – Hydrogen Storage, 2024). In depleted oil and natural gas reservoirs, the best cycles for subterranean hydrogen storage depend on factors including supply and demand balance and storage efficiency. Furthermore, research is also being conducted to optimize storage operations and reduce hydrogen loss throughout the storage operations life cycle (Epelle et al., 2022; Lysyy et al., 2021). Based on their geological features, abandoned mines can be used to store hydrogen using injection-production cycles similar to those used for other storage facilities.

There are some of the factors we need to consider for all types of UHS.

- Optimization of injection-production cycles: based on patterns of energy use and storage capacity, optimize the length and frequency of injection and withdrawal cycles.
- Safety precautions: follow safety procedures during the injection and withdrawal stages to guarantee the storage medium's integrity and stop leakage.
- Impact on the environment: evaluate and reduce any possible environmental impacts linked to the storage of hydrogen in each geological formation.
- Regulatory compliance: assure adherence to legal specifications for the storage and use of hydrogen.
- Monitoring and maintenance: establish processes for ongoing maintenance and monitoring to guarantee the storage system's long-term dependability and integrity.
- Integration with energy infrastructure: integrate hydrogen storage devices for smooth operation and increased efficiency with the current energy infrastructure.

The geology and hydrogeological properties of the storage medium, as well as the local legal framework and safety concerns, will determine the specifics of each injection-production cycle. Thorough engineering studies and site-specific assessments are essential for a successful operation of UHS.

6.4 Activities for underground hydrogen storage project: from identification to well engineering

The activities involved in the UHS project range from well engineering to project identification. As shown in Fig. 6.5, tasks related to well engineering, operational planning, and subsurface project identification for a UHS project involve detailed procedure consisting of subsurface project identification, operational planning, and well engineering.

6.4.1 Subsurface project identification

The process of identifying subsurface projects for hydrogen storage involves assessing and choosing subterranean formations or reservoirs that provide the best circumstances for the safe and effective storage of hydrogen. The following steps are involved in the process:

- **Feasibility study**: To analyze the technical, financial, environmental, and regulatory elements of the UHS project, conduct screening tests, feasibility studies, and preliminary assessments.
- **Site selection**: Apply engineering, hydrogeological, geological, and geophysical criteria to locate, assess, and choose appropriate subsurface formations, locations, and reservoirs.
- **Resource assessment**: To evaluate storage capacity, injectivity, confinement, performance, and viability, conduct comprehensive reservoir characterization, modeling, and simulation studies.

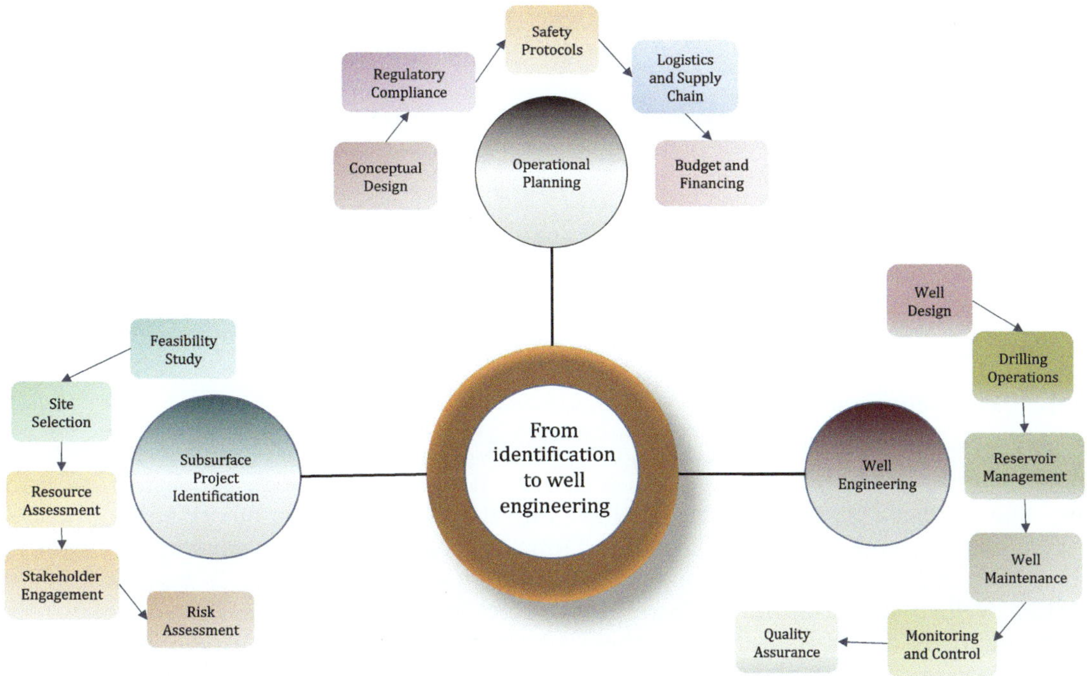

FIGURE 6.5 From identification to well engineering.

- **Stakeholder engagement**: To handle issues, get permissions, gain permits, and promote cooperation, become involved with communities, landowners, regulatory bodies, and partners.
- **Risk assessment**: To identify, evaluate, and address potential technical, safety, environmental, social, and economic risks, conduct thorough risk assessments, hazard evaluations, impact studies, and mitigation plans.

6.4.2 Operational planning

For an UHS project, operational planning includes developing conceptual designs, strategies for regulatory compliance, safety procedures, logistics management, and financial frameworks to guarantee successful implementation and operation. Following are the key steps involved in planning.

- **Conceptual design**: Create engineering concepts, conceptual designs, system architectures, and operational frameworks for the UHS project, encompassing surface installations, storage facilities, and infrastructure.
- **Regulatory compliance**: To ensure adherence to relevant laws, regulations, standards, and requirements, establish regulatory compliance plans, permits procedures, environmental evaluations, and legal frameworks.

- **Safety protocols**: To guarantee safe, secure, and dependable operations throughout the project life cycle, implement safety protocols, rules, procedures, training programs, and best practices.
- **Logistics and supply chain**: Explain supply chain management, procurement tactics, locating equipment, material handling, transportation, and delivery schedules in order to carry out project execution.
- **Budget and financing**: To assist with project development, financing, and investment decisions, prepare budget estimates, financial models, funding strategies, investment plans, cost-benefit assessments, and economic evaluations.

6.4.3 Well engineering

The design of wells is a crucial aspect in UHS, encompassing the wellbore's stability, as well as the selection and integration of critical components such as, casing and tubing. The wellbore is the actual hole drilled into the earth to access the underground reservoir. Casing and tubing are also fundamental components of well design (Ramesh Kumar et al., 2023). Casing refers to the large-diameter pipe that is installed in the drilled well to stabilize the wellbore and protect it from collapsing. It also isolates the wellbore from groundwater and other subsurface formations to prevent any cross-contamination. Tubing, on the other hand, is a smaller-diameter pipe that is run into the well after the casing is in place. It serves as the conduit through which the hydrogen is injected into or extracted from the underground reservoir. In Aquifers, the wellbore must be designed to withstand the high pressures and temperatures encountered at greater depths. This involves using thicker and stronger casing to ensure the wellbore remains stable and secure. The wellbore's stability in these conditions is also maintained using specialized drilling techniques and fluids. For shallower depleted reservoirs, the wellbore might be part of an existing infrastructure, but it still requires careful assessment and potential modification to suit hydrogen storage. The focus here is on adapting the wellbore to ensure it can safely handle the specific pressures and environmental conditions associated with hydrogen.

The UHS project requires precise drilling operations, such as casing design and directed procedures, in addition to specific well design intended for uses like as production, observation, and injection. The focus is on reservoir management measures to improve safety and efficiency of storage, in addition to regular well maintenance and intervention tactics. Rigid quality assurance, regulatory compliance, and operational standards serve as the foundation for optimal performance and safety, which are ensured through continuous monitoring through innovative technology. Following are the details of the process In designing wells for UHS projects, the focus on structural design, material selection, and specific components such as casings, tubing, and compressors is crucial for the functionality and safety of the storage system.

- **Well design**: Create engineering requirements, drilling schedules, completion plans, construction methods, and well designs that are specifically suited for UHS applications. These may include production, observation, monitoring, and injection wells.

- **Drilling operations**: To drill wells safely and effectively, carry out drilling operations, well construction, wellbore integrity, casing design, cementing, drilling fluids management, directional drilling, and well control procedures.
- **Reservoir management**: To optimize storage capacity, efficiency, and containment, apply reservoir management, pressure management, fluid dynamics, injection methods, production optimization, well testing, and performance monitoring.
- **Well maintenance**: To guarantee operational dependability, lifespan, and performance, carry out well maintenance, workover operations, corrective measures, well interventions, integrity evaluations, and surveillance activities.
- **Monitoring and control**: To continually monitor well conditions, reservoir behavior, and system performance, install, run, and maintain monitoring systems, sensors, instrumentation, control systems, data collection, real-time monitoring, and automation technologies.
- **Quality assurance**: Maintaining engineering standards, regulatory compliance, and operational excellence requires the establishment of quality assurance, quality control, inspection protocols, testing methods, compliance monitoring, and verification processes.

The UHS project encompasses a thorough and methodical approach to well engineering, operational planning, and subsurface project identification. To ensure the successful development, implementation, and operation of an UHS facility, it is imperative to address a number of critical activities, including feasibility, site selection, resource assessment, stakeholder engagement, risk assessment, conceptual design, regulatory compliance, safety protocols, logistics, supply chain, budgeting, financing, well design, drilling operations, reservoir management, well maintenance, monitoring, control, and quality assurance. Effective project management requires collaboration between interdisciplinary teams, specialists, stakeholders, and partners at every stage of the project life cycle to efficiently handle technical problems, minimize risks, maximize performance, and accomplish project objectives.

Globally, several nations have shown interest in development of hydrogen hubs. There are numerous ongoing projects on the development of UHS. In Europe, European Hydrogen Backbone Project (see Fig. 6.6) is being developed where in 33 EU network operators are participating encompassing 24 EU countries. The goal is to develop five pan U hydrogen supply and import corridors through development phases in 2030 and 2040 as shown in Fig. 6.4. Numerous research initiatives, like HyCAVmobil in Germany (EWE, 2022) and HyPSTER in France (HyPSTER, 2021), are planned to begin testing in an effort to demonstrate quick cycling in large-scale hydrogen storage. Additional studies conducted in the Netherlands (Hystock, 2021), Germany (Garbe, 2023), and France (HDF Energy, 2024) are examining the feasibility of converting natural gas salt caverns into hydrogen storage facilities. Other forms of underground storage sites (such as depleted gas fields, aquifers, and lined hard rock caves) are also being researched (IEA 50, 2023). In Sweden, a test facility for storing hydrogen in walled hard rock caves began operations in

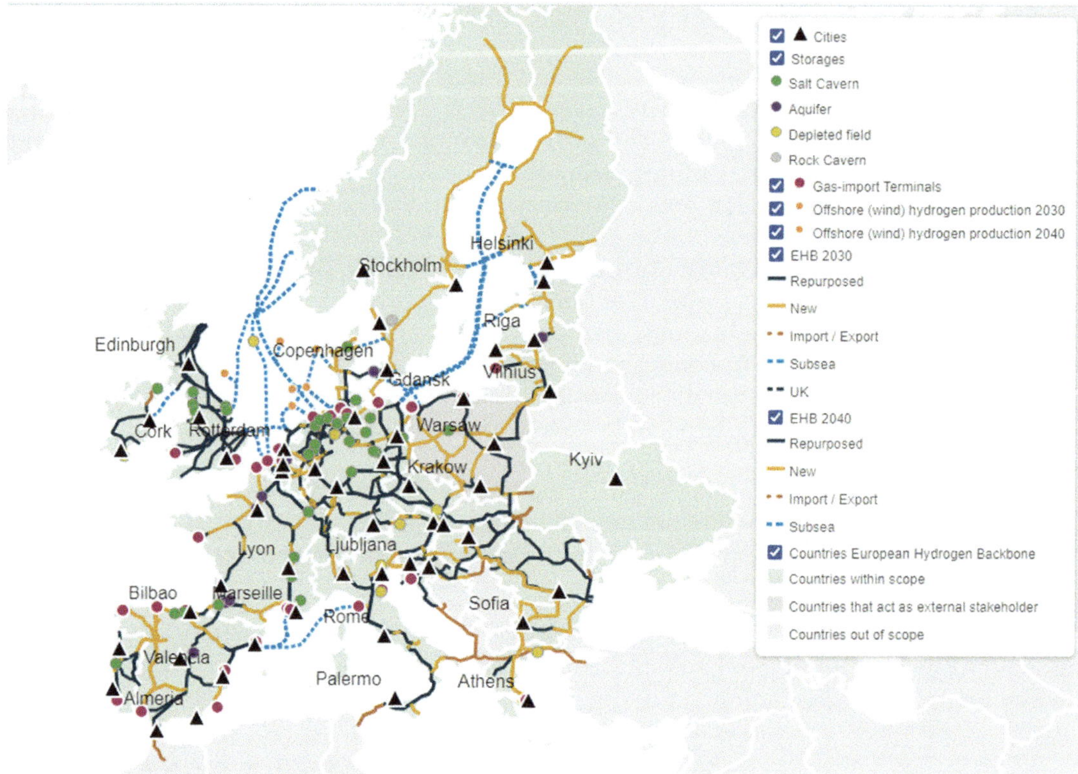

FIGURE 6.6 European hydrogen backbone map (EHB, 2024). Map lines delineate study areas and do not necessarily depict accepted national boundaries.

2022 (SSAB, 2022). Global bulk storage capacity is expected to increase from 0.5 TWh to 70 TWh by 2030 under the NZE Scenario (IEA 50, 2023).

6.5 Conclusion and outlook

In light of the intermittent nature of renewable power sources, the chapter addresses the possibilities of subsurface UHS as a solution to the problems associated with storing hydrogen, a clean energy carrier. Aquifers, salt caverns, rock caverns, and abandoned mines are just a few of the UHS kinds that are covered. It also emphasizes the benefits of UHS, such as its enormous storage capacity, safety, and longevity. In order to fully fulfill UHS's potential for large-scale, long-term energy storage, it will be necessary to address technological, safety, environmental, and regulatory constraints.

Here are the findings and recommendations for subsurface UHS going forward:

- UHS is a viable option for long-term energy storage since it has many benefits, including a huge storage capacity, durability, and safety.

- Aquifers, salt caverns, rock caverns, and abandoned mines are some examples of UHS. Each form of UHS has its own opportunities and difficulties, necessitating extensive technical, safety, and environmental considerations.
- Ongoing research and innovation are necessary to address technical issues like hydrogen embrittlement and storage efficiency, as well as to develop cutting-edge monitoring systems and safety procedures, in order to fully fulfill the promise of UHS.
- To solve regulatory and safety problems, provide good technical designs, and establish a strong regulatory framework for UHS, industry, government, and academic institutions must collaborate together.
- To fully realize its potential for large-scale, long-term energy storage in support of sustainable, low-carbon energy systems, and to overcome the challenges and hazards associated with subsurface storage, UHS will need to continue its research, innovation, and collaboration.

References

Aghaei, H., Al-Yaseri, A., Toorajipour, A., Shahsavani, B., Yekeen, N., & Edlmann, K. (2023). Host-rock and cap-rock wettability during hydrogen drainage: Implications of hydrogen subsurface storage. *Fuel*, *351*, 129048. Available from https://doi.org/10.1016/j.fuel.2023.129048.

Ajayi, T., Gomes, J. S., & Bera, A. (2019). A review of CO_2 storage in geological formations emphasizing modeling, monitoring and capacity estimation approaches. *Petroleum Science*, *16*(5), 1028–1063. Available from https://doi.org/10.1007/s12182-019-0340-8, http://www.springerlink.com/content/120897/.

Alamooti, A. M., & Malekabadi, F. K. (2018). An introduction to enhanced oil recovery. *Fundamentals of enhanced oil and gas recovery from conventional and unconventional reservoirs* (pp. 1–40). Elsevier Inc. Available from https://www.sciencedirect.com/book/9780128130278/fundamentals-of-enhanced-oil-and-gas-recovery-from-conventional-and-unconventional-reservoirs, https://doi.org/10.1016/B978-0-12-813027-8.00001-1.

Alms, K., Ahrens, B., Graf, M., & Nehler, M. (2023). Linking geological and infrastructural requirements for large-scale underground hydrogen storage in Germany. *Frontiers in Energy Research*, *11*. Available from https://doi.org/10.3389/fenrg.2023.1172003, http://www.frontiersin.org/Energy_Research/.

Alva, G., Lin, Y., & Fang, G. (2018). An overview of thermal energy storage systems. *Energy*, *144*, 341–378. Available from https://doi.org/10.1016/j.energy.2017.12.037, http://www.elsevier.com/inca/publications/store/4/8/3/.

Amid, A., Mignard, D., & Wilkinson, M. (2016). Seasonal storage of hydrogen in a depleted natural gas reservoir. *International Journal of Hydrogen Energy*, *41*(12), 5549–5558. Available from https://doi.org/10.1016/j.ijhydene.2016.02.036, http://www.journals.elsevier.com/international-journal-of-hydrogen-energy/.

Armaroli, N., & Balzani, V. (2011). The hydrogen issue. *ChemSusChem*, *4*(1), 21–36. Available from https://doi.org/10.1002/cssc.201000182, http://www.interscience.wiley.com/jpages/1864-5631.

Badawy, A. M., & Ganat, T. A. A. O. (2021). Rock properties and reservoir engineering: A practical view. *Rock Properties and Reservoir Engineering: A Practical View, 2022*. Available from https://doi.org/10.1007/978-3-030-87462-9_1.

Berna-Escriche, C., Vargas-Salgado, C., Alfonso-Solar, D., & Escrivá-Castells, A. (2022). Hydrogen production from surplus electricity generated by an autonomous renewable system: Scenario 2040 on Grand Canary Island, Spain. *Sustainability*, *14*(19), 11884. Available from https://doi.org/10.3390/su141911884.

Cai, J., Chen, Y., Liu, Y., Li, S., & Sun, C. (2022). Capillary imbibition and flow of wetting liquid in irregular capillaries: A 100-year review. *Advances in Colloid and Interface Science*, *304*, 102654. Available from https://doi.org/10.1016/j.cis.2022.102654.

Camargo, J. T., White, J. A., Hamon, F. P., Fakeye, V., Buscheck, T. A., & Huerta, N. (2024). Managing reservoir dynamics when converting natural gas fields to underground hydrogen storage. *International Journal of Hydrogen Energy*, *49*, 1261–1273. Available from https://doi.org/10.1016/j.ijhydene.2023.09.165, https://www.sciencedirect.com/science/journal/03603199.

Chen, Y., Jin, X., Zeng, L., Zhong, Z., Mehana, M., Xiao, W., Pu, W., Regenauer-Lieb, K., & Xie, Q. (2023). Role of large-scale underground hydrogen storage and its pathways to achieve net-zero in China. *Journal of Energy Storage, 72*, 108448. Available from https://doi.org/10.1016/j.est.2023.108448.

Chen, Z., Zhou, F., & Rahman, S. S. (2014). Effect of cap rock thickness and permeability on geological storage of CO_2: Laboratory test and numerical simulation. *Energy Exploration and Exploitation, 32*(6), 943–964. Available from https://doi.org/10.1260/0144-5987.32.6.943, http://eea.sagepub.com/.

Civan, F. (2007). *Petrophysics – Flow functions and parameters* (pp. 101–124). Elsevier BV. Available from https://doi.org/10.1016/b978-075067738-7/50005-1.

Colas, E., Klopries, E.-M., Tian, D., Kroll, M., Selzner, M., Bruecker, C., Khaledi, K., Kukla, P., Preuße, A., Sabarny, C., Schüttrumpf, H., & Amann, F. (2023). Overview of converting abandoned coal mines to underground pumped storage systems: Focus on the underground reservoir. *Journal of Energy Storage, 73*, 109153. Available from https://doi.org/10.1016/j.est.2023.109153.

Crotogino, F. (2016). *Larger scale hydrogen storage storing energy: With special reference to renewable energy sources* (pp. 411–429). Elsevier Inc. Available from http://www.sciencedirect.com/science/book/9780128034408, https://doi.org/10.1016/B978-0-12-803440-8.00020-8.

Curiale, J. A., & Curtis, J. B. (2016). Organic geochemical applications to the exploration for source-rock reservoirs – A review. *Journal of Unconventional Oil and Gas Resources, 13*, 1–31. Available from https://doi.org/10.1016/j.juogr.2015.10.001, http://www.sciencedirect.com/science/journal/22133976.

D3.1_Overview-of-all-known-underground-storage-technologies (n.d.).

De, S. K., & Prabu, V. (2017). Experimental studies on humidified/water influx O_2 gasification for enhanced hydrogen production in the context of underground coal gasification. *Journal of Hydrogen Energy, 42*(20), 14089–14102. Available from https://doi.org/10.1016/j.ijhydene.2017.04.112, http://www.journals.elsevier.com/international-journal-of-hydrogen-energy/.

Delshad, M., Umurzakov, Y., Sepehrnoori, K., Eichhubl, P., & Fernandes, B. R. B. (2022). Hydrogen storage assessment in depleted oil reservoir and saline aquifer. *Energies, 15*(21), 8132. Available from https://doi.org/10.3390/en15218132.

Dias, W., Roehl, D., Mejia, C., & Sotomayor, P. (2023). Cavern integrity for underground hydrogen storage in the Brazilian pre-salt fields. *Journal of Hydrogen Energy, 48*(69), 26853–26869. Available from https://doi.org/10.1016/j.ijhydene.2023.03.292, http://www.journals.elsevier.com/international-journal-of-hydrogen-energy/.

EHB. (2024). European hydrogen backbone maps. https://www.ehb.eu/page/european-hydrogen-backbone-maps.

EWE. (2022). Status quo – The path to the future of energy. https://www.ewe.com/en/shaping-the-future/weg-in-die-energiezukunft/status-quo.

Elberry, A. M., Thakur, J., & Veysey, J. (2021). Seasonal hydrogen storage for sustainable renewable energy integration in the electricity sector: A case study of Finland. *Journal of Energy Storage, 44*, 103474. Available from https://doi.org/10.1016/j.est.2021.103474.

Elberry, A. M., Thakur, J., Santasalo-Aarnio, A., & Larmi, M. (2021). Large-scale compressed hydrogen storage as part of renewable electricity storage systems. *Journal of Hydrogen Energy, 46*(29), 15671–15690. Available from https://doi.org/10.1016/j.ijhydene.2021.02.080, http://www.journals.elsevier.com/international-journal-of-hydrogen-energy/.

Engie. (2021). H_2 in the underground: Are salt caverns the future of hydrogen storage? http://innovation.engie.com/en/news/news/did-you-know-/hydrogen-underground-storage-salt-caverns/25906.

Epelle, E. I., Obande, W., Udourioh, G. A., Afolabi, I. C., Desongu, K. S., Orivri, U., Gunes, B., & Okolie, J. A. (2022). Perspectives and prospects of underground hydrogen storage and natural hydrogen. *Sustainable Energy & Fuels, 6*(14), 3324–3343. Available from https://doi.org/10.1039/d2se00618a.

Fanchi, J. R. (2002). *Measures of rock–fluid interactions* (pp. 108–132). Elsevier BV. Available from https://doi.org/10.1016/b978-075067522-2/50007-0.

Feldmann, F., Hagemann, B., Ganzer, L., & Panfilov, M. (2016). Numerical simulation of hydrodynamic and gas mixing processes in underground hydrogen storages. *Environmental Earth Sciences, 75*(16). Available from https://doi.org/10.1007/s12665-016-5948-z, http://www.springerlink.com/content/121380/.

Flanigan, O. (1995). *Characteristics of underground storage* (pp. 54–67). Elsevier BV. Available from https://doi.org/10.1016/b978-088415204-0/50006-6.

Garbe, A. (2023). H2CAST Etzel – Making energy transition work.

Gautier, D. L. (2004). *Oil and natural gas resource assessment: Geological methods* (pp. 607–615). Elsevier BV. Available from https://doi.org/10.1016/b0-12-176480-x/00266-7.

Geologic CO_2 Sequestration Technology and Cost Analysis—Technical Support Document. (2008).

van Gessel (2023). Serge task 42: Underground hydrogen storage. https://www.ieahydrogen.org/task/task-42-underground-hydrogen-storage/.

Golf-Racht, T. D. V. (1982). *Physical properties of rocks. Developments in Petroleum Science.* (12). Elsevier. Available from https://doi.org/10.1016/S0376-736108.

HDF Energy (2024). Game-changing hydrogen solutions: Decarbonizing the power, heavy mobility and industrial sectors. https://www.hdf-energy.com/.

Haddad, P. G., Ranchou-Peyruse, M., Guignard, M., Mura, J., Casteran, F., Ronjon-Magand, L., Senechal, P., Isaure, M. P., Moonen, P., Hoareau, G., Dequidt, D., Chiquet, P., Caumette, G., Cezac, P., & Ranchou-Peyruse, A. (2022). Geological storage of hydrogen in deep aquifers – An experimental multidisciplinary study. *Energy and Environmental Science, 15*(8), 3400–3415. Available from https://doi.org/10.1039/d2ee00765g, http://pubs.rsc.org/en/journals/journal/ee.

Hangx, S. J. T., Spiers, C. J., & Peach, C. J. (2010). Mechanical behavior of anhydrite caprock and implications for CO_2 sealing capacity. *Journal of Geophysical Research: Solid Earth, 115*(7). Available from https://doi.org/10.1029/2009JB006954, http://onlinelibrary.wiley.com/journal/10.1002/(ISSN)2169-9356.

Henkel, S., Pudlo, D., & Gaupp, R. (2013). Research sites of the H_2STORE project and the relevance of lithological variations for hydrogen storage at depths. *Energy Procedia, 40*, 25–33. Available from https://doi.org/10.1016/j.egypro.2013.08.004, http://www.sciencedirect.com/science/journal/18766102.

HyPSTER. (2021). First EU-supported large scale green H_2 underground storage demonstrator. https://hypster-project.eu/about-the-project/.

Hystock. (2021). Demonstration hydrogen storage A8. https://www.hystock.nl/en/about-hystock/demonstration-hydrogen-storage-a8.

Hyne, N. J. (2001). *Nontechnical guide to petroleum geology, exploration, drilling, and production.* Penn Well Corporation.

IEA. (2021). Proving the viability of underground hydrogen storage. https://www.iea.org/articles/proving-the-viability-of-underground-hydrogen-storage.

IEA 50. (2023). Hydrogen. https://www.iea.org/energy-system/low-emission-fuels/hydrogen.

Ilgen, A. G., Aman, M., Espinoza, D. N., Rodriguez, M. A., Griego, J. M., Dewers, T. A., Feldman, J. D., Stewart, T. A., Choens, R. C., & Wilson, J. (2018). Shale-brine-CO_2 interactions and the long-term stability of carbonate-rich shale caprock. *International Journal of Greenhouse Gas Control, 78*, 244–253. Available from https://doi.org/10.1016/j.ijggc.2018.07.002, http://www.elsevier.com/wps/find/journaldescription.cws_home/709061/description#description.

Ineris. (2022). State of knowledge on the storage of hydrogen in salt caverns. https://www.ineris.fr/en/state-knowledge-storage-hydrogen-salt-caverns.

Jafari Raad, S. M., Leonenko, Y., & Hassanzadeh, H. (2022). Hydrogen storage in saline aquifers: Opportunities and challenges. *Renewable and Sustainable Energy Reviews, 168*, 112846. Available from https://doi.org/10.1016/j.rser.2022.112846.

Kadkhodaie, A., Kadkhodaie, R. (n.d.). Chapter 4 – Pore type and pore size distribution. Reservoir characterization of tight gas sandstones. 2022. https://doi.org/10.1016/B978-0-323-90180-2.00006-2.

Kapusta, K., Stańczyk, K., Wiatowski, M., & Chećko, J. (2013). Environmental aspects of a field-scale underground coal gasification trial in a shallow coal seam at the experimental mine barbara in Poland. *Fuel, 113*, 196–208. Available from https://doi.org/10.1016/j.fuel.2013.05.015.

Klemme, H. D. (1977). *World oil and gas reserves from analysis of giant fields and petroleum basins (PROVINCES)* (pp. 217–260). Elsevier BV. Available from https://doi.org/10.1016/b978-0-08-021735-2.50022-7.

Krishna, R., Titus, E., Salimian, M., Okhay, O., Rajendran, S., Rajkumar, A., Sousa, J. M. G., Ferreira, A. L. C., Campos, J., & Gracio, J. (2012). *Hydrogen storage for energy application.* InTech. Available from 10.5772/51238.

Laadel, N.-E., El Mansori, M., Kang, N., Marlin, S., & Boussant-Roux, Y. (2022). Permeation barriers for hydrogen embrittlement prevention in metals – A review on mechanisms, materials suitability and efficiency. *International Journal of Hydrogen Energy, 47*(76), 32707–32731. Available from https://doi.org/10.1016/j.ijhydene.2022.07.164.

Lackey, G., Freeman, G. M., Buscheck, T. A., Haeri, F., White, J. A., Huerta, N., & Goodman, A. (2023). Characterizing hydrogen storage potential in U.S. underground gas storage facilities. *Geophysical Research*

Letters, 50(3). Available from https://doi.org/10.1029/2022GL101420, http://agupubs.onlinelibrary.wiley.com/hub/journal/10.1002/(ISSN)1944-8007/.

Li, H., Qin, B., Jiang, Y., Zhao, Y., & Shi, W. (2022). Data-driven optimal scheduling for underground space based integrated hydrogen energy system. *IET Renewable Power Generation, 16*(12), 2521–2531. Available from https://doi.org/10.1049/rpg2.12386, https://ietresearch.onlinelibrary.wiley.com/journal/17521424.

Linde. (2024). Storing hydrogen in underground salt caverns. https://www.lindehydrogen.com/technology/hydrogen-storage.

Liu, X., Shi, X., Li, Y., Li, P., Zhao, K., Ma, H., Yang, C. (2021). Maximum gas production rate for salt cavern gas storages. *Energy. 234*, 121211, Available from https://doi.org/10.1016/j.energy.2021.121211.

Londe, L. F. (2021). Society of petroleum engineers undefined four ways to store large quantities of hydrogen. Paper presented at the Abu Dhabi International Petroleum Exhibition & Conference, Abu Dhabi, UAE, November 2021. Paper Number: SPE-208178-MS. https://onepetro.org/SPEADIP/21ADIP/conference/1-21ADIP, https://doi.org/10.2118/208178-MS.

Lyons, W. C., & Plisga, G. J. (2004). *Reservoir engineering. Standard handbook of petroleum and natural gas engineering* (pp. 1–225). Gulf Professional Publishing. Available from https://doi.org/10.1016/B978-075067785-1/50017-8.

Lysyy, M., Fernø, M., & Ersland, G. (2021). Seasonal hydrogen storage in a depleted oil and gas field. *Journal of Hydrogen Energy, 46*(49), 25160–25174. Available from https://doi.org/10.1016/j.ijhydene.2021.05.030, http://www.journals.elsevier.com/international-journal-of-hydrogen-energy/.

Lyu, X., Yun, L., Xu, J., Liu, H., Yu, X., Peng, P., Ouyang, M., & Luo, Y. (2023). Sealing capacity evolution of gypsum salt caprocks under multi-cycle alternating stress during operations of underground gas storage. *Journal of Petroleum Science and Engineering, 220*, 111244. Available from https://doi.org/10.1016/j.petrol.2022.111244.

Magoon, L. B. (2004). *Petroleum system: Nature's distribution system for oil and gas* (pp. 823–836). Elsevier BV. Available from https://doi.org/10.1016/b0-12-176480-x/00251-5.

Mazarei, M., Davarpanah, A., Ebadati, A., & Mirshekari, B. (2019). The feasibility analysis of underground gas storage during an integration of improved condensate recovery processes. *Journal of Petroleum Exploration and Production Technology, 9*(1), 397–408. Available from https://doi.org/10.1007/s13202-018-0470-3, http://www.springerlink.com/content/2190-0566/.

McPhee, C., Reed, J., & Zubizarreta, I. (2015). Capillary pressure. *Developments in Petroleum Science, 64*. Available from https://doi.org/10.1016/B978-0-444-63533-4.00009-3, http://www.elsevier.com/wps/find/bookdescription.cws_home/BS_DPS/description#description.

Meda, U. S., Bhat, N., Pandey, A., Subramanya, K. N., & Lourdu Antony Raj, M. A. (2023). Challenges associated with hydrogen storage systems due to the hydrogen embrittlement of high strength steels. *Journal of Hydrogen Energy, 48*(47), 17894–17913. Available from https://doi.org/10.1016/j.ijhydene.2023.01.292, http://www.journals.elsevier.com/international-journal-of-hydrogen-energy/.

van der Meer, L. G. H., & Obdam, A. (2008). Is carbon dioxide in case of natural gas storage a feasible cushion gas? https://co2-cato.org/publications/carbon-dioxide-case-natural-gas-storage/

Miocic, J., Heinemann, N., Edlmann, K., Scafidi, J., Molaei, F., & Alcalde, J. (2022). Underground hydrogen storage: A review. *Geological Society, London, Special Publications, 528*(1), 73–86. Available from https://doi.org/10.1144/sp528-2022-88.

Muhammed, N. S., Haq, B., Al Shehri, D., Al-Ahmed, A., Rahman, M. M., & Zaman, E. (2022). A review on underground hydrogen storage: Insight into geological sites, influencing factors and future outlook. *Energy Reports, 8*, 461–499. Available from https://doi.org/10.1016/j.egyr.2021.12.002, http://www.journals.elsevier.com/energy-reports/.

Muhammed, N. S., Haq, M. B., Al Shehri, D. A., Al-Ahmed, A., Rahman, M. M., Zaman, E., & Iglauer, S. (2023). Hydrogen storage in depleted gas reservoirs: A comprehensive review. *Fuel, 337*. Available from https://doi.org/10.1016/j.fuel.2022.127032, http://www.journals.elsevier.com/fuel/.

Muhammed, N. S., Haq, B., & Al Shehri, D. A. (2023). Hydrogen storage in depleted gas reservoirs using nitrogen cushion gas: A contact angle and surface tension study. *Journal of Hydrogen Energy, 48*(98), 38782–38807. Available from https://doi.org/10.1016/j.ijhydene.2023.06.208, http://www.journals.elsevier.com/international-journal-of-hydrogen-energy/.

Møller, K. T., Jensen, T. R., Akiba, E., & Li, Hw (2017). Hydrogen – A sustainable energy carrier. *Progress in Natural Science: Materials International, 27*(1), 34–40. Available from https://doi.org/10.1016/j.pnsc.2016.12.014, https://www.sciencedirect.com/journal/progress-in-natural-science-materials-international/issues.

Najjar, Y. S. H. (2013). Hydrogen safety: The road toward green technology. *Journal of Hydrogen Energy, 38*(25), 10716–10728. Available from https://doi.org/10.1016/j.ijhydene.2013.05.126, http://www.journals.elsevier.com/international-journal-of-hydrogen-energy/.

Ni, M. (2006). An overview of hydrogen storage technologies. *Energy Exploration and Exploitation, 24*(3), 197–209. Available from https://doi.org/10.1260/014459806779367455, http://docserver.ingentaconnect.com/deliver/connect/mscp/01445987/v24n3/s5.pdf?expires = 1168407179&id = 34505847&titleid = 871&accname = Innodata + -+ CAB + Abstracts&checksum = 5986B5035EA68F841764C7AA55459E72.

Nuclear Energy Agency. (2014). Natural analogues for safety cases of repositories in rock salt: Salt club workshop proceedings. https://www.oecd-nea.org/jcms/pl_19454/natural-analogues-for-safety-cases-of-repositories-in-rock-salt-salt-club-workshop-proceedings?details = true.

Okoroafor, E. R., Kim, T. W., Nazari, N., Watkins, H. Y., Saltzer, S. D., & Kovscek, A. R. (2022). Assessing the underground hydrogen storage potential of depleted gas fields in Northern California, *Paper presented at the SPE annual technical conference and exhibition*. Houston, Texas, USA, October 2022. Paper Number: SPE-209987-MS. https://www.onepetro.org/conferences/SPE/97SPE.

Osman, A. I., Mehta, N., Elgarahy, A. M., Hefny, M., Al-Hinai, A., Al-Muhtaseb, A. H., & Rooney, D. W. (2022). Hydrogen production, storage, utilisation and environmental impacts: A review. *Environmental Chemistry Letters, 20*(1), 153–188. Available from https://doi.org/10.1007/s10311-021-01322-8, http://springerlink.metapress.com/app/home/journal.asp?wasp = d86tgdwvtg0yvw9gvkwp&referrer = parent&backto = browsepublicationsresults,140,541.

Otto, M., Chagoya, K. L., Blair, R. G., Hick, S. M., & Kapat, J. S. (2022). Optimal hydrogen carrier: Holistic evaluation of hydrogen storage and transportation concepts for power generation, aviation, and transportation. *Journal of Energy Storage, 55*, 105714. Available from https://doi.org/10.1016/j.est.2022.105714.

PHMSA. (2021). 2023 12 26 Fact sheet: Underground natural gas storage caverns. https://www.phmsa.dot.gov/pipeline/underground-natural-gas-storage/fact-sheet-underground-natural-gas-storage-caverns.

Panfilov, M. (2015). Underground and pipeline hydrogen storage. *Compendium of hydrogen energy: Hydrogen storage, distribution and infrastructure* (Volume 2, pp. 91–115). Elsevier. Available from https://www.sciencedirect.com/book/9781782423621, https://doi.org/10.1016/B978-1-78242-362-1.00004-3.

Patanwar, Y. K., Kim, H. M., Deb, D., & Gujjala, Y. K. (2024). Underground storage of hydrogen in lined rock caverns: An overview of key components and hydrogen embrittlement challenges. *Journal of Hydrogen Energy, 50*, 116–133. Available from https://doi.org/10.1016/j.ijhydene.2023.08.342, https://www.sciencedirect.com/science/journal/03603199.

Peppas, A., Kottaridis., & Politi, C. (2023). Potential application of RES and underground H_2 storage in abandoned mines. *Materials Proceedings, 15*. Available from https://doi.org/10.3390/materproc2023015030.

Perera, M. S. A. (2023). A review of underground hydrogen storage in depleted gas reservoirs: Insights into various rock-fluid interaction mechanisms and their impact on the process integrity. *Fuel, 334*, 126677. Available from https://doi.org/10.1016/j.fuel.2022.126677.

Portarapillo, M., & Di Benedetto, A. (2021). Risk assessment of the large-scale hydrogen storage in salt caverns. *Energies, 14*(10), 2856. Available from https://doi.org/10.3390/en14102856.

Preuster, P., Alekseev, A., & Wasserscheid, P. (2017). Hydrogen storage technologies for future energy systems. *Annual Review of Chemical and Biomolecular Engineering, 8*, 445–471. Available from https://doi.org/10.1146/annurev-chembioeng-060816-101334, http://www.annualreviews.org/journal/chembioeng.

Puntervold, T., Strand, S., Mamonov, A., & Piñerez, I. D. T. (2022). *Enhanced oil recovery by smart water injection in sandstone reservoirs. Recovery improvement* (pp. 109–184). Elsevier. Available from https://www.sciencedirect.com/book/9780128233634, https://doi.org/10.1016/B978-0-12-823363-4.00002-9.

Quintos Fuentes, J. E., & Santos, D. M. F. (2023). Technical and economic viability of underground hydrogen storage. *Hydrogen (Switzerland), 4*(4), 975–1000. Available from https://doi.org/10.3390/hydrogen4040057, https://www.mdpi.com/journal/hydrogen.

Ramesh Kumar, K., Honorio, H., Chandra, D., Lesueur, M., & Hajibeygi, H. (2023). Comprehensive review of geomechanics of underground hydrogen storage in depleted reservoirs and salt caverns. *Journal of Energy Storage, 73*, 108912. Available from https://doi.org/10.1016/j.est.2023.108912.

Raza, A., Arif, M., Glatz, G., Mahmoud, M., Al Kobaisi, M., Alafnan, S., & Iglauer, S. (2022). A holistic overview of underground hydrogen storage: Influencing factors, current understanding, and outlook. *Fuel, 330*, 125636. Available from https://doi.org/10.1016/j.fuel.2022.125636.

Reitenbach, V., Ganzer, L., Albrecht, D., & Hagemann, B. (2015). Influence of added hydrogen on underground gas storage: A review of key issues. *Environmental Earth Sciences, 73*(11), 6927–6937. Available from https://doi.org/10.1007/s12665-015-4176-2, http://www.springerlink.com/content/121380/.

Ruiz Maraggi, L. M., & Moscardelli, L. G. (2023). Modeling hydrogen storage capacities, injection and withdrawal cycles in salt caverns: Introducing the GeoH_2 salt storage and cycling app. *Journal of Hydrogen Energy, 48*(69), 26921–26936. Available from https://doi.org/10.1016/j.ijhydene.2023.03.293, http://www.journals.elsevier.com/international-journal-of-hydrogen-energy/.

SSAB. (2022). HYBRIT: A unique, underground, fossil-free hydrogen gas storage facility is being inaugurated in Luleå. https://www.ssab.com/en/news/2022/06/hybrit-a-unique-underground-fossilfree-hydrogen-gas-storage-facility-is-being-inaugurated-in-lule.

Sekar, L. K., Kiran, R., Okoroafor, E. R., & Wood, D. A. (2023). Review of reservoir challenges associated with subsurface hydrogen storage and recovery in depleted oil and gas reservoirs. *Journal of Energy Storage, 72*. Available from https://doi.org/10.1016/j.est.2023.108605, http://www.journals.elsevier.com/journal-of-energy-storage/.

Selley, R. C. (1998). *Elements of petroleum geology*. Gulf Professional Publishing.

Seo, S. K., Yun, D. Y., & Lee, C. J. (2020). Design and optimization of a hydrogen supply chain using a centralized storage model. *Applied Energy, 262*. Available from https://doi.org/10.1016/j.apenergy.2019.114452, https://www.journals.elsevier.com/applied-energy.

Smart Energy International. (2022). Underground rock cavern hydrogen storage facility inaugurated in Luleå https://www.smart-energy.com/storage/underground-rock-cavern-hydrogen-storage-facility-inaugurated-in-lulea/.

Speight, J. G. (2019). *Recovery, storage, and transportation* (pp. 149–186). Elsevier BV. Available from https://doi.org/10.1016/b978-0-12-809570-6.00005-9.

Tao, Y., He, Y., Zhao, Z., Wu, D., & Deng, Q. (2023). Sealing of oil-gas reservoir caprock: Destruction of shale caprock by micro-fractures. *Frontiers in Earth Science, 10*. Available from https://doi.org/10.3389/feart.2022.1065875, https://www.frontiersin.org/journals/earth-science.

Tarkowski, R. (2019). Underground hydrogen storage: Characteristics and prospects. *Renewable and Sustainable Energy Reviews, 105*, 86–94. Available from https://doi.org/10.1016/j.rser.2019.01.051, https://www.journals.elsevier.com/renewable-and-sustainable-energy-reviews.

Thiyagarajan, S. R., Emadi, H., Hussain, A., Patange, P., & Watson, M. (2022). A comprehensive review of the mechanisms and efficiency of underground hydrogen storage. *Journal of Energy Storage, 51*, 104490. Available from https://doi.org/10.1016/j.est.2022.104490.

Tiab, D., & Donaldson, E. C. (2016). *Porosity and permeability* (pp. 67–186). Elsevier BV. Available from https://doi.org/10.1016/b978-0-12-803188-9.00003-6.

Tutorial 5 – Hydrogen Storage in depleted reservoirs—PFLOTRAN documentation. (2024).

Uniper. (2022). Gas storage technology. https://www.uniper.energy/energy-storage-uniper/gas-storage-technology.

Veziroglu, N. T., Sherif, S. A., & Barbir, F. (2005). *Hydrogen energy solutions. Environmental solutions* (pp. 143–180). Elsevier Inc. Available from http://www.sciencedirect.com/science/book/9780120884414, https://doi.org/10.1016/B978-012088441-4/50008-3.

Webb, J. A., & White, S. (2013). Karst in deserts. *Treatise on Geomorphology, 6*. Available from http://www.sciencedirect.com/science/referenceworks/9780080885223. https://doi.org/10.1016/B978-0-12-374739-6.00138-X.

Wheaton, R. (2016). *Basic rock and fluid properties* (pp. 5–57). Elsevier BV. Available from https://doi.org/10.1016/b978-0-08-101019-8.00002-8.

Zeng, L., Sarmadivaleh, M., Saeedi, A., Chen, Y., Zhong, Z., & Xie, Q. (2023). Storage integrity during underground hydrogen storage in depleted gas reservoirs. *Earth-Science Reviews, 247*, 104625. Available from https://doi.org/10.1016/j.earscirev.2023.104625.

Zivar, D., Kumar, S., & Foroozesh, J. (2021). Underground hydrogen storage: A comprehensive review. *Journal of Hydrogen Energy, 46*(45), 23436–23462. Available from https://doi.org/10.1016/j.ijhydene.2020.08.138, http://www.journals.elsevier.com/international-journal-of-hydrogen-energy/.

Zohuri, B. (2018). *Hydrogen storage processes and technologies* (pp. 257–279). Springer Nature America, Inc. Available from https://doi.org/10.1007/978-3-319-93461-7_8.

CHAPTER

7

Physics of hydrogen flow in porous media: two-phase (or multiphase) flow and transport of hydrogen in porous geological storage

Birger Hagemann, Sebastian Hogeweg and Julia Michelsen

Department Reservoir Engineering, Clausthal University of Technology, Institute of Subsurface Energy Systems, Clausthal-Zellerfeld, Germany

7.1 Introduction

The storage of pure hydrogen and hydrogen gas mixtures in porous geological formations was rarely done in the past. The different characteristics of hydrogen, such as density and viscosity, compared to natural gas or methane could lead to an unexpected flow behavior in the subsurface. Both aquifers and depleted hydrocarbon reservoirs can be used to develop underground hydrogen pore storages. However, as illustrated in Fig. 7.1, the governing processes during the development period will differ. Aquifers are initially saturated only by water or brine and a gas cap is created during the development. Depleted hydrocarbon reservoirs, where the focus is on gas reservoirs, can have high residual gas saturation but exhibit depleted reservoir pressures.

The development of a gas cap in aquifers requires the displacement of water. This two-phase displacement process could be inefficient because hydrogen has a low viscosity and very low density in contrast to water. Consequences could be the occurrence of strongly pronounced viscous fingering and gravity override of the water phase (Carden & Paterson, 1979; Paterson, 1983). The two-phase flow in porous media is a complex process in which the combination of pore scale and macroscale effects must be considered. This is discussed in more detail in Section 7.4.

In depleted gas reservoirs, some residual gas remains in the reservoir. If the influx of aquifer water into the reservoir was weak, the residual gas saturation almost corresponds

Subsurface Hydrogen Energy Storage
DOI: https://doi.org/10.1016/B978-0-443-24071-3.00008-X

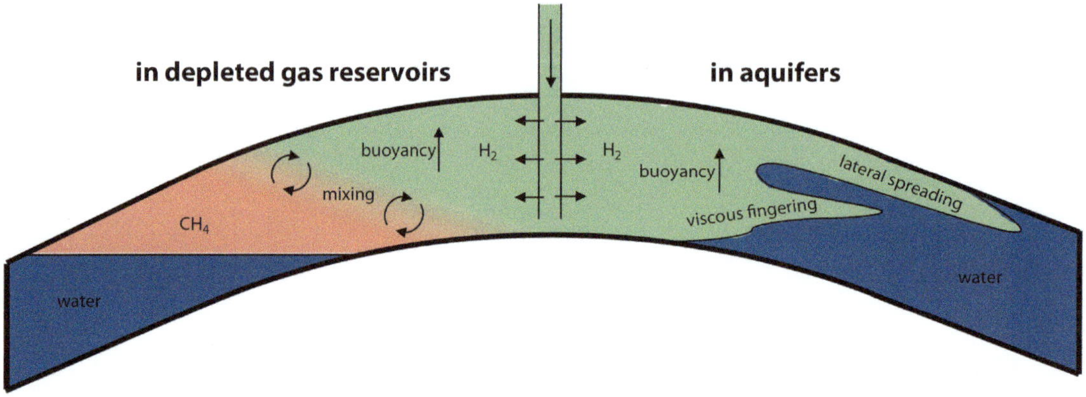

FIGURE 7.1 Sketch of the relevant flow characteristics during underground hydrogen storage.

to the initial gas saturation and only the pressure was depleted. A challenge could be the displacement of the residual gas by hydrogen. Different conversion schemes are suggested in the literature for the transformation of gas storages from one gas to another, which can also be applied for the development of an UHS (Tek, 1989). A transformation by cyclic injection and production using the same wells could result in a highly contaminated gas production, especially in the first years of operation. In other transformation schemes, hydrogen is injected on one edge of the reservoir. This pushes the residual gas to the other side where it can be produced using other wells. This sweeping strategy potentially results in a more pure hydrogen withdrawal. The low density of hydrogen can also be utilized for this purpose. In a storage layer with a large thickness, hydrogen can be injected through perforations at the top of the structure and the natural gas can be withdrawn at the bottom through other wells or perforation intervals. As different gases are miscible under storage conditions, gas mixing with these methods is unavoidable. The mixing process between two gases, which is influenced by advective or dispersive flow but also by molecular diffusion, is described in more detail in Section 7.5.

Two-phase flow and gas mixing processes must also be taken into account during storage operation. Two-phase processes are particularly relevant at the contact between gas cap and aquifer and can lead to gas trapping or a premature water production via the storage wells. On the other hand, gas mixing processes can lead to unwanted contamination of the withdrawn gas.

In this chapter, the relevant fluid properties (Section 7.2) and flow geometry and regimes (Section 7.3) in the subsurface are discussed first. The nonmiscible and miscible displacement processes mentioned above are then explained in Sections 7.4 and 7.5.

7.2 Fluid properties of hydrogen

When hydrogen or gases (gas mixture containing hydrogen) are injected into the subsurface, it needs to be considered that thermodynamic properties but also energy densities are different from those of conventionally stored gas (natural gas). For the flow processes in the subsurface, especially the fluid transport properties, need to be taken into account. As hydrogen is compressible, the fluid properties are strongly influenced by pressure and temperature. As a result, the transport properties need to be regarded under storage conditions which are typically in the range of 20–200 bar and 20°C–100°C. Under these conditions, hydrogen is in supercritical state but still behaves gas-like. Hydrogen, as the lightest chemical element overall, comes with low viscosity and less compressibility than most other gases. An example of the trend of compressibility (Z-factor) and viscosity versus pressure in comparison to relevant natural gas components is depicted in Fig. 7.2 and Fig. 7.3.

For most gases, the Z-factor at increased pressures is below one, yielding increased compressibility, while the compressibility of hydrogen worsens with increasing pressure. Methane has the highest compressibility between 150 and 200 bar (at 50°C). For hydrogen there is no optimum in this respect. Nevertheless, the storage capacity increases with rising pressure, but the increase is lower than for an ideal gas. The low density and viscosity of hydrogen can have an influence on the forces acting in two-phase flow processes. In terms of viscosity, hydrogen comes up with the lowest values and is approximately 1–1.5 times smaller than the viscosity of methane. While the density mainly impacts phenomena such as gravity override, the viscosity differences could lead to unstable displacements characterized by viscous fingering. The larger interfacial tension (IFT) between hydrogen and water/brine compared to other gases could have an impact on capillary effects and thus relative permeability curves what alters the two-phase displacement further (discussed in Section 7.4). The mixing of the components in the gas phase (e.g., with the cushion gas) is strongly governed by molecular diffusion and mechanical dispersion. In this context, the high diffusivity of hydrogen compared to other gases must be considered more closely (discussed in Section 7.5).

For the two-phase flow processes also, the solubility of hydrogen in the formation water or brine needs to be considered. At 100 bar and 25°C approximately 1.7 Sm^3 of H_2 are soluble in 1 m^3 of water. The hydrogen solubility increases with pressure roughly following Henry's law (Chabab et al., 2024). In Fig. 7.4, the solubility is shown versus brine salinity at different temperatures.

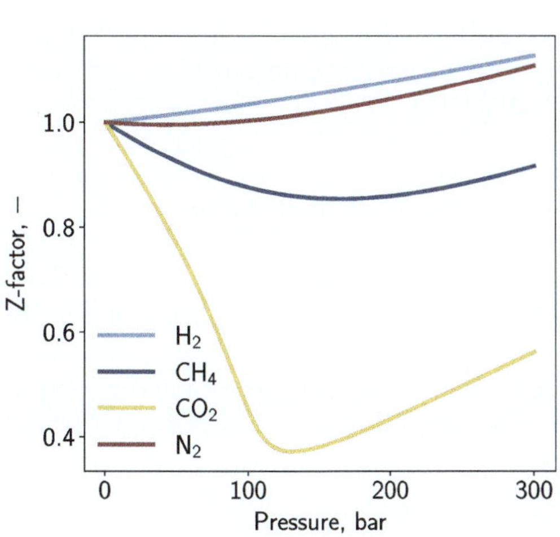

FIGURE 7.2 Compressibility factor of relevant gases at 50°C, calculated based on Peng and Robinson (1976).

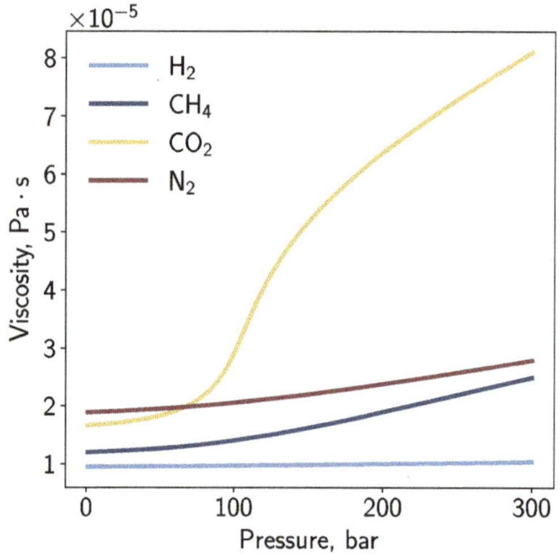

FIGURE 7.3 Viscosity of relevant gases at 50°C, calculated based on Peng and Robinson (1976) and viscosity model according to Stiel and Thodos (1961).

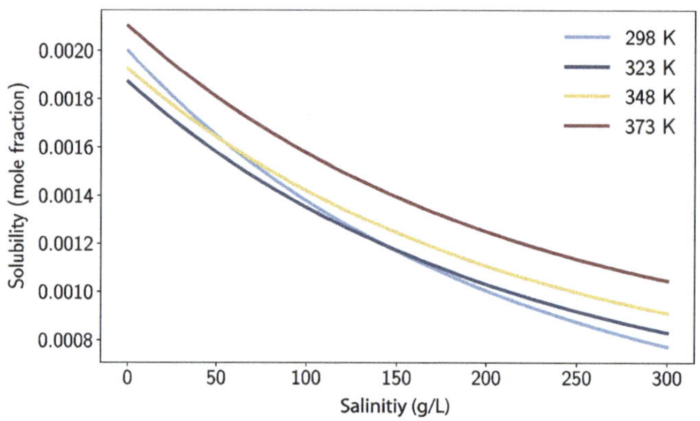

FIGURE 7.4 Hydrogen solubility versus salinity at different temperatures calculated by the Chabab et al. (2024) correlation.

With increasing salinity less hydrogen becomes soluble in brine. At a salinity of 234 g/L the solubility is decreased to approximately 0.7 Sm^3/m^3 of water. Compared to methane, the hydrogen solubility is slightly lower (Chabab et al., 2021).

7.3 Reservoir geometry, flow regimes, and rock properties

For the physics of flow, it is first necessary to consider the geometric conditions, flow regimes and properties of the porous media. Fig. 7.5 shows a typical structure of a pore storage and its characteristics.

In general, the reservoir consists of the following components: (1) storage formation(s) separated into the gas cap and aquifer; (2) overburden/caprock, and (3) observation horizon. Several wells are available for the operation of the storage. These can be divided into storage and observation wells. The storage wells are used to inject and withdraw the gas, while the observation wells

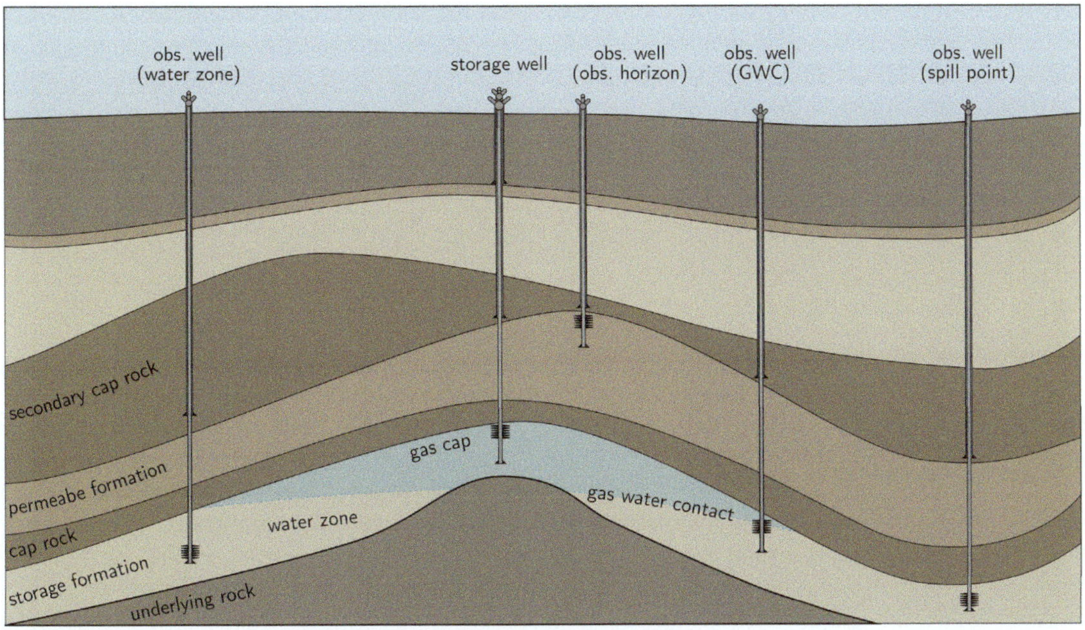

FIGURE 7.5 Schematic structure of a pore storage (Katz & Lee, 1990). *Source: Adapted from Katz, D. L. & Lee, R. L. (1990). Natural Gas Engineering Production and Storage. New York: McGraw-Hill Publishing Co.*

are used to monitor the storage facility. These are typically used both in the storage formation to monitor the pressure and spatial spreading of the gas cap (e.g., at the possible spill point) and in overlying permeable rock layers to monitor possible communication with the storage horizon.

The various flow regimes that occur in the storage layers originating from the wells are described below. This refers to the number of phases (single-phase vs two-phase flow) and the acting forces (viscous forces vs inertial forces) on the one hand but also to the flow geometry (horizontal and radial flow) on the other hand.

7.3.1 Flow regimes

Different flow regimes can occur in the rock layers of a pore storage depending on the pressure gradient, rock properties and fluid properties: pre-Darcy flow, Darcy flow, Forchheimer flow, and turbulent flow. In the majority of the high permeable rock layers, laminar flow takes place through the porous medium. When the flow is dominated by viscous forces, this type of flow can be described using Darcy's law (Darcy, 1856) on the continuum scale. Assuming that the reservoir rock is dry or that only immobile connate water is present, Darcy's law for single-phase flow is as follows:

$$u = -\frac{K}{\mu}(\nabla P - \rho g)$$

here, the Darcy velocity u in m/s is described as a function of the permeability of the rock (K in m^2), the dynamic viscosity of the fluid (μ in Pa • s) and the pressure gradient (∇P in Pa/m). In addition, a flow potential is created by the density of the fluid (ρ in kg/m^3) and the gravity acceleration (g in m/s^2).

Storage formations within which the gas can flow are usually thin with a height of a few meters up to some tens of meters. In comparison, the horizontal dimension is very large, up to several kilometers. The main flow direction is therefore horizontal. Considering the region around a well, the flow conditions are radial. The Darcy equation can be rearranged to calculate the radial flow around a well:

$$u = -\frac{2\pi rKh}{\mu}\frac{dP}{dr}$$

where h is the height of the formation in m and r is the radius in m. The equation can be used to describe the inflow and outflow of storage wells. During injection, the gas decelerates due to the increasing cross-section for the flow. During withdrawal, in contrast, the gas is accelerated when it gets closer to the well as the flow cross-section decreases. The highest flow velocity is therefore in the immediate vicinity of the well. Gas storages have to provide high injection and withdrawal rates, usually one or two orders of magnitude higher than during the depletion of a reservoir. Consequently, it is possible that Darcy's law is no longer valid in the near-wellbore region, since a pressure gradient greater than the linear relationship is present due to the effects of inertia or even turbulent flow. An estimate can be made using the Reynolds number, which describes the relationship between inertial forces and viscous forces. For flow in porous media, it can be calculated as follows (Bear, 2018):

$$\mathrm{Re} = \frac{V\sqrt{\frac{K}{\phi T}}}{\nu}$$

where V is the average fluid velocity in m/s, K is the permeability in m^2, Φ is the porosity, T is the tortuosity of the porous medium, and ν is the kinematic viscosity in m^2/s. Darcy's law is valid for Reynolds numbers less than 1 (Bear, 2018). Sometimes it is still applied in the transition range up to 10. Above 10, the inertial forces dominate and the application of Darcy's law would lead to large deviations. In such cases, the Forchheimer equation can be applied, which is an extension of the Darcy equation (Forchheimer, 1901):

$$-\frac{dP}{dx} = \frac{\mu}{K}u + \frac{\rho}{k_2}u^2$$

here an additional pressure drop which is caused by the inertial forces is taken into account by a quadratic term and the non-Darcy permeability coefficient (k_2 in m). The Darcy or Forchheimer equations can be applied based on numerical or analytical methods to calculate flow rates in the vicinity of storage wells during injection and withdrawal.

In low-permeability intermediate layers, a deviation from Darcy's law in the other direction is also possible. This flow regime is referred to as pre-Darcy flow and can be expected at very low flow rates (Dejam et al., 2017). Pre-Darcy flow is characterized by a threshold in the pressure gradient which must be exceeded before a flow occurs. This behavior is due to the strong interaction between solid surfaces and fluid in the very narrow pore space. Solid–fluid surface forces and the resulting non-Newtonian fluid viscosity can play a role in this flow regime (Dejam et al., 2017).

As gas storages are operated with high injection and withdrawal rates, the main driving force during the operation will be compression and expansion of the gas cap. Thereby, a certain amount of gas remains always in the reservoir as cushion gas. However, it cannot be assumed that the formation water is immobile in all regions of the storage. A mobile water phase may also be present in the edges of the gas cap (transition to the aquifer) and depending on the strength of the water drive in some cases even near the wells. In this case, the flow behavior is characterized by the competing movement of two phases in the pore space. Interactions with oil in former oil reservoirs are also possible, resulting in a three-phase system. However, this will not be discussed further here as the focus for UHS is on depleted gas reservoirs and aquifers.

During gas injection, the aquifer water is forced back by the increasing pressure of the gas cap. During withdrawal, in contrast, water displaces the gas due to the decreasing pressure of the gas cap and reduces the size of the gas cap. For the description of these effects, Darcy's law can be extended to two phases:

$$u_i = -\frac{Kk_{ri}}{\mu_i}(\nabla P_i - \rho g)$$

where the subscript i refers to the fluid phase (gas or water). In addition to the parameters described above, the relative permeability curves (k_{ri}) are required here to describe the two-phase flow behavior. Relative permeability is usually a function of the saturation. It is used to reflect the ability of a phase to flow in the presence of another phase in the pore space relative to its ability to flow when it occupies the entire pore phase. Due to capillary pressure in the pore space the pressures of the two phase are not equal. Both, relative permeability and capillary pressure, are discussed in more detail in Section 7.4.

7.3.2 Rock properties

Suitable storage rocks must predominantly possess two characteristics: porosity and permeability. A certain porosity is necessary to provide the required storage volume for the gas. The rock pores must be sufficiently interconnected to provide a flow path for the gas. This rock property is referred to as permeability. Highly permeable sandstone formations offer ideal conditions for constructing a storage facility. However, since it is also necessary to consider which rock formations are available at the desired location, carbonate rocks are also used for gas storage in many cases. Fig. 7.6 and Fig. 7.7 show two histograms for the rock properties of the European gas storage facilities.

It can be seen that the storage formations of the majority of gas storages have a permeability between 30 mD and 2D with an average of around 1D. For the porosity, the majority is within the range of 0.12–0.3, with an average of around 0.21.

Another rock parameter is rock compressibility. On the one hand, elastic deformation of the rock changes the porosity and permeability depending on the pore pressure. On the

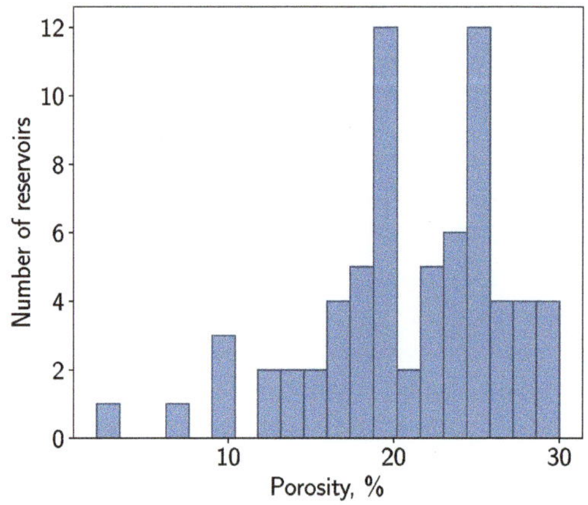

FIGURE 7.6 Porosity of the storage layer(s) of the European gas storage facilities.

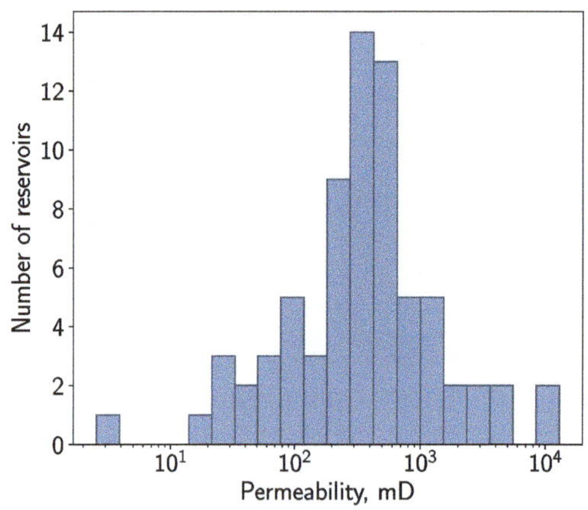

FIGURE 7.7 Permeability of the storage layer(s) of the European gas storage facilities.

other hand, the cyclically changing pore pressure also leads to a breathing effect (swelling when the storage is full and shrinking when it is empty).

7.3.3 Flow boundaries

Storage formations are often stratified due to their depositional origin. The individual layers have different rock properties and thus contribute more or less to the storage volume and flow path. At the top and bottom (possibly also between several storage layers) low-permeability layers act as a no flow boundary for the advective gas flow. Due to the buoyancy forces that cause gas to rise, the low-permeable layer above the storage (referred to as caprock) is particularly important. It is important that these water-saturated rock layers have a capillary threshold pressure that is high enough so that it is not exceeded during storage operation. Capillary threshold pressure is the pressure difference between the nonwetting (gas) and wetting phase (water) that must be exceeded so that the nonwetting phase can penetrate into a porous medium completely saturated with the wetting phase to displace the wetting phase. Suitable caprocks must also be water-wet when in contact with hydrogen and have a capillary threshold pressure of at least some tens of bars.

However, this does not mean that there is a complete barrier for hydrogen. Hydrogen can dissolve in the formation water and penetrate the low-permeability layers, for example, the caprock, through diffusive transport in the aqueous phase. This was investigated by experimental measurements of the diffusive flux through potential caprock samples (Michelsen et al., 2022; Salina Borello et al., 2024). The measurements have shown effective diffusion coefficients in the range of 8×10^{-11} to 6×10^{-8} m^2/s. Such low diffusion coefficients lead to very low diffusive rates compared to advective transport rates, and over time the concentration gradients become smaller, further reducing the diffusive rate. Even if this transport process does not disappear completely, it is often referred to as a once-only loss, which can be in the range of up to 2% of the storage capacity (Carden & Paterson, 1979).

7.4 Two-phase flow mechanism

In aquifers a gas cap has to be created during the development period, hence, the aquifer water has to be displaced. Depending on the external factors, a continuous or step-wise expansion of the gas cap is conceivable. In depleted gas fields, a gas cap is already in place. However, the operation of UHS, which will be done in a cyclic way with alternating periods of injection, withdrawal and idle, will nevertheless lead to water movements at the edges of the reservoir. Depending on the strength of the water drive, the size of the gas cap can still change, which means that the storage operation is characterized by the two-phase flow behavior. Under certain circumstances, the water phase can reach the storage wells during the withdrawal periods. The physical processes which are important for the flow and alternating displacement between two almost immiscible fluids depends on several factors which are focused in this section. The section is separated into wettability and IFT, relative permeability and hysteresis and displacement instabilities.

7.4.1 Wettability and interfacial tension

Gas and water are almost immiscible fluids that form a two-phase system within the porous rock. On pore scale, the phases are separated by abrupt interfaces as illustrated for a strong water-wet system in Fig. 7.8.

The interaction between the phases is related to forces at the fluid–solid (surface tension) and fluid–fluid (IFT) interfaces. For a gas–water systems usually water tends more to adhere to the solid surfaces than gas (Ahmed, 2006). This means that the system is water-wet where water is rather present in smaller pores while the gas phase will inhibit larger pores or channels (Ahmed, 2006). In addition to the adhesive forces, the behavior is influenced by the IFT between gas and water. The combination of these forces leads to a pressure difference across the gas–water interface which is referred to as capillary pressure. The

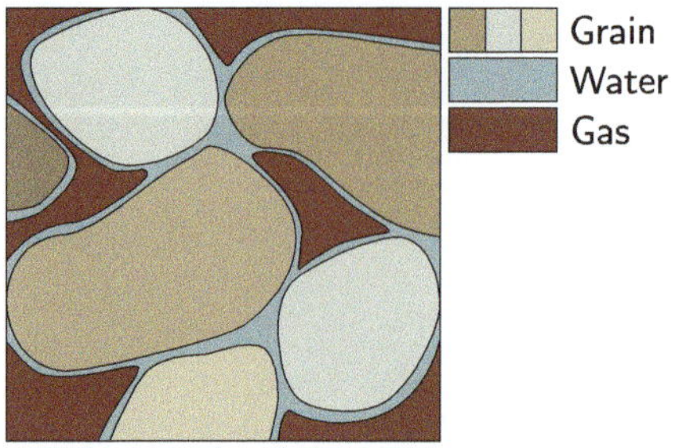

FIGURE 7.8 Sketch of fluid distribution on pore scale: for a gas–water system the wetting phase is water and the nonwetting phase is gas.

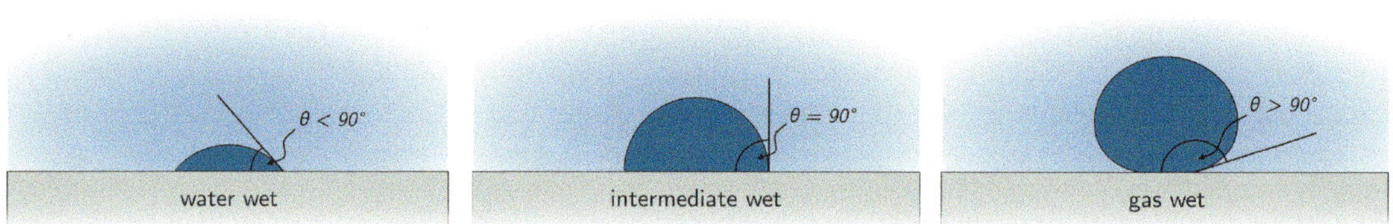

FIGURE 7.9 Illustration of the contact angle for water-wet, intermediate-wet, and gas-wet systems.

balance of these forces creates a curvature at the fluid interfaces. Based on this, the capillary pressure can be expressed by the Young–Laplace equation (Laplace, 1805; Young, 1805):

$$P_c = \sigma\left(\frac{1}{r_1} + \frac{1}{r_2}\right)$$

where P_c is the capillary pressure in Pa, σ is the IFT in N/m and r_1 and r_2 are the principal radii of the interface curvature in m.

The configuration of the phases in the porous medium results from the thermodynamic equilibrium of the system. The solid–liquid and solid–gas phase boundaries are predetermined because the pore walls are fixed. Since an equilibrium of the surface energies occurs in the thermodynamic equilibrium, the angle at which the fluid–fluid interfaces meet a pore wall results from the following balance equation (Young, 1805):

$$\gamma_{sg} - \gamma_{sw} - \gamma_{wg}\cos(\theta) = 0$$

where γ_{sg} is the solid–gas surface energy in J/m^2, γ_{sw} is the solid–water surface energy in J/m^2 and γ_{wg} is the water–gas surface energy in J/m^2. The angle θ (in °), known as the contact angle, describes the wettability of surfaces or pore walls in porous media. Fig. 7.9 depicts different types of wettability for the system solid–water–gas.

If the angle between the solid surface and the interface between water and gas is less than 90°, the system is water-wet. An angle of 90° means intermediate wettability and an angle greater than 90° means gas-wet. Instead of using the surface energies, the Young equation can be used to calculate the contact angle as a function of surface tension and IFT (Young, 1805):

$$\cos\theta = \frac{\sigma_{sg} - \sigma_{sw}}{\sigma_{wg}}$$

where σ_{wg} is the water–gas IFT in N/m, σ_{sw} is the solid–water surface tension in N/m and σ_{sg} is solid–gas surface tension in N/m.

In recent years, a number of studies have been carried out on the wettability of storage rocks in the presence of hydrogen as a gas phase. A list of studies on IFT between hydrogen or hydrogen-containing gas mixtures and water or brine is shown in Table 7.1.

The measurements show that the IFT ranges from 24 to 82 mN/m and is dependent on pressure, temperature and salinity of the water phase. From the studies it can be concluded that IFT decreases with increasing pressure and increasing temperature. Comparisons to other gases show that that for hydrogen–water/brine systems the IFT is slightly higher than for methane- and carbon dioxide–water/brine systems (Isfehani et al., 2023; Mirchi et al., 2022).

If in the Young–Laplace equation only the IFT is increased, this results in a larger contact angle and consequently a shift in the direction to gas-wet. In order to investigate this behavior in more detail, the contact angle of H_2–water–rock systems was also measured in various studies. In Table 7.2 a selection of studies is listed.

The measurements confirm that reservoir rocks are typically water-wet also when the gas phase is hydrogen. Most measurements show a strong to weakly water-wet behavior. Only a few measurements (e.g., with calcite at 100 bar by Esfandyari et al., 2022) show a very slight gas-wet contact angle. Generally, it was observed that the water/brine contact angle increases with increasing pressure (Esfandyari et al., 2022; Iglauer et al., 2021). Comparisons between hydrogen and methane show very similar contact angles (Hashemi et al., 2022). A water-wet behavior is advantageous in terms of effective gas permeability, as this means that high gas mobility can be expected at low water saturations during injection and withdrawal of hydrogen.

7.4.2 Relative permeability and hysteresis

The pore scale effects mentioned above also influence the flow of the fluid phases. Not the entire pore space is available for each phase, and consequently, the flow of the phases

TABLE 7.1 Summary of interfacial tension measurements for hydrogen–water/brine systems.

References	Used fluids	Temp., °C	Pressure, bar	Method	Interfacial tension, mN/m
Esfandyari et al. (2022)	H_2 and brine/distilled water	20–80	10–100	Captive drop	24–82
Mirchi et al. (2022)	H_2/H_2–CH_4 mixture and water/brine	22–60	69	Rising/captive bubble	53.55–70.36
Isfehani et al. (2023)	H_2–CO_2 mixture and water/brine	50–80	34–207	Pendent drop	32.1–67.8
Hosseini et al. (2022)	H_2 and water/brine	25–150	27–345	Pendent drop	46.97–80.77
Higgs et al. (2022)	H_2 and water/brine	25	69–207	Captive bubble, pendent drop	56.5–72.45

TABLE 7.2 Summary of contact angle measurements for hydrogen–water/brine–rock systems.

References	Used fluids	Temp., °C	Pressure, bar	Method	Rock	Brine contact angle, °
Mirchi et al. (2022)	$H_2/H_2–CH_4$ mixtures and brine	22–60	69	Pendent drop, rising/ captive bubble	Berea sandstone and Indiana limestone	52.42–71.1
Higgs et al. (2022)	H_2 and brine	25	69–207	Pendent drop, captive bubble, μCT	Bentheimer sandstone	27–59.75
Hashemi et al. (2022)	$CH_4/H_2–CH_4$ mixtures and brine	30–50	20–100	Captive bubble	Bentheimer sandstone	25–45
Esfandyari et al. (2022)	H_2 and water/ brine	20–80	10–100	Captive bubble	Quartz, granite, gypsum, basalt, calcite, dolomite, shale, anhydrite	17–93
Iglauer et al. (2021)	H_2 and brine	23–70	1–250	Tilted plate	Quartz	Mostly 50–70

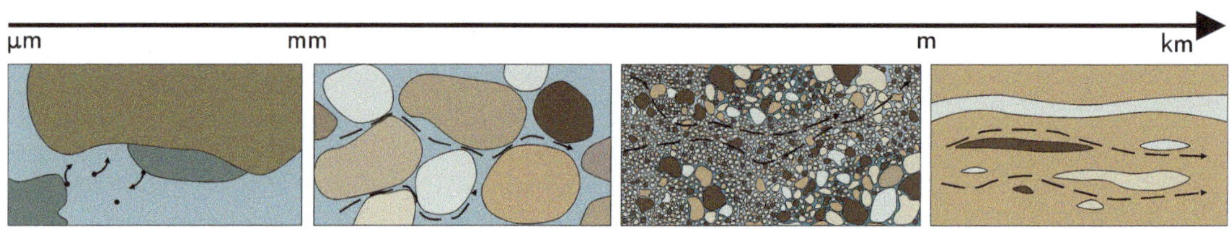

FIGURE 7.10 From pore scale to Darcy scale (Kobus & Haar, 1996). Source: *Adapted from Kobus, H., & de Haar, U. (1996).* Perspektiven der Wasserforschung Ergebnisse aus der 50. Sitzung der Kommission am 10. und 11. März 1994. Mitteilung 14 der Senatskommission für Wasserforschung, *Wiley-VCH.*

interferes with each other. To be able to flow each phase needs to have a continuous pathway through the porous medium (Ahmed, 2006).

On the macroscopic (Darcy) scale the two-phase system can be averaged over a representative elementary volume (REV) (cf. Fig. 7.10).

The pore space occupied by each phase divided through the total pore space is referred to as saturation. The ability for each phase to flow through the porous medium is described by the concept of relative permeability which was derived from laboratory experiments (Ahmed, 2006). The relative permeability is the ratio between the effective permeability for one fluid phase and the absolute permeability as a function of saturation.

As a consequence of the necessity for a continuous flow path for a phase, the relative permeability will become zero at some critical value for the saturation which is still larger than zero. This saturation is referred to as residual saturation (Helmig, 1997). The displacement processes when the wetting phase (water) displaces the nonwetting phase (gas) and vice versa have different characteristics (Ahmed, 2006). During imbibition the wetting phase penetrates into the larger pores or channels while during drainage the nonwetting phase displaces the wetting phase in the smaller pores (Helmig, 1997). As a result, the relative permeability curves as a function of saturation have different shapes for the wetting and for the nonwetting phase. In both cases some amount of the displaced fluid phase will be left behind. In Fig. 7.11 relative permeability curves are shown for a typical water-wet gas–water system.

The displacement problem between two immiscible fluids on macroscopic scale was solved mathematically by Buckley and Leverett (1942) in a simplified form. They derived an analytical solution for the saturation in a one-dimensional domain when the displacing fluid is injected from one side. The solution consists of a displacement front followed by a rarefaction wave (cf. Fig. 7.12). The value of saturation behind the displacement front depends on the mobility ratio between the phases. The characteristics of this behavior were also confirmed by laboratory flooding experiments (Buckley & Leverett, 1942).

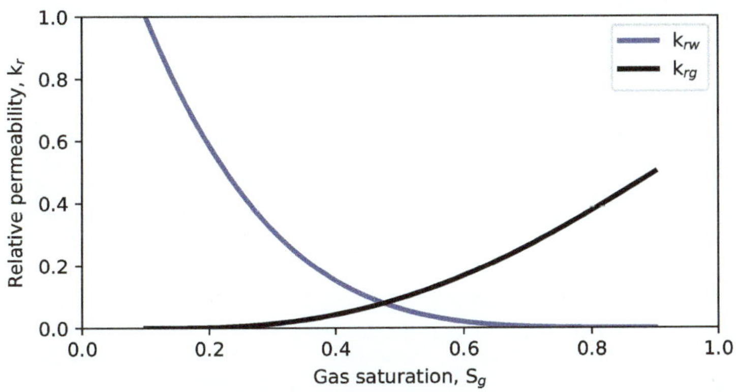

FIGURE 7.11 Relative permeability as a function of water saturation for a typical water-wet gas–water system.

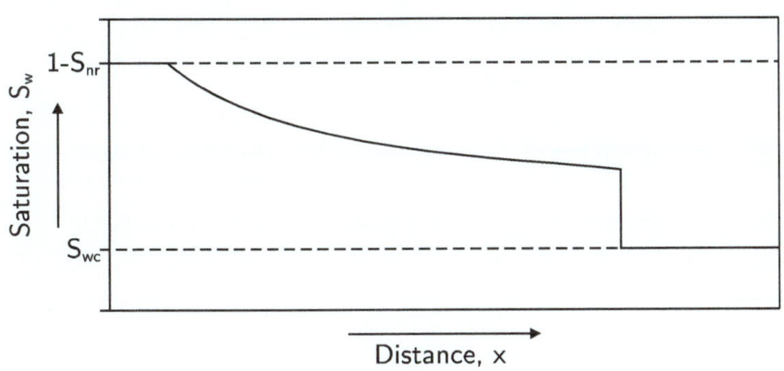

FIGURE 7.12 Saturation versus distance in a two-phase displacement process: the example shows the displacement of gas by water.

In recent years, a number of studies have measured relative permeability curves for hydrogen with water or brine as the second phase. Some of these studies are summarized in Table 7.3.

Drainage relative permeability curves were measured by Rezaei et al. (2022), Yekta et al. (2018), and Higgs et al. (2023). The measurements carried out by Rezaei et al. (2022) in unsteady-state conditions showed a dependence on rock type and pressure. The comparison with methane shows that the relative permeability of hydrogen is lower at the same saturation. However, the measurements of Higgs et al. (2023) with Bentheimer sandstone under steady-state conditions showed very similar results compared to nitrogen and methane. No comparison with other gases was made in the study by Yekta et al. (2018). In the studies by Boon and Hajibeygi (2022), Lysyy et al. (2022), and Higgs et al. (2024) drainage and imbibition relative permeability curves were measured under steady-state conditions. The rocks used are Bentheimer and Berea sandstones. These studies showed hysteresis effects for the relative permeability curves, which means that the curves for drainage and imbibition are not identical (cf. Fig. 7.13). Hysteresis effects can appear in relative permeabilities and capillary pressure curves as a consequence of the behavior of the wettability. One explanation for this are deviations between advancing and receding contact angles that arise in a dynamic system. Various studies have shown that also surface geometry, heterogeneity and roughness and adsorption of chemical species play a role in hysteresis (Elhaj et al., 2020). In addition, the

TABLE 7.3 Summary of relative permeability measurements for hydrogen—water/brine—rock systems.

References	Used gases	Temp., °C	Pressure, bar	Method	Rock	Main findings
Rezaei et al. (2022)	H_2, CH_4, N_2	80	1–207	Unsteady state (drainage only)	Sandstone and carbonate rock	Dependency on pressure and rock type, H_2 relative permeability lower than for CH_4 at the same saturation
Higgs et al. (2023)	H_2, CH_4, N_2	25	21	Steady-state (drainage only)	Bentheimer sandstone	H_2 relative permeability very similar to other gases
Yekta et al. (2018)	H_2	20, 45	55, 100	Steady-state (drainage only)	Vosges sandstone	Repeatable results at different conditions
Lysyy et al. (2022)	H_2	30	30	Steady-state (primary drainage, imbibition and secondary drainage)	Berea sandstone	Strong hysteresis effects were observed
Boon and Hajibeygi (2022)	H_2	18	100	Steady-state (drainage and imbibition)	Berea sandstone	Gravity segregation and channeling/viscous fingering were observed
Higgs et al. (2024)	H_2	25	21	Steady-state (drainage and imbibition)	Bentheimer sandstone	Very low (end point) relative permeability for H_2

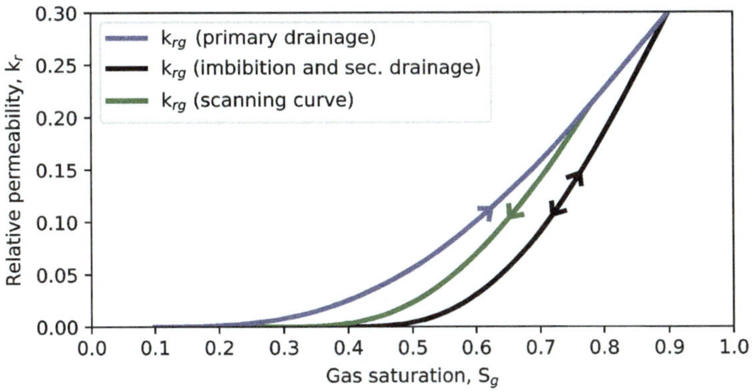

FIGURE 7.13 Relative permeability of the gas phase for alternating drainage and imbibition processes as a function of gas saturation (hysteresis).

saturation history in the porous medium is important. A pore wall which was in contact with oil is more likely to be oil-wet while a pore wall which was in contact with water is water-wet. As a consequence, wettability and capillary pressure and relative permeability curves are a function of the operation history of a reservoir and might change during the operation of an UHS.

Depending on the saturation at which the drainage or imbibition process is interrupted during a reciprocal displacement, the curve can also regress differently (scanning curve in Fig. 7.13). Lysyy et al. (2022) also measured the secondary drainage curve, which showed a deviating behavior compared to the primary one. Higgs et al. (2024) confirm a very low (end point) hydrogen relative permeability as also measured in other studies before. The measurements carried out by Boon and Hajibeygi (2022) in a medical X-ray CT scanner showed that phenomena such as gravity segregation and channeling or viscous fingering already play a role on the core scale. This will be discussed later in this section.

The impact of relative permeability hysteresis on the storage operation was investigated by numerical simulation studies (Bo et al., 2023; Ershadnia et al., 2023; Pan et al., 2023). The three studies have in common that an aquifer scenario is simulated in which the gas cap must first be developed by injecting cushion gas. Also, all of the results have the same tendency. Relative permeability hysteresis has a strong influence on the hydrogen injectivity and productivity compared to simulation scenarios where drainage and imbibition follows the same curve. The consideration of relative permeability hysteresis results in lower hydrogen recovery and a larger water extraction from the storage.

7.4.3 Displacement instabilities

In a three-dimensional system the stability of the displacement front needs be considered additionally. Two-phase flow in porous media is always a combination of viscous, gravitational and capillary forces. All three forces can have stabilizing or destabilizing effects. Dimensionless numbers are often used to compare the forces with each other. For porous media flow the capillary number is used to relate the viscous forces to capillary forces and the Bond number is often used to relate gravitational forces to capillary forces:

$$Ca = \frac{\mu V}{\sigma}$$

$$Bo = \frac{\Delta \rho g a^2}{\sigma}$$

where μ is the dynamic viscosity, V is the characteristic flow velocity, σ is IFT, ρ is the density, g is the gravitational acceleration and a is the typical pore size. The capillary number shows that viscous forces are dominating at a relatively high flow velocity while capillary forces are dominating at low flow velocity. The Bond number shows the difference in density of the phases as an influencing factor. Thus, with a large difference in density, the gravitational forces can predominate, while with a small difference in density, the capillary forces are greater.

The following describes various situations relevant to UHS and whether the combination of different forces can lead to stable or instable displacement. Factors, which are important in this context, are the density difference between the phases, the mobility ratio, the permeability distribution, and the positioning of storage wells. The low density and viscosity of hydrogen could make the displacement instable what results in a low volumetric displacement efficiency. Two physical effects which could arise and lower the efficiency are viscous fingering and gravity segregation.

Viscous fingering arises due to a contrast in the mobilities of the displacing and displaced fluid. An assessment can be done by using the mobility ratio (Terry, 2001):

$$M = \frac{k_{r1}\mu_2}{k_{r2}\mu_1}$$

where k_{r1} is relative permeability of the displacing fluid at the average saturation behind the displacement front, k_{r2} is relative permeability of the displaced fluid at the average saturation ahead the displacement front and μ_1 and μ_2 are the viscosities respectively of the displacing and displaced fluid. In general it is accepted that mobility ratios smaller than one result in a stable displacement (Ho & Webb, 2006). In contrast for mobility ratios higher than one the front becomes instable and fingers occur in the displacement front. A rough estimation for hydrogen storages results in mobility ratios in the order of 2–5 for the displacement of water. Consequently, an instable displacement is expected.

The cause for viscous fingering are small perturbations which result from microscopic heterogeneities although the porous medium can be homogenous at macroscopic scale (Kueper & Frind, 1988). As destabilization forces are stronger than stabilization forces the small perturbations continue to propagate as large fingers. In Fig. 7.14, the formation of fingers is shown in a simulation of a gas–water displacement problem. The fingers are created by permeability anomalies (four rectangles on the left side) and then continue to grow in length during the further displacement process.

The spreading of fingers can be conducted by several phenomena. Fingers tend to spread perpendicular to the main flow direction what can either result in joining of two fingers or a splitting at the tip (Ho & Webb, 2006). Additionally, a shielding can occur which means that

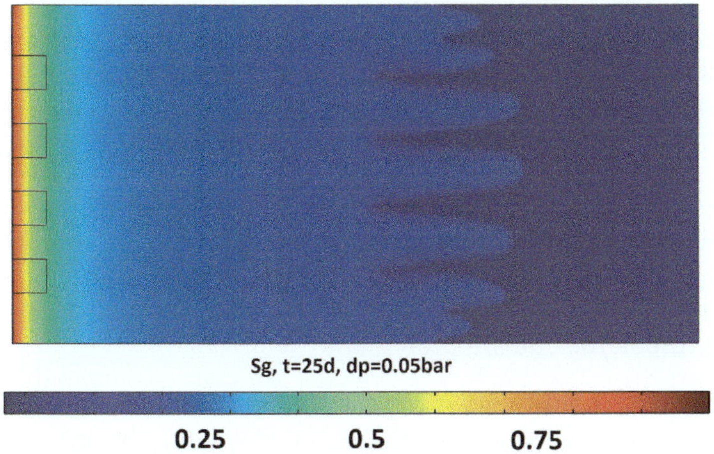

FIGURE 7.14 Simulation of a gas–water displacement characterized by viscous fingers in a two-dimensional horizontal cross section where the four squares represent a permeability anomaly.

only a smaller number of fingers continues to propagate and the remaining fingers become slower or stop (Ho & Webb, 2006). The result is a tree-like structure (Paterson, 1983). The factors influencing this process are viscosity differences, density differences, and surface tension (Kueper & Frind, 1988). However, underground structures also have permeability heterogeneities at macroscopic scale which result in different flow potentials. Thereby viscous fingers tend to propagate in the direction of high flow potential (Kueper & Frind, 1988). Viscous fingering should not be confused with channeling on macroscopic scale. Channeling occurs in heterogenous porous media where fingers are created due to different flow potentials.

Gravity segregation arises due to different densities and mobilities of the displacing and displaced fluid (Tek, 1989). When a less dense displacing fluid prefers to flow at the top of the structure the behavior is referred to as density override. In contrast, when a more dense displacing fluid prefers to flow at the bottom of the structure, it is referred to as density underride. This effect was investigated mathematically by Dietz (1953). Based on some assumptions he derived a balance for the viscous and gravitational forces which separate an instable from a stable displacement. His findings allow the following conclusions for UHS where the less dense and more mobile gas phase is usually above the more dense and less mobile water phase. The downdip displacement of water by gas during injection becomes instable at a critical velocity (Tek, 1989). Above this velocity gravity override occurs. However, the up dip displacement of gas by water during withdrawal is stable at any velocity (Tek, 1989). In UHS especially in aquifers the instable displacement could be associated with several problems. At first fingers could spread laterally much further than expected and some of them could pass the spill point of the structure (Paterson, 1983). If this occurs the gas is carried away by buoyancy forces and consequently lost (Paterson, 1983). Secondly, a low volumetric sweep efficiency is obtained which results in a small storage volume. Thirdly, the fingers provide a large contact area between hydrogen and the water phase where a dissolution could take place (Paterson, 1983). The dissolved amount of hydrogen becomes also unrecoverable. Finally, mobile fluids could be left behind which will be produced in the subsequent withdrawal period.

The injection rate provides a possible control for the extension of fingers. This was investigated in a simulation study by Hagemann et al. (2015). Fig. 7.15 and Fig. 7.16 show the results for a low injection rate and a high injection rate. It can be seen that at a high rate, a possible spill point is already reached at a much lower injected gas volume.

7.5 Gas mixing behavior

The mixing between hydrogen and other gases plays a major role when a depleted gas reservoir is transformed to an UHS or when an alternative cushion gas is used. In the literature, for example, nitrogen and carbon dioxide are proposed to be used as cushion gases (Oldenburg, 2003; Pfeiffer & Bauer, 2015; Tek, 1989). The principle difference compared to the mechanism of gas–water flow is the fact that there is no abrupt transition between the different gases at pore scale. In a one-dimensional displacement, this leads, on the one hand, to a complete or piston-like displacement by advection where no initial gas is left behind the front (Bedrikovetsky, 1993). On the other hand, the miscibility leads to a smearing out of the front due to diffusive or dispersive flux at the front (Tek, 1989). The relationship between the two transport processes can be described using the Peclet number. The Peclet number for mass transport is defined as (Bear, 2018):

$$Pe = \frac{LV}{D}$$

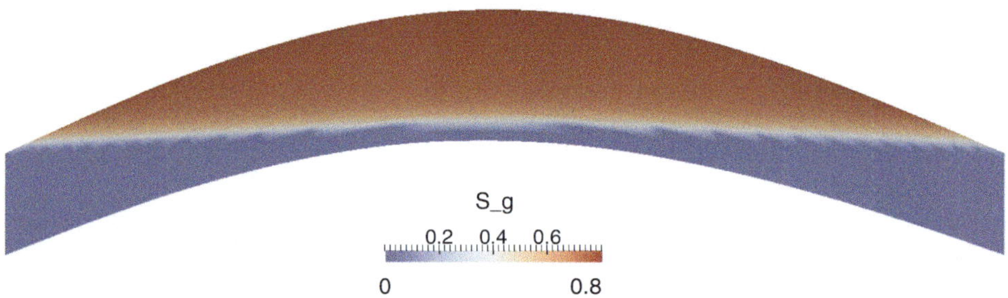

FIGURE 7.15 Simulation of gas injection into a two-dimensional anticline structure at a low injection rate.

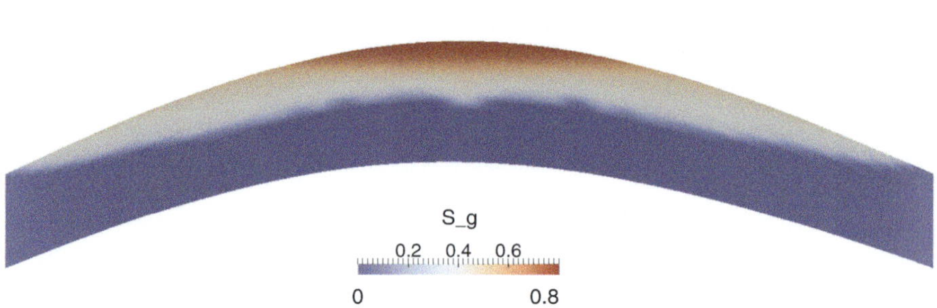

FIGURE 7.16 Simulation of gas injection into a two-dimensional anticline structure at a high injection rate.

where L is the characteristic length in m, V is the average flow velocity in m/s and D is the effective diffusion coefficient in m^2/s. In Fig. 7.17, the behavior of a miscible displacement is shown for different Peclet numbers. A high Peclet number means that the diffusive/dispersive flux has a low influence and the displacement front is piston-like. In contrast, a small Peclet number means that the diffusive flux has a significant influence which results in a strong smearing out of the front. In a more-dimensional system the overall process is influenced by heterogeneities and anisotropies of the porous medium and mobility ratios, density differences, molecular diffusion and mechanical dispersion between the different gases (Tek, 1989).

In the following sections, various factors influencing the mixing behavior are discussed.

7.5.1 Mobility and gravity effects

Similar to the gas–water displacement also the gas–gas displacement process can become instable due to differences in mobility and density of the displacing and displaced gas. Mobility differences occur mainly due to different dynamic viscosities. Hydrogen has a very low viscosity which results in a mobility ratio around 1.5 for the system H_2–CH_4 and 4 for the system H_2–CO_2 under typical storage conditions. This could result in an instable displacement when hydrogen is injected to displace another gas. However, this effect is much less than in a gas–water displacement because the miscibility leads to a high dispersion of the front which acts as stabilizing force (Ho & Webb, 2006). In Feldmann et al. (2016), the stability of gas–gas displacement processes has been studied by numerical simulations. It was shown that in a homogeneous reservoir without gravitational effects the displacement of methane by hydrogen has a stable displacement front (cf. Fig. 7.18).

The impact of density differences could be more relevant during gas–gas displacements. Differences in the density of hydrogen and other gases can be in the range of ~80% for the system H_2–CH_4 and ~95% for the system H_2–CO_2 under typical storage conditions. The effect can have a negative influence when the injection aims to displace another gas but instead gravity override occurs. However, the effect can be also used to keep different gases segregated, for example, when carbon dioxide is used as cushion gas

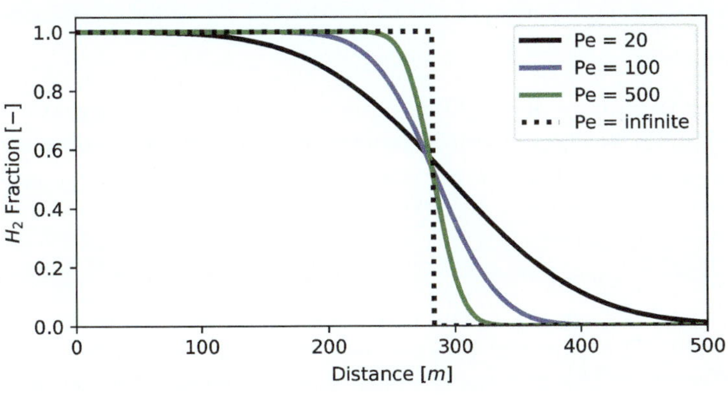

FIGURE 7.17 H_2 mole fraction in a one-dimensional displacement at different Peclet numbers.

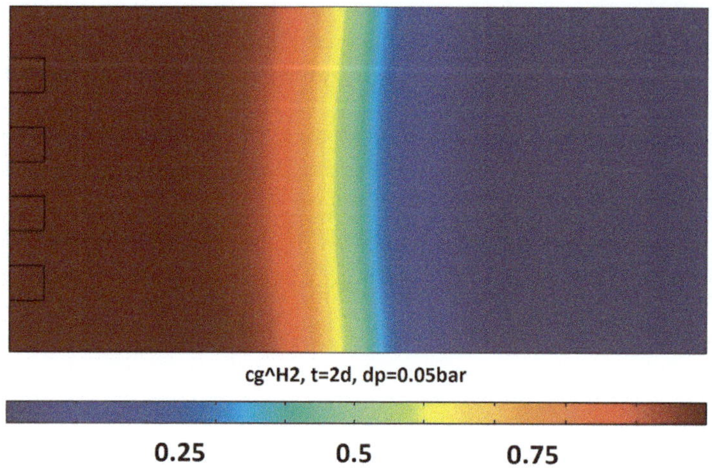

FIGURE 7.18 Simulation result for a gas–gas displacement process in a two-dimensional horizontal cross-section (Feldmann et al., 2016).

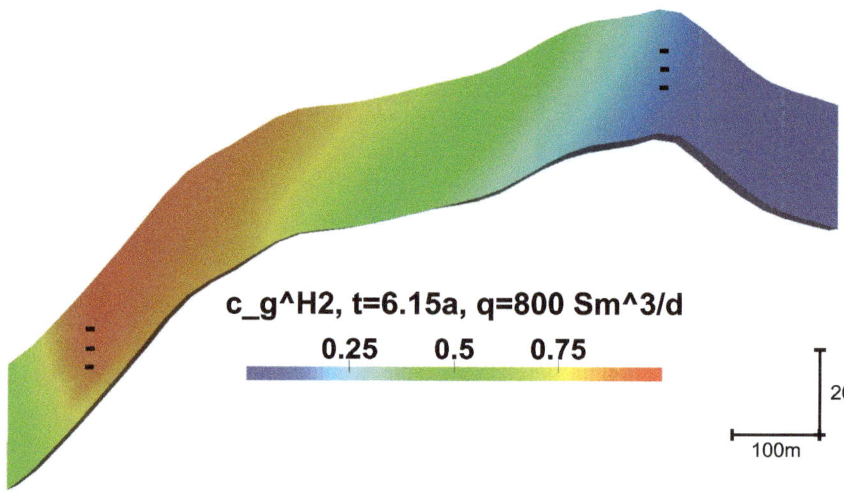

FIGURE 7.19 Simulation result for a gas–gas displacement process in a two-dimensional vertical cross-section (Feldmann et al., 2016).

(Tek, 1989). Simulations studies by Feldmann et al. (2016) have shown that a slight gravity overriding takes place when hydrogen is injected into a reservoir which initially contains natural gas (cf. Fig. 7.19).

7.5.2 Molecular diffusion

Molecular diffusion is generally considered as a slow process when compared to advective/convective transport (Tek, 1989). This physical process is driven by chemical potential and occurs without pressure differences. The process refers to the movement of molecules driven by the inherent tendency to equalize concentration gradients, which arises from the random motion known as Brownian motion. It can take place under both stationary and unsteady conditions. The difference between the two conditions is that stationary diffusion occurs at a constant rate, while the unsteady diffusion rate is a function of time. In both conditions diffusion can be described using Fick's laws. According to Fick's first law, the diffusive flux is directly proportional to the gradient in concentration (Fick, 1855).

The diffusivity in a porous medium is typically lower compared to free diffusion of gases, as the gas encounters less space and must traverse a longer distance within the porous medium. Various factors, such as porosity, tortuosity, and the existence of other fluid phases within the pores, influence the effective diffusivity. In the case of binary systems, the diffusive flux of gas components in a porous medium can be formulated by the following equation:

$$J^k_{\mathrm{diff}} = -\rho D^{AB}_{\mathrm{eff}} \nabla c^k$$

In the given equation, ${J_{\mathrm{diff}}}^k$ is the flux of component k resulting from diffusion, measured in mol/m^2/s, ρ denotes the molar density of the gas in mol/m^3, while ${D_{\mathrm{eff}}}^{AB}$ represents the effective diffusion coefficient for the binary system, in m^2/s. ∇c^k is the gradient of the mole fraction for component k. The effective molecular diffusion coefficient in a porous medium can be expressed as follows:

$$D^{AB}_{\mathrm{eff}} = \phi S_g \tau D^{AB}_{\mathrm{bulk}}$$

where Φ is the porosity, S_g the gas saturation, τ the tortuosity factor and ${D_{\mathrm{bulk}}}^{\mathrm{AB}}$ the bulk diffusion coefficient in m^2/s. For the description of diffusion in porous media the tortuosity of the pore structure needs to be considered. The model of Millington and Quirk (1961) estimates the tortuosity factor of a porous medium based on its porosity and saturation:

$$\tau = \phi^{\frac{1}{3}} S_i^{\frac{7}{3}}$$

Knudsen diffusion occurs when the average distance covered by gas molecules, known as the mean free path, becomes comparable to or larger than the pore sizes of the porous medium. Consequently, the molecules collide more frequently with the walls of the pores rather than with each other. The extent of Knudsen diffusion can be quantified by the Knudsen number, denoted as Kn. A Knudsen number higher than 10 indicates the dominance of Knudsen diffusion, where collisions between molecules and the pore walls are the primary factor. To determine whether Knudsen diffusion is relevant during the hydrogen flow in underground storages the Knudsen number is calculated using the equation provided by Bear (2013):

$$Kn = \frac{\lambda}{l_{\mathrm{pm}}}$$

where λ represents the mean free path of molecules in m and l_{pm} denotes the characteristic length dimension of the void space, specifically the pore diameter, also in m. A mean free path of 6.6×10^{-8} m for the hydrogen molecules was assumed. Considering a typical storage rock the mean pore diameter is around 5×10^{-5} m. The estimated Knudsen number is 1.32×10^{-3}. Therefore it can be concluded that Knudsen diffusion does not have a significant influence on the diffusion process in storage rocks. However, it could become relevant in low-permeability layers above and below.

Different methods have been devised to measure effective molecular diffusion within porous materials. An overview of diffusion measurements reported in the literature, along with their corresponding results, is provided in Table 7.4.

Chen et al. (1977) conducted diffusion experiments using modified Wicke–Kallenbach setups on porous samples. In their experiments the binary diffusion of methane and nitrogen through sandstone samples was measured and effective diffusion coefficients were determined. The resulting effective molecular diffusion coefficients vary between 2.59×10^{-7} and 2.00×10^{-5} m^2/s for dry rock samples.

A similar approach was used by Pandey et al. (1974) for the diffusion measurement of helium and nitrogen through porous rock samples. The diffusion was measured at steady state but also at unsteady state conditions. At steady state conditions, the diffusion coefficients lie between 2.14×10^{-8} and 1.19×10^{-6} m^2/s for dry rock samples. The diffusion coefficients at unsteady-state conditions are between 1.67×10^{-10} and 1.88×10^{-7} m^2/s.

Strauch et al. (2023) investigated the diffusion of hydrogen through different potential storage site rocks. In their study sandstone, rock salt, and claystone samples were used. An experimental setup, which comprises two chambers and a rock sample, was developed to determine hydrogen breakthrough times and hydrogen diffusion coefficients. The determined effective diffusion coefficients in these experiments range from 10^{-9} to 10^{-8} m^2/s. It has been observed that water decreases the diffusion rate in all conducted experiments.

In Arekhov et al. (2023), an experimental setup was constructed, which enables the measurement of effective multicomponent diffusion coefficients at high pressure and temperature conditions. For the diffusion measurements three different reservoir rocks from potential UHS sites were used. The determined diffusion coefficients vary between 1.49×10^{-7} and 1.51×10^{-6} m^2/s.

TABLE 7.4 Effective gas–gas diffusion measurements in the literature.

References	Used gases	Temp., °C	Pressure, bar	Method	Material	Effective diffusion coefficient, m^2/s
Pandey et al. (1974)	He, N_2	24–42	1–5	Steady and unsteady state	Carbonate, sandstone, and shale samples	1.19×10^{-4} to 1.67×10^{-8}
Chen et al. (1977)	CH_4, N_2	35	1	Steady state	Sandstone samples	2.59×10^{-7} to 2.00×10^{-5}
Strauch et al. (2023)	H_2, Air	Room temperature	1	Unsteady state	Sandstone, claystone, rock salt samples	10^{-9} to 10^{-8}
Arekhov et al. (2023)	H_2, CH_4	28	5–40	Unsteady state	Sandstone samples	8.38×10^{-7} to 1.51×10^{-6}
Michelsen, Langanke, et al. (2023)	H_2, CH_4	25–100	10–200	Pseudo steady-state	Sandstone and limestone samples	5.0×10^{-9} to 3.71×10^{-7}

Michelsen, Langanke, et al. (2023) performed more than 30 effective diffusion measurements on sandstone and limestone rock samples using a pseudo-steady measurement method. The measurements have been done at storage conditions with methane and hydrogen. The determined effective diffusion coefficients are in the range between 5.0×10^{-9} and 3.71×10^{-7} m^2/s.

It can be concluded that the diffusion coefficients determined in the literature show a large range. In addition, there are only a few diffusion experiments in literature, where hydrogen was used. Only two studies exist where effective diffusion coefficients were measured for hydrogen under relevant subsurface conditions (Arekhov et al., 2023; Michelsen, Langanke, et al., 2023).

7.5.3 Mechanical dispersion

Mechanical dispersion is a mixing process of fluids due to their movement and current in porous media. Dispersion is driven by variations in the flow velocity which can occur on different scales. These variations are mainly caused by the pore-size distribution, tortuosity and heterogeneity of the porous medium. On microscopic scale larger pores and pore throats lead to higher velocities than smaller pores. In addition, the nonuniform velocity profile along the pore cross-section induces velocity variations (Ho & Webb, 2006).

The effect has to be introduced into macroscale models to compensate the error made by the assumption of an averaged Darcy velocity. Consequently, this effect arises during the upscaling to an REV.

Mechanical dispersion can be described by assuming a linear relation to the flow velocity (Scheidegger, 1961):

$$J^k_{\text{disp}} = -\rho D^k_{\text{disp}} \nabla c^k$$

$$D^k_{\text{disp}} = \phi S_g \left(\frac{V V^T}{\|V\|} (a_L - a_T) + \|V\| a_T \right)$$

where $J_{\text{disp}}{}^k$ is the dispersive flux of component k in mol/m^2/s, ρ is the molar density of gas in mol/m^3, α_L in the longitudinal (parallel to the main flow direction) dispersivity in m and α_T is the transverse (perpendicular to the main flow direction) dispersivity in m, V is the flow velocity in m/s and ∇c^k is the gradient of the mole fraction of component k in 1/m.

Information about the dispersion measurements in literature is summarized in Table 7.5.

Kurihara et al. (2000) and Yang et al. (2023) conducted core flooding experiments to measure the dispersion between different gases in Bandera and Berea sandstone samples on the scale of a few centimeters. In the studies by Michelsen, Thaysen, et al. (2023) and Kurihara et al. (2000) dispersion was measured using slimtube coils filled with an artificial porous medium. These studies investigated the process of dispersion between different gases on the scale of a few meters. Carriere et al. (1985) and Laille et al. (1986) investigated the mixing behavior of gases on reservoir scale by performing tracer tests in the field.

TABLE 7.5 Gas dispersivity measurements in the literature.

References	Used gases	Temp., °C	Pressure, bar	Method	Material	Dispersivity, m
Yang et al. (2023)	H_2, N_2	N/A	50	Core flooding	Berea sandstone sample	3.10×10^{-4}
Mohammed et al. (2019)	N_2, CO_2, CH_4	35–40	103	Core flooding	Bandera gray sandstone sample	6.37×10^{-4}
Kurihara et al. (2000)	CH_4, Ar	N/A	N/A	Slimtube displacement	Glass beads	1.20×10^{-2}
Michelsen, Thaysen, et al. (2023)	H_2, CH_4	25–100	50–200	Slimtube displacement	Glass beads	1.8×10^{-2} to 6.0×10^{-2}
Carriere et al. (1985)	H_2 tracer with natural gas	N/A	N/A	Field test	Porous reservoir	2
Laille et al. (1986)	H_2 tracer with natural gas	N/A	N/A	Field test	Porous reservoir	25

In Table 7.5 it can be recognized that the measurements of the dispersivity gave results which deviate by several orders of magnitude. As the process is strongly influenced by the considered scale, laboratory measurements cannot be transferred directly to reservoir scale (Tek, 1989). Only two studies have been performed with hydrogen under the relevant gas storage conditions (Michelsen, Thaysen, et al., 2023; Yang et al., 2023). In Michelsen, Thaysen, et al. (2023) it was shown that the dispersivity depends on pressure, temperature, and flow velocity. It was concluded that the fluid properties, in this case the gas properties, have an important influence on the dispersivity. The tracer tests, which have been performed on reservoir scale, have shown that the longitudinal dispersivity is between 2 and 25 m (Carriere et al., 1985; Laille et al., 1986; Tek, 1989). The transverse dispersivity is usually one or two orders of magnitude less. Assuming flow velocities of several meters per day, which are common in gas storages, the longitudinal mechanical dispersion coefficient will be around 5×10^{-4} m^2/s. Hence, the mixing by mechanical dispersion could be much more pronounced than only by molecular diffusion. However, in a gas storage the flow velocity varies over time depending on whether gas is being injected or withdrawn. Moreover, it can vary greatly locally since it is much higher near the wellbore than further away. The flow velocity also depends on the permeability of the storage layer, as it is higher in layers with high permeability than in layers with a lower permeability. Fig. 7.20 illustrates the relation between molecular diffusion and mechanical dispersion in porous media to the flow velocity. When the flow velocity is low, molecular diffusion primarily governs the mixing process. In contrast, at higher velocities the mixing is primarily controlled by mechanical dispersion. The transitional region between these two extremes is influenced by both processes simultaneously.

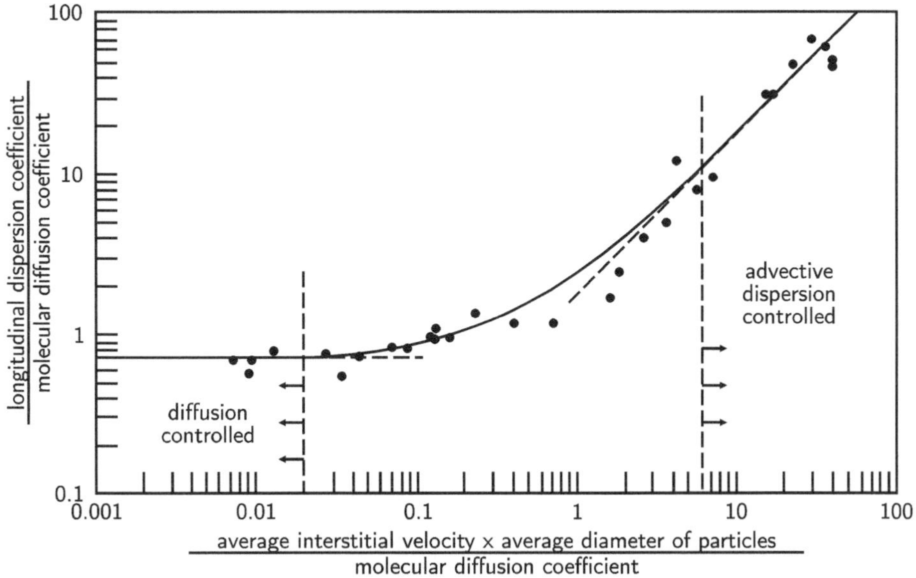

FIGURE 7.20 Diffusive and dispersive mixing regimes in porous media (Perkins & Johnston, 1963). Source: *Adapted from Perkins, T. K., & Johnston, O. C. (1963). A review of diffusion and dispersion in porous media.* Society of Petroleum Engineers Journal, *3(01), 70–84, https://doi.org/10.2118/480-pa.*

7.6 Conclusions

Storage rock layers of UHSs are characterized by a small thickness in relation to large lateral extent. Through the storage rock predominantly laminar porous media fluid flow takes place. This flow process can be described on macroscopic scale using Darcy's law. In the immediate vicinity of storage wells, however, it is possible that the flow velocity is high and inertial effects play a role or even a turbulent flow takes place. Since porous subsurface formations usually have formation water and there can be a contact to an aquifer at the edges of the gas cap, two-phase flow processes must be taken into account. In addition, mixing processes between gases need to be considered when the injected hydrogen or gas with admixed hydrogen is injected into reservoirs with residual natural gas or an alternative gas is used as cushion gas.

Over the past few years, a large number of laboratory measurements have been carried out to better understand the fluid properties, two-phase flow characteristics, and transport rates for hydrogen-containing systems. It has been shown that there are some differences to natural gas. Effects revealed by the investigations are deviations in the IFT with water or brine, differences in the wetting behavior and consequently differences in the relative permeability curves, which are often used for numerical simulation of two-phase flow processes. It has also been shown that diffusion rates of systems with hydrogen are significantly higher than for methane. This concerns the gas–gas mixing but also the transport in low-permeability layers where hydrogen diffuses as dissolved gas in the water phase. Simulation studies on macroscopic scale have shown that instabilities could occur due to high mobility ratios or large density differences. This could lead to an incomplete displacement with gas trapping, density override or viscous fingering affecting the technical and economic efficiency of UHS.

References

Ahmed, T. (2006). *Reservoir engineering handbook*. Gulf Professional Publishing.

Arekhov, V., Zhainakov, T., Clemens, T., & Wegner, J. (2023). Measurement of effective hydrogen-hydrocarbon gas diffusion coefficients in reservoir rocks. *Society of Petroleum Engineers – SPE EuropEC – Europe Energy Conference featured at the 84th EAGE Annual Conference and Exhibition, EURO 2023*. Available from https://doi.org/10.2118/214451-MS.

Bear, J. (2018). *Modeling phenomena of flow and transport in porous media* (1). Springer. Available from https://doi.org/10.1007/978-3-319-72826-1.

Bear, J. (2013). Dynamics of fluids in porous media. Courier Corporation.

Bedrikovetsky, P. (1993). Petroleum Engineering & Development Studies*Mathematical theory of oil and gas recovery, with applications to ex-USSR oil and gas fields* (4). Kluwer. Available from https://doi.org/10.1007/978-94-017-2205-6.

Boon, M., & Hajibeygi, H. (2022). Experimental characterization of [Formula: see text]/water multiphase flow in heterogeneous sandstone rock at the core scale relevant for underground hydrogen storage (UHS). *Scientific Reports, 12*(1), 14604. Available from https://doi.org/10.1038/s41598-022-18759-8.

Bo, Z., Boon, M., Hajibeygi, H., & Hurter, S. (2023). Impact of experimentally measured relative permeability hysteresis on reservoir-scale performance of underground hydrogen storage (UHS). *International Journal of Hydrogen Energy, 48*(36), 13527–13542. Available from https://doi.org/10.1016/j.ijhydene.2022.12.270.

Buckley, S. E., & Leverett, M. C. (1942). Mechanism of fluid displacement in sands. *Transactions of the AIME, 146* (01), 107–116. Available from https://doi.org/10.2118/942107-g.

Carden, P., & Paterson, L. (1979). Physical, chemical and energy aspects of underground hydrogen storage. *International Journal of Hydrogen Energy, 4*(6), 559–569. Available from https://doi.org/10.1016/0360-3199(79)90083-1.

Carriere, J. F., Fasanino, G., & Tek, M. R. (1985). Mixing in underground storage reservoirs. *Proceedings – SPE Annual Technical Conference and Exhibition, 1985-September*. Available from https://doi.org/10.2118/14202-ms.

Chabab, S., Cruz, J. L., Poulain, M., Ducousso, M., Contamine, F., Serin, J. P., & Cézac, P. (2021). Thermodynamic modeling of mutual solubilities in gas-laden brines systems containing CO_2, CH_4, N_2, O_2, H_2, H_2O, NaCl, $CaCl_2$, and KCl: Application to degassing in geothermal processes. *Energies, 14*(17). Available from https://doi.org/10.3390/en14175239.

Chabab, S., Kerkache, H., Bouchkira, I., Poulain, M., Baudouin, O., Moine, É., Ducousso, M., Hoang, H., Galliéro, G., & Cézac, P. (2024). Solubility of H_2 in water and NaCl brine under subsurface storage conditions: Measurements and thermodynamic modeling. *International Journal of Hydrogen Energy, 50*, 648–658. Available from https://doi.org/10.1016/j.ijhydene.2023.10.290.

Chen, L. L., Katz, D. L., & Tek, M. R. (1977). Binary gas diffusion of methane-nitrogen through porous solids. *AIChE Journal, 23*(3), 336–341. Available from https://doi.org/10.1002/aic.690230317.

Darcy, H. (1856). *Les fontaines publiques de la ville de Dijon: Exposition et application des principes à suivre et des formules à employer dans les questions de distribution d'eau* (1). Victor Dalmont.

Dejam, M., Hassanzadeh, H., & Chen, Z. (2017). Pre-Darcy flow in porous media. *Water Resources Research, 53*(10), 8187–8210. Available from https://doi.org/10.1002/2017WR021257.

Dietz, D. N. (1953). A theoretical approach to the problem of encroaching and by-passing edge water. *Proceedings of the Koninklijke Nederlandse Akademie van Wetenschappen, B56*, 83–92.

Elhaj, M. A., Hossain, E., Imtiaz, S. A., & Naterer, G. F. (2020). Hysteresis of wettability in porous media: A review. *Journal of Petroleum Exploration and Production Technology, 10*(5), 1897–1905. Available from https://doi.org/10.1007/s13202-020-00872-x.

Ershadnia, R., Singh, M., Mahmoodpour, S., Meyal, A., Moeini, F., Hosseini, S. A., Sturmer, D. M., Rasoulzadeh, M., Dai, Z., & Soltanian, M. R. (2023). Impact of geological and operational conditions on underground hydrogen storage. *International Journal of Hydrogen Energy, 48*(4), 1450–1471. Available from https://doi.org/10.1016/j.ijhydene.2022.09.208.

Esfandyari, H., Sarmadivaleh, M., Esmaeilzadeh, F., Ali, M., Iglauer, S., & Keshavarz, A. (2022). Experimental evaluation of rock mineralogy on hydrogen-wettability: Implications for hydrogen geo-storage. *Journal of Energy Storage, 52*. Available from https://doi.org/10.1016/j.est.2022.104866.

Feldmann, F., Hagemann, B., Ganzer, L., & Panfilov, M. (2016). Numerical simulation of hydrodynamic and gas mixing processes in underground hydrogen storages. *Environmental Earth Sciences, 75*(16). Available from https://doi.org/10.1007/s12665-016-5948-z.

Fick, A. (1855). Über Diffusion: Poggendorffs Annalen der Physik und Chemie. *The London, Edinburgh, and Dublin Philosophical Magazine and Journal of Science*, 30–39.

Forchheimer, P. (1901). Wasserbewegung. *Zeitschrift des Vereines deutscher Ingenieure, 45*, 1782–1788.

Hagemann, B., Rasoulzadeh, M., Panfilov, M., Ganzer, L., & Reitenbach, V. (2015). Mathematical modeling of unstable transport in underground hydrogen storage. *Environmental Earth Sciences, 73*(11), 6891–6898. Available from https://doi.org/10.1007/s12665-015-4414-7.

Hashemi, L., Boon, M., Glerum, W., Farajzadeh, R., & Hajibeygi, H. (2022). A comparative study for H_2–CH_4 mixture wettability in sandstone porous rocks relevant to underground hydrogen storage. *Advances in Water Resources, 163*. Available from https://doi.org/10.1016/j.advwatres.2022.104165.

Helmig, R. (1997). *Multiphase flow and transport processes in the subsurface: A contribution to the modeling of hydrosystems* (1). Springer.

Higgs, S., Da Wang, Y., Sun, C., Ennis-King, J., Jackson, S. J., Armstrong, R. T., & Mostaghimi, P. (2022). In-situ hydrogen wettability characterisation for underground hydrogen storage. *International Journal of Hydrogen Energy, 47*(26), 13062–13075. Available from https://doi.org/10.1016/j.ijhydene.2022.02.022.

Higgs, S., Da Wang, Y., Sun, C., Ennis-King, J., Jackson, S. J., Armstrong, R. T., & Mostaghimi, P. (2023). Comparative analysis of hydrogen, methane and nitrogen relative permeability: Implications for underground hydrogen storage. *Journal of Energy Storage, 73*. Available from https://doi.org/10.1016/j.est.2023.108827.

Higgs, S., Da Wang, Y., Sun, C., Ennis-King, J., Jackson, S. J., Armstrong, R. T., & Mostaghimi, P. (2024). Direct measurement of hydrogen relative permeability hysteresis for underground hydrogen storage. *International Journal of Hydrogen Energy, 50*, 524–541. Available from https://doi.org/10.1016/j.ijhydene.2023.07.270.

Hosseini, M., Fahimpour, J., Ali, M., Keshavarz, A., & Iglauer, S. (2022). H_2 – brine interfacial tension as a function of salinity, temperature, and pressure; implications for hydrogen geo-storage. *Journal of Petroleum Science and Engineering, 213*. Available from https://doi.org/10.1016/j.petrol.2022.110441.

Ho, C., & Webb, S. (2006). *Gas transport in porous media* (20). Springer. Available from https://doi.org/10.1007/1-4020-3962-X.

Iglauer, S., Ali, M., & Keshavarz, A. (2021). Hydrogen wettability of sandstone reservoirs: Implications for hydrogen geo-storage. *Geophysical Research Letters, 48*(3). Available from https://doi.org/10.1029/2020GL090814.

Isfehani, Z. D., Sheidaie, A., Hosseini, M., Fahimpour, J., Iglauer, S., & Keshavarz, A. (2023). Interfacial tensions of (brine + H_2 + CO_2) systems at gas geo-storage conditions. *Journal of Molecular Liquids, 374*. Available from https://doi.org/10.1016/j.molliq.2023.121279.

Katz, D. L., & Lee, R. L. (1990). *Natural Gas Engineering Production and Storage*. New York: McGraw-Hill Publishing Co.

Kobus, H., & Haar, U. de (1996). Perspektiven der Wasserforschung Ergebnisse aus der 50. Sitzung der Kommission am 10. und 11. März 1994. Mitteilung 14 der Senatskommission für Wasserforschung. Wiley-VCH.

Kueper, B. H., & Frind, E. O. (1988). An overview of immiscible fingering in porous media. *Journal of Contaminant Hydrology*, 2(2), 95–110. Available from https://doi.org/10.1016/0169-7722(88)90001-0.

Kurihara, M., Liang, J., Fujimoto, F., Nghiem, L., & Sammon, P. (2000). Development and application of underground gas storage simulator. *Proceedings of the SPE Asia Pacific Conference on Integrated Modelling for Asset Management*, 339–352. Available from https://doi.org/10.2118/59438-ms.

Laille, J. P., Coulomb, C., & Tek, M. R. (1986). Underground storage in Cerville-Velaine, France: A case history in conversion and inert gas injection as cushion substitute. *Proceedings – SPE Annual Technical Conference and Exhibition*. Available from https://doi.org/10.2118/15588-ms.

Laplace, P. S. (1805). Traité De Mécanique Céleste: Supplément au dixième livre du traité de méchanique céleste sur l'action capillaire. De l'Imprimerie de Crapelet.

Lysyy, M., Føyen, T., Johannesen, E. B., Fernø, M., & Ersland, G. (2022). Hydrogen relative permeability hysteresis in underground storage. *Geophysical Research Letters, 49*(17). Available from https://doi.org/10.1029/2022GL100364.

Michelsen, J., Hagemann, B., Ganzer, L., & Hujer, W. (2022). Measurement of hydrogen diffusion through caprock samples. *6th International Conference on Fault and Top Seals 2022*. Available from https://doi.org/10.3997/2214-4609.202243061.

Michelsen, J., Langanke, N., Hagemann, B., Hogeweg, S., Ganzer, L. (2023). Diffusion measurements with hydrogen and methane through reservoir rock samples. 36th International Symposium of the Society of Core Analysts.

Michelsen, J., Thaysen, E. M., Hogeweg, S., Hagemann, B., Hassanpouryouzband, A., Langanke, N., Edlmann, K., & Ganzer, L. (2023). Hydrogen reservoir flow behaviour: Measurements of molecular diffusion, mechanical dispersion and relative permeability. Available from https://www.hyuspre.eu/wp-content/uploads/2023/04/HyUSPRe_D4.4_Methodology-and-results-of-H2-reservoir-flow-behavior-experiments_2023.03.31.pdf.

Millington, R. J., & Quirk, J. P. (1961). Permeability of porous solids. *Transactions of the Faraday Society, 57*, 1200–1207. Available from https://doi.org/10.1039/TF9615701200.

Mirchi, V., Dejam, M., & Alvarado, V. (2022). Interfacial tension and contact angle measurements for hydrogen – methane mixtures/brine/oil-wet rocks at reservoir conditions. *International Journal of Hydrogen Energy, 47*(82), 34963–34975. Available from https://doi.org/10.1016/j.ijhydene.2022.08.056.

Mohammed, N., Abubakar, A. J., Enyi, G. C., & Ghavami, G. N. (2019). Flow characteristics through gas alternating gas injection during enhanced gas recovery. *Society of Petroleum Engineers – SPE Gas and Oil Technology Showcase and Conference 2019, GOTS 2019*. Available from https://doi.org/10.2118/198658-ms.

Oldenburg, C. M. (2003). Carbon dioxide as cushion gas for natural gas storage. *Energy and Fuels, 17*(1), 240–246. Available from https://doi.org/10.1021/ef020162b.

Pandey, G. N., Tek, M. R., & Katz, D. L. (1974). Diffusion of fluids through porous media with implications in petroleum geology. *American Association of Petroleum Geologists Bulletin, 58*(2), 291–303. Available from https://doi.org/10.1306/83D913DA-16C7-11D7-8645000102C1865D.

Pan, B., Liu, K., Ren, B., Zhang, M., Ju, Y., Gu, J., Zhang, X., Clarkson, C. R., Edlmann, K., Zhu, W., & Iglauer, S. (2023). Impacts of relative permeability hysteresis, wettability, and injection/withdrawal schemes on underground hydrogen storage in saline aquifers. *Fuel, 333*. Available from https://doi.org/10.1016/j.fuel.2022.126516.

Paterson, L. (1983). The implications of fingering in underground hydrogen storage. *International Journal of Hydrogen Energy, 8*(1), 53–59. Available from https://doi.org/10.1016/0360-3199(83)90035-6.

Peng, D. Y., & Robinson, D. B. (1976). A new two-constant equation of state. *Industrial and Engineering Chemistry Fundamentals, 15*(1), 59–64. Available from https://doi.org/10.1021/i160057a011.

Perkins, T. K., & Johnston, O. C. (1963). A review of diffusion and dispersion in porous media. *Society of Petroleum Engineers Journal, 3*(01), 70–84. Available from https://doi.org/10.2118/480-pa.

Pfeiffer, W. T., & Bauer, S. (2015). Subsurface porous media hydrogen storage – Scenario development and simulation. *Energy Procedia, 76*, 565–572. Available from https://doi.org/10.1016/j.egypro.2015.07.872.

Rezaei, A., Hassanpouryouzband, A., Molnar, I., Derikvand, Z., Haszeldine, R. S., & Edlmann, K. (2022). Relative permeability of hydrogen and aqueous brines in sandstones and carbonates at reservoir conditions. *Geophysical Research Letters, 49*(12). Available from https://doi.org/10.1029/2022GL099433.

Salina Borello, E., Bocchini, S., Chiodoni, A., Coti, C., Fontana, M., Panini, F., Peter, C., Pirri, C. F., Tawil, M., Mantegazzi, A., Marzano, F., Pozzovivo, V., Verga, F., & Viberti, D. (2024). Underground hydrogen storage safety: Experimental study of hydrogen diffusion through caprocks. *Energies, 17*(2). Available from https://doi.org/10.3390/en17020394.

Scheidegger, A. E. (1961). General theory of dispersion in porous media. *Journal of Geophysical Research, 66*. Available from https://doi.org/10.1029/JZ066i010p03273.

Stiel, L. I., & Thodos, G. (1961). The viscosity of nonpolar gases at normal pressures. *AIChE Journal, 7*(4), 611–615. Available from https://doi.org/10.1002/aic.690070416.

Strauch, B., Pilz, P., Hierold, J., & Zimmer, M. (2023). Experimental simulations of hydrogen migration through potential storage rocks. *International Journal of Hydrogen Energy, 48*(66), 25808–25820. Available from https://doi.org/10.1016/j.ijhydene.2023.03.115.

Tek, M. (1989). *Underground storage of natural gas: Theory and practice* (171). Springer Science & Business Media. Available from 10.1007/978-94-009-0993-9.

Terry, R. E. (2001). *Encyclopedia of physical science and technology.*

Young, T. (1805). An essay on the cohesion of fluids. *Philosophical Transactions of the Royal Society of London.*

Yang, K., Kobeissi, S., Ling, N., Li, M., Esteban, L., May, E. F., & Johns, M. L. (2023). Measurement of hydrogen dispersion in rock cores using benchtop NMR. *International Journal of Hydrogen Energy, 48*(45), 17251–17260. Available from https://doi.org/10.1016/j.ijhydene.2023.01.197.

Yekta, A. E., Manceau, J.-C., Gaboreau, S., Pichavant, M., & Audigane, P. (2018). Determination of hydrogen–-water relative permeability and capillary pressure in sandstone: Application to underground hydrogen injection in sedimentary formations. *Transport in Porous Media, 122*(2), 333–356. Available from https://doi.org/10.1007/s11242-018-1004-7.

CHAPTER

8

Physiochemical parametric considerations for optimal underground hydrogen storage

Nawal Noshad[1], *Shwetank Krishna*[2], *Gerhard Thonhauser*[2], *Syahrir Ridha*[3], *Aymn Abdulrahman*[4] *and Suhaib Umer Ilyas*[4]

[1]Department of Chemical Engineering, University of Gujrat, Gujrat, Pakistan [2]Chair of Drilling and Completion Engineering, Department Geo Energy, Montanuniversität Leoben, Leoben, Austria [3]Institute of Hydrocarbon Recovery, Universiti Teknologi PETRONAS, Seri Iskandar, Perak Darul Ridzuan, Malaysia [4]Chemical Engineering Department, University of Jeddah, Jeddah, Kingdom of Saudi Arabia

8.1 Introduction

Underground hydrogen storage (UHS) is a relatively new and promising technology that is now attracting the attention of researchers because of the potential of hydrogen to be used as an energy vector (Gahleitner, 2013; Mouli-Castillo et al., 2021) to build a large-scale hydrogen economy (Lord et al., 2014; Tarkowski, 2019). This attentiveness towards hydrogen as an energy source is supported by the environmental concerns associated with fossil fuels and their scarcity in comparison to rising energy demands (Höök & Tang, 2013), as well as excess and deficits of renewable energy (an alternative energy source) caused by utilizing sources such as solar, wind, etc. that have limitations of seasonal dependability and geographical constraints (Heinemann, Alcalde, et al., 2021).

Hydrogen can be stored underground in large amounts in geological formations such as salt caverns (Lankof & Tarkowski, 2020; Ozarslan, 2012), deep aquifers (Sainz-Garcia et al., 2017), depleted hydrocarbon regions (Amid et al., 2016; Hemme & van Berk, 2018), etc., each storage type having its own scope of application (Tarkowski, 2019). Salt caverns are artificially drilled within salt domes or bedded salt deposits by controlled freshwater injection (Muhammed et al., 2022). Currently, only salt caverns store pure hydrogen as far

Subsurface Hydrogen Energy Storage
DOI: https://doi.org/10.1016/B978-0-443-24071-3.00009-1

as UHS is concerned. The viscoplastic characteristics of the rock salt ensure the tightness of hydrogen storage and long-term mechanical stability (Wei et al., 2017), making injection and extraction flexible. Furthermore, the plastic properties of salt protect against cracks and fractures, and the saline environment prevents the microbial consumption of hydrogen (Khaledi et al., 2016). An aquifer is found hundreds of feet deep and is a permeable and porous rock filled with water that may be fresh or saline (Muhammed et al., 2022). Saline aquifers have a large potential storage capacity, and UHS in deep aquifers is the most cost-effective option among the three major storage formations (Lord et al., 2014). Depleted hydrocarbon reservoirs are by far the most common means of UHS. However, there have been several difficulties in storing pure hydrogen in aquifers and depleted reservoirs because of the geochemical and microbial reaction-dependent risks associated with UHS.

In UHS, hydrogen (H_2) is injected into the subsurface and stored by a seal layer with extremely low permeability, that is, the caprock, and can be withdrawn when needed (Iglauer, 2022). This is significant because H_2 needs to be recoverable, while carbon dioxide (CO_2) is often stored underground permanently. Thus, understanding fluid flow in porous media is essential for two-way flow. Drainage and imbibition are the two processes that take place during injection and withdrawal. The decrease in wetting phase saturation when H_2 (i.e., the nonwetting phase) replaces brine (i.e., the wetting phase) as it is injected into the storage formation is known as drainage. Conversely, imbibition occurs during withdrawal when the nonwetting phase is displaced by the wetting phase, thus increasing the wetting phase saturation in the pore space. The different trapping mechanisms in the two processes of imbibition and drainage affect the mobility of H_2 and result in capillary pressure and relative permeability hysteresis (discussed in Section 8.2.3.4) and cause loss of H_2. Capillary pressure and relative permeability are both dependent on wettability (the ability of a fluid to adhere to a solid surface in the presence of immiscible or partially miscible fluids), that is, a function of contact angles (Hashemi, Blunt, et al., 2021) (refer to Section 8.2.3.1). Although hydrogen productivity loss is inevitable as recovered H_2 will always be less than the stored H_2, the deliverability of UHS can be enhanced by controlling several factors such as storage depth, sweep efficiency, injection, and production rates during the injection and withdrawal, respectively (Buscheck et al., 2023).

In underground storage, different trapping mechanisms such as structural, residual, dissolution, and mineral trapping play a vital role in understanding H_2 immobilization and its consequences, for example, H_2 loss during withdrawal (Raza et al., 2022). Structural or hydrodynamic trapping (Nilsen et al., 2012) refers to the trapped H_2 that has been accumulated beneath the caprock. The extremely low permeability of the caprock (typically shale or mudstone) prevents the buoyant H_2 gas from escaping through this sealing layer. Residual trapping (Baz et al., 2016; Rahman et al., 2016; Taku Ide et al., 2007), or capillary trapping, is the entrapment of H_2 in the pores of the reservoir rock by capillary forces and causes a significant amount of loss during withdrawal. However, the trapped H_2 may be recoverable with successive cycles. Residual trapping is also highly dependent on the wettability of the rock, as it determines the mobility of the fluid. Dissolution trapping (Emami-Meybodi et al., 2015; Suekane et al., 2008) occurs when H_2 dissolves into the brine present in the formation. While it is an efficient storage mechanism, it causes H_2 loss until an equilibrium between H_2 and brine is established. This means that dissolution will

TABLE 8.1 Cushion gas percentages concerning storage type.

Storage type	Cushion gas percentage (%)	Injection period (days)	Withdrawal period (days)
Saline aquifers	50–80	200–250	100–150
Depleted hydrocarbon reservoirs	50–70	200–250	100–150
Salt caverns	20–30	20–40	20–40

From Amid et al. (2016), Buscheck et al. (2023), and CEDIGAZ, 2017. Underground Gas Storage in the World – Part, 2017.

decrease with successive injection and withdrawal cycles. Mineral trapping (Xu et al., 2003) deals with any potential geochemical reactions that might occur upon the interaction of H_2 with in-situ fluids or the reservoir rock. The efficiency of these trapping mechanisms is influenced by several site characteristics, such as caprock properties, brine salinity, porosity, permeability heterogeneity, and the configuration of the injection well (Al-Khdheeawi et al., 2018).

Another essential aspect of H_2 storage and withdrawal is the amount of cushion gas. A cushion gas is injected into the storage formation prior to the injection of the working gas, that is, H_2, to maintain operational pressure, limit pressure swings for efficient injection and withdrawal cycles, and prevent water inflow into the storage medium during withdrawal (Tarkowski et al., 2021). It also assists in efficiently utilizing the maximum storage capacity of the formation. The cushion gas to working gas ratio in the formation depends on the type of storage and its geological parameters, such as reservoir depth and permeability, and it decreases with subsequent cycles. (Heinemann, Scafidi, et al., 2021). Knowing the appropriate volume of cushion gas required helps evaluate costs associated with injection and storage accurately. The percentage of cushion gas of the total gas stored in the three main storages is given in Table 8.1 (Amid et al., 2016; Buscheck et al., 2023; CEDIGAZ, 2017).

Cushion gases such as nitrogen (N_2) and CO_2 make the displacement of water or brine much more efficient because of their higher viscosity and density compared to H_2. N_2 also has the advantage of exhibiting good wetting behavior. In depleted hydrocarbon reservoirs, the available natural gas can be used as cushion gas (Feldmann et al., 2016).

The injection of H_2 comes with several complexities involving the displacement of in-situ fluids during UHS operations and the interactions of the injected gas with these fluids as well as with the reservoir rock. These factors are particularly essential to be considered for UHS because of the low density and viscosity of H_2 compared to that of water or brine present in the storage formation. This difference in densities results in inefficient displacement, viscous fingering, and gravity override of the water phase, and thus the loss of H_2.

8.2 Influencing physiochemical parameters concerning different types of underground hydrogen storage

The flow behavior of H_2 in the pore structure of the subsurface storage formation and its interaction with the solid (reservoir rock) and fluids (water, brine, oil, etc.) are important to study when dealing with a two-way hydrogen flow in UHS. It is also vital to understand the behavior of hydrogen in the pore network of the sediments at underground pressures and temperatures. These pore-scale effects (i.e., solid, fluid, and solid-fluid properties) help govern the reservoir fluid dynamics in addition to storage capacity and performance, containment security, and potential hydrogen injection and withdrawal rates (Pan, Yin, Ju, et al., 2021). All these factors, including storage type and size, are significantly influenced by several physiochemical parameters (Pan, Yin, Ju, et al., 2021), particularly in porous media, that is, saline aquifers and depleted hydrocarbon reservoirs, that provide long-term storage compared to salt caverns, which have limitations of short-duration storage (Buscheck et al., 2023; Heinemann et al., 2021). These parameters that greatly influence UHS at the pore scale are discussed and elaborated on in this chapter.

8.2.1 Solid-based properties

8.2.1.1 Absolute permeability

Absolute permeability (k_a) measures the ease of fluid flow through the porous medium (reservoir rock) when it is 100% saturated by this fluid. It is an intrinsic property not affected by fluid or fluid-solid interactions (Raats, 1973). Darcy's Law (given in Eq. 8.1) calculates absolute permeability, provided that the subsurface flow is laminar (having Reynold number less than 1–10), one phase is present at complete pore space saturation, and no reaction occurs between the rock and fluid.

$$k_a = \frac{Q\mu\Delta L}{A\Delta P} = \frac{V\mu\Delta L}{\Delta P} \tag{8.1}$$

where Q represents the volumetric flow rate of the fluid, μ is the fluid viscosity, ΔL is the distance between the injection and withdrawal well, A is the cross-sectional area, V is the fluid seepage velocity, and ΔP is the pressure differential. As the flowrate increases, the relationship between the differential pressure and seepage velocity becomes nonlinear because the inertial forces become more significant (Chen et al., 2015; Mustapha et al., 2015). In this case, Darcy's law is corrected by the addition of the seepage velocity cubic term and inertia force, presented in Eq. (8.2).

$$\Delta P = \frac{\mu}{k_a}V + \frac{\gamma}{\mu}\rho^2 V^3 \tag{8.2}$$

As the flow rate continues to rise, resulting in an increase in the Reynolds number, non-Darcy flow is attained. In such instances, the Forchheimer equation, provided in Eq. (8.3), becomes applicable (Chen et al., 2015).

$$\Delta P = \frac{\mu}{k_{fh}}V + \beta\rho V^2 \tag{8.3}$$

where k_{fh} is the Forchheimer permeability, β is the non-Darcy flow coefficient, and ρ is the fluid density. Here, note that $k_a \neq k_{fh}$ because of the weak inertia transition region. The primary explanation for this occurrence is believed to be viscous dissipation, wherein kinetic energy transforms into internal energy through work done against viscous stress.

Under low Reynolds numbers (Re), the dissipation is induced by viscous forces and remains consistent. As Reynolds numbers increase, the dissipation escalates due to heightened inertial forces, subsequently leading to alterations in the absolute permeability of the medium.

Absolute permeability is sensitive to changes in pore pressure or effective stress ($\sigma_{\text{effective}}$) (Palmer & Mansoori, 1998; Zheng et al., 2015), that is, the difference between overburden and formation pressures. The effect of $\sigma_{\text{effective}}$ on k_a is caused by sample consolidation and a possible change in pore structure (Chen et al., 2019), the impact of which is more apparent in low-permeability reservoirs compared to medium- to high-permeability reservoirs (Zhong et al., 2020). Fig. 8.1 describes the relationship between $\sigma_{\text{effective}}$ and k_a (Zhong et al., 2020). It can be observed that k_a decreases with $\sigma_{\text{effective}}$ during pressure loading for some rock samples (collected from permeability stress sensitivity tests from a tight sandstone reservoir in the Huanxian area of the Ordos Basin). Samples 1, 2, 3, and 4 correspond to L159-1, M217-1, M109-1, and H82-1 from Zhong et al. (2020). However, the response of k_a to $\sigma_{\text{effective}}$ is much larger with fractured reservoirs (Ju et al., 2017; Zheng et al., 2015). In UHS operations, k_a will increase during H_2 injection, that is, the unloading cycle, and decrease during H_2 withdrawal, that is, the loading cycle. Understanding the relationship between the two factors also enables accurate reservoir scale predictions before the UHS operation, subject to specific conditions. Therefore, it can be deduced that injecting hydrogen into or recovering it from highly permeable storage sites such as limestone and sandstone reservoirs is more energy-efficient (Sander et al., 2017).

Nonetheless, flow capacity, that is, the product of the reservoir's absolute permeability and thickness, is a much better indicator of hydrogen productivity than either of these parameters alone, which might be insufficient to determine the reservoir's productivity for hydrogen recovery (Okoroafor et al., 2022).

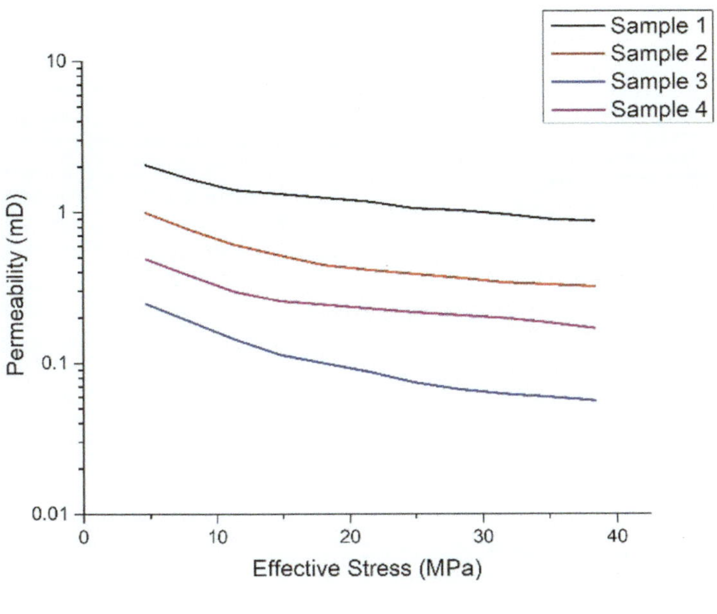

FIGURE 8.1 Absolute permeability as a function of effective stress for loading cycle. Source: *From Zhong, X., Zhu, Y., Liu, L., Yang, H., Li, Y., Xie, Y., & Liu, L. (2020). The characteristics and influencing factors of permeability stress sensitivity of tight sandstone reservoirs.* Journal of Petroleum Science and Engineering, 191. *https://doi.org/10.1016/j.petrol.2020.107221.*

8.2.1.2 *Effective porosity*

Effective porosity ($\phi_{effective}$) is the ratio of the connected pore volume to the bulk volume of the rock. It is different from absolute porosity (which considers total pore volume), as dead-end or disconnected pores do not aid the flow of H_2. Effective porosity is used to find the maximum hydrogen storage capacity (C) using Eq. (8.4).

$$C = L \times T \times W \times \phi_{effective} \times S_{H_2} \tag{8.4}$$

where L, T, and W are the formation length, thickness, and width, respectively, and S_{H_2} is the H_2 saturation, that is, the ratio of the pore volume filled with hydrogen to the total pore volume. We can observe from the equation that the maximum hydrogen capacity is directly proportional to the effective porosity. $\phi_{effective}$ and thus C is much greater and easier to measure (via helium pycnometry) in conventional rocks such as limestone and sandstone compared to shale (Arif et al., 2021) or coal (Zhang et al., 2016), that is, unconventional rocks, in which UHS prediction has high uncertainties (Pan, Yin, Ju, et al., 2021).

Effective porosity is also sensitive to effective stress to a certain extent (Zheng et al., 2015). Several empirical and theoretical models have been proposed to study the stress sensitivity of porosity and permeability (Katsube et al., 1991; Shi & Wang, 1986). As illustrated in Fig. 8.2, and similar to the relationship between k_a and $\sigma_{effective}$, $\phi_{effective}$ decreases with $\sigma_{effective}$ (with a faster rate than permeability (Li et al., 2014)) in the loading cycle. Samples 1, 2, 3, and 4 in Fig. 8.2 are limestone samples taken from the Cobourg formation in Ontario, Canada (Hu et al., 2020). The dependence on effective porosity signifies the importance of analyzing effective stress and its implications before UHS operations.

The containment security of hydrogen for UHS should also be properly examined before operation. Attention should be given to the reactivity of hydrogen with the storage material, as these reactions often take place over time and are likely to go unnoticed at the

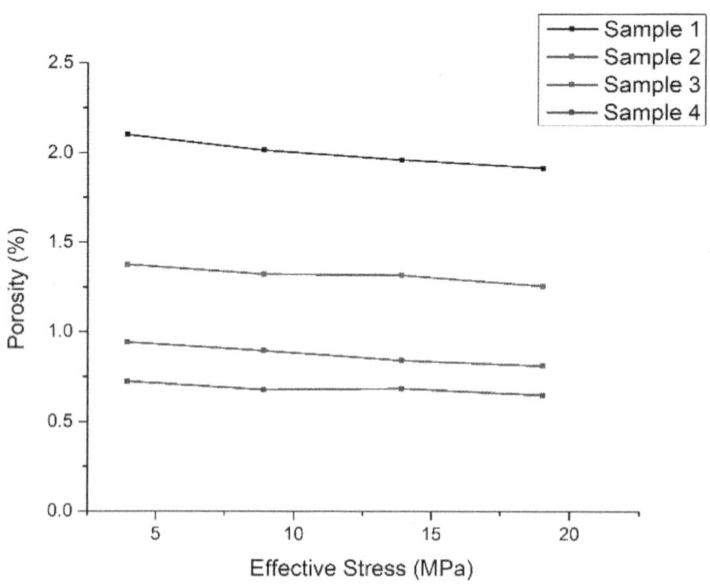

FIGURE 8.2 Effective porosity as a function of effective stress. Source: *From Hu, Z., Klaver, J., Schmatz, J., Dewanckele, J., Littke, R., Krooss, B. M., & Amann-Hildenbrand, A. (2020). Stress sensitivity of porosity and permeability of Cobourg limestone.* Engineering Geology, 273. *https://doi.org/10.1016/j.enggeo.2020.105632.*

initial stages. These reactions reduce the effective porosity and, thus, the storage capacity due to the phenomenon of expansion, which is attributed to the penetration of hydrogen into the solid phase (Al-Yaseri et al., 2023).

8.2.2 Fluid-based properties

8.2.2.1 *Fluid density*

One of the primary reasons to develop underground H_2 storage is the challenge of storing this low-density fluid on a large scale on the surface. H_2 has a density as low as 0.089 kg/m (Tarkowski, 2019). Fluid density (ρ) changes with temperature and pressure, which increases with formation depth owing to geothermal and hydrostatic gradients (Iglauer, 2018). ρ_{H_2} increases substantially with pressure and only slightly with temperature, as also observed in Fig. 8.3 (Pan, Yin, Ju, et al., 2021). However, ρ_{H_2} is less sensitive to temperature and pressure effects than ρ_{CH_4}, and ρ_{CO_2} which means that ρ_{H_2} is generally low even at high pressures.

Additionally, ρ_{H_2} is much less than ρ_{H_2O} at similar thermophysical conditions, which would result in viscous fingering, strong gravity segregation (Han et al., 2016; Taku Ide et al., 2007) and a rapid upward movement of H_2. H_2, having a lower density, accumulates beneath the caprock and above the high-density fluid (brine). The accumulated H_2 will likely penetrate through the caprock depending on several factors, including the entry capillary pressure, which is also interrelated to the rock-fluid interfacial tension (a low-density fluid makes it even more significant). This enhances the risk of leakage; thus, caprock integrity should be ensured (Pan, Yin, Ju, et al., 2021).

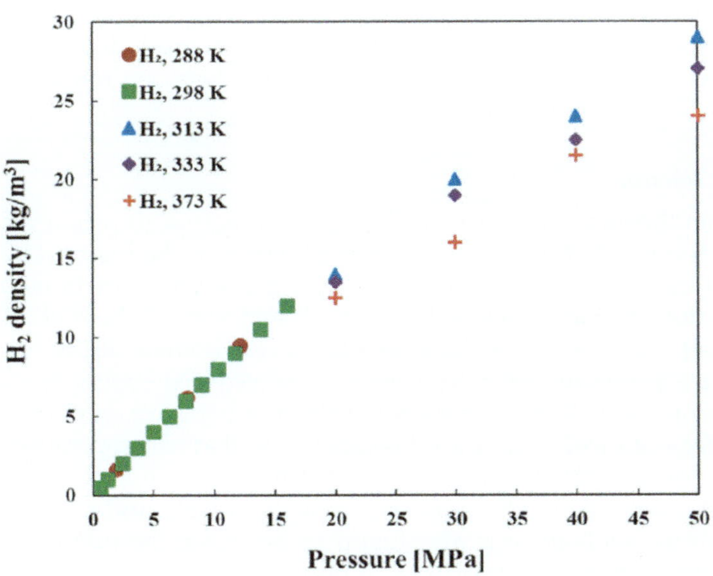

FIGURE 8.3 H_2 density as a function of temperature and pressure. Source: *From Pan, B., Yin, X., Ju, Y., & Iglauer, S. (2021). Underground hydrogen storage: Influencing parameters and future outlook.* Advances in Colloid and Interface Science, 294. *https://doi.org/10.1016/j.cis.2021.102473.*

Fluid densities are also used to predict contact angles and, thus, rock surface wettability (further elaborated in Section 8.2.3.1). Contact angles are significant in determining fluid distribution properties and storage capacities, amongst other aspects (Pan et al., 2019).

8.2.2.2 Fluid viscosity

Fluid viscosity (μ), that is, the ratio of shear stress between two fluid layers to the velocity gradient, measures fluid flow resistance and is incredibly important for H_2 injection and withdrawal. The fluid viscosity ratio (which is high for water and H_2 at similar thermophysical conditions) determines the fluid-fluid interfacial tension as well as viscous fingering (Chang et al., 2019; Wang et al., 2013), which can lead to problems like untimely surface leakage during H_2 injection, low H_2 sweep efficiency, etc. (Carden & Paterson, 1979). The low viscosity of H_2, as well as its high diffusivity, cause high mobility. Thus, there are more uncertainties in H_2 flow through porous media saturated with brine. This makes it hard for H_2 to replace brine, and the risk of viscous fingering increases. Even a tiny leakage can accelerate the formation of fingerings and cause severe H_2 loss (Perera, 2023). Viscous fingering reduces gas deliverability by decreasing the sweep efficiency (during injection and withdrawal) and increasing the loss of H_2 by dissolution and diffusion into brine (Buscheck et al., 2023). This effect can be minimized by lowering the injection rate. This will allow the capillary and gravitation forces to dominate the viscous forces and cause a more stable injection (Feldmann et al., 2016).

In Fig. 8.4, the effect of the CH_4 and H_2 mixture is represented (Hassanpouryouzband et al., 2020). It depicts that with the increase in H_2 concentration, the effect of temperature and pressure on H_2 viscosity is minimal as compared to its lower concentration. This finding is consistent with Heinemann, Alcalde, et al. (2021), who observed a similar trend where the viscosity of H_2 was only slightly affected by temperature and pressure within the range of subsurface storage conditions. The study also compared the viscosities of H_2, CO_2, and CH_4, revealing that H_2 has significantly lower viscosities than CO_2 and CH_4, especially at elevated temperatures and pressures. This relates to the fluid density—a dense gas will have more frequent and stronger collisions and friction between the molecules, thus increasing viscosity.

8.2.2.3 Fluid-fluid interfacial tension

The interfacial tension between the two fluids (γ_{FF}) is the energy required to produce a boundary surface between two immiscible fluids as a result of adhesion or the intermolecular interaction between the molecules on both sides (Israelachvili, 2022). There are limited studies available with regard to the interfacial tension between H_2 and water (Chow et al., 2018; Pan et al., 2020), the result of one of which is illustrated in Fig. 8.5 (Perera, 2023).

γ_{FF} is a function of temperature, pressure, and fluid composition (Arif, Al-Yaseri, et al., 2016; Arif, Barifcani, et al., 2016; Pan et al., 2020), although the effect of pressure on interfacial tension between H_2 and water is almost insignificant compared to that of temperature. The interfacial tension was lessened with the increase in temperature and pressure. However, different patterns are observed for different gases, such as CH_4 and CO_2. In another study (Ye et al., 2008), brine molality was also found to influence the interfacial tension of hydrogen and brine, which increased linearly with brine concentration.

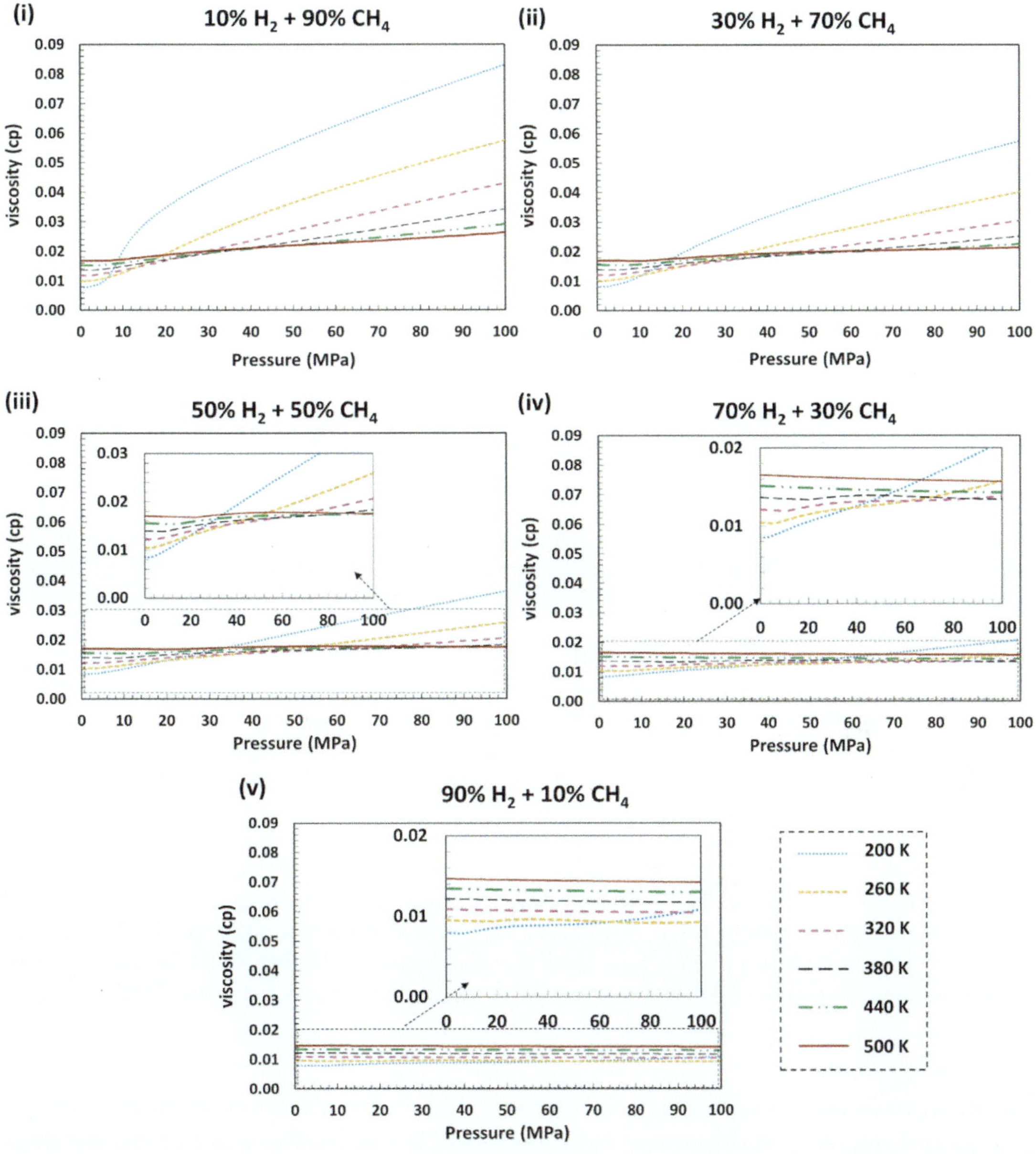

FIGURE 8.4 H_2 viscosity as a function of temperature and pressure when mixed in different concentration with CH_4. Source: *From Hassanpouryouzband, A., Joonaki, E., Edlmann, K., Heinemann, N., & Yang, J. (2020). Thermodynamic and transport properties of hydrogen containing streams.* Scientific Data, *7(1). https://doi.org/10.1038/s41597-020-0568-6.*

8.2.2.4 Solubility

While H_2 solubility in brine or other fluids enhances the H_2 trapping capacity, it is a major cause of hydrogen loss in UHS during withdrawal, which is also attributed to the

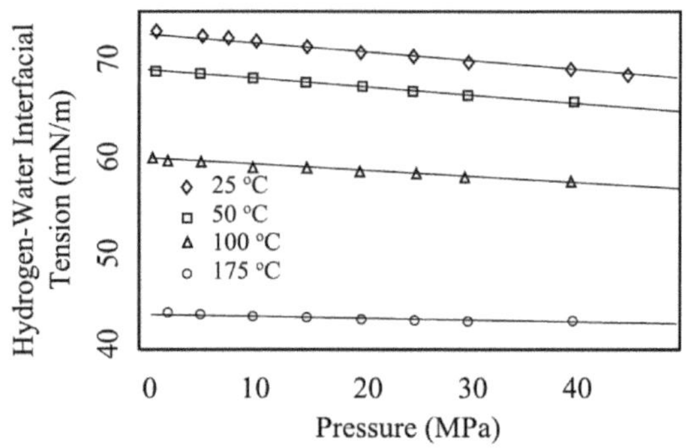

FIGURE 8.5 H_2-water interfacial tension. Source: *From Perera, M. S. A. (2023). A review of underground hydrogen storage in depleted gas reservoirs: Insights into various rock-fluid interaction mechanisms and their impact on the process integrity.* Fuel, 334. *https://doi.org/10.1016/j.fuel.2022.126677.*

low density of H_2 (0.089 kg/m) (Tarkowski, 2019). H_2 loss will take place until an equilibrium is established between H_2 and brine. H_2 solubility in UHS also affects storage and fluid mobility (Chabab et al., 2020). It influences the rock-fluid interaction as it affects the chemistry of the pore fluid. H_2 dissolved in brine may also alter thermodynamic equilibrium conditions that can lead to changes in pH and increased chemical reactivity or mineral trapping. When H_2 dissolves in brine at a certain pressure, it moves without any bubbles; however, bubbles form when this pressure is released. H_2 solubility in brine is much lower because of weak Van der Waals forces than CO_2, which makes strong hydrogen bonds (Perera, 2023).

H_2 dissolution is much greater in subsurface conditions than in ambient conditions. H_2 solubility in brine is a function of pressure, temperature, and salinity, as illustrated in Fig. 8.6A–C (Muhammed et al., 2022); it increases with an increase in pressure and temperature and decreases with an increase in brine salinity (Chabab et al., 2020; Li et al., 2018). A similar effect of temperature and pressure has been observed for oleic and other nonaqueous liquids such as pure hydrocarbons, heavy oils, alcohols, etc., although they may have one magnitude greater H_2 solubility than aqueous liquids (Zhao & Jin, 2019; Saajanlehto et al., 2014). Higher dissolution with increased temperatures is because of the elevated kinetic energy of the system, which allows the vibrating hydrogen to break the liquid molecules more easily. Similarly, high pressures essentially push the gas into the liquid, thus increasing dissolution, and the decreasing solubility with increasing salinity is associated with the bonding between reservoir rock and NaCl (Perera, 2023).

8.2.2.5 Diffusivity

Diffusivity is another essential parameter that describes the amount of gas that diffuses across a rock surface and determines the amount of H_2 lost. Similar to solubility, the diffusivity of H_2 into brine decreases with progressing injection and withdrawal cycles as it equilibrates. It can be observed from Fig. 8.7A and B (Muhammed et al., 2022) that diffusivity is a function of temperature, pressure, and the type of fluid (Zhao et al., 2021). In hydrocarbons, the diffusivity of H_2 decreased with an increase in temperature, while in water, there was a direct relation between diffusivity and temperature (Vivian & King, 1964; Zhao & Jin, 2019). It is suggested that this is attributed to the effect of temperature on Brownian motion, molecular momentum, and intermolecular forces (Bird, 2002).

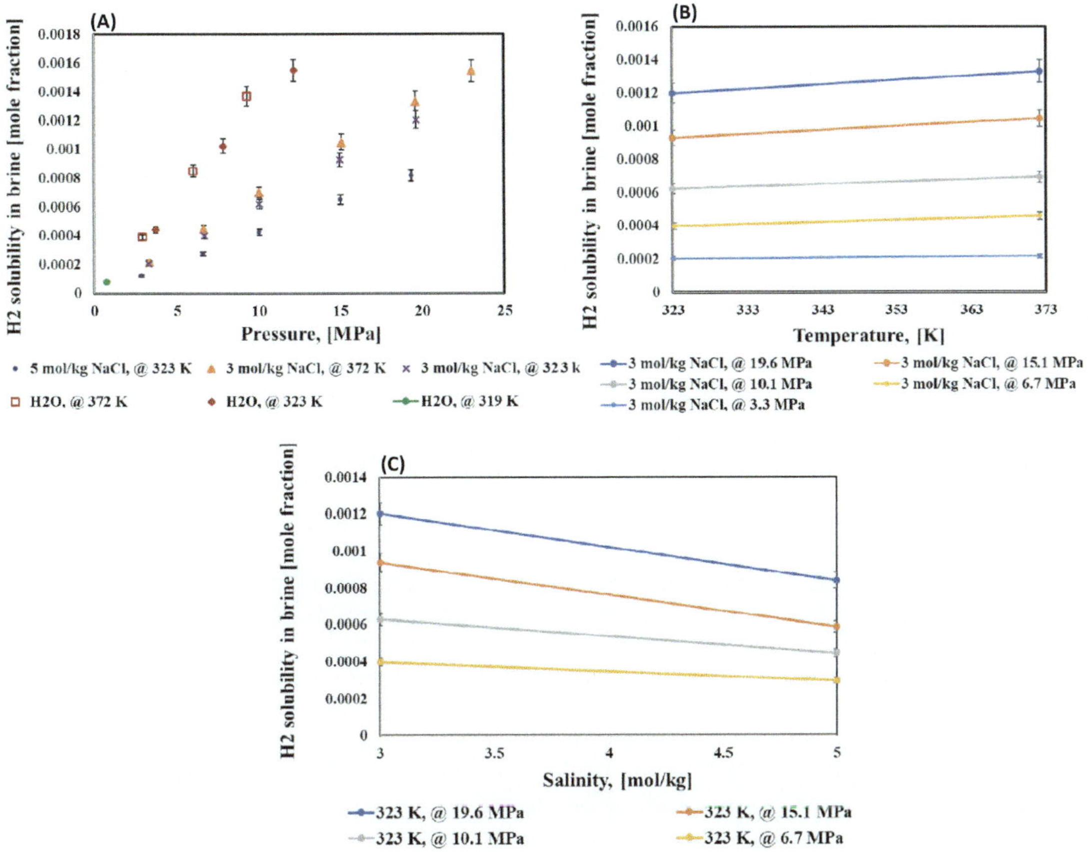

FIGURE 8.6 H_2 solubility as a function of (A) Pressure, (B) Temperature, and (C) Salinity (Muhammed et al., 2022).

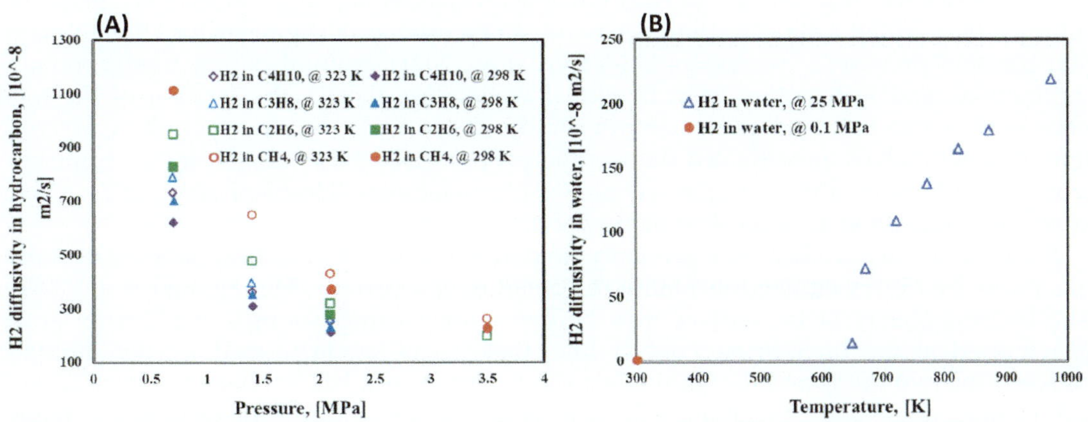

FIGURE 8.7 H_2 diffusivity as a function of pressure and temperature for (A) different hydrocarbons and (B) water (Muhammed et al., 2022).

Data from available literature indicates that H_2 losses attributed to diffusion during UHS would be much more significant in aquifers than in depleted hydrocarbon reservoirs, as diffusivity is enhanced in the presence of water (Muhammed et al., 2022). However, diffusive loss of H_2 is difficult to avoid because a large caprock area is exposed to the accumulated H_2, as these reservoirs usually have a much greater lateral length than their thickness (Perera, 2023).

8.2.3 Solid-fluid properties

8.2.3.1 Wettability

Wettability is the relative affinity of a liquid with a solid surface in the presence of another fluid (i.e., immiscible or partially miscible). In natural porous media, wettability may vary over the surface based on surface roughness, liquid adsorption on mineral or rock, etc. (Bradford & Leij, 1995; Eral, 't Mannetje, & Oh, 2013). In fact, wettability heterogeneity has a substantial impact on fluid displacement mechanisms and phase distributions during injection (Al-Khdheeawi et al., 2018). Depending on the rock-fluid system's wetting behavior, fluid distribution governs several other factors, for example, relative permeability and capillary pressure at mesoscopic and reservoir scales, plume migration, and the trapping capacity of the injected gas (Iglauer, 2022; Bird, 2002; Eral, 't Mannetje, & Oh, 2013; Bradford & Leij, 1995). H_2 wettability is an essential parameter for UHS as it can evaluate injection and withdrawal rates, containment security, and storage capacity (Al-Khdheeawi et al., 2018). It depends on the characteristics of the reservoir rock and the pore fluid, as well as hydrogen properties (Perera, 2023). Oil, water (brine), or gas can be used as a wetting phase, depending on the system and its interactive forces (Yekeen et al., 2020). Aquifer materials such as quartz, sulfates, and carbonates are strongly water-wet, while coal, talc, graphite, etc. may be neutrally wet to oil-wet. Water-wet clays, for instance, are found to enhance the structural and residual trapping of hydrogen, which has the drawback of lower withdrawal efficiencies (Pan et al., 2023). Moreover, gases like nitrogen that exhibit a good wetting behavior can be used as cushion gas for UHS to maintain constant pressure in reservoirs (Al-Yaseri et al., 2021). Natural porous media containing both oil and water often have fractional wettability, that is, the fraction of the total pore surface area that is oil- or water-wet (Bradford & Leij, 1995). However, there are very few studies that have investigated the wetting properties of the hydrogen-water/brine system, particularly in-situ hydrogen wettability characteristics (Hashemi et al., 2021; Higgs et al., 2022; Iglauer et al., 2021; Pan et al., 2021).

Rock surface wettability may generally be defined in terms of hydrophobicity or aqueous phase hydrophilicity and its ability to spread over a surface (Muhammed et al., 2022). The hydrophobicity of the surface may be determined using Young's equilibrium angle (θ_y) formed at the three-phase contact line, that is, the angle formed by the interface between the two fluids and the solid when the reservoir rock has not been completely wetted. Contact angle and contact angle hysteresis are both significant for wettability characterization (Pan et al., 2019). The contact angle may change under the influence of an external force, such as gravity. The difference between the advancing (the highest possible)

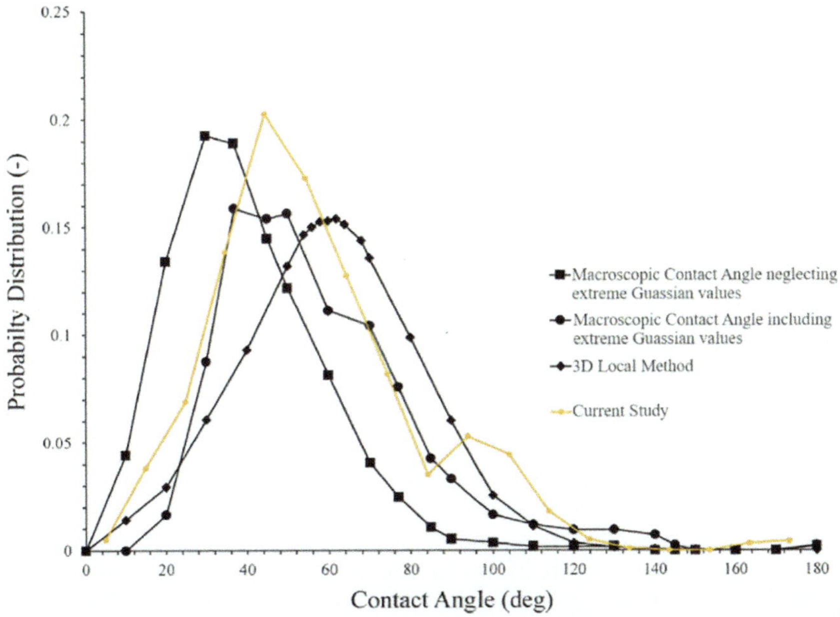

FIGURE 8.8 Distribution of contact angles (compared to another study). Source: *From Higgs, S., Wang, Y. D., Sun, C., Ennis-King, J., Jackson, S. J., Armstrong, R. T., & Mostaghimi, P. (2023). Direct measurement of hydrogen relative permeability hysteresis for underground hydrogen storage.* International Journal of Hydrogen Energy. *https://doi.org/10.1016/j.ijhydene.2023.07.270.*

contact angle and the receding (the lowest possible) contact angle is known as contact angle hysteresis (Makkonen, 2017).

The contact angle wettability assessment method has the advantage of providing rapid measurements for varying operating conditions and is simple, reliable, and versatile (Arif et al., 2021; Yekeen et al., 2020). The rock wettability and contact angles of various reservoir rocks from different origins are presented by Slobod and Blum (1952). The Devonian limestone exhibited an apparent contact angle of 39 degrees, indicating strong water-wetness behavior. The Yates Sandstone displayed an angle of 71 degrees, indicating intermediate water wetness. Alundum (RA-1139) showed a contact angle of 67 degrees, suggesting weak water wetness. Additionally, the synthetic clearfork and tensleep sand exhibited contact angles of 73 and 81 degrees, respectively, indicating intermediate water wetness for both samples. Fig. 8.8 represents the distribution of contact angles from a study (Higgs et al., 2023) (in orange), compared to another study (Higgs et al., 2022), both of which had similar trends, having mean contact angles of 57 and 53 degrees, respectively.

8.2.3.2 Solid-fluid interfacial tension

Solid-fluid interfacial tension (γ_{SF}) is an essential parameter that has an effect on the fluid spreading behavior (Perera, 2023) and thus the distribution and migration of H_2 in the subsurface (Arif, Barifcani, et al., 2016), and it is directly related to wettability by Young's equation, as given in Eq. (8.5).

$$\cos\theta = \frac{\gamma_{rock-H_2} - \gamma_{rock-H_2O}}{\gamma_{H_2-H_2O}} \tag{8.5}$$

where γ_{rock-H_2}, γ_{rock-H_2O}, and $\gamma_{H_2-H_2O}$ are solid-H_2, solid-water and H_2-water interfacial tensions respectively.

γ_{SF} is a function of temperature, pressure, and acid content, which is illustrated in Fig. 8.9A and B (Perera, 2023). These figures depict that γ_{rock-H_2} interfacial tension reduced with temperature and pressure for quartz, while at constant temperature and pressure, γ_{rock-H_2} decreased with an increase in acid concentration. The increase in pressure causes an increase in gas density, which naturally strengthens the rock-H_2 intermolecular forces. This is also why γ_{rock-H_2} is more significant than the interfacial tension between rock and CH_4 or CO_2, as H_2 has a lesser density.

An increase in temperature results in increased mobility; the high kinetic energy of the gas molecules gives them less time to interact with the rock surface, which results in lower γ_{rock-H_2}. The increase in organic acid concentration, on the other hand, has to do with the increase in the probability of the organic acid molecules adsorbing onto the rock (Ye et al., 2008). Consequently, the wettability of the gas on the rock increases as its surface becomes more hydrophobic. While this causes the repulsion of water, the non-polar molecule H_2 attaches itself and spreads onto the rock surface (Pan, Yin, & Iglauer, 2021).

8.2.3.3 *Capillary pressure*

Capillary pressure (P_c), that is, the pressure between two immiscible fluids in a thin tube, is dependent on capillary forces, which facilitate multiphase flow in a porous

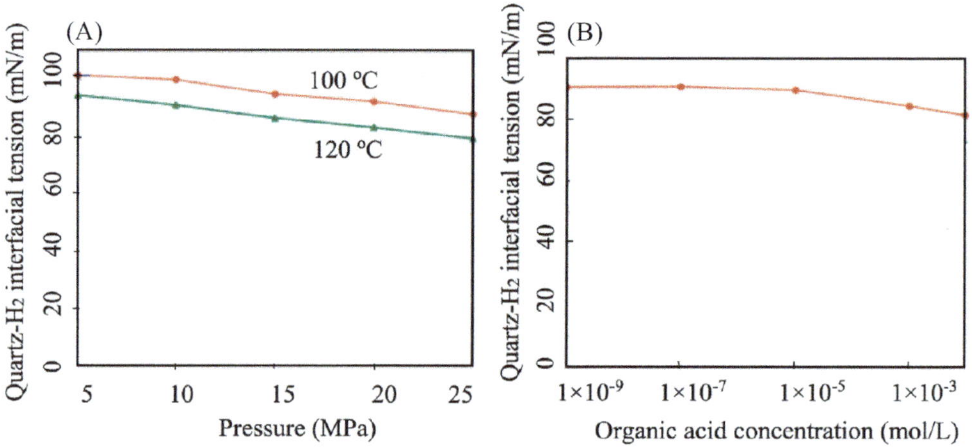

FIGURE 8.9 Solid-H_2 interficial tensions as a function of (A) temperature and pressure (B) organic acid concentration (Perera, 2023).

medium. Capillary pressure is formed as a consequence of the interaction of fluid and solid forces and can be found using Eq. (8.6) (Muhammed et al., 2022).

$$P_c = \frac{2\gamma\cos\theta}{r} \tag{8.6}$$

where γ, θ and r are the interfacial tension, contact angle, and radius of the largest pore, respectively.

To achieve a larger H_2 storage capacity, capillary pressure for larger H_2 saturation or smaller brine saturation will be needed as H_2 displaces the in-situ brine in the pore space (Pan, Yin, Ju, et al., 2021). The amount of pressure required to inject the H_2 (and thus displace the brine) is determined by P_c. This establishes P_c as an important parameter to determine UHS feasibility. Fig. 8.10 illustrates the very limited data available on capillary pressure with regard to rock-H_2-water systems (Raza et al., 2022).

After gas injection, the H_2 plume is split into discontinuous clusters when spontaneous imbibition by brine displaces the injected H_2 in the storage medium (Hu et al., 2018). This process will subsequently increase the storage capacity through the residual trapping mechanism (Hu et al., 2017; Iglauer & Wülling, 2016). This trapping should be minimized to avoid loss in the withdrawal phase, as H_2 clusters are more likely to be remobilized (Raza et al., 2022; Tanino & Blunt, 2013).

P_c also governs structural trapping and thus containment security and the storage capacity of H_2 (Guiltinan et al., 2018; Iglauer et al., 2015; Ren et al., 2014). The buoyancy pressure (P_B) can be estimated using Eq. (8.7), assuming that H_2 and water (brine) both exist as a continuous phase.

$$P_B = (\rho_{brine} - \rho_{H_2})gh = \Delta\rho gh \tag{8.7}$$

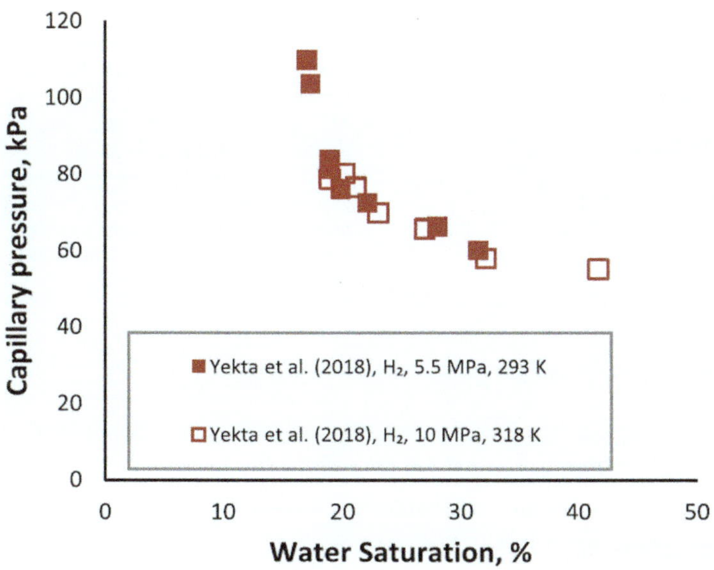

FIGURE 8.10 Capillary pressure as a function of water saturation in sandstone for H_2-water system. Source: *From Raza, A., Arif, M., Glatz, G., Mahmoud, M., Al Kobaisi, M., Alafnan, S., & Iglauer, S. (2022). A holistic overview of underground hydrogen storage: Influencing factors, current understanding, and outlook.* Fuel, 330. *https://doi.org/10.1016/j.fuel.2022.125636.*

8.2.3.4 *Relative permeability*

Relative permeability (k_r) is the ratio of effective permeability (k_e), that is, the fluid conductance of a fluid in the presence of other fluids, to absolute permeability (k_a), and has a great influence on multiphase fluid flow through porous media. Relative permeability is one of the two factors that have a major impact on storage efficiency by increasing the loss of H_2 (Zivar, Kumar, & Foroozesh, 2021). k_r is a function of several parameters, including surface wettability, saturation, contact angle, and pore size distribution (Burnside & Naylor, 2014; Krevor et al., 2012; Ruprecht et al., 2014), and is an important parameter to determine injection and withdrawal rates (Higgs et al., 2022). Limited data is available for the relative permeability of H_2 and water. One data set available where k_r of both water and H_2 were measured in reservoir conditions is illustrated in Fig. 8.11.

Relative permeability hysteresis, which refers to the change in relative permeability of fluids after displacement (Gao et al., 2023), is found to have a significant impact on the cyclic UHS operation (Lysyy et al., 2022; Zhang et al., 2023). k_r may depend on saturation according to the history of the system. The process of hysteresis is observed during drainage, where H_2 (nonwetting phase) is injected to saturate the reservoir, followed by an imbibition process, where gas withdrawal reduces saturation. The entrapment of the nonwetting phase is a consequence of capillary pressure hysteresis. Capillary pressure and relatively permeability hysteresis occur because the amount of pressure required for H_2 to replace brine during drainage is greater than the pressure required for brine to replace H_2 in the imbibition process. Relative permeability hysteresis leads to less H_2 injection and withdrawal as a result of H_2 entrapment (Jadhawar & Saeed, 2024). However, it is also found that this withdrawn gas has a higher purity (Pan et al., 2023). Fig. 8.12 illustrates the relative permeability for the two processes (drainage and imbibition) in the gas–water system (Buscheck et al., 2023) (data collected from Fatemi and Sohrabi (2018)), where k_{rg}

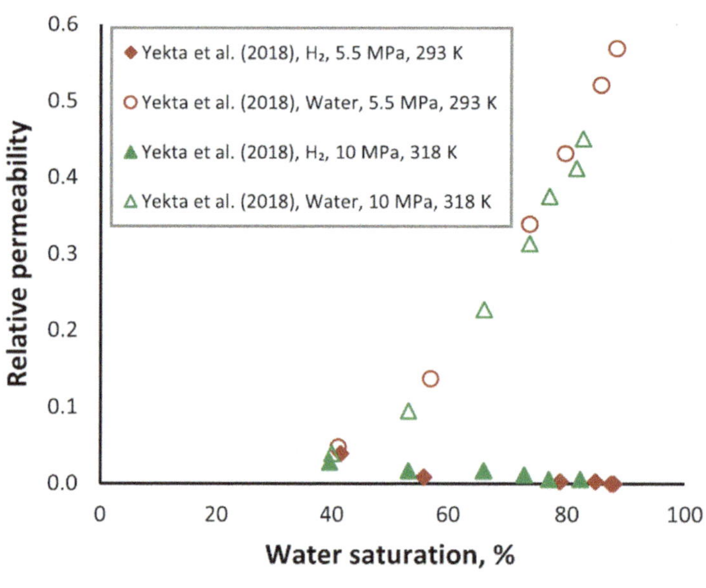

FIGURE 8.11 Relative permeability as a function of water saturation in sandstone for H_2-water system. Source: *From Raza, A., Arif, M., Glatz, G., Mahmoud, M., Al Kobaisi, M., Alafnan, S., & Iglauer, S. (2022). A holistic overview of underground hydrogen storage: Influencing factors, current understanding, and outlook.* Fuel, 330. *https://doi.org/10.1016/j.fuel.2022.125636.*

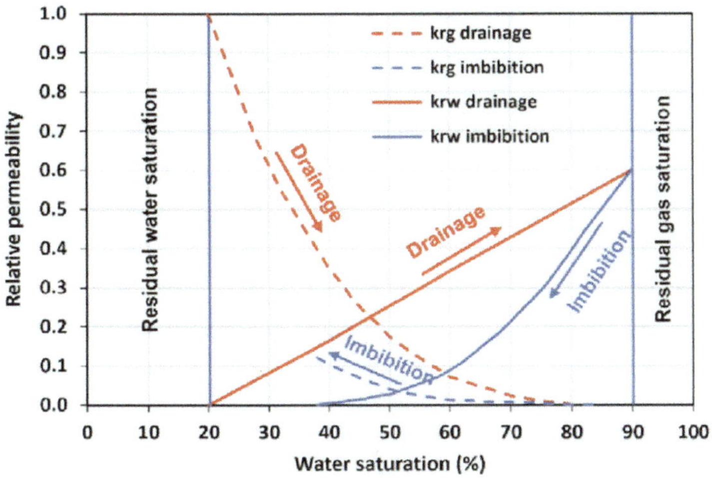

FIGURE 8.12 Imbibition and drainage relative permeability as a function of wetting phase saturation for gas–water system. Source: *From Buscheck, T. A., Goodman, A., Lackey, G., De Toledo Camargo, J., Huerta, N., Haeri, F., Freeman, G. M., & White, J. A. (2023). Underground storage of hydrogen and hydrogen/methane mixtures in porous reservoirs: Influence of reservoir factors and engineering choices on deliverability and storage operations.* International Journal of Hydrogen Energy. *https://doi.org/10.1016/j.ijhydene.2023.07.073.*

and k_{rw} are the relative permeabilities of gas and water, respectively. The difference in the k_r curves is because of the different trapping mechanisms of H_2 for both of these processes.

8.2.3.5 Mobility ratio

Mobility ratio (M) is the unstable displacement of a more viscous fluid (water) by a less viscous fluid (H_2). It can also be defined as the product of absolute permeability (k_a) and relative permeability (k_r) divided by the fluid viscosity (μ), as given in Eq. (8.8).

$$M = \frac{\left(\frac{k_a \times k_{r1}}{\mu_1}\right)}{\left(\frac{k_a \times k_{r2}}{\mu_2}\right)} = \frac{k_{r1}\mu_2}{k_{r2}\mu_1} \tag{8.8}$$

The mobility ratio is used to investigate the injection and withdrawal efficiency of H_2 and H_2-formation brine interfacial stability (Al-Abri et al., 2012; Phukan et al., 2020). A larger M would result in low H_2 sweep efficiency and inefficient displacement of formation brine due to viscous fingering (Berg & Ott, 2012). Generally, lesser mobility ratios ($M < 1$) provide more stable displacement fronts. Sweep efficiencies (both areal and vertical) decrease as M increases for a given amount of fluid injected. In other words, high values of M cause flow instability. As a result, viscous fingers are formed (Al-Abri et al., 2012). M is dependent on temperature, pressure, water saturation, and the gas type, as shown in Fig. 8.13 (Pan, Yin, Ju, et al., 2021). Temperature, pressure, and water saturation were all observed to have an inverse relationship with M.

8.2.3.6 Adsorption–desorption

Gas adsorption and desorption are important phenomena, particularly where porous media are concerned. Adsorption is the adhesion of fluid or dissolved solid molecules onto a surface, while desorption involves the removal of adsorbed molecules at high

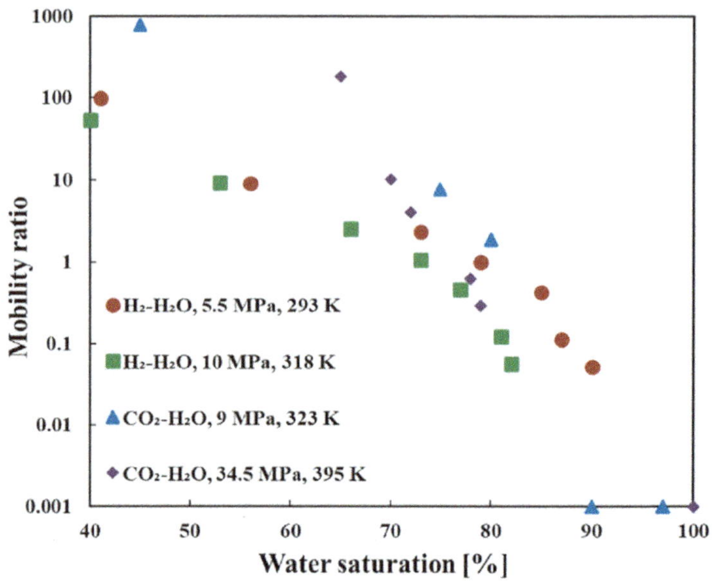

FIGURE 8.13 Mobility ratio as a function of temperature, pressure, and water saturation for H_2/CO_2-water-sandstone system. Source: *From Pan, B., Yin, X., Ju, Y., & Iglauer, S. (2021). Underground hydrogen storage: Influencing parameters and future outlook.* Advances in Colloid and Interface Science, 294. *https://doi.org/10.1016/j.cis.2021.102473.*

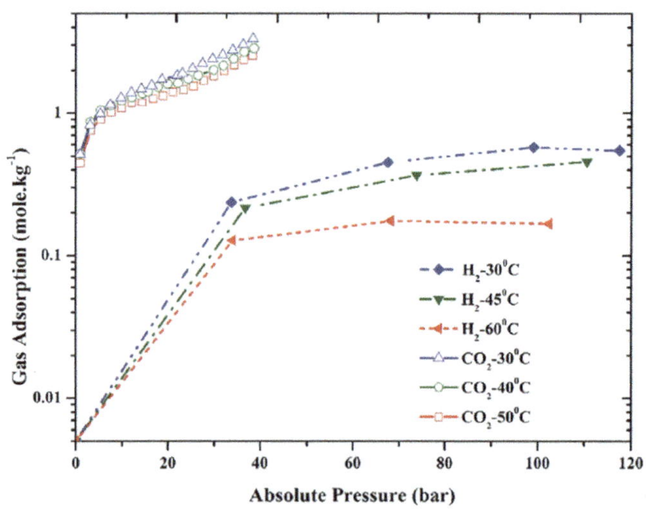

FIGURE 8.14 H_2 adsorption on subbituminous coal as a function of temperature and pressure (compared with CO_2). Source: *From Iglauer, S., Abid, H., Al-Yaseri, A., & Keshavarz, A. (2021). Hydrogen adsorption on sub-bituminous coal: Implications for hydrogen geo-storage.* Geophysical Research Letters, 48 *(10). https://doi.org/10.1029/2021GL092976.*

temperatures and pressures. Both of these phenomena play a crucial role in organic-rich materials such as coal or shale. Adsorption capacity is subject to organic matter content and pore structure (Rani et al., 2019).

Adsorption as a function of pressure and temperature is depicted in Fig. 8.14 (Iglauer et al., 2021), which demonstrates the suitability of higher pressures for higher adsorption rates for H_2 compared with adsorption rates of CO_2 (the higher rate of adsorption for CO_2 was associated with the low zeta potential of the reservoir rock used in the study, i.e., coal). Larger rock surface areas also encourage more adsorption (Bardelli et al., 2014).

Furthermore, Van der Waals forces play an important role in H_2 adsorption. Weak Van der Waals forces between H_2 and the rock, combined with low temperatures and high pressures, are optimal conditions for large H_2 storage densities in UHS via adsorption (Andersson & Grönkvist, 2019). However, high pressure would only be useful up to a certain limit due to the space occupied by the preadsorbed gas (Klebanoff et al., 2014). Temperature, on the other hand, had a weaker effect on the adsorption profile, with a marginally smaller amount adsorbed at a higher temperature. It is important to note that adsorption, being an exothermic process, may affect the degree of adsorption of H_2 despite the weak Van der Waals forces (Raza et al., 2022). Furthermore, the rate of gas desorption is usually lower than the adsorption rate (Yang et al., 2016), which could result in rock deformation and changes in permeability, and could even impact withdrawal activities (Wei et al., 2019).

8.3 Prospects and challenges

UHS, while promising, is a relatively new technology with very little preliminary work and data available. A lot of the research done so far has used carbon geo-sequestration (CGS) projects or natural gas reservoirs as a point of reference to gauge several aspects of UHS and its implications, simply because of the lack of data available on H_2. However, comparison with and extrapolating data from these sources has very obvious limitations, as H_2, CO_2, and CH_4 (or other fluids) all have different physical and chemical properties. These distinctions are particularly apparent in fluid properties and solid-fluid interactions.

Moreover, even from the available literature regarding the parameters that significantly influence UHS operations, these studies are limited to specific reservoir rocks and conditions (for a given property). For instance, insufficient information is available to accurately determine the diffusion loss through hydrogen under deep reservoir conditions or the effect of cushion gas and residually trapped hydrogen in subsurface conditions. The lack of datasets in some contexts also proves to be a challenge to do further research, despite its potential. For example, the use of Henry's Law to determine Henry's constant (for solubility comparison) has been limited because the partial pressure above the liquid has not been reported in the literature. Similarly, the Einstein relationship has not been used to relate diffusivity to viscosity, as neither of these properties have been reported for the same thermophysical conditions.

8.4 Summary and recommendations

UHS is a favorable solution for large-scale H_2 storage in formations such as salt caverns, aquifers, and depleted hydrocarbon reservoirs. UHS operations depend on several factors, such as rock properties (k_a and $\phi_{effective}$), fluid properties (ρ, μ, γ_{FF}, solubility, and diffusivity), and solid-fluid interactions (wettability, γ_{SF}, P_c, k_r, M, and adsorption–desorption). These properties play a significant role in determining UHS performance, storage capacity, containment security, etc.

However, very scarce data is available regarding these influencing physiochemical properties, subject to specific conditions only. Conclusions drawn from the underground storage of other gases cannot be used to navigate the operations of UHS in all scenarios. These properties need to be assessed in several different porous geologic media as well as under different geothermal conditions, taking into consideration multiple influencing factors at a time. Therefore, UHS research demands more investment in research for accurate evaluation of the process so that UHS applications can not only be enhanced but also analyzed for its feasibility.

References

Al-Abri, A., Sidiq, H., & Amin, R. (2012). Mobility ratio, relative permeability and sweep efficiency of supercritical CO_2 and methane injection to enhance natural gas and condensate recovery: Coreflooding experimentation. *Journal of Natural Gas Science and Engineering*, *9*, 166–171. Available from https://doi.org/10.1016/j.jngse.2012.05.011.

Al-Khdheeawi, E. A., Vialle, S., Barifcani, A., Sarmadivaleh, M., & Iglauer, S. (2018). Effect of wettability heterogeneity and reservoir temperature on CO_2 storage efficiency in deep saline aquifers. *International Journal of Greenhouse Gas Control*, *68*, 216–229. Available from https://doi.org/10.1016/j.ijggc.2017.11.016, http://www.elsevier.com/wps/find/journaldescription.cws_home/709061/description#description.

Al-Yaseri, A., Al-Mukainah, H., Yekeen, N., & Al-Qasim, A. S. (2023). Experimental investigation of hydrogen-carbonate reactions via computerized tomography: Implications for underground hydrogen storage. *International Journal of Hydrogen Energy*, *48*(9), 3583–3592. Available from https://doi.org/10.1016/j.ijhydene.2022.10.148, http://www.journals.elsevier.com/international-journal-of-hydrogen-energy/.

Al-Yaseri, A., Wolff-Boenisch, D., Fauziah, C. A., & Iglauer, S. (2021). Hydrogen wettability of clays: Implications for underground hydrogen storage. *International Journal of Hydrogen Energy*, *46*(69), 34356–34361. Available from https://doi.org/10.1016/j.ijhydene.2021.07.226, http://www.journals.elsevier.com/international-journal-of-hydrogen-energy/.

Amid, A., Mignard, D., & Wilkinson, M. (2016). Seasonal storage of hydrogen in a depleted natural gas reservoir. *International Journal of Hydrogen Energy*, *41*(12), 5549–5558. Available from https://doi.org/10.1016/j.ijhydene.2016.02.036.

Andersson, J., & Grönkvist, S. (2019). Large-scale storage of hydrogen. *International Journal of Hydrogen Energy*, *44*(23), 11901–11919. Available from https://doi.org/10.1016/j.ijhydene.2019.03.063.

Arif, M., Al-Yaseri, A. Z., Barifcani, A., Lebedev, M., & Iglauer, S. (2016). Impact of pressure and temperature on CO_2-brine-mica contact angles and CO_2-brine interfacial tension: Implications for carbon geo-sequestration. *Journal of Colloid and Interface Science*, *462*, 208–215. Available from https://doi.org/10.1016/j.jcis.2015.09.076, http://www.elsevier.com/inca/publications/store/6/2/2/8/6/1/index.htt.

Arif, M., Barifcani, A., & Iglauer, S. (2016). Solid/CO_2 and solid/water interfacial tensions as a function of pressure, temperature, salinity and mineral type: Implications for CO_2-wettability and CO_2 geo-storage. *International Journal of Greenhouse Gas Control*, *53*, 263–273. Available from https://doi.org/10.1016/j.ijggc.2016.08.020.

Arif, M., Mahmoud, M., Zhang, Y., & Iglauer, S. (2021). X-ray tomography imaging of shale microstructures: A review in the context of multiscale correlative imaging. *International Journal of Coal Geology*, *233*. Available from https://doi.org/10.1016/j.coal.2020.103641.

Bardelli, F., Mondelli, C., Didier, M., Vitillo, J. G., Cavicchia, D. R., Robinet, J. C., Leone, L., & Charlet, L. (2014). Hydrogen uptake and diffusion in Callovo–Oxfordian clay rock for nuclear waste disposal technology. *Applied Geochemistry*, *49*, 168–177. Available from https://doi.org/10.1016/j.apgeochem.2014.06.019.

Baz, H., Noureldin, M., Allinson, W. G., & Cinar, Y. (2016). A field-scale investigation of residual and dissolution trapping of CO_2 in a saline formation in Western Australia. *International Journal of Greenhouse Gas Control*, *46*, 86–99. Available from https://doi.org/10.1016/j.ijggc.2015.12.032.

Berg, S., & Ott, H. (2012). Stability of CO_2–brine immiscible displacement. *International Journal of Greenhouse Gas Control*, *11*, 188–203. Available from https://doi.org/10.1016/j.ijggc.2012.07.001.

Bird, R. B. (2002). Transport phenomena. *Applied Mechanics Reviews, 55*(1). Available from https://doi.org/10.1115/1.1424298.

Bradford, S. A., & Leij, F. J. (1995). Fractional wettability effects on two-and three-fluid capillary pressure-saturation relations. *Journal of Contaminant Hydrology, 20*(1-2), 89–109. Available from https://doi.org/10.1016/0169-7722(95)00027-S.

Burnside, N. M., & Naylor, M. (2014). Review and implications of relative permeability of CO_2/brine systems and residual trapping of CO_2. *International Journal of Greenhouse Gas Control, 23*, 1–11. Available from https://doi.org/10.1016/j.ijggc.2014.01.013.

Buscheck, T. A., Goodman, A., Lackey, G., De Toledo Camargo, J., Huerta, N., Haeri, F., Freeman, G. M., & White, J. A. (2023). Underground storage of hydrogen and hydrogen/methane mixtures in porous reservoirs: Influence of reservoir factors and engineering choices on deliverability and storage operations. *United States International Journal of Hydrogen Energy*. Available from https://doi.org/10.1016/j.ijhydene.2023.07.073, http://www.journals.elsevier.com/international-journal-of-hydrogen-energy/.

Carden, P., & Paterson, L. (1979). Physical, chemical and energy aspects of underground hydrogen storage. *International Journal of Hydrogen Energy, 4*(6), 559–569. Available from https://doi.org/10.1016/0360-3199(79)90083-1.

CEDIGAZ (2017). Underground Gas Storage in the World – Part. 1. Report Published on 7-July-2017.

Chabab, S., Théveneau, P., Coquelet, C., Corvisier, J., & Paricaud, P. (2020). Measurements and predictive models of high-pressure H_2 solubility in brine ($H_2O + NaCl$) for underground hydrogen storage application. *International Journal of Hydrogen Energy, 45*(56), 32206–32220. Available from https://doi.org/10.1016/j.ijhydene.2020.08.192.

Chang, C., Kneafsey, T. J., Zhou, Q., Oostrom, M., & Ju, Y. (2019). Scaling the impacts of pore-scale characteristics on unstable supercritical CO_2-water drainage using a complete capillary number. *International Journal of Greenhouse Gas Control, 86*, 11–21. Available from https://doi.org/10.1016/j.ijggc.2019.04.010.

Chen, T., Feng, X. T., Cui, G., Tan, Y., & Pan, Z. (2019). Experimental study of permeability change of organic-rich gas shales under high effective stress. *Journal of Natural Gas Science and Engineering, 64*, 1–14. Available from https://doi.org/10.1016/j.jngse.2019.01.014, http://www.elsevier.com/wps/find/journaldescription.cws_home/716470/description#description.

Chen, Y. F., Zhou, J. Q., Hu, S. H., Hu, R., & Zhou, C. B. (2015). Evaluation of Forchheimer equation coefficients for non-Darcy flow in deformable rough-walled fractures. *Journal of Hydrology, 529*, 993–1006. Available from https://doi.org/10.1016/j.jhydrol.2015.09.021, http://www.elsevier.com/inca/publications/store/5/0/3/3/4/3.

Chow, Y. T. F., Maitland, G. C., & Trusler, J. P. M. (2018). Interfacial tensions of ($H_2O + H_2$) and ($H_2O + CO_2 + H_2$) systems at temperatures of (298–448) K and pressures up to 45 MPa. *Fluid Phase Equilibria, 475*, 37–44. Available from https://doi.org/10.1016/j.fluid.2018.07.022, https://www.journals.elsevier.com/fluid-phase-equilibria.

Emami-Meybodi, H., Hassanzadeh, H., Green, C. P., & Ennis-King, J. (2015). Convective dissolution of CO_2 in saline aquifers: Progress in modeling and experiments. *International Journal of Greenhouse Gas Control, 40*, 238–266. Available from https://doi.org/10.1016/j.ijggc.2015.04.003, http://www.elsevier.com/wps/find/journaldescription.cws_home/709061/description#description.

Eral, H. B., 't Mannetje, D. J. C. M., & Oh, J. M. (2013). Contact angle hysteresis: A review of fundamentals and applications. *Colloid and Polymer Science, 291*(2), 247–260. Available from https://doi.org/10.1007/s00396-012-2796-6.

Fatemi, S. M., & Sohrabi, M. (2018). Relative permeabilities hysteresis for oil/water, gas/water and gas/oil systems in mixed-wet rocks. *Journal of Petroleum Science and Engineering, 161*, 559–581. Available from https://doi.org/10.1016/j.petrol.2017.11.014, http://www.sciencedirect.com/science/journal/09204105/81.

Feldmann, F., Hagemann, B., Ganzer, L., & Panfilov, M. (2016). Numerical simulation of hydrodynamic and gas mixing processes in underground hydrogen storages. *Environmental Earth Sciences, 75*(16). Available from https://doi.org/10.1007/s12665-016-5948-z.

Gahleitner, G. (2013). Hydrogen from renewable electricity: An international review of power-to-gas pilot plants for stationary applications. *International Journal of Hydrogen Energy, 38*(5), 2039–2061. Available from https://doi.org/10.1016/j.ijhydene.2012.12.010.

Gao, J., Kong, D., Peng, Y., Zhou, Y., Liu, Y., & Zhu, W. (2023). Pore-scale mechanisms and hysteresis effect during multi-cycle injection and production process in underground hydrogen storage reservoir. *Energy, 283*. Available from https://doi.org/10.1016/j.energy.2023.129007.

Guiltinan, E. J., Espinoza, D. N., Cockrell, L. P., & Cardenas, M. B. (2018). Textural and compositional controls on mudrock breakthrough pressure and permeability. *Advances in Water Resources, 121*, 162–172. Available from https://doi.org/10.1016/j.advwatres.2018.08.014.

Han, J., Lee, M., Lee, W., Lee, Y., & Sung, W. (2016). Effect of gravity segregation on CO_2 sequestration and oil production during CO_2 flooding. *Applied Energy, 161*, 85–91. Available from https://doi.org/10.1016/j.apenergy.2015.10.021.

Hashemi, L., Blunt, M., & Hajibeygi, H. (2021). Pore-scale modelling and sensitivity analyses of hydrogen-brine multiphase flow in geological porous media. *Scientific Reports, 11*(1). Available from https://doi.org/10.1038/s41598-021-87490-7.

Hashemi, L., Glerum, W., Farajzadeh, R., & Hajibeygi, H. (2021). Contact angle measurement for hydrogen/brine/sandstone system using captive-bubble method relevant for underground hydrogen storage. *Advances in Water Resources, 154*. Available from https://doi.org/10.1016/j.advwatres.2021.103964.

Hassanpouryouzband, A., Joonaki, E., Edlmann, K., Heinemann, N., & Yang, J. (2020). Thermodynamic and transport properties of hydrogen containing streams. *Scientific Data, 7*(1). Available from https://doi.org/10.1038/s41597-020-0568-6.

Heinemann, N., Alcalde, J., Miocic, J. M., Hangx, S. J. T., Kallmeyer, J., Ostertag-Henning, C., Hassanpouryouzband, A., Thaysen, E. M., Strobel, G. J., Schmidt-Hattenberger, C., Edlmann, K., Wilkinson, M., Bentham, M., Stuart Haszeldine, R., Carbonell, R., & Rudloff, A. (2021). Enabling large-scale hydrogen storage in porous media – The scientific challenges. *Energy & Environmental Science, 14*(2), 853–864. Available from https://doi.org/10.1039/d0ee03536j.

Heinemann, N., Scafidi, J., Pickup, G., Thaysen, E. M., Hassanpouryouzband, A., Wilkinson, M., Satterley, A. K., Booth, M. G., Edlmann, K., & Haszeldine, R. S. (2021). Hydrogen storage in saline aquifers: The role of cushion gas for injection and production. *International Journal of Hydrogen Energy, 46*(79), 39284–39296. Available from https://doi.org/10.1016/j.ijhydene.2021.09.174.

Hemme, C., & van Berk, W. (2018). Hydrogeochemical modeling to identify potential risks of underground hydrogen storage in depleted gas fields. *Applied Sciences, 8*(11). Available from https://doi.org/10.3390/app8112282.

Higgs, S., Wang, Y. D., Sun, C., Ennis-King, J., Jackson, S. J., Armstrong, R. T., & Mostaghimi, P. (2023). Direct measurement of hydrogen relative permeability hysteresis for underground hydrogen storage. *International Journal of Hydrogen Energy*. Available from https://doi.org/10.1016/j.ijhydene.2023.07.270, http://www.journals.elsevier.com/international-journal-of-hydrogen-energy/.

Higgs, S., Wang, Y. D., Sun, C., Ennis-King, J., Jackson, S. J., Armstrong, R. T., & Mostaghimi, P. (2022). In-situ hydrogen wettability characterisation for underground hydrogen storage. *International Journal of Hydrogen Energy, 47*(26), 13062–13075. Available from https://doi.org/10.1016/j.ijhydene.2022.02.022.

Hu, R., Wan, J., Kim, Y., & Tokunaga, T. K. (2017). Wettability effects on supercritical CO_2–brine immiscible displacement during drainage: Pore-scale observation and 3D simulation. *China International Journal of Greenhouse Gas Control, 60*, 129–139. Available from https://doi.org/10.1016/j.ijggc.2017.03.011, http://www.elsevier.com/wps/find/journaldescription.cws_home/709061/description#description.

Hu, R., Wan, J., Yang, Z., Chen, Y. F., & Tokunaga, T. (2018). Wettability and flow rate impacts on immiscible displacement: A theoretical model. *Geophysical Research Letters, 45*(7), 3077–3086. Available from https://doi.org/10.1002/2017GL076600, http://onlinelibrary.wiley.com/journal/10.1002/(ISSN)1944-8007/issues?year = 2012.

Hu, Z., Klaver, J., Schmatz, J., Dewanckele, J., Littke, R., Krooss, B. M., & Amann-Hildenbrand, A. (2020). Stress sensitivity of porosity and permeability of Cobourg limestone. *Engineering Geology, 273*. Available from https://doi.org/10.1016/j.enggeo.2020.105632.

Höök, M., & Tang, X. (2013). Depletion of fossil fuels and anthropogenic climate change—A review. *Energy Policy, 52*, 797–809. Available from https://doi.org/10.1016/j.enpol.2012.10.046.

Iglauer, S. (2018). Optimum storage depths for structural CO_2 trapping. *International Journal of Greenhouse Gas Control, 77*, 82–87. Available from https://doi.org/10.1016/j.ijggc.2018.07.009.

Iglauer, S. (2022). Optimum geological storage depths for structural H_2 geo-storage. *Journal of Petroleum Science and Engineering, 212*. Available from https://doi.org/10.1016/j.petrol.2021.109498.

Iglauer, S., Abid, H., Al-Yaseri, A., & Keshavarz, A. (2021). Hydrogen adsorption on sub-bituminous coal: Implications for hydrogen geo-storage. *Geophysical Research Letters, 48*(10). Available from https://doi.org/10.1029/2021GL092976, http://agupubs.onlinelibrary.wiley.com/hub/journal/10.1002/(ISSN)1944-8007/.

Iglauer, S., Ali, M., & Keshavarz, A. (2021). Hydrogen wettability of sandstone reservoirs: Implications for hydrogen geo-storage. *Geophysical Research Letters, 48*(3). Available from https://doi.org/10.1029/2020GL090814.

Iglauer, S., Al-Yaseri, A. Z., Rezaee, R., & Lebedev, M. (2015). CO_2 wettability of caprocks: Implications for structural storage capacity and containment security. *Geophysical Research Letters, 42*(21), 9279–9284. Available from https://doi.org/10.1002/2015GL065787, http://onlinelibrary.wiley.com/journal/10.1002/(ISSN)1944-8007/issues?year = 2012.

Iglauer, S., & Wülling, W. (2016). The scaling exponent of residual nonwetting phase cluster size distributions in porous media. *Geophysical Research Letters, 43*(21), 11–260. Available from https://doi.org/10.1002/2016GL071298, http://agupubs.onlinelibrary.wiley.com/hub/journal/10.1002/(ISSN)1944-8007/.

Israelachvili, J. N. (2022). *Surface forces* (pp. 793–816). Informa UK Limited. Available from http://doi.org/10.1201/9780367811815-58.

Jadhawar, P., & Saeed, M. (2024). Mechanistic evaluation of the reservoir engineering performance for the underground hydrogen storage in a deep North Sea aquifer. *International Journal of Hydrogen Energy, 50*, 558–574. Available from https://doi.org/10.1016/j.ijhydene.2023.07.272.

Ju, Y., Gong, W., Chang, C., Xie, H., Xie, L., & Liu, P. (2017). Three-dimensional characterisation of multi-scale structures of the Silurian Longmaxi shale using focused ion beam-scanning electron microscopy and reconstruction technology. *Journal of Natural Gas Science and Engineering, 46*, 26–37. Available from https://doi.org/10.1016/j.jngse.2017.07.015.

Katsube, T. J., Mudford, B. S., & Best, M. E. (1991). Petrophysical characteristics of shales from the Scotian shelf. *Geophysics, 56*(10), 1681–1689. Available from https://doi.org/10.1190/1.1442980.

Khaledi, K., Mahmoudi, E., Datcheva, M., & Schanz, T. (2016). Stability and serviceability of underground energy storage caverns in rock salt subjected to mechanical cyclic loading. *International Journal of Rock Mechanics and Mining Sciences, 86*, 115–131. Available from https://doi.org/10.1016/j.ijrmms.2016.04.010.

Klebanoff, L. E., Ott, K. C., Simpson, L. J., O'Malley, K., & Stetson, N. T. (2014). Accelerating the understanding and development of hydrogen storage materials: A review of the five-year efforts of the three DOE hydrogen storage materials centers of excellence. *Metallurgical and Materials Transactions E, 1*(2), 81–117. Available from https://doi.org/10.1007/s40553-014-0011-z.

Krevor, S. C. M., Pini, R., Zuo, L., & Benson, S. M. (2012). Relative permeability and trapping of CO_2 and water in sandstone rocks at reservoir conditions. *Water Resources Research, 48*(2). Available from https://doi.org/10.1029/2011WR010859.

Lankof, L., & Tarkowski, R. (2020). Assessment of the potential for underground hydrogen storage in bedded salt formation. *International Journal of Hydrogen Energy, 45*(38), 19479–19492. Available from https://doi.org/10.1016/j.ijhydene.2020.05.024.

Li, D., Beyer, C., & Bauer, S. (2018). A unified phase equilibrium model for hydrogen solubility and solution density. *International Journal of Hydrogen Energy, 43*(1), 512–529. Available from https://doi.org/10.1016/j.ijhydene.2017.07.228.

Li, Y., Tang, D., Xu, H., Meng, Y., & Li, J. (2014). Experimental research on coal permeability: The roles of effective stress and gas slippage. *Journal of Natural Gas Science and Engineering, 21*, 481–488. Available from https://doi.org/10.1016/j.jngse.2014.09.004.

Lord, A. S., Kobos, P. H., & Borns, D. J. (2014). Geologic storage of hydrogen: Scaling up to meet city transportation demands. *International Journal of Hydrogen Energy, 39*(28), 15570–15582. Available from https://doi.org/10.1016/j.ijhydene.2014.07.121, http://www.journals.elsevier.com/international-journal-of-hydrogen-energy/.

Lysyy, M., Føyen, T., Johannesen, E. B., Fernø, M., & Ersland, G. (2022). Hydrogen relative permeability hysteresis in underground storage. *Geophysical Research Letters, 49*(17). Available from https://doi.org/10.1029/2022GL100364, http://agupubs.onlinelibrary.wiley.com/hub/journal/10.1002/(ISSN)1944-8007/.

Makkonen, L. (2017). A thermodynamic model of contact angle hysteresis. *The Journal of Chemical Physics, 147*(6). Available from https://doi.org/10.1063/1.4996912.

Mouli-Castillo, J., Heinemann, N., & Edlmann, K. (2021). Mapping geological hydrogen storage capacity and regional heating demands: An applied UK case study. *Applied Energy, 283*. Available from https://doi.org/10.1016/j.apenergy.2020.116348.

Muhammed, N. S., Haq, B., Al Shehri, D., Al-Ahmed, A., Rahman, M. M., & Zaman, E. (2022). A review on underground hydrogen storage: Insight into geological sites, influencing factors and future outlook. *Energy*

Reports, *8*, 461–499. Available from https://doi.org/10.1016/j.egyr.2021.12.002, http://www.journals.elsevier.com/energy-reports/.

Mustapha, H., De Langavant, L., & Ann Giddins, M. (2015). Darcy and non-darcy flows in fractured gas reservoirs. *Society of Petroleum Engineers – SPE reservoir characterisation and simulation conference and exhibition*, RCSC 2015, 180–193. doi: 10.2118/175596-ms.

Nilsen, H. M., Syversveen, A. R., Lie, K. A., Tveranger, J., & Nordbotten, J. M. (2012). Impact of top-surface morphology on CO_2 storage capacity. *International Journal of Greenhouse Gas Control*, *11*, 221–235. Available from https://doi.org/10.1016/j.ijggc.2012.08.012, http://www.elsevier.com/wps/find/journaldescription.cws_home/709061/description#description.

Okoroafor, E. R., Saltzer, S. D., & Kovscek, A. R. (2022). Toward underground hydrogen storage in porous media: Reservoir engineering insights. *International Journal of Hydrogen Energy*, *47*(79), 33781–33802. Available from https://doi.org/10.1016/j.ijhydene.2022.07.239, http://www.journals.elsevier.com/international-journal-of-hydrogen-energy/.

Ozarslan, A. (2012). Large-scale hydrogen energy storage in salt caverns. *International Journal of Hydrogen Energy*, *37*(19), 14265–14277. Available from https://doi.org/10.1016/j.ijhydene.2012.07.111.

Palmer, I., & Mansoori, J. (1998). How permeability depends on stress and pore pressure in coalbeds: A new model. *SPE Reservoir Engineering (Society of Petroleum Engineers)*, *1*(6), 539–543. Available from https://doi.org/10.2118/52607-pa.

Pan, B., Gong, C., Wang, X., Li, Y., & Iglauer, S. (2020). The interfacial properties of clay-coated quartz at reservoir conditions. *Fuel*, *262*. Available from https://doi.org/10.1016/j.fuel.2019.116461.

Pan, B., Li, Y., Xie, L., Wang, X., He, Q., Li, Y., Hejazi, S. H., & Iglauer, S. (2019). Role of fluid density on quartz wettability. *Journal of Petroleum Science and Engineering*, *172*, 511–516. Available from https://doi.org/10.1016/j.petrol.2018.09.088.

Pan, B., Liu, K., Ren, B., Zhang, M., Ju, Y., Gu, J., Zhang, X., Clarkson, C. R., Edlmann, K., Zhu, W., & Iglauer, S. (2023). Impacts of relative permeability hysteresis, wettability, and injection/withdrawal schemes on underground hydrogen storage in saline aquifers. *Fuel*, *333*. Available from https://doi.org/10.1016/j.fuel.2022.126516.

Pan, B., Yin, X., & Iglauer, S. (2021). Rock-fluid interfacial tension at subsurface conditions: Implications for H_2, CO_2 and natural gas geo-storage. *International Journal of Hydrogen Energy*, *46*(50), 25578–25585. Available from https://doi.org/10.1016/j.ijhydene.2021.05.067.

Pan, B., Yin, X., Ju, Y., & Iglauer, S. (2021). Underground hydrogen storage: Influencing parameters and future outlook. *Advances in Colloid and Interface Science*, *294*. Available from https://doi.org/10.1016/j.cis.2021.102473.

Perera, M. S. A. (2023). A review of underground hydrogen storage in depleted gas reservoirs: Insights into various rock-fluid interaction mechanisms and their impact on the process integrity. *Fuel*, *334*. Available from https://doi.org/10.1016/j.fuel.2022.126677.

Phukan, R., Gogoi, S. B., & Tiwari, P. (2020). Effects of CO_2-foam stability, interfacial tension and surfactant adsorption on oil recovery by alkaline-surfactant-alternated-gas/CO_2 flooding. *Colloids and Surfaces A: Physicochemical and Engineering Aspects*, *597*. Available from https://doi.org/10.1016/j.colsurfa.2020.124799.

Raats, P. A. C. (1973). Dynamics of fluids in porous media. *Soil Science Society of America Journal*, *37*(4). Available from https://doi.org/10.2136/sssaj1973.03615995003700040004x.

Rahman, T., Lebedev, M., Barifcani, A., & Iglauer, S. (2016). Residual trapping of supercritical CO_2 in oil-wet sandstone. *Journal of Colloid and Interface Science*, *469*, 63–68. Available from https://doi.org/10.1016/j.jcis.2016.02.020.

Rani, S., Padmanabhan, E., & Prusty, B. K. (2019). Review of gas adsorption in shales for enhanced methane recovery and CO_2 storage. *Journal of Petroleum Science and Engineering*, *175*, 634–643. Available from https://doi.org/10.1016/j.petrol.2018.12.081, http://www.sciencedirect.com/science/journal/09204105/81.

Raza, A., Arif, M., Glatz, G., Mahmoud, M., Al Kobaisi, M., Alafnan, S., & Iglauer, S. (2022). A holistic overview of underground hydrogen storage: Influencing factors, current understanding, and outlook. *Fuel*, *330*. Available from https://doi.org/10.1016/j.fuel.2022.125636.

Ren, B., Sun, Y., & Bryant, S. (2014). Maximizing local capillary trapping during CO_2 injection. *Energy Procedia*, *63*, 5562–5576. Available from https://doi.org/10.1016/j.egypro.2014.11.590.

Ruprecht, C., Pini, R., Falta, R., Benson, S., & Murdoch, L. (2014). Hysteretic trapping and relative permeability of CO_2 in sandstone at reservoir conditions. *International Journal of Greenhouse Gas Control, 27*, 15–27. Available from https://doi.org/10.1016/j.ijggc.2014.05.003.

Saajanlehto, M., Uusi-Kyyny, P., & Alopaeus, V. (2014). Hydrogen solubility in heavy oil systems: Experiments and modeling. *Fuel, 137*, 393–404. Available from https://doi.org/10.1016/j.fuel.2014.08.015.

Sainz-Garcia, A., Abarca, E., Rubi, V., & Grandia, F. (2017). Assessment of feasible strategies for seasonal underground hydrogen storage in a saline aquifer. *International Journal of Hydrogen Energy, 42*(26), 16657–16666. Available from https://doi.org/10.1016/j.ijhydene.2017.05.076.

Sander, R., Pan, Z., & Connell, L. D. (2017). Laboratory measurement of low permeability unconventional gas reservoir rocks: A review of experimental methods. *Australia Journal of Natural Gas Science and Engineering, 37*, 248–279. Available from https://doi.org/10.1016/j.jngse.2016.11.041, http://www.elsevier.com/wps/find/journaldescription.cws_home/716470/description#description.

Shi, Y., & Wang, C. Y. (1986). Pore pressure generation in sedimentary basins: Overloading versus aquathermal. *Journal of Geophysical Research: Solid Earth, 91*(B2), 2153–2162. Available from https://doi.org/10.1029/jb091ib02p02153.

Slobod, R. L., & Blum, H. A. (1952). Method for determining wettability of reservoir rocks. *Journal of Petroleum Technology, 4*(01), 1–4. Available from https://doi.org/10.2118/137-g.

Suekane, T., Nobuso, T., Hirai, S., & Kiyota, M. (2008). Geological storage of carbon dioxide by residual gas and solubility trapping. *International Journal of Greenhouse Gas Control, 2*(1), 58–64. Available from https://doi.org/10.1016/s1750-5836(07)00096-5.

Taku Ide, S., Jessen, K., & Orr, F. M. (2007). Storage of CO_2 in saline aquifers: Effects of gravity, viscous, and capillary forces on amount and timing of trapping. *International Journal of Greenhouse Gas Control, 1*(4), 481–491. Available from https://doi.org/10.1016/S1750-5836(07)00091-6, http://www.elsevier.com/wps/find/journaldescription.cws_home/709061/description#description.

Tanino, Y., & Blunt, M. J. (2013). Laboratory investigation of capillary trapping under mixed-wet conditions. *Water Resources Research, 49*(7), 4311–4319. Available from https://doi.org/10.1002/wrcr.20344.

Tarkowski, R. (2019). Underground hydrogen storage: Characteristics and prospects. *Renewable and Sustainable Energy Reviews, 105*, 86–94. Available from https://doi.org/10.1016/j.rser.2019.01.051.

Tarkowski, R., Uliasz-Misiak, B., & Tarkowski, P. (2021). Storage of hydrogen, natural gas, and carbon dioxide – Geological and legal conditions. *International Journal of Hydrogen Energy, 46*(38), 20010–20022. Available from https://doi.org/10.1016/j.ijhydene.2021.03.131.

Vivian, J. E., & King, C. J. (1964). Diffusivities of slightly soluble gases in water. *AIChE Journal, 10*(2), 220–221. Available from https://doi.org/10.1002/aic.690100217.

Wang, Y., Zhang, C., Wei, N., Oostrom, M., Wietsma, T. W., Li, X., & Bonneville, A. (2013). Experimental Study of crossover from capillary to viscous fingering for supercritical CO_2–water displacement in a homogeneous pore Network. *Environmental Science & Technology, 47*(1), 212–218. Available from https://doi.org/10.1021/es3014503.

Wei, M., Liu, J., Elsworth, D., Li, S., & Zhou, F. (2019). Influence of gas adsorption induced non-uniform deformation on the evolution of coal permeability. *International Journal of Rock Mechanics and Mining Sciences, 114*, 71–78. Available from https://doi.org/10.1016/j.ijrmms.2018.12.021.

Wei, T. Y., Lim, K. L., Tseng, Y. S., & Chan, S. L. I. (2017). A review on the characterization of hydrogen in hydrogen storage materials. *Renewable and Sustainable Energy Reviews, 79*, 1122–1133. Available from https://doi.org/10.1016/j.rser.2017.05.132.

Xu, T., Apps, J. A., & Pruess, K. (2003). Reactive geochemical transport simulation to study mineral trapping for CO_2 disposal in deep arenaceous formations. *Journal of Geophysical Research: Solid Earth, 108*(2). Available from http://onlinelibrary.wiley.com/journal/10.1002/(ISSN)2169-9356.

Yang, Z., Wang, W., Dong, M., Wang, J., Li, Y., Gong, H., & Sang, Q. (2016). A model of dynamic adsorption-diffusion for modeling gas transport and storage in shale. *Fuel, 173*, 115–128. Available from https://doi.org/10.1016/j.fuel.2016.01.037, http://www.journals.elsevier.com/fuel/.

Ye, Z., Zhang, F., Han, L., Luo, P., Yang, J., & Chen, H. (2008). The effect of temperature on the interfacial tension between crude oil and gemini surfactant solution. *Colloids and Surfaces A: Physicochemical and Engineering Aspects, 322*(1-3), 138–141. Available from https://doi.org/10.1016/j.colsurfa.2008.02.043.

Yekeen, N., Padmanabhan, E., Sevoo, T. A., Kanesen, K. A., & Okunade, O. A. (2020). Wettability of rock/CO_2/brine systems: A critical review of influencing parameters and recent advances. *Journal of Industrial and*

Engineering Chemistry, 88, 1–28. Available from https://doi.org/10.1016/j.jiec.2020.03.021, http://www.sciencedirect.com/science/journal/1226086X.

Zhang, H., Zhang, Y., Al Kobaisi, M., Iglauer, S., & Arif, M. (2023). Effect of cyclic hysteretic multiphase flow on underground hydrogen storage: A numerical investigation. *International Journal of Hydrogen Energy*. Available from https://doi.org/10.1016/j.ijhydene.2023.08.169, http://www.journals.elsevier.com/international-journal-of-hydrogen-energy/.

Zhang, Y., Lebedev, M., Sarmadivaleh, M., Barifcani, A., Rahman, T., & Iglauer, S. (2016). Swelling effect on coal micro structure and associated permeability reduction. *Fuel, 182*, 568–576. Available from https://doi.org/10.1016/j.fuel.2016.06.026.

Zhao, X., & Jin, H. (2019). Investigation of hydrogen diffusion in supercritical water: A molecular dynamics simulation study. *International Journal of Heat and Mass Transfer, 133*, 718–728. Available from https://doi.org/10.1016/j.ijheatmasstransfer.2018.12.164.

Zhao, X., Jin, H., Chen, Y., & Ge, Z. (2021). Numerical study of H_2, CH_4, CO, O_2 and CO_2 diffusion in water near the critical point with molecular dynamics simulation. *Computers & Mathematics with Applications, 81*, 759–771. Available from https://doi.org/10.1016/j.camwa.2019.11.012.

Zheng, J., Zheng, L., Liu, H. H., & Ju, Y. (2015). Relationships between permeability, porosity and effective stress for low-permeability sedimentary rock. *International Journal of Rock Mechanics and Mining Sciences, 78*, 304–318. Available from https://doi.org/10.1016/j.ijrmms.2015.04.025, http://www.elsevier.com/inca/publications/store/2/5/6/index.htt.

Zhong, X., Zhu, Y., Liu, L., Yang, H., Li, Y., Xie, Y., & Liu, L. (2020). The characteristics and influencing factors of permeability stress sensitivity of tight sandstone reservoirs. *Journal of Petroleum Science and Engineering, 191*. Available from https://doi.org/10.1016/j.petrol.2020.107221.

Zivar, D., Kumar, S., & Foroozesh, J. (2021). Underground hydrogen storage: A comprehensive review. *International Journal of Hydrogen Energy, 46*(45), 23436–23462. Available from https://doi.org/10.1016/j.ijhydene.2020.08.138.

CHAPTER

9

Geochemical and geomechanical synergies during underground hydrogen storage schemes

Prashant Jadhawar and Hosein Doryani

School of Engineering, University of Aberdeen, Aberdeen, United Kingdom

9.1 Introduction

Exploiting renewable energy resources is an undeniable path towards achieving the net zero carbon emissions target. However, there is a significant mismatch between the production and usage of these resources. One way to overcome this issue is to store the energy generated by renewable energy sources to be used during times of high consumption and low production (Osman et al., 2022; Saeed & Jadhawar, 2024a, 2024b). Hydrogen is considered to be a suitable energy carrier due to its high energy potential. Hydrogen can be generated through various forms of processing. Gray/brown hydrogen is generated through natural gas steam reformation and coal gasification. Blue hydrogen is generated by natural gas steam reformation, where most of the CO_2 produced is stored underground. In green hydrogen production, through water electrolysis, hydrogen is generated by the use of renewable energy sources like wind and solar energy. In the process of electrolysis, CO_2 is not generated. In turquoise hydrogen production, through CH_4 pyrolysis, natural gas is thermally decomposed and H_2 and solid carbon are generated (Osman et al., 2022).

Hydrogen can be stored above the surface (surface storage) or below the surface (underground storage). Surface hydrogen storage can be further categorized into two storage types: physical storage and material-based storage. In physical storage, the hydrogen's physical state is changed to be suitable for storage. These involve compressed (gas) hydrogen, cryogenic (liquid) hydrogen, and cryo-compressed hydrogen. In material-based hydrogen storage, hydrogen is adsorbed on the surface of the adsorbent. Surface storage of hydrogen is associated with different concerns and challenges,

Subsurface Hydrogen Energy Storage
DOI: https://doi.org/10.1016/B978-0-443-24071-3.00010-8

including operational safety and leakage concerns and limited storage capacity. Hence, subsurface hydrogen storage can be an attractive solution to overcome these limitations (Amirthan & Perera, 2022). Underground hydrogen storage sites include depleted oil and gas reservoirs, saline aquifers, salt caverns, and abandoned mines. Depleted hydrocarbon reservoirs, with high storage capacities and impermeable caprocks, offer a favorable solution for the storage of hydrogen. In such reservoirs, already-available infrastructure can be utilized to efficiently store hydrogen (Jahanbakhsh et al., 2024). Moreover, previous experiences at large-scale CO_2 or CH_4 storage sites can help minimize the risks associated with underground hydrogen storage. Various factors, including reservoir type and geology, gas storage energy density, technology efficiency, dynamic flow paths, sustainable underground space management, periods of storage site operation, and field operational parameters like injection and production scenarios, need to be taken into account to find out how a specific reservoir could be suitable for storage practices. Hence, the knowledge and technology associated with subsurface CO_2 and CH_4 storage cannot be directly linked to hydrogen storage (Tarkowski, Uliasz-Misiak, & Tarkowski, 2021). Accordingly, the unique properties of hydrogen might trigger new physiochemical processes that do not exist in CO_2 or natural gas storage. In this section, we will discuss how the differences between CO_2, CH_4, and H_2 properties can limit the applicability of the technologies developed for natural gas storage and hydrogen storage.

Hydrogen exhibits a greater diffusivity than methane. H_2 diffusivity is approximately 3 times higher than CH_4 diffusivity. Specifically, in pure water at 25°C, the diffusion coefficient for hydrogen is $5.13 \times 10^{-9}\ m^2/s$, while for CH_4, it is $1.85 \times 10^{-9}\ m^2/s$. Typically, the diffusion of an aqueous species is characterized by its diffusion coefficient in pure water (m^2/s) (Zivar et al., 2020). However, in porous systems where solids are present, the impact of porosity on diffusion also needs to be considered. This involves adjusting the diffusion coefficient with tortuosity to derive an effective diffusion coefficient (as shown in Eq. (9.1) (Saeed & Jadhawar, 2024a, 2024b)), where d_m' represents the effective diffusion coefficient, d_m is the diffusion coefficient in pure water, and τ is the tortuosity (Song & Zhang, 2013). Tortuosity is defined as the ratio of the effective diffusion mass fluxes within a porous medium to the ideal diffusion mass fluxes in solutions where there is no rock matrix. At standard temperature, available measured data suggest that effective H_2 diffusivity could be 10 times higher than CH_4 diffusivity.

However, data regarding the effective diffusion coefficients of hydrogen relative to underground storage conditions with higher temperatures and pressures is currently lacking (Hemme & van Berk, 2018):

$$d_m' = \frac{d_m}{\tau^2} \tag{9.1}$$

Considering that hydrogen has a significantly lower molecular weight and higher diffusivity compared to other natural gases, the likelihood of its leakage to the surface is higher than for other types of gases. In terms of viscosity, hydrogen exhibits a lower viscosity compared to methane. The lower viscosity of hydrogen compared to methane indicates that hydrogen has higher mobility. This increased mobility could result in lower residual hydrogen within a porous medium containing both hydrogen and methane.

Consequently, a higher efficiency of hydrogen withdrawal is anticipated later on. The occurrence of fluid coning is directly influenced by viscosity. Consequently, during hydrogen withdrawal, the lower viscosity of hydrogen reduces the probability of coning problems. Additionally, hydrogen density is also lower than that of methane. Under standard conditions, the density of hydrogen is eight times smaller than that of methane, and this disparity grows by 20%–30% under storage conditions (Zivar et al., 2020). Because hydrogen is less dense than natural gas, storing the same mass of hydrogen gas necessitates higher pressure. Hydrogen-pure water solubility is lower than that of methane at standard conditions. The differing properties of hydrogen compared to methane could pose risks when storing hydrogen in fields that were originally designed for natural gas storage. Table 9.1 shows the comparison of the properties of H_2 and CH_4.

Depleted gas reservoirs provide good potential for UHS. There is sufficient data and knowledge about these reservoirs (reservoir geology, rock characteristics, storage capacity, gas injectivity, and any potential leakage possibility). However, the differing properties of H_2 compared to CH_4 and CO_2 could pose risks when storing hydrogen in fields that were originally designed for natural gas storage. These risks are associated with those reactions that can occur between the minerals, facilities, fluids, microorganisms, and any other existing component in the underground thermodynamic conditions.

Table 9.2 shows the general information and comparison between different aspects of CO_2 storage, underground gas storage (UGS), and UHS. It can be observed that to properly assess the feasibility of a successful underground storage project, different parameters should be considered. Some of the main parameters that need attention are operational conditions, bio-geochemical reactions, leakage risks associated with the type of gas injected, potential hazards that might occur in underground storage sites, and the integrity of the materials, like wells, pipelines, and surface facilities. Failure to carefully analyze these aspects might lead to the unsuccessful or catastrophic implementation of storage sites.

TABLE 9.1 Physical and thermodynamic properties of H_2 and CH_4 (Zivar et al., 2021).

Properties	H_2	CH_4
Molecular weight	2.016	16.043
Density @ 25°C and 1 atm	0.082 kg/m^3	0.657 kg/m^3
Viscosity @ 25°C and 1 atm	0.89×10^{-5} Pa/s	1.1×10^{-5} Pa/s
Solubility in pure water @ 25°C and 1 atm	7.9×10^{-4} (molkg/w H_2 (g))	1.4×10^{-3} (mol kg/w CH_4(g))
Normal boiling point	−253°C	−165°C
Critical pressure	12.8 atm	45.79 atm
Critical temperature	−239.95°C	−82.3°C
Heating value	120–142 kJ/g	205–55.5 kJ/g
Diffusion in pure water @ 25°C	5.13×10^{-9} m^2/s	1.85×10^{-9} m^2/s

TABLE 9.2 Comparison between different aspects of CO_2 storage, underground gas storage and underground hydrogen storage (Kumar et al., 2023; Muhammed et al., 2022).

	CO_2 storage	UGS	UHS
Operating conditions	Continuous injection of CO_2 is carried out into depleted reservoirs and salt caverns. There is potential to utilize CO_2 as a cushion gas for Underground Hydrogen Storage (UHS). Upon complete injection of CO_2, the pressure is maintained at a constant level.	The cyclic process of injecting and producing natural gas occurs over extended periods of time.	The cyclic injection and production of H_2 occur at a higher frequency compared to natural gas. Hydrogen exhibits greater mobility compared to other stored gases, potentially introducing operational challenges due to limited practical experience. Continued cycles of intergranular swelling and shrinkage can lead to gradual rock damage.
Bio-geochemical reactions	The interaction between the host rock and supercritical CO_2 can change the properties of the rocks, impacting their ductility and cohesion.	Depending on the selected reservoir site, stored methane may undergo reactions that release various gases, such as sulfides, which can lead to the degradation of the rocks.	Biochemical reactions involving reservoir rocks and hydrogen (H_2) can alter the properties of the rocks and contribute to degradation. These interactions may also affect the integrity of the underground seal. In the case of caverns, salt rock is commonly considered to be inert to hydrogen.
Leakage risks	Leakage may arise from seal failures and the opening of faults. In CCS scenarios with a corrosive environment and low gas-water interfacial tension, the porosity and permeability of rocks can be altered. Additionally, failures in top of formations in salt caverns can result in leakage, influenced by operational limits and heterogeneities within the caverns.	Because the IFT between water and methane is high, the likelihood of leakage through the caprock is reduced.	The dissolution of minerals can alter the properties of both the reservoir and the caprock. This process may result in permanent deformation and a reduced threshold for fault slip, potentially leading to the propagation or growth of fractures. In salt caverns, hydrogen percolation, hydraulic fracturing at the cavern roof, and salt creep, especially in heterogeneous geological domains, can accelerate fault slip, ultimately leading to hydrogen leakage.
Potential hazards	Subsidence, uplift, induced seismicity, and fracture propagation may occur in CO_2 storage operations. Operational conditions and the selected reservoir site influence the impacts.	Subsidence, uplift, induced seismicity, and fracture propagation may occur in underground gas storage operations. Operational conditions and the selected reservoir site influence the impacts.	Rapid cyclic loading can expedite induced seismicity, compromise the integrity of the caprock, and potentially result in permanent subsidence or uplift on the surface. Inelastic deformation, such as creep, can accelerate movement along preexisting faults.

(*Continued*)

TABLE 9.2 (Continued)

	CO_2 storage	UGS	UHS
Well integrity mechanisms	Corrosion, cement carbonation, and elastomer degradation can occur due to the presence of carbonic acid.	The risk of corrosion, cement carbonation, and hydrogen blistering is relatively low, contingent upon the chosen reservoir site.	There is a high risk of steel corrosion, hydrogen blistering, and sulfidation, which can also lead to elastomer degradation due to the presence of H_2.

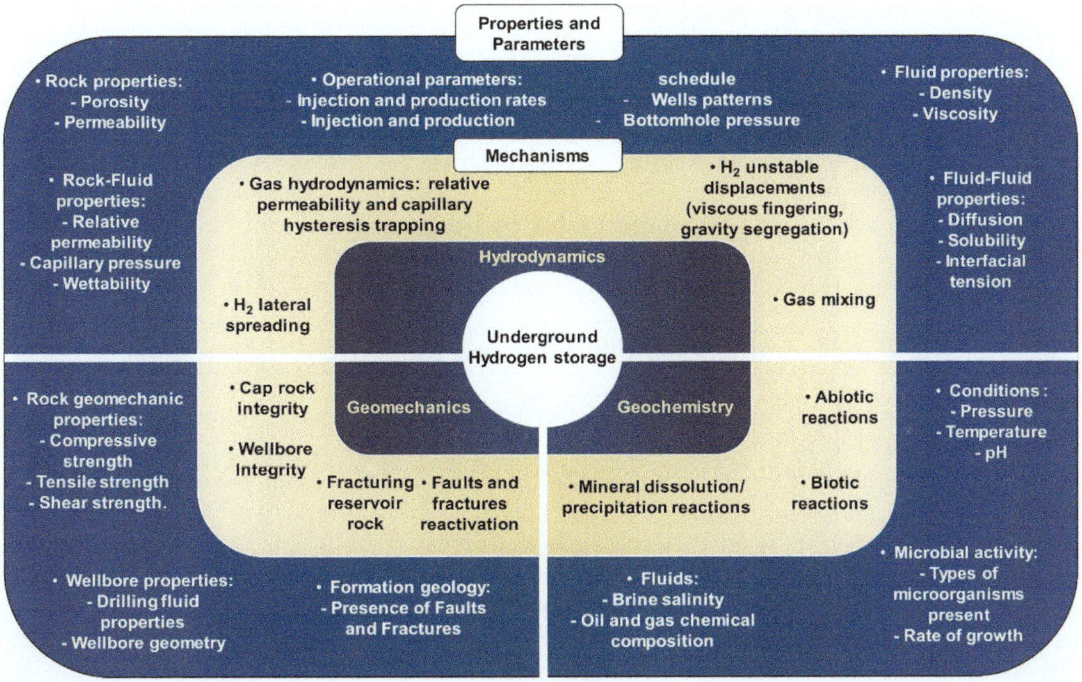

FIGURE 9.1 Aspects of underground hydrogen storage modeling (Saeed & Jadhawar, 2024a).

Fig. 9.1 shows the aspects, mechanisms, properties, and parameters influencing the UHS process. The main aspects of underground hydrogen storage analysis are hydrodynamics, geochemistry, and geomechanics. Different mechanisms are involved in each of these aspects.

The study of hydrodynamics at UHS involves examining the flow of hydrogen gas and other fluids through the porous media at the underground storage site. It also focuses on the geological formation's capacity for storing hydrogen gas during injection and the efficiency of recovery during withdrawal periods (Jadhawar & Saeed, 2023, 2024; Saeed, Jadhawar & Bagala, 2023; Saeed & Jadhawar, 2024a, 2024b). Various factors can influence this hydrodynamic process and can be divided into two main types: reservoir-specific factors and operational factors. Reservoir-specific factors include porosity, permeability, and pressure, among others. Operational factors encompass hydrogen gas injection and withdrawal rates, storage cycles, and the number of operational wells involved in the UHS process (Jadhawar & Saeed, 2023; Saeed & Jadhawar, 2024a, 2024b). Chemical reactions may occur between the stored hydrogen gas, cushion gas, reservoir fluids, and reservoir minerals in underground hydrogen storage. These interactions can lead to the transformation of hydrogen into other gases, such as CH_4 and H_2S, potentially causing losses in stored hydrogen (Saeed, Jadhawar, & Bagala, 2023). Furthermore, interactions with other gases in the reservoir may lower the purity of the hydrogen gas. Changes in the porosity and permeability of the geological formation due to mineral dissolution or precipitation are also possible outcomes of H_2 geochemical activity during underground storage (Saeed & Jadhawar, 2024a, 2024b; Saeed, Jadhawar & Bagala, 2023). Geomechanical studies are mainly concerned with caprock, wellbore, and reservoir rock integrity. Geochemical and hydrodynamic mechanisms have a direct influence on the geomechanical properties. In other words, due to hydrodynamic and geochemical processes, the geomechanical properties of the geological formation may shift, raising the risk of structural collapse or faulting and compromising the stability of the cap rock. As a result, for a successful implementation of underground hydrogen storage, it is essential to consider comprehensive cumulative geomechanical, geochemical, and hydrodynamic effects (Saeed & Jadhawar, 2024a, 2024b).

9.2 Fundamentals of fluid–fluid and rock–fluid interactions

Fluid–fluid and rock–fluid interactions are critical parameters for effective screening of UHS practices. These factors dictate the sealing capabilities of caprock, the distribution and saturation of H_2 at the pore scale within the rock, and the displacement of fluids. The main fluid–fluid and rock–fluid properties and associated interactions that are related to UHS are solubility, fluid–fluid IFT, and rock wettability (Perera, 2023).

Solubility plays a pivotal role in understanding losses and gas entrapment in underground formations. Typically, gases are retained or trapped underground through various mechanisms, including structural, capillary, dissolution (solubility), mineralization, and adsorption processes. Structural trapping arises when gas becomes confined within the pores or fractures of a rock formation, influenced by factors like shape, size, and connectivity of the pore network. This mechanism is dependent on reservoir rock properties such as permeability, porosity, and pore throat dimensions. Capillary trapping occurs when gas is captured in tiny bubbles due to capillary forces at the gas-liquid interface, dependent on pore fluid properties like surface tension and contact angle (Perera, 2023).

Solubility trapping happens when gas dissolves into the fluid residing in the pores and remains trapped due to its low solubility in the liquid phase. Mineralization trapping happens when the injected gas has interactions with the original minerals and fluids existing in the formation. Both dissolution and mineralization trapping rely on pore fluid properties such as salinity, pH, and temperature.

Adsorption trapping occurs when gas molecules adhere to mineral grain surfaces within the rock formation due to attractive forces, depending on mineral surface properties like surface area, roughness, chemistry, and pore size distribution.

In UHS, trapping mechanisms like solubility (due to fluid—fluid interactions) and mineralization (fluid—rock interaction) can result in the permanent loss of the injected gas, rendering them undesirable. For instance, if a gas exhibits high solubility, a larger portion of the injected gas will dissolve into the fluids, potentially diminishing the amount of gas trapped within the pore spaces. Conversely, if the solubility of the gas is low, a lower amount of it will dissolve into the existing reservoir fluids, leading to an increase in the amount of gas trapped in the pore space. Hence, evaluating and knowing the solubility behavior of H_2 gas in reservoir fluids, such as brine, within a three-phase region is a critical factor. It determines the extent to which the injected gas may be lost through dissolution into the fluids and how much of it will still exist as a distinct gas phase. At UHS conditions, temperature, pressure, and reservoir salinity play pivotal roles in determining the solubility of H_2. Increasing pressure and temperature typically lead to increased dissolution of H_2, whereas higher brine salinity tends to decrease H_2 solubility.

At the interface between H_2 and other fluids in the reservoir, a phase boundary exists. An essential feature of this phase boundary is the interfacial tension (IFT) among the various phases existing in the reservoir. The IFT of H_2 in the presence of reservoir fluids governs the fluid behavior within the reservoir. Therefore, this parameter is deemed significant for estimating gas storage efficiency and is vital for appropriately designing injection and withdrawal schemes. To measure the IFT at the fluid—fluid interface, the pendant drop method is frequently employed. In this method, the profile and shape of a fluid droplet (liquid) suspended from a needle within a chamber containing another fluid (gas) are analyzed. Pressure, temperature, salinity, the presence of organic acids, and the existence of cushion gas are considered to be the parameters that have effects on the IFT between H_2 and reservoir fluids (Raza et al., 2022).

To select a suitable site for UHS, wettability, which is defined as the one specific fluid's ability to stay on the surface of a solid in the presence of other immiscible fluids, is a critical parameter. The dynamics of the fluid flow in porous media are affected by the wettability. During UHS, it involves the influence of wettability on how hydrogen moves within the reservoir and also its possible movement in the caprock. Moreover, wettability influences the storage capacity and security of the formation. Wettability, obtained through contact angle measurements, serves as a fundamental parameter in conducting studies on multiphase flow in porous media. It is particularly important for estimating upscaled capillary pressure and relative permeability curves (Kumar et al., 2023; Perera, 2023; Raza et al., 2022; Saeed & Jadhawar, 2024a,b).

The absorption and desorption characteristics of hydrogen are significant factors in fluid-rock interactions, especially in reservoirs containing high volumes of rocks characterized by high surface areas, such as coal and clayey rock. Given that reservoir rocks often contain a substantial amount of clay, mainly during the cementation phase, it is crucial to investigate the adsorption and desorption characteristics of hydrogen in clay. It's worth noting that the hydrogen adsorption capacity of coal is approximately ten times lower than that of CO_2. This discrepancy occurs because hydrogen lacks an external electric dipole moment, and its quadrupole moment is also significantly smaller than that of CO_2. Consequently, hydrogen adsorption onto coal predominantly occurs through induced dipole moment interactions generated by dispersive Van der Waals forces (Borello, 2024; Perera, 2023).

The small molecular size of hydrogen makes it highly buoyant and diffusive through any medium. As a result, the average effective diffusion coefficients of hydrogen in water and reservoir rocks are considerably greater than those of other gases. As hydrogen is stored, diffusive losses gradually diminish over time. This is mainly due to the fact that the surrounding pores become saturated with hydrogen, thereby reducing the hydrogen concentration gradient across the reservoir. Also, temperature and pressure affect the rate of diffusion. Specifically, temperature directly impacts hydrogen diffusivity by altering molecular momentum, Brownian motion, and intermolecular forces. Diffusion decreases with increasing pressure due to the closer molecular proximity of water at higher pressure, making it challenging for hydrogen to move through the fluid inside the pore. Additionally, the diffusion characteristics of hydrogen vary depending on the hydrocarbon type, thus varying from one reservoir to another. Ideally, reservoirs with a thickness equivalent to their width would be optimal to minimize diffusive hydrogen loss; however, such reservoirs are rare. Typical geological reservoirs have significantly greater lateral dimensions than thickness, leaving them susceptible to diffusive hydrogen loss. Consequently, a substantial area of the caprock is exposed to accumulated hydrogen beneath it (due to the buoyancy of hydrogen), making it challenging to prevent diffusive hydrogen loss through the caprock (Saeed & Jadhawar, 2024a,b; Song & Zhang, 2013; Zivar et al., 2020).

Effective storage of hydrogen within a depleted oil and gas reservoir is largely contingent upon the integrity of the caprock, which serves as a barrier, impeding the flow of hydrogen into the upper rock layers and, ultimately, into the atmosphere. Naturally present water-saturated clayey rock formations can form caprocks with thicknesses of up to hundreds of meters, making them well-suited for containing injected hydrogen within underground reservoirs. These flow restrictions can occur when the caprock consists of pores of sufficiently small size, such as those found in claystone, and when high capillary entry pressures are present. Essentially, the pressure disparity between the wetting and nonwetting phases creates interactive forces between the fluids and the surfaces of the rock grains, known as capillary pressure (P_c). During the injection of hydrogen into the reservoir, H_2 displaces the reservoir's resident fluid. Due to gravity segregation (caused by the density difference between the resident fluid and H_2), H_2 will move upward to form a plume underneath the caprock and above the resident fluid. The ability and possibility of the accumulated H_2 to move into the caprock, which is located on top of the H_2 plume, depends on the capillary pressure threshold, which is the initial entry pressure required for H_2 to enter the caprock pore space. Capillary entry pressure, P_{ce}, is dependent on the rock-fluid parameters, including IFT, contact angle, and pore size. P_{ce} is calculated using Eq. (9.2) (Perera, 2023).

$$P_{ce} = \frac{2\gamma_{brine-hydrogen}\cos\theta}{r} \tag{9.2}$$

where $\gamma_{brine-hydrogen}$ is the IFT between brine and hydrogen, θ is the contact angle, and r is the radius of the largest pores in the caprock.

As discussed, rock-fluid and fluid-fluid interactions are important parameters that need to be considered during UHS. These interactions would affect the static and kinetic properties of the formations and the reactions that might take place within the formations.

9.2.1 Geochemical evaluations of hydrogen storage in porous media

Different types of underground sources are available for UHS. These include depleted oil and gas reservoirs, saline aquifers, and salt caverns. Fig. 9.2 shows the relevant information associated with these types of storage sites, with a focus on microbial activities. These types of storage sites have different temperatures and salinity ranges. Hence, in terms of geochemical studies, it is essential to provide enough information for each type of storage site selected for UHS. As it is illustrated in Fig. 9.3, information on H_2 storage in depleted gas reservoirs and deep aquifers is very low. Microbial activity and geochemical aspects of UHS need much more effort and require more attention.

As discussed in the previous section (Fig. 9.1), in geochemical modeling, it is essential to account for several parameters, such as the reservoir's pressure and temperature, the composition of the resident fluid, salinity, and pH levels, as well as the types of microorganisms present and their growth rates.

In subsurface environments, hydrogen consumption and generation typically result from two processes: (Saeed & Jadhawar, 2024a, 2024b) abiotic and (Osman et al., 2022) biotic. Components of abiotic reactions consist of nonliving elements within porous media, including water, rock minerals, and gases. Abiotic reactions are generally the result of geochemical rock-fluid interactions. Biotic processes or microbial activities involve living components such as bacteria present in subsurface porous media. The operational temperature required for abiotic processes can reach up to 600°C. In contrast, biotic processes operate at lower temperatures, closer to the typical prevailing temperatures in storage sites. Fig. 9.3 shows a schematic of the geochemical reactions involved in UHS. Generally, two major groups of microbial life exist within the subsurface environment. These single-cell groups are bacteria and archaea. The biochemical and geochemical reactions taking place in subsurface are mainly due to the activity of

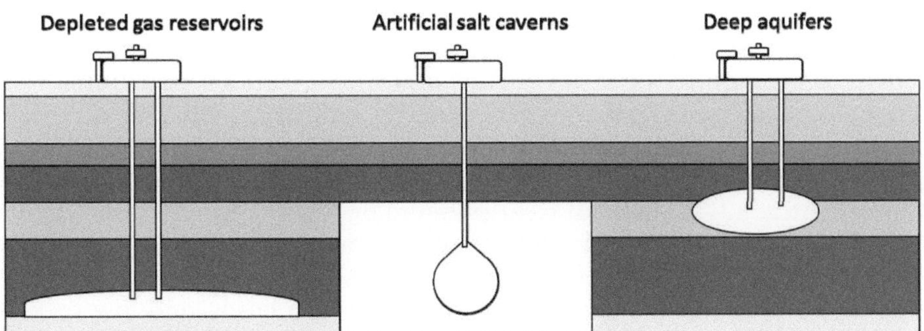

	Depleted gas reservoirs	Artificial salt caverns	Deep aquifers
Temperature range	high range	mostly 20 – 35°C	7 – 174 °C
Salinity	high range	in sump up to saturation	5 – 52.000 ppm
Experience for H_2 storage	low	high	very low
Knowledge on microbial activity during H_2 storage	low	very low	low

FIGURE 9.2 General information on underground hydrogen storage specific to depleted gas reservoirs, salt caverns, and aquifers (Dopffel, Jansen, & Gerritse, 2021).

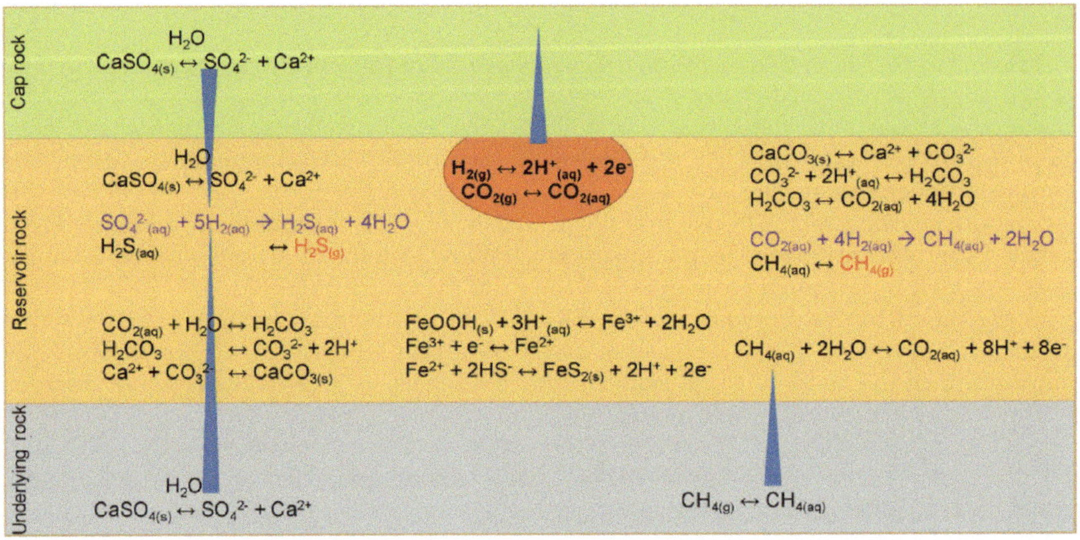

FIGURE 9.3 Schematic of typical geochemical activity during underground hydrogen storage (Hemme & van Berk, 2018).

these groups. Microorganisms have been discovered in rocks located several kilometers beneath the Earth's surface, with cell counts ranging from 104 to 108 cells per gram of rock. The total number of microorganisms in the subsurface is estimated to be $2-6 \times 10^{29}$. The origin of these microbes in the subsurface can be both indigenous, introduced by natural transport or sedimentation processes, and anthropogenically introduced by human activities such as drilling, pumping, or mining (Dopffel, Jansen, & Gerritse, 2021).

Microbes require specific environmental conditions and essential resources to remain active. These include water, an energy source, and various essential elements like carbon, nitrogen, phosphorus, and trace elements. Microbial activity can occur across a range of conditions, from extreme to ambient, with cell numbers and diversity typically lower under extreme conditions. Temperature tolerance for microbial life spans from $-15^{\circ}C$ to approximately $121^{\circ}C$, without clear pressure or brine salinity thresholds. The pH range for microbial activity is typically between pH 0 and 11, with the highest diversity and cell numbers observed between pH 6 and 7. Additionally, factors such as high concentrations of toxic chemicals, low water activity, high radiation levels, and low rock permeability can significantly influence microbial presence and reduce their abundance. It is essential to evaluate all these parameters for a comprehensive assessment of subsurface microbial activity. Microbes can derive energy through the oxidation of electron donors while simultaneously reducing an electron acceptor. These electron donors, found in both water and rock formations, include simple organic compounds like various acids such as acetate or formate, as well as complex organic compounds like plant material and hydrocarbons, in addition to inorganic compounds. Molecular H_2 is regarded as one of the most crucial electron donors for microbial respiration in subsurface environments. Due to its low reduction potential, H_2 can be utilized by a wide range of metabolically diverse organisms.

Specific enzymes known as hydrogenases, located within microbial cells, facilitate the splitting of H_2 into protons and electrons. These electrons can then be employed to chemically store cellular energy, which is subsequently utilized to drive various reactions within the cell, including processes like CO_2 and N_2 fixation.

Hydrogen generation during hydrogen storage operations hasn't been widely documented in the literature, leaving uncertainty about its favorability. However, one potential process for hydrogen generation involves nitrogenases, which could produce hydrogen gas as a by-product of nitrogen fixation. In this process, 1 mol of hydrogen and 2 mol of ammonia are generated through the consumption of 1 mol of nitrogen. Nitrogenases can become important in scenarios where nitrogen gas is utilized as a cushion gas (Zivar et al., 2001).

Several biotic processes are known for consuming hydrogen, including methanogenesis, acetogenesis, sulfate reduction, and iron (III) reduction (Dopffel, Jansen, & Gerritse, 2021; Saeed & Jadhawar, 2024a, 2024b; Zivar et al., 2020).

Methanogenesis occurs in the presence of archaea microorganisms under optimal conditions of 90 bar pressure and temperatures ranging from 30°C to 40°C in subsurface environments. It's worth noting that archaea activity has been observed at even higher temperatures. In this process, hydrogen is converted to methane and water (Saeed, Jadhawar, & Bagala, 2023).

$$CO_2 + 4H_2 \leftrightarrow CH_4 + 2H_2O$$

Methanogenesis can have an influence on the overall performance of UHS through possible gas mixing and reducing the H_2 injectivity (and productivity) due to pore plugging, which is triggered by the reduction in porosity and permeability.

Acetogenesis is the process by which acetogenic bacteria produce acetate. This can transpire through either the reduction of carbon dioxide or the reduction of organic acids to form acetate. During an acidogenic metabolic reaction, homoacetogenic bacteria consume hydrogen and carbon dioxide, resulting in the production of methane and acetate. Acetate can then be readily converted to acetic acid through protonation (Saeed, Jadhawar, & Bagala, 2023):

$$2CO_2 + 4H_2 \leftrightarrow CH_3COOH + 2H_2O$$

Sulfate-reducing bacteria (SRB) are microorganisms found in subsurface hydrocarbon reservoirs that consume hydrogen and produce hydrogen sulfide. In hydrocarbon reservoirs, when incompatible water is injected into the reservoirs, these bacteria are activated. SRB are anaerobic and highly adaptable microorganisms that can pose serious risks to injection and/or production facilities due to the generation of hydrogen. In the process involving SRB, water is generated as a byproduct. Besides the disadvantages associated with the production of H2S, the generation of water could pose another issue as it can increase water saturation, thereby reducing the saturation of hydrogen and its productivity. It is expected that the activity of SRB might reduce in depleted hydrocarbon reservoirs and in an acidic aqueous environment. The process of hydrogen consumption by SRB in the presence of sulfate ions, resulting in hydrogen sulfide generation, is described by the following reaction (Saeed, Jadhawar, & Bagala, 2023):

$$SO_4^{2-} + 5H_2 \leftrightarrow H_2S + 4H_2O$$

SRB, sourced from anhydrite ($CaSO_4$) mineral dissolution, utilize H_2 for H_2S production, using the energy gained for their cellular growth. The main concern with SRB lies in its potential for facility corrosion. They can thrive in various gas storage sites, particularly natural gas reservoirs. SRB corrodes metallic casing both chemically, through H_2S production, and electrically through withdrawal. Their survival depends on reservoir conditions like temperature, salinity, brine complexity, pH, and underground pressure. However, pressure has a lesser impact on microbial activity, particularly in thermophiles and hyperthermophiles.

Each of the microbial metabolisms described above involves two distinct types of reactions: catabolic reactions, which involve the breakdown of complex molecules into simpler ones, and anabolic reactions, which involve the synthesis of complex molecules from simpler ones. The catabolic reaction provides energy for microbial growth, while reproduction occurs through the anabolic reaction. The complete microbial metabolism results from the combination of these two reactions. Table 9.3 shows the catabolic and anabolic reactions involved in microbial metabolisms during hydrogen storage.

Iron-reducing bacteria (IRB) are another type of microorganism found in subsurface reservoirs that consume hydrogen. In this process, Fe_2O_3 (iron oxide) reacts with hydrogen. This type of reaction is usually called the ferric reaction, and it is generally associated with the interaction of microorganisms with minerals in the reservoir. Iron-reducers can be either heterotrophs, which utilize organic carbon as their food source, or autotrophs, bacteria capable of synthesizing their own food. IRB depend on the presence of iron oxides, organic carbon, and minerals such as smectite and chlorite to facilitate metal oxidation processes and reduce iron (III) phases.

$$3Fe_2^{3+} + H_2 \rightarrow 2Fe_3^{2+}O_4 + H_2O$$

TABLE 9.3 Metabolic process of each microbe during underground hydrogen storage (Shojaee et al., 2024).

Metabolism	Reactions
Methanogenesis	***Catabolic:***
	$HCO_3 + 4H_{2(aq)} + H \rightarrow CH_{4(aq)} + 3H_2O$
	Anabolic:
	$HCO_3^- + 2.1H_{2(aq)} + 0.2NH_4^- + 0.8H^+ \rightarrow CH_{1.8}O_{0.5}N_{0.2} + 2.5H_2O$
Acetogenesis	***Catabolic:***
	$2HCO_3^- + 4H_{2(aq)} + H^+ \rightarrow C_2H_3O_2^- + 4H_2O$
	Anabolic:
	$HCO_3^- + 2.1H_{2(aq)} + 0.2NH_4^- + 0.8H^+ \rightarrow CH_{1.8}O_{0.5}N_{0.2} + 2.5H_2O$
Sulfate reduction	***Catabolic:***
	$SO_4^2 + 4H_{2(aq)} + H \rightarrow HS + 4H_2O$
	Anabolic:
	$0.35C_2H_3O_2^- + 0.3CO_2 + 0.2NH_4^+ + 0.7H_{2(aq)} + 0.15H^+ \rightarrow CH_{1.8}O_{0.5}N_{0.2} + 0.8H_2O$

During this reaction within sandstone reservoirs, water can be released and occupy the interstitial pore space of sandstone as a byproduct, leading to excessive water saturation and mineral dissolution.

These biotic reactions were not immediately observed in the reservoir. Their nature is progressive and is kinetically controlled by the bacteria's activity that consumes hydrogen, leading to reduction reactions. It should be noted that there are different side effects associated with microbial activities during UHS. Some of these side effects include gas mixture changes, souring and H2S formation, and steel corrosion. Each of these side effects can have adverse effects on the performance of hydrogen storage site integrity and productivity.

9.2.2 Geomechanical aspects of hydrogen-mineral interaction during the hydrogen storage in geoformations

Insufficient understanding of subsurface reservoir and stress conditions, particularly in faulted reservoirs, can result in injection-induced seismicity. In the worst-case scenario of triggering an earthquake, this not only jeopardizes the stability of subsurface resource storage operations but also has societal repercussions, raising public concerns about such subsurface activities. Injecting fluids at high pressure diminishes frictional strength along fault planes and reduces effective overburden stress conditions in the fluid-pressurized zone. Consequently, there's an increase in driving shear stress and a decrease in frictional coefficient due to lubrication in sandstones. Moreover, the induced stress change may activate remote faults, contingent upon the disparity between the injection and diffusion rates of the injected fluid in the reservoir.

The subsurface contains several types of rocks and different soil, making it highly heterogeneous. Each material point in the subsurface is at equilibrium with the overburden pressure and the horizontal stress, and therefore subject to the stress tensor σ, expressed in the $x-y-z$ plane as (Kumar et al., 2023):

$$\sigma = \begin{matrix} \sigma_{xx} & \sigma_{xy} & \sigma_{xz} \\ \sigma_{yx} & \sigma_{yy} & \sigma_{yz} \\ \sigma_{zx} & \sigma_{zy} & \sigma_{zz} \end{matrix}$$

When the stress tensor is transformed in such a way that the shear stress components (σ_{xy}, σ_{xz}, σ_{yx}, σ_{yz}, σ_{zx}, σ_{zy}) become zero, it yields the principal stresses (σ_{xx}, σ_{yy}, σ_{zz}). Those are commonly used to describe the stress state, along with the Von Mises stress and the hydrostatic stress, two different metrics representing, respectively, the deviatoric and mean stress. Fig. 9.4 shows an illustration of the stresses that exist in depleted reservoirs during H_2 injection and production.

Fluids inside the interconnected pores within the rocks also exert pressure. This pressure is normally referred to as pore pressure (P_p). During H_2 injection and production, the pore pressure changes. The variations in pore pressure can be obtained by using the fluid conversion of momentum equation (Eq. 9.3). In this equation, time and spatial variations in the velocity (left-hand side of the equation) are related to the pressure

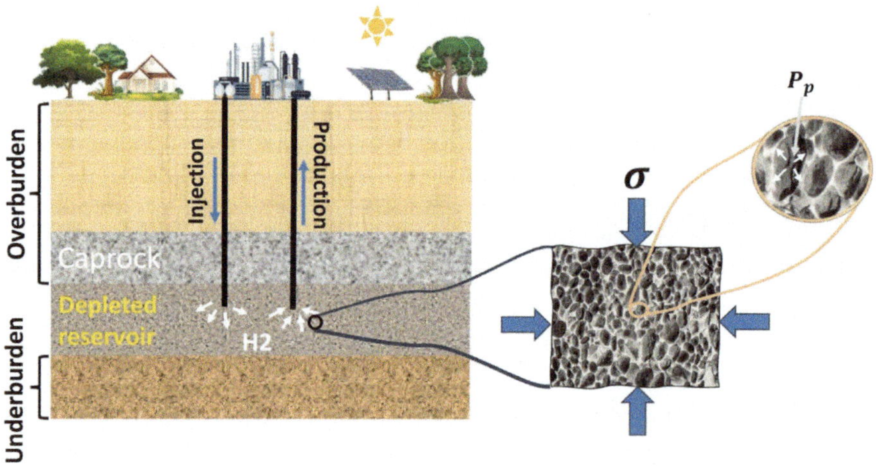

FIGURE 9.4 Schematic representation of the stresses related to the H_2 injection and withdrawal in depleted hydrocarbon reservoirs (Kumar et al., 2023).

gradient and viscous stress of the fluid (terms in the right-hand side of the equation, respectively).

$$\rho\left(\frac{\partial v}{\partial t} + v.\nabla v\right) = -\nabla P + \mu \nabla^2 v \tag{9.3}$$

where ρ is fluid density, v is the fluid velocity, and ∇P is pressure gradient of the fluid.

Taking into account the formation stresses and fluid pressure would yield effective stress, which is defined as below (Terzaghi et al., 1996):

$$\sigma' = \sigma - \alpha P_{\mathrm{p}} I \tag{9.4}$$

In this equation, α is the Biot coefficient, which is related to the compressibility of the pores within the rocks (Biot & Willis, 1957). α can be calculated by Eq. (9.5).

$$\alpha = 1 - \frac{K_{\mathrm{T}}}{K_{\mathrm{S}}} \tag{9.5}$$

where K_{T} and K_{S} represent the moduli of rock and grains, respectively.

When H_2 is injected into the porous medium, pore pressure will increase. A change in the pore pressure would result in a change in the effective stress, and it would initially associated with the elastic strain of the rocks. Constitutive models are used for determining the elastic strain of rocks because of effective stress. These models have the general form expressed as $\sigma = C/\varepsilon_{\mathrm{el}}$, where $\varepsilon_{\mathrm{el}}$ is the elastic strain and C is a function of Young's modulus and Poisson's ratio. More details regarding the constitutive models will be presented in subsequent sections.

9.2.3 Cumulative effects of geochemical and geomechanics on the underground hydrogen storage system

The interaction of injected fluids with reservoir rocks can have enduring consequences for reservoir performance and safety. The dissolution or deposition of minerals along fault planes is pivotal in assessing the potential for slippage and the load-bearing capacity of the faulted surface.

While the microbial presence and its impact on oil and gas recovery in subsurface hydrocarbon reservoirs are well-explored, there is considerable diversity within these microbial communities. The subset of microbes responsive to hydrocarbons differs from those interacting with H_2, which is a fundamental and straightforward electron donor (Teng et al., 2019; Xu et al., 2021). Although the activity of these microbes under reservoir conditions in the presence of H_2 is not extensively documented, laboratory experiments have shown their sensitivity to temperature, pressure, and pH. These microbes not only consume H_2 to produce gases like CH_4, H_2S, and organic acids but also form biofilms. These biofilms can modify reservoir wettability and reduce accessible pore spaces after successive H_2 injection and production cycles. Similar to the attention that has been devoted to the effects of biofilms on H_2 flow aspects, there is a need to consider the impacts of biofilms on the geomechanical properties of underground reservoirs (Boon et al., 2024; Kumar et al., 2023).

It is anticipated that changes in reservoir wettability due to microbial activity will alter how H_2 and formation brine interact with the existing fault plane, potentially resulting in mechanical instability. Iron-reducing bacteria have the ability to substitute iron in minerals in the presence of H_2, which could impact the reservoir's stability, depending on the prevalence of iron-bearing minerals. Additionally, H_2 consumption can decrease reservoir pressure, leading to gradual subsidence due to increased effective stress (Mirchi et al., 2023; Corina et al., 2022).

Fig. 9.5 shows the processes with respect to geomechanics in a depleted gas reservoir during UHS. These processes may lead to geomechanical hazards and need to be considered. Fault

FIGURE 9.5 Geomechanical processes involved in underground hydrogen storage in depleted gas reservoirs (Kumar et al., 2023)

reactivation, land uplift or subsidence, caprock leakage, wellbore integrity, and microbial activities are among the most important interplaying processes that need to be investigated thoroughly. It should be noted that most of these processes are common among different types of gas storage sites for different types of gas injections. However, the presence of H_2 during UHS, along with microbial activities and microorganisms, might affect the rock material's integrity.

9.3 Modeling and simulation

Modeling and simulation of complex processes like UHS requires extensive analysis of influencing phenomena as well as the mechanisms and parameters involved. As discussed in previous sections, in UHS modeling, the most important aspects that need to be considered are hydrodynamics, geochemistry, and geomechanics (Okoroafor et al., 2023; Zeng et al., 2022). In each of these aspects, different influencing mechanisms exist. As discussed in Section 9.2.3, it should be noted that some of these mechanisms and properties are directly or indirectly connected to each other. Fully coupled simulations can also incorporate additional interdependent processes or effects, such as thermal changes and geochemical reactions. Integrating these interconnected processes into fully coupled simulations would yield more precise results when modeling UHS; however, this would come at an extra cost to the simulation time (Saeed & Jadhawar, 2024a, 2024b).

9.3.1 Geochemical reactions in underground hydrogen storage

Geochemical models utilize a set of mathematical equations derived from the conceptual analysis of different principal variables. These equations represent different natural processes within a system. In complex natural underground systems, numerous processes exist that influence the extent of geochemical reactions. These include fluid flow, chemical reactions, biological interactions, heat transfer, and chemical mass transport (Hagemann, 2018; Thiyagarajan et al., 2022). In most cases, these processes are all coupled to each other, making the analysis of the system more complex. Through fluid flow, precipitation and dissolution reactions might occur, by which porosity and permeability alteration in the reservoir might be induced. These can change how fluid will flow underground and also influence the velocity of the components existing in the system. Fluid flow will also have an impact on the advective or dispersive transport of the chemical components. Chemical reactions would have a significant influence on the partitioning of the species in aqueous or other phases present. Biological interactions involve those interactions where living microorganisms play a part in them. These microorganisms, usually serve as chemical reaction catalysts (Giustetto et al., 2015; Perera, 2023). Chemical mass transport could occur either by advection or dispersion. In advection, chemical species are carried along with the movement of the carrier fluids. Whereas dispersion involves the movement of microscopic particles or species through molecular diffusion or microscopic variations in the speed of each component within a very limited space.

In geochemical modeling, thermodynamics is coupled with chemical reaction kinetics to investigate how a geological system is responding during a specified process that might be considered to occur in a reservoir (Tremosa et al., 2023; Zhan et al., 2024).

As discussed, during UHS, geochemical reactions may involve injected fluid (H_2), cushion gas, and resident fluids inside the porous media. These reactions can be biotic or abiotic. Abiotic reactions are those reactions in which no living organism (such as bacteria) is involved. During UHS, these reactions can be classified as equilibrium reactions, mineral dissolution/precipitation reactions, and ion exchange reactions. The most important abiotic reactions during UHS are listed in Table 9.4. Equilibrium reactions have a general form as Eq. (9.6) (Saeed & Jadhawar, 2024a, 2024b):

$$aR + bS \rightarrow cP \tag{9.6}$$

TABLE 9.4 Equilibrium reaction and equilibrium constant for main reactions during underground hydrogen storage (Hemme & van Berk, 2018).

Equilibrium reaction	logK at 25°C, 1 bar
$KAlSi_3O_8 + 8H_2O = K^+ + Al(OH)^{4-} + 3H_4SiO_4$	− 20.573
$NaAlSi_3O_8 + 8H2O = Na^+ + Al(OH)^{4-} + 3H_4SiO_4$	− 18.002
$Al_2Si_2O_5(OH)_4 + 6H^+ = H_2O + 2H_4SiO_4 + 2Al_3^+$	7.435
$SiO_2 + 2H_2O = H_4SiO_4$	3.98
$CaCO_3 = CO_3^{2-} + Ca^{2+}$	8.48
$FeS_2 + 2H^+ + 2e^- = Fe_2^+ + 2HS^-$	− 18.479
$K_{0.6}Mg_{0.25}Al_{2.3}Si_{3.5}O_{10}(OH)_2 + 11.2H_2O = 0.6K^+ + 0.25Mg^{2+}$	− 40.267
$+ 2.3Al(OH)^{4-}3.5H_4SiO_4 + 1.2H^+$	
$NaAlCO_3(OH)_2 + 3H^+ = Al^{3+} + HCO^{3-} + Na^+ + 2H_2O$	4.35
$FeS + H^+ = Fe^{2+} + HS^-$	− 4.648
$CaMg(CO_3)_2 = Ca^{2+} + Mg^{2+} + 2CO_3^{2-}$	− 17.09
$NaHCO_3 = HCO_3^- + Na^+$	− 0.11
$CaSO_4 = Ca^{2+} + SO_4^{2-}$	− 4.39
$NaCl = Cl^- + Na^+$	1.570
$CaSO_4 \cdot 2H_2O = Ca^{2+} + SO_4^{2-} + 2H_2O$	− 4.58
$S + 2H^+ + 2e^- = H_2S$	4.882
$BaSO_4 = Ba^{2+} + SO_4^{2-}$	− 9.97
$FeOOH + 3H^+ = Fe^{3+} + 2H_2O$	− 1.0
$H_2 = H_2$	− 3.1050
$CO_2 = CO_2$	− 1.468
$CH_4 = CH_4$	− 2.8502
$H_2S = H^+ + HS^-$	− 7.9759
$N_2 = N_2$	− 3.1864

where a, b, and c are stoichiometric coefficients of the equilibrium reaction. This reaction is controlled by K, which is the equilibrium constant of the reaction and is calculated using Eq. (9.7):

$$K = \frac{\alpha_P^c}{\alpha_R^a \cdot \alpha_S^b} \tag{9.7}$$

where a is the activity of chemical species in the equation and can be calculated using species concentration and activity coefficient as shown in Eq. (9.8).

$$\alpha_i = \gamma_i \cdot C_i \tag{9.8}$$

Activity coefficients can be obtained by the WATEEQ Debye–Huckel model (Saeed & Jadhawar, 2024a, 2024b).

Table 9.4 shows the equilibrium reactions that might occur during UHS. In a hydrogen geologic storage system, some of the reactions that may not involve hydrogen directly can influence the pH and chemistry of the solution, subsequently affecting the solubility of hydrogen in water. Higher ionic strength in a solution, such as from increased salinity, typically causes a salting-out effect for many gases, which decreases their solubility in the solution (Saeed & Jadhawar, 2024a, 2024b). As mentioned, the rate of kinetic reactions is influenced by temperature, pressure, and salinity, so assuming constant reaction rates in modeling may produce inaccurate outcomes. Additionally, conducting geochemical experiments that explore hydrogen reactions with different reservoir minerals, brine compositions, and conditions is necessary to calibrate and validate reaction rates, which in turn improves the geochemical modeling of UHS (Saeed & Jadhawar, 2024a, 2024b).

The majority of the studies focused on modeling the geochemical reactions during UHS have utilized two main geochemical modeling softwares, namely PHREEQC (pH-Redox EQuilibrium written in the C programming language) and Geochemist's Workbench (GWB). The PHREEQC software, developed by the US Geological Survey, is an advanced geochemical modeling software that enables the modeling of aqueous, mineral, ion exchange, and surface complexation reactions. In GWB, which is developed by Aqueous Solutions LLC, various types of geochemical reactions can be simulated. These reactions include aqueous, kinetic mineral dissolution and precipitation, as well as reactive and colloidal transport (Hemme & van Berk, 2018; Saeed & Jadhawar, 2024a, 2024b).

9.3.2 Geomechanical modeling

When modeling the geomechanical alterations during UHS, it is imperative to evaluate the stresses induced on the reservoir rock due to hydrogen gas injection and withdrawal. Additionally, it is crucial to analyze the rock's response to these applied stresses in terms of strength, strain, and elasticity.

Two of the most common techniques for performing geomechanical modeling of underground hydrogen storage are one-way coupling and two-way coupling. Underground hydrogen storage practices involve the injection of hydrogen, hydrogen flow inside porous media, and the production of hydrogen. In one-way coupling, the first fluid flow equations are solved. The distribution of fluid saturation and pressure is obtained. Then, this information is applied in geomechanical equations and simulations to investigate the effects of changes in pore pressure and fluid saturation resulting from fluid flow on possible rock deformations (Gelencsér et al., 2023; Zeng et al., 2023). In two-way coupling, the effects of fluid flow and rock deformation are investigated simultaneously. In this approach, the effects of fluid flow, which are considered to be variations in pore pressure and fluid saturations, would have an instantaneous influence on the stress-induced behavior of rocks. In the same manner, the stress-strain properties of the rock would have a direct influence on the fluid flow inside the porous media (Bai and Tahmasebi, 2022; Burtonshaw et al., 2022; Xue et al., 2023). The two-way coupling method is considered to be more accurate as it includes simultaneous coupling effects of rock mechanics and fluid flow properties. However, the computational cost of the two-way coupling method is higher than that of the one-way coupling method. Fully coupled simulations can also incorporate additional interdependent processes or effects, such as thermal changes and geochemical reactions. Integrating these interconnected processes into fully coupled simulations would yield more precise results when modeling UHS; however, this would come at an extra cost to the simulation time (Epelle et al., 2022; Zivar et al., 2020).

To investigate the hydro-mechanical responses of subsurface formations, it's vital to explore the processes rocks undergo during cyclic loading. Constitutive models are designed to facilitate this understanding, enabling the projection of rocks' long-term behaviors. These models serve a critical role in anticipating how rocks will react under varied operational conditions, which are often difficult to replicate in laboratory settings. Fig. 9.6 shows a schematic of the effects of constant stress applied to the rock over time.

In the first region, the transient region, rock's behavior is identified as elastic and viscoelastic (time-dependent elastic behavior). The subsequent phase is the steady-state domain, which can be subdivided into time-dependent inelastic strain and time-independent inelastic strain, commonly known as plasticity. Inelastic strains like plasticity, viscoplasticity, and creep are frequently encountered in subsurface formations. Plasticity entails permanent deformation that occurs immediately (time-independent) when stress levels

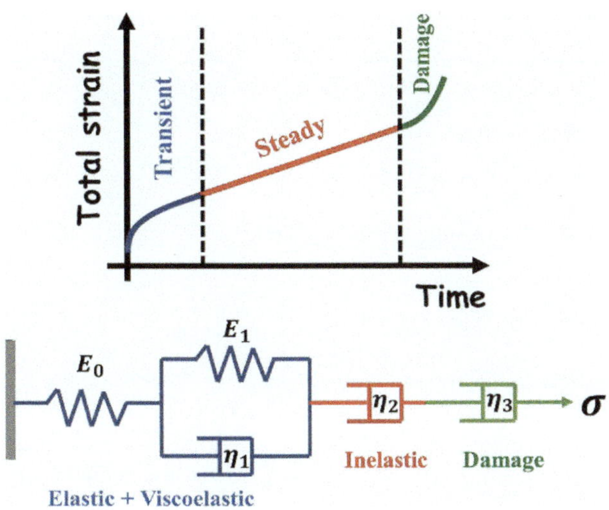

FIGURE 9.6 Schematic representation of the rock strain while being under constant stress (Kumar et al., 2023)

surpass a specific yield threshold (Grgic et al., 2022; Patanwar et al., 2024). Likewise, viscoplasticity entails permanent deformation when stress levels reach the yield threshold. However, in the case of viscoplasticity, the rate of stress application also significantly influences the outcome. Hence, viscoplastic behavior is referred to as having time-dependent deformation characteristics. Another form of time-dependent inelastic deformation is referred to as creep deformation, where the material undergoes continuous deformation under a constant and sustained external load, regardless of the stress levels (Khaledi et al., 2016; Maury Fernandez et al., 2024). Finally, damage and strain may occur, creating microcracks that can ultimately result in rock failure. The last stage must be avoided in order to maintain safe operation at UHS. It should be noted that each reservoir contains different types of rocks that follow a different stress-strain pattern. It is important to take into account such heterogeneity in the constitutive models for accurate geomechanical modeling (Kumar et al., 2023).

Table 9.5 shows examples of constitutive models that can be applied during UHS. These models follow the specific physical behavior of rock material and its response to the applied stress. Depending on the rock type and its stress-strain behavior, a specific type of constitutive model should be applied to perform a geomechanical simulation.

To measure rock deformation accurately, it is essential to integrate models based on various physical phenomena like fault slip laws, viscoelasticity, plasticity, and creep. Field-scale simulators can utilize these models to calculate permanent deformation, stress distribution, and fault movements, with calibration from laboratory experiments. Incorporating fault slip models derived from lab-scale experiments into field simulators can help assess the potential for fault slippage and induced seismicity in heavily faulted reservoirs (Amirthan & Perera, 2023; Tenthorey et al., 2024). When utilizing constitutive models derived from laboratory experiments, which typically operate on a timescale of days or weeks, in field-scale simulations spanning years or decades, interpretation of the

TABLE 9.5 Constitutive model examples during underground hydrogen storage (Kumar et al., 2023).

Physics	Constitutive mode	Type of rock
Viscoelasticity	General Kelvin model: $\frac{\varepsilon}{\sigma} = \frac{1}{E_1} + \frac{1}{E_2}\left[1 - \exp\left(-\frac{E_2 t}{\eta_K}\right)\right]$ Fractional maxwell model: $\frac{\varepsilon}{\sigma} = \frac{1}{E_H} + \frac{1}{\eta_A^\alpha}\frac{t^\alpha}{\Gamma(1+\alpha)}$	Sandstone
Plasticity	MCC Yield surface: $F(\sigma, \varepsilon_v^p) = q^2 + M^2 p'(p' - p_c'(\varepsilon_v^p)) = 0$ Generalized plasticity: $d\sigma = \frac{1}{H_{vp}}(n_{gL/U} \otimes n)$ General cyclic plasticity model: $f^y = (q - (p+p_c)\alpha)^2 - (p+p_c)^2(M^y)^2$	Sandstone
Creep	Power law: $\dot{\varepsilon}^{cr} = Ae^{-\frac{Q}{RT}}\sigma^{n-1}st^{m-1}$ Linear relationship: $\varepsilon_{cr} = \frac{\sigma - \sigma_{BP}}{\eta}\Delta t$	Sandstone
Viscoplasticity	Yield surface: $F(\sigma, \varepsilon_v^p) = q^2 + M^2 p'(p' - p_c'(\varepsilon_v^p)) = 0$	Shale
Plasticity	General cyclic plasticity model: $f^y = (q - (p+p_c)\alpha)^2 - (p+p_c)^2(M^y)^2$	Carbonate
Creep	Yield surface: $F(\sigma, \varepsilon_v^p) = q^2 + M^2 p'(p' - p_c'(\varepsilon_v^p)) = 0$	Carbonate
Fault stability model	Byerlee law: $\tau^{\max} = \left|\mu_c + \mu_f.\sigma_n\right|$	Fracture models

results must be carried out with caution. These models are often developed based on external loading experiments conducted on rock samples. Hence, it's crucial to validate the applicability of these constitutive models in scenarios involving cyclic injection and fluid production (Lysyy et al., 2023). Confidence in these models can be enhanced by using them to forecast reservoir behavior over longer timescales and comparing the predictions with field-scale observations.

9.4 Challenges and knowledge gap

UHS is a promising approach toward achieving net zero carbon emission targets. However, it is associated with several technical, safety, and economic challenges. Interdisciplinary research and investigations are required for safe operations during UHS. H_2 has unique thermophysical properties, including low viscosity, low density, small molecular size, and high diffusivity, which may lead to potential failures in successfully delivering UHS projects. Fluid–fluid and rock–fluid interactions during H_2 injection and withdrawal pose new geochemical and geomechanical challenges (Heinemann et al., 2021; Navaid, Emadi, & Watson, 2023; Raza et al., 2022). These challenges include, but are not limited to, H_2 consumption, reduction in H_2 purity, land uplift or subsidence, fault reactivation, reducing reservoir, caprock, and well integrity, and adverse impacts on surface and subsurface facilities. To prevent such issues, comprehensive analysis of H_2 cyclic injection and production should be carried out (Epelle et al., 2022; Kumari & Ranjith, 2023; Ma et al., 2024; Wei et al., 2023).

Current knowledge on the geochemistry and geomechanics of UHS is limited. Further analytical and experimental research should be conducted to investigate the geochemical reactions taking place in the presence of H_2. More extensive hydrodynamic characterization of H_2 flow in underground reservoirs should be carried out. Experimental geomechanical studies should be conducted to better understand the impacts that H_2 might have on the integrity of the formations and wells. H_2 storage Numerical and analytical studies should encompass coupled methods that consider the simultaneous effects of different processes that might occur during H_2 storage. An appropriate methodology should be developed to select potential UHS sites.

9.5 Conclusions

The geochemical reactivity of hydrogen and its effect on the migration of hydrogen in different types of reservoirs are critical. It is crucial to understand the reactivity of hydrogen with rock-forming minerals, mineralogical transformations (for different rock and reservoir types) due to the presence of hydrogen, and how they modify the pore structure of the rock and affect the storage properties. The geochemical changes upon hydrogen injection are susceptible to varying with reservoir conditions, the potential chemical reactions, and the respective changes in flow properties under the in-situ conditions anticipated at the storage site under short- and long-term conditions. In-situ migration of hydrogen coupled to geochemical alterations of the rocks or the brine components interaction to study the influence of the hydrogen reactions on the reservoir operations has direct influence on underground hydrogen storage. Several challenges related to geochemical and geomechanical modeling and simulation still exist to mitigate for properly implementing an underground hydrogen storage project. Extended research and development work is required to understand the geochemical and geomechanical interactions within the reservoirs for storing hydrogen underground.

References

Amirthan, T., & Perera, M. S. A. (2022). The role of storage systems in hydrogen economy: A review. *Journal of Natural Gas Science and Engineering.*, *108*. Available from https://doi.org/10.1016/j.jngse.2022.104843.

Amirthan, T., & Perera, M. S. A. (2023). Underground hydrogen storage in Australia: A review on the feasibility of geological sites. *International Journal of Hydrogen Energy*, *48*(11), 4300–4328. Available from https://doi.org/10.1016/j.ijhydene.2022.10.218, http://www.journals.elsevier.com/international-journal-of-hydrogen-energy/.

Bai, T., & Tahmasebi, P. (2022). Coupled hydro-mechanical analysis of seasonal underground hydrogen storage in a saline aquifer. *Journal of Energy Storage.*, *50*. Available from https://doi.org/10.1016/j.est.2022.104308.

Biot, M. A., & Willis, D. G. (1957). The elastic coefficients of the theory of consolidation. *Journal of Applied Mechanics, Transactions ASME*, *24*(4), 594–601. Available from https://doi.org/10.1115/1.4011606, https://asmedigitalcollection.asme.org/appliedmechanics.

Boon, M., Buntic, I., Ahmed, K., Dopffel, N., Peters, C., & Hajibeygi, H. (2024). Microbial induced wettability alteration with implications for underground hydrogen storage. *Scientific Reports*, *14*(1). Available from https://doi.org/10.1038/s41598-024-58951-6.

Borello, S. (2024). Underground hydrogen storage safety: Experimental study of hydrogen diffusion through caprocks. *Energies (Basel)*, *17*. Available from https://doi.org/10.3390/EN17020394/S1.

Burtonshaw J. E. J. Paluszny A. Thomas R. N. Zimmerman R. W. 2022 The influence of hydraulic fluid properties on induced seismicity during underground hydrogen storage, *56th U.S. Rock Mechanics/Geomechanics Symposium*, American Rock Mechanics Association (ARMA): United Kingdom.

Corina, A., Zikovic, V., Soustelle, V., & Ter Heege, J. (2022). Review of current practices and existing experimental data for well and rock materials under cyclic hydrogen injection and withdrawal. H2020 HyUSPRe project report D. 2022;5.

Dopffel, N., Jansen, S., & Gerritse, J. (2021). Microbial side effects of underground hydrogen storage – Knowledge gaps, risks and opportunities for successful implementation. *International Journal of Hydrogen Energy*, *46*(12), 8594–8606. Available from https://doi.org/10.1016/j.ijhydene.2020.12.058, http://www.journals.elsevier.com/international-journal-of-hydrogen-energy/.

Epelle, E. I., Obande, W., Udourioh, G. A., Afolabi, I. C., Desongu, K. S., Orivri, U., Gunes, B., & Okolie, J. A. (2022). Perspectives and prospects of underground hydrogen storage and natural hydrogen. *Sustainable Energy & Fuels*, *6*(14), 3324–3343. Available from https://doi.org/10.1039/d2se00618a.

Gelencsér, O., Árvai, C., Mika, L. T., Breitner, D., LeClair, D., Szabó, C., Falus, G., & Szabó-Krausz, Z. (2023). Effect of hydrogen on calcite reactivity in sandstone reservoirs: Experimental results compared to geochemical modeling predictions. *Journal of Energy Storage*, *61*. Available from https://doi.org/10.1016/j.est.2023. 106737.

Giustetto, R., Gonella, D., Bianciotto, V., Lumini, E., Voyron, S., Costa, E., & Diana, E. (2015). Transfiguring biodegradation of frescoes in the Beata Vergine del Pilone Sanctuary (Italy): Microbial analysis and minerochemical aspects. *International Biodeterioration and Biodegradation*, *98*, 6–18. Available from https://doi.org/10.1016/j.ibiod.2014.10.020, http://www.elsevier.com/inca/publications/store/4/0/5/8/9/9.

Grgic, D., Al Sahyouni, F., Golfier, F., Moumni, M., & Schoumacker, L. (2022). Evolution of gas permeability of rock salt under different loading conditions and implications on the underground hydrogen storage in salt caverns. *Rock Mechanics and Rock Engineering*, *55*(2), 691–714. Available from https://doi.org/10.1007/s00603-021-02681-y, http://www.springer.com/journal/603.

Hagemann, B. (2018). *Numerical and analytical modeling of gas mixing and bio-reactive transport during underground hydrogen storage*. Cuvillier Verlag.

Heinemann, N., Alcalde, J., Miocic, J. M., Hangx, S. J. T., Kallmeyer, J., Ostertag-Henning, C., Hassanpouryouzband, A., Thaysen, E. M., Strobel, G. J., Schmidt-Hattenberger, C., Edlmann, K., Wilkinson,

M., Bentham, M., Stuart Haszeldine, R., Carbonell, R., & Rudloff, A. (2021). Enabling large-scale hydrogen storage in porous media-the scientific challenges. *Energy and Environmental Science*, *14*(2), 853–864. Available from https://doi.org/10.1039/d0ee03536j, http://pubs.rsc.org/en/journals/journal/ee.

Hemme, C., & van Berk, W. (2018). Hydrogeochemical modeling to identify potential risks of underground hydrogen storage in depleted gas fields. *Applied Sciences (Switzerland)*, *8*(11). Available from https://doi.org/10.3390/app8112282, https://www.mdpi.com/2076-3417/8/11/2282/pdf.

Jadhawar P. Saeed M. 2023 Fluid-fluid and rock-fluid parametric investigation of the underground storage of hydrogen in a Deep North Sea aquifer, *SPE Offshore Europe Conference Proceedings*, Society of Petroleum Engineers (SPE): United Kingdom. https://www.onepetro.org/conferences/SPE/15OE, https://doi.org/10.2118/215598-MS.

Jadhawar, P., & Saeed, M. (2024). Mechanistic evaluation of the reservoir engineering performance for the underground hydrogen storage in a deep North Sea aquifer. *International Journal of Hydrogen Energy*, *50*, 558–574. Available from https://doi.org/10.1016/j.ijhydene.2023.07.272, https://www.sciencedirect.com/science/journal/03603199.

Jadhawar, P., & Saeed, M. (2023). Optimizing the operational efficiency of the underground hydrogen storage scheme in a deep North Sea aquifer through compositional simulations. *Journal of Energy Storage*, *73*. Available from https://doi.org/10.1016/j.est.2023.108832.

Jahanbakhsh, A., Potapov-Crighton, A. L., Mosallanezhad, A., Kaloorazi, N. T., & Maroto-Valer, M. M. (2024). Underground hydrogen storage: A UK perspective. *Renewable and Sustainable Energy Reviews.*, *189*. Available from https://doi.org/10.1016/j.rser.2023.114001.

Khaledi, K., Mahmoudi, E., Datcheva, M., & Schanz, T. (2016). Stability and serviceability of underground energy storage caverns in rock salt subjected to mechanical cyclic loading. *International Journal of Rock Mechanics and Mining Sciences*, *86*, 115–131. Available from https://doi.org/10.1016/j.ijrmms.2016.04.010, http://www.elsevier.com/inca/publications/store/2/5/6/index.htt.

Kumar, K. R., Honorio, H., Chandra, D., Lesueur, M., & Hajibeygi, H. (2023). Comprehensive review of geomechanics of underground hydrogen storage in depleted reservoirs and salt caverns. *Journal of Energy Storage*, *73*. Available from https://doi.org/10.1016/j.est.2023.108912.

Kumari, W. G. P., & Ranjith, P. G. (2023). An overview of underground hydrogen storage with prospects and challenges for the Australian context. Geoenergy *Science and Engineering*, 231. Available from https://doi.org/10.1016/j.geoen.2023.212354, https://www.sciencedirect.com/journal/geoenergy-science-and-engineering.

Lysyy, M., Liu, N., Solstad, C. M., Fernø, M. A., & Ersland, G. (2023). Microfluidic hydrogen storage capacity and residual trapping during cyclic injections: Implications for underground storage. *International Journal of Hydrogen Energy*, *48*(80), 31294–31304. Available from https://doi.org/10.1016/j.ijhydene.2023.04.253, http://www.journals.elsevier.com/international-journal-of-hydrogen-energy/.

Ma, N., Zhao, W., Wang, W., Li, X., & Zhou, H. (2024). Large scale of green hydrogen storage: Opportunities and challenges. *International Journal of Hydrogen Energy*, *50*, 379–396. Available from https://doi.org/10.1016/j.ijhydene.2023.09.021.

Maury Fernandez, D., Emadi, H., Hussain, A., & Thiyagarajan, S. R. (2024). A holistic review on wellbore integrity challenges associated with underground hydrogen storage. *International Journal of Hydrogen Energy*, *57*, 240–262. Available from https://doi.org/10.1016/j.ijhydene.2023.12.230, https://www.sciencedirect.com/science/journal/03603199.

Mirchi, V., Dejam, M., Alvarado, V., & Akbarabadi, M. (2023). Effect of cushion gas on hydrogen/brine flow behavior in oil-wet rocks with application to hydrogen storage in depleted oil and gas reservoirs. *Energy & Fuels*, *37*(19), 15231–15243. Available from https://doi.org/10.1021/acs.energyfuels.3c02884.

Muhammed, N. S., Haq, B., Al Shehri, D., Al-Ahmed, A., Rahman, M. M., & Zaman, E. (2022). A review on underground hydrogen storage: Insight into geological sites, influencing factors and future outlook. *Energy Reports*, *8*, 461–499. Available from https://doi.org/10.1016/j.egyr.2021.12.002, http://www.journals.elsevier.com/energy-reports/.

Navaid, H. B., Emadi, H., & Watson, M. (2023). A comprehensive literature review on the challenges associated with underground hydrogen storage. *International Journal of Hydrogen Energy*, *48*(28), 10603–10635. Available from https://doi.org/10.1016/j.ijhydene.2022.11.225, http://www.journals.elsevier.com/international-journal-of-hydrogen-energy/.

Okoroafor, E. R., Sampaio, L., Gasanzade, F., Perez Claro, Y., Zhou, J. D., Saltzer, S. D., Bauer, S., & Kovscek, A. R. (2023). Intercomparison of numerical simulation models for hydrogen storage in porous media using different codes. *Energy Conversion and Management, 292*. Available from https://doi.org/10.1016/j.enconman.2023.117409, https://www.journals.elsevier.com/energy-conversion-and-management.

Osman, A. I., Mehta, N., Elgarahy, A. M., Hefny, M., Al-Hinai, A., Al-Muhtaseb, A. H., & Rooney, D. W. (2022). Hydrogen production, storage, utilisation and environmental impacts: a review. *Environmental Chemistry Letters, 20*(1), 153–188. Available from https://doi.org/10.1007/s10311-021-01322-8, http://springerlink.metapress.com/app/home/journal.asp?wasp = d86tgdwvtg0yvw9gvkwp&referrer = parent&backto = browsepublicationsresults.

Patanwar, Y. K., Kim, H. M., Deb, D., & Gujjala, Y. K. (2024). Underground storage of hydrogen in lined rock caverns: An overview of key components and hydrogen embrittlement challenges. *International Journal of Hydrogen Energy, 50*, 116–133. Available from https://doi.org/10.1016/j.ijhydene.2023.08.342, https://www.sciencedirect.com/science/journal/03603199.

Perera, M. S. A. (2023). A review of underground hydrogen storage in depleted gas reservoirs: Insights into various rock-fluid interaction mechanisms and their impact on the process integrity. *Fuel, 334*. Available from https://doi.org/10.1016/j.fuel.2022.126677.

Raza, A., Arif, M., Glatz, G., Mahmoud, M., Kobaisi, M. A., Alafnan, S., & Iglauer, S. (2022). A holistic overview of underground hydrogen storage: Influencing factors, current understanding, and outlook. *Fuel, 330*. Available from https://doi.org/10.1016/j.fuel.2022.125636.

Saeed, M., & Jadhawar, P. (2024a). Optimizing underground hydrogen storage in aquifers: The impact of cushion gas type. *International Journal of Hydrogen Energy, 52*, 1537–1549. Available from https://doi.org/10.1016/j.ijhydene.2023.08.352, https://www.sciencedirect.com/science/journal/03603199.

Saeed, M., & Jadhawar, P. (2024b). Modelling underground hydrogen storage: A state-of-the-art review of fundamental approaches and findings. *Gas Science and Engineering, 121*. Available from https://doi.org/10.1016/j.jgsce.2023.205196, https://www.sciencedirect.com/science/journal/29499089.

Saeed, M., Jadhawar, P., & Bagala, S. (2023). Geochemical effects on storage gases and reservoir rock during underground hydrogen storage: A depleted north sea oil reservoir case study. *Hydrogen, 4*(2), 323–337. Available from https://doi.org/10.3390/hydrogen4020023.

Shojaee, A., Ghanbari, S., Wang, G., & Mackay, E. (2024). Interplay between microbial activity and geochemical reactions during underground hydrogen storage in a seawater-rich formation. *International Journal of Hydrogen Energy, 50*, 1529–1541. Available from https://doi.org/10.1016/j.ijhydene.2023.10.061.

Song, J., & Zhang, D. (2013). Comprehensive review of caprock-sealing mechanisms for geologic carbon sequestration. *Environmental Science and Technology, 47*(1), 9–22. Available from https://doi.org/10.1021/es301610p, http://pubs.acs.org/journal/esthag.

Tarkowski, R., Uliasz-Misiak, B., & Tarkowski, P. (2021). Storage of hydrogen, natural gas, and carbon dioxide – Geological and legal conditions. *International Journal of Hydrogen Energy, 46*(38), 20010–20022. Available from https://doi.org/10.1016/j.ijhydene.2021.03.131, http://www.journals.elsevier.com/international-journal-of-hydrogen-energy/.

Teng, Y., Xu, Y., Wang, X., & Christie, P. (2019). Function of biohydrogen metabolism and related microbial communities in environmental bioremediation. *Frontiers in Microbiology, 10*. Available from https://doi.org/10.3389/fmicb.2019.00106.

Tenthorey, E., Hsiao, W. M., Puspitasari, R., Giddins, M. A., Pallikathekathil, Z. J., Dandekar, R., Suriyanto, O., & Feitz, A. J. (2024). Geomechanics of hydrogen storage in a depleted gas field. *Australia International Journal of Hydrogen Energy, 60*, 636–649. Available from https://doi.org/10.1016/j.ijhydene.2024.02.189, https://www.sciencedirect.com/science/journal/03603199.

Terzaghi, K., Ralph, R.B., Peck, B., & Mesri, G. (1996). Soil mechanics in engineering practice.

Thiyagarajan, S. R., Emadi, H., Hussain, A., Patange, P., & Watson, M. (2022). A comprehensive review of the mechanisms and efficiency of underground hydrogen storage. *Journal of Energy Storage, 51*. Available from https://doi.org/10.1016/j.est.2022.104490, http://www.journals.elsevier.com/journal-of-energy-storage/.

Tremosa, J., Jakobsen, R., & Gallo, Y. L. (2023). Assessing and modeling hydrogen reactivity in underground hydrogen storage: A review and models simulating the Lobodice town gas storage. *Frontiers in Energy Research, 11*. Available from https://doi.org/10.3389/fenrg.2023.1145978.

Wei, Y., Guangyao, L., Zhong, L., Mengqi, H., Jingen, D., & Zelin, M. (2023). Progress and challenges of underground hydrogen storage technology. *Acta Petrolei Sinica, 44*.

Xu, Y., Teng, Y., Dong, X., Wang, X., Zhang, C., Ren, W., Zhao, L., Luo, Y., & Greening, C. (2021). Genome-resolved metagenomics reveals how soil bacterial communities respond to elevated H_2 availability. *Soil Biology and Biochemistry, 163*. Available from https://doi.org/10.1016/j.soilbio.2021.108464.

Xue, Y., Liu, J., Liang, X., Li, X., Wang, S., Ma, Z., Zhang, S., & Jiao, X. (2023). Influence mechanism of brine-gas two-phase flow on sealing property of anisotropic caprock for hydrogen and carbon energy underground storage. *International Journal of Hydrogen Energy, 48*(30), 11287–11302. Available from https://doi.org/10.1016/j.ijhydene.2022.05.173, http://www.journals.elsevier.com/international-journal-of-hydrogen-energy/.

Zeng, L., Keshavarz, A., Xie, Q., & Iglauer, S. (2022). Hydrogen storage in Majiagou carbonate reservoir in China: Geochemical modelling on carbonate dissolution and hydrogen loss. *International Journal of Hydrogen Energy, 47*(59), 24861–24870. Available from https://doi.org/10.1016/j.ijhydene.2022.05.247, http://www.journals.elsevier.com/international-journal-of-hydrogen-energy/.

Zeng, L., Keshavarz, A., Nilesh Kumar, J., Al-Yaseri, A., Sarmadivaleh, M., Xie, Q., & Iglauer, S. (2023). Geochemical modelling of hydrogen wettability on Quartz: Implications for underground hydrogen storage in sandstone reservoirs. *Journal of Molecular Liquids., 371*. Available from https://doi.org/10.1016/j.molliq.2022.121076.

Zhan, S., Zeng, L., Al-Yaseri, A., Sarmadivaleh, M., & Xie, Q. (2024). Geochemical modelling on the role of redox reactions during hydrogen underground storage in porous media. *International Journal of Hydrogen Energy, 50*, 19–35. Available from https://doi.org/10.1016/j.ijhydene.2023.06.153, https://www.sciencedirect.com/science/journal/03603199.

Zivar, D., Kumar, S., & Foroozesh, J. (2020). Underground hydrogen storage: A comprehensive review. Available from https://doi.org/10.1016/j.ijhydene.2020.08.138.

CHAPTER

10

Microbial considerations/aspects of underground hydrogen storage

Tausif Ahmad[1], *Narendra Kumar*[2] and *Muhammad Mubashir*[3]

[1]Physical Science and Engineering Division, King Abdullah University of Science and Technology (KAUST), Thuwal, Saudi Arabia [2]Ali I. Al-Naimi Petroleum Engineering Research Center, King Abdullah University of Science and Technology, Thuwal, Saudi Arabia [3]Water Technologies Innovation Institute & Research Advancement, Saudi Water Authority, WTIIRA-SWA, Jubail, Saudi Arabia

10.1 Introduction

One potential strategy for storing hydrogen (H_2) on a large scale is the utilization of geological formations for underground storage. Various options are available for subsurface storage of H_2, including depleted oil and gas reservoirs, deep aquifers, abandoned mines, artificial salt, and hard rock caverns (Caglayan et al., 2020; Flesch et al., 2018; Heinemann et al., 2018). The majority of underground structures are not devoid of life and instead serve as habitats for a wide range of varied microbiological species (Varjani & Gnansounou, 2017). In a situation of underground hydrogen storage (UHS), the "microbial biosphere" will come into direct contact with the stored H_2, and the activity of these microorganisms will significantly impact the functioning of several operational systems dependent on the stored H_2. For the feasibility of H_2 storage, the presence of microorganisms in reservoirs is one of the most crucial factors that must be considered. In the subsurface, microbes can either be anthropogenically introduced by the subsurface exploration of human activities such as drilling, pumping, or mining, or they can be indigenous through natural transport or sedimentation processes. Most of the time, there is not enough study available on the immediate and long-term consequences of subsurface operations on natural ecosystems. A long-lasting transformation of the microbial diversity and activity will result from human actions in the subterranean, which modifies its condition (Vigneron

Subsurface Hydrogen Energy Storage
DOI: https://doi.org/10.1016/B978-0-443-24071-3.00011-X

et al., 2017). The microbial interaction with H_2 depends on decisive variables such as temperature, substrate availability, pH, and anaerobic environments, with a crucial role played by the hydrogenase enzymes in catalyzing these interactions. The irreversible damage caused by the microbial activity on the geological H_2 storage is the permanent loss of H_2 characteristics resulting from its conversion into by-products such as methane (CH_4) and hydrogen sulfide (H_2S) (Fig. 10.1). The ongoing proliferation of microorganisms within the reservoir has the potential to alter the quality of the stored H_2 and induce tubular corrosion as well as pore blockage, leading to a reduction in the injectivity of H_2.

Bacteria and archaea are the predominant categories of unicellular microorganisms, which play a pivotal role in facilitating a wide array of biochemical and geological phenomena (Hoehler & Jørgensen, 2013). Microbial entities have been discovered beneath the rock formations at considerable depths, spanning many Kilometers. The concentration of these microorganisms has been reported to range from 10^4 to 10^8 cells in a gram of rock, culminating in an anticipated aggregate count of approximately 2×10^{29}–6×10^{29} cells (Dutta et al., 2018; Magnabosco et al., 2018). Microbial life exhibits a temperature tolerance spanning from $-15°C$ to $121°C$. The pH range within which microbial life can thrive extends from zero (0) to 11, with notable variations in the cell abundance and the diversity observed, particularly within the pH range of 6–7. Furthermore, no discernible constraints pertaining to pressure or brine salinity impede microbial survival (Flesch et al., 2018). Additional factors, such as elevated levels of hazardous substances, reduced water activity, increased radiation exposure, and decreased rock permeability, can exert a substantial impact and diminish the abundance of microbes (Halim et al., 2014; Jagger, 1983). All these aspects must be considered while estimating microbial activity in underground habitats.

Depleted hydrocarbon reservoirs appear as suitable options for UHS. Given their ubiquitous presence in the upper lithosphere, microorganisms are also anticipated to be found within subsurface geological formations that are being considered for potential H_2 storage. The high microbial activity possibly oxidizes H_2, which is a highly energetic substrate in the environment. This leads to H_2 quantity and quality loss owing to the occurrence of the unwanted reaction causing the generation of H_2S, CH_4, and other phenomena.

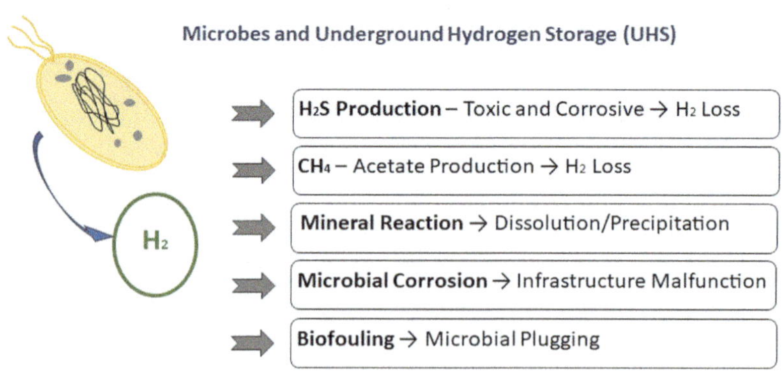

FIGURE 10.1 Microbial interaction with hydrogen during underground hydrogen storage.

The ingestion of H_2 in the subsurface is influenced by two mechanisms, namely chemical reactions involving biotic and abiotic components (Gregory et al., 2019; Reitenbach et al., 2015). The abiotic components of an ecosystem encompass nonliving entities, including water, rock minerals, and gases. In contrast, the biotic components consist of living organisms, such as bacteria (Panfilov, 2016). The required operational temperature for an abiotic process may reach a maximum of 600°C. However, the biotic process exhibits a lower temperature range that aligns closely with the prevailing temperatures typically found in the storage sites. The abiotic process takes place under high-temperature conditions, which are not suitable for storage sites. However, a study documented methanation, which is the process of H_2 conversion to CH_4, occurring at a temperature of 50°C in the Chimaera seep located in Turkey (Etiope et al., 2011).

The level of microbial activity observed in man-made salt caverns is comparatively lower in comparison to other types of storage sites. The limited surface area of the cavern inhibits the formation of biofilm and subsequent blockage. The presence of salt and sulfur within the salt cavern with high salinity levels leads to the imposition of osmotic stress on cells, hence causing a significant reduction in microbial diversity (Yekta et al., 2018). Halobacterium is an organism that has undergone significant evolutionary adaptations, enabling it to thrive in environments with high salinity levels, reaching concentrations as high as 100–150 g/L. The presence of sulfate in salt caves with brine indicates a significant risk of microbiological H_2S formation, as evidenced by the liquid–gas interphase (Dopffel et al., 2021).

The existing works in this area are limited and sometimes yield conflicting or incomplete findings on the microbiological, permeability, and physicochemical conditions under which a specific observation was made. For instance, at the Altmark hydrogen storage test site, the conditions included a temperature of 80°C, a NaCl salinity of 7.2 M, and a pressure of 20 MPa. The geological composition consisted of sandstone containing clay minerals, anhydrite, and dolomite. The gas composition was comprised of 22.6% H_2, 4.3% He, 8.3% CO_2, and 64.8% N_2. It was hypothesized that the precipitation of minerals filling the pores was attributed to microbial activity, as biofilms were observed. However, no changes in permeability were reported (Thaysen et al., 2023). Similarly, the studies on the gas storage site located in Ketzin mention permeability changes due to microbial activity. The research conducted on the Lobodice storage site in the Czech Republic focused on porous medium storage. It depicted that at a relatively low temperature of 35°C, approximately 45%–60% of the H_2 present in the stored town gas underwent microbiological conversion into CH_4 or H_2S. A study conducted by Šmigáň et al. showed a reduction in reservoir pressure leads to a rise in the abundance of methanogenic microbial cells (Šmigáň et al., 1990).

The laboratory cultivation of methanogens has been seen to facilitate the conversion of H_2 into CH_4. Another example of porous media storage may be found at the previous town gas storage site in Ketzin, Germany. This case study documented various phenomena, including total gas volume reduction, corrosion, alterations in permeability, and modifications in gas composition. Specifically, a decrease in carbon monoxide (CO) levels was accompanied by an increase in H_2, CH_4, and carbon dioxide (CO_2) concentrations (Thiyagarajan et al., 2022). Divergent results were documented about the storage of town gas in the saline aquifer at Beynes, France, with no instances of operational challenges or H_2 losses being observed (Strobel et al., 2020).

10.2 Microbial impact on subterranean hydrogen storage

UHS presents an assuring opportunity for addressing a key challenge in the realm of renewable energy—the disparity between energy generation and usage arising from the seasonal variations in atmospheric conditions. The excess electricity produced renewably has the potential to be transformed into hydrogen, serving as an energy carrier through the water electrolysis process (Khajooie et al., 2023). Subsequently, this hydrogen can be reintegrated into the energy grid during periods of high demand. Various subsurface opportunities exist in fabricated salt caverns, emptied reservoirs of oil and gas, solid rock caverns, deep aquifers, and disused mine formations for storing H_2 (Fig. 10.2) (Dopffel et al., 2021). With their inherent chemical characteristics and the extensive operational knowledge derived from natural gas management, man-made cavities within salt formations are suggested as highly effective storage sites (Schwab et al., 2022). Table 10.1 summarizes a concise overview of the primary microbial outlooks encountered during subsurface H_2 storage, as well as the possible associated issues.

10.3 Microbial metabolism of hydrogen

The chemical inactivity of H_2 at reservoir pressures and temperatures is mostly attributed to the strong electron-proton bonding within the H_2 atom. However, it is worth noting that certain microorganisms found in rocks can utilize H_2 under specific circumstances. Microorganisms can derive energy through the process of oxidizing electron donors while simultaneously

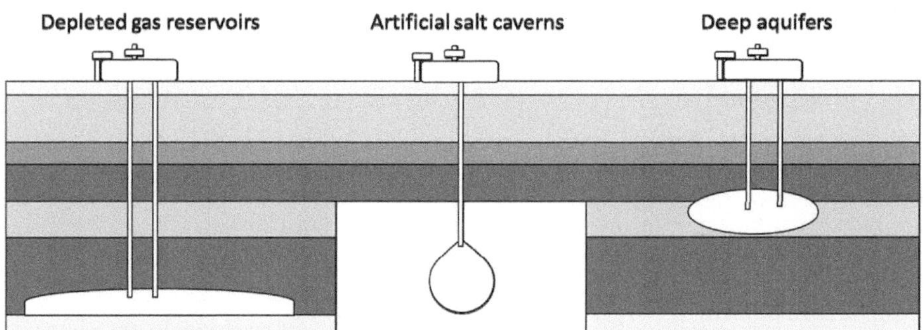

	Depleted gas reservoirs	Artificial salt caverns	Deep aquifers
Temperature range	high range	mostly 20 – 35°C	7 – 174 °C
Salinity	high range	in sump up to saturation	5 – 52.000 ppm
Experience for H_2 storage	low	high	very low
Knowledge on microbial activity during H_2 storage	low	very low	low

FIGURE 10.2 Diagram of the three main subterranean hydrogen storage methods showing the critical microbial activity parameters. Source: *With permission from Dopffel, N., Jansen, S., & Gerritse, J. (2021). Microbial side effects of underground hydrogen storage – Knowledge gaps, risks and opportunities for successful implementation.* International Journal of Hydrogen Energy, 46*(12), 8594–8606. https://doi.org/10.1016/j.ijhydene.2020.12.058.*

TABLE 10.1 List of various microorganism species influencing underground hydrogen storage UHS (Gregory et al., 2014).

Microbes	Type	Phylum	Species	Possibilities in aquifers
Prokaryote	Bacteria	β-Proteobacteria	*Rhodoferax ferrireducens*	Iron reduction
			Gallionella spp.	Iron oxidation & formation of iron deposits that induced plugging
		γ-Proteobacteria	*Escherichia coli*	Pathogenic bacteria
		Actinobacteria	*Rhodococcus* spp.	Degradation of hydrocarbons and polychlorinated biphenyls
Prokaryote	Archaea	Crenarchaeota	*Nitrosopumilus maritimus*	Ammonia oxidation
		Euryarchaeota	*Methanosaeta* spp.	Methanogenesis
Eukaryotic	Protists	Rhizaria	*Cercomonas* spp.	Grazing bacterial biofilms
Virus		Group IV virus (N.B. not a phylum)	Hepatits A virus	Pathogenic virus

From Gregory et al. (2014). Microbial communities in UK aquifers: Current understanding and future research needs. Quarterly Journal of Engineering Geology and Hydrogeology, *47(2), 145–157. https://doi.org/10.1144/qjegh2013-059.*

reducing an electron acceptor. The water and/or rock can contain electron donors, such as various acids (e.g., formate or acetate) as simple organic molecules or more complicated organic compounds like plant material and hydrocarbons. Additionally, inorganic substances can also serve as electron donors. Molecular H_2 has significant importance as an electron donor for microbial respiration in underground environments. Due to its notably low reduction potential, H_2 serves as a viable energy source for several groups of organisms with distinct metabolic characteristics (Gregory et al., 2019). From a technical standpoint, it can be stated that bacteria do not engage in the direct consumption of H_2. Instead, they derive their sustenance from the energy generated through a redox reaction facilitated by the bacterium involving H_2 and other entities.

10.3.1 Bacterial metabolism

Bacterial metabolism can be classified into two distinct types. One of the metabolic processes is respiratory metabolism, often known as respiration, which involves the initiation of redox processes within the microorganism's body to harness the energy released during these reactions. Another metabolic process is constructive metabolism, sometimes known as nutrition. This metabolic process is alternatively referred to as carbon fixation or biomass formation. The concept of respiratory metabolism can be shown by Sabatier's reaction.

$$4H_2 + CO_2 = CH_4 + 2H_2O \tag{10.1}$$

Initially, the bacterial cell facilitates the dissolution of H_2 and CO_2 molecules. CO_2 undergoes dissociation, resulting in the formation of COH^+ cations. These cations have the ability to take electrons, making them potential electron acceptors. Bacterial proteins play a role in promoting the separation of an electron from a H_2 atom. The interaction between the free electron and cations is characterized by a multifaceted series of events, ultimately leading to the production of CH_4, water, and the liberation of energy. The aforementioned process is typically denoted as respiratory metabolism or, in a more concise manner, bacterial respiration. Bacteria, in conjunction with H_2, necessitate the utilization of electron acceptors, particularly cations. CO_2 is not the sole material employed for this purpose; other cations such as sulfates found in reservoir water or iron-III ($Fe_3{}^+$) commonly present in reservoir rocks are also utilized. The occurrence of redox reactions is a consequence of respiratory metabolism, wherein H_2 acts as the reductant by donating its electrons. The bacterial-mediated processes exhibit rapid kinetics and facilitate the conversion of H_2 into various chemical compounds.

The purpose of constructive metabolism is to facilitate the generation of biomass within the bacterial population. The process involves the uptake of carbon from external sources, whether organic or nonorganic, and its subsequent conversion into organic compounds that are utilized for the synthesis of biomass. Certain bacteria only utilize inorganic CO_2 as their source of carbon. These bacteria are referred to as autotrophic, as they possess the ability to synthesize organic compounds through their own metabolic processes. There are individuals that utilize organic matter, whereas another group can utilize both organic and nonorganic substances. The cellular mechanisms involved in converting CO_2 into organic compounds are collectively referred to as carbon fixation. The majority of methanogenic, acetogenic, and sulfate-reducing bacteria (SRB) exhibit autotrophic characteristics. Iron-reducing bacteria (IRB) employ a combination of organic materials and CO_2 as substrates for the generation of biomass. The process of carbon fixation results in the augmentation of bacterial biomass and size. Once the cellular mass surpasses a crucial threshold, it undergoes cellular division, known as a dichotomy, which is preceded by the process of DNA replication. On a global scale, the rate of biomass production is proportional to the growth of bacterial populations. There is a significant interdependence between respiration and biomass production. It is true that bacteria are unable to generate biomass in the absence of concurrent respiration, and they are also unable to respire if there is a lack of carbon available for fixation. In an abiotic context, most reactions that involve the oxidation of H_2 tend to exhibit slowness; nevertheless, these processes are accelerated by microbial enzymes that serve as catalysts (Hoehler & Jørgensen, 2013). In respiratory metabolism, methanogenesis, acetogenesis, sulfate reduction, and iron-III reduction are among the commonly observed hydrogenotrophic biotic processes that can occur within subterranean gas storage wells and are schematically represented in Fig. 10.3.

10.3.1.1 Methanogenesis

All methanogenic archaea exhibit obligate anaerobic characteristics. There are three taxonomic groups that can be identified, specifically Methanobacteriales, Methanococcales, and Methanomicrobiales, which combined comprise a total of more than 40 species. The dimensions of several of them display values that are less than 0.45 μm. The presence of oxygen at a concentration of 0.004% entirely inhibits the growth of the organisms. The preferred conditions for methanogenesis growth include a pressure of 90 bar (1305.34 Psi) and

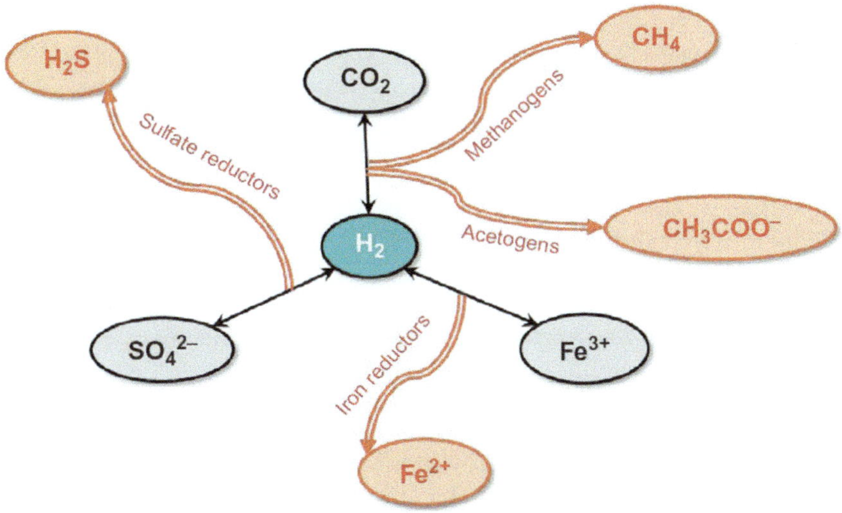

FIGURE 10.3 Hydrogenotrophic microorganisms in underground hydrogen storage. Source: *With permission from Panfilov, M. (2016).* Underground and pipeline hydrogen storage *(pp. 91–115). Elsevier BV. https://doi.org/10.1016/b978-1-78242-362-1.00004-3.*

a temperature range of 30°C–40°C, which is classified as mesophilic (Reitenbach et al., 2015). However, it is noteworthy that these organisms can still exhibit growth even at a higher temperature of 122°C (Thaysen et al., 2021). The archaea bacterium, belonging to the methanogenic bacteria family, exhibits robust growth under these specific temperature conditions. In contrast to other species, this particular organism possesses the ability to inhabit underground environments characterized by significant depths devoid of oxygen (Reitenbach et al., 2015). The reaction of methanogenesis can be exemplified by Sabatier's reaction (Eq. 10.1).

The Lobodice Town gas storage project serves as a prominent illustration of the conversion process from H_2 to CH_4, facilitated by methanogenesis. This project involves the substantial transformation of H_2 into CH_4 (Šmigáň et al., 1990). The group of acetogenic bacteria includes several species, such as *Clostridium aceticum, Clostridium thermoautotrophicum, Sporomusa sphaeroides, Butyribacterium methylotrophicum, Acetogenium kivui, S. ovata, Acetobacterium woodii*, and various other species.

10.3.1.2 Acetogenesis

Acetogenic bacteria exhibit maximum activity under identical pressure and temperature conditions as methanogens. During the acetogenesis process, the conversion of acetate to acetic acid can occur via the reaction depicted below (Panfilov, 2016).

$$4H_2 + 2CO_2 = CH_3COOH + 2H_2O \tag{10.2}$$

Acetogenesis, a metabolic route led by anaerobic bacteria (acetogen bacteria) occurring in various environments, plays a crucial role in the carbon cycle. This biological process can potentially cause huge economic losses and other harmful environmental consequences.

10.3.1.3 *Sulfate reduction*

SRB are a type of microorganism that utilize H_2 as a source of energy and generate H_2S within subterranean hydrocarbon reservoirs. The preferred temperature range for the activity of SRB is approximately 38°C (Bernardez et al., 2013). However, there have been reports of SRB activity occurring at higher temperatures, reaching up to 110°C (Zivar et al., 2021). The identification of sulfate-reducing Archaea, known as Archaeoglobus, was first reported in 1987 (Panfilov, 2016). SRB is a type of anaerobic and versatile bacteria that has the potential to inflict significant harm to injection and/or production facilities because of H_2S production. Water is produced as a byproduct during the SRB process. In addition to the drawbacks associated with production, the generation of water may pose an additional disadvantage due to its potential to increase water saturation. This, in turn, might lead to reduced H_2 saturation, limiting the available area for H_2 storage.

The bacterial activity in hydrocarbon reservoirs is documented, particularly in cases where water that is incompatible with the formation is injected. An increase in production has been recorded in the Las Heras-Cerro Grande oilfield located in the Gulf of San Jorge Basin, Argentina, because of water flooding. This rise can be attributed to the presence of active SRB within the reservoir (Cavallaro et al., 2005). The phenomenon of H_2 consumption by SRB in the presence of sulfate ions, resulting in the formation of H_2S, can be described by the following chemical reaction (Hemme & van Berk, 2018; Liebscher et al., 2016).

$$4H_2 + SO_4^{2-} + 2H^+ = 4H_2O + H_2S \tag{10.3}$$

10.3.1.4 *Iron-III reduction*

IRB are known to contribute to the consumption of H_2 in subterranean reservoirs. The reduction of iron-III is restricted to the presence of iron oxides and minerals containing iron, like chlorite and smectite. Additionally, organic carbon must be accessible, as dissimilatory iron-reducing bacteria (DIRB) operate as obligate heterotrophs, meaning they generate cellular carbon from organic compounds (Hernsdorf et al., 2017). The bacterial species under consideration are Geobacter metallireducens and Shewanella putrefaciens. The iron component (Fe_3O_4) undergoes a reaction with H_2 in this process. The interaction between the IRB and minerals found in rocks has the potential to produce a significant amount of water. The reaction is presented as follows:

$$H_2 + Fe_2O_3 = H_2O + Fe_3O_4 \tag{10.4}$$

The water production per unit mass of H_2 is twice as high in the reactions of SRB and IRB compared to the reactions of methanogenesis or acetogenesis. The presence of an abundance of water may serve as an indication of the occurrence of processes involving SRB and IRB. The relationships between the different variables in microbial activities are depicted through a workflow format in Fig. 10.4.

10.3.2 Hydrogen oxidation due to microbial activities

The H_2 oxidizing processes can be classified based on their H_2 threshold and standard free energy change ($\Delta G^{\circ\prime}$), which are two valuable parameters for comparing the probability and sequence of reactions (Table 10.2). The H_2 threshold is a parameter that establishes the

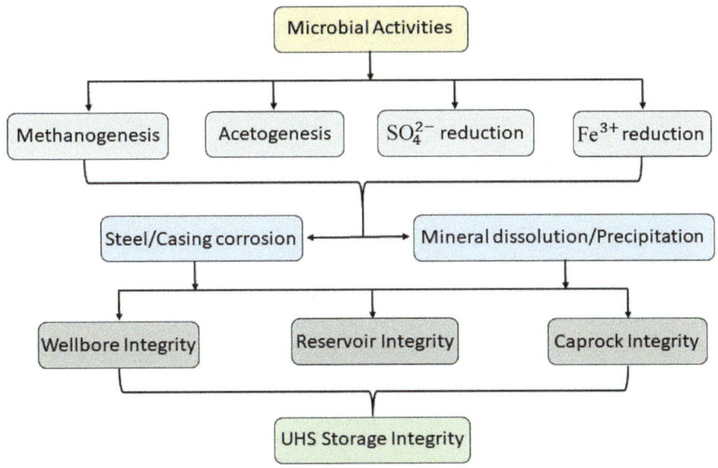

FIGURE 10.4 Illustrated is the sequential arrangement of connections between the identified variables that influence storage integrity during underground hydrogen storage.

minimum concentration of H_2 at which it ceases to be consumed. Assuming all other variables are held constant, it may be inferred that the microbial community exhibiting the lowest threshold value for H_2 is likely to achieve the highest level of success in outcompeting other populations for H_2 (Lovley & Goodwin, 1988). The $\Delta G^{\circ\prime}$ value represents the thermodynamic propensity of a reaction under standard conditions, specifically at standard temperature and pressure, pH 7, and with all reactants present at a concentration of 1 M. In conditions characterized by high pressure and temperature and low nutritional levels, the $\Delta G^{\circ\prime}$ values can be employed to ascertain the relative energy favorability of different reactions. As seen by the data presented in Table 10.2, it is observed that a decrease in $\Delta G^{\circ\prime}$ values (indicating an increase in accessible free energy) is typically associated with a decrease in H_2 thresholds.

Hydrogenotrophic sulfate reduction, hydrogenotrophic methanogenesis, and homoacetogenesis are metabolic processes that necessitate the highest threshold concentration of H_2. These processes are also characterized by having the lowest "ΔG°," as shown in Table 10.2. The microorganisms capable of reducing sulfur species, including sulfate, sulfide, and thiosulfate, are collectively referred to as sulfur species-reducing microorganisms (SSRM) and are involved in the process of sulfur species reduction (SSR) (Findlay, 2016). The ability of sulfur to be directly respired is constrained due to its minimal solubility (1.6×10^{-7}), thus necessitating cell adherence to sulfur particles for the process to occur. Nonetheless, sulfur readily combines with sulfide generated during sulfate reduction, giving rise to readily metabolizable polysulfides. Iron oxides are found abundantly in numerous sediments and aquifers; however, they are generally absent in carbon-rich oil fields because they have undergone reduction over millions of years and are not renewed (Lovley, 1991). Also, merely detecting iron reduction by bacteria in a laboratory, where they are provided with a medium enriched for DIRB, does not automatically imply that these bacteria will carry out iron reduction in their natural surroundings. The cell numbers frequently fall within the range of low to intermediate (10–100 cells/mL), potentially encompassing nonhydrogenotrophic species.

TABLE 10.2 Oxidation processes that consume H_2, along with their respective free energy yields ($\Delta G^{\circ\prime}$) and observed H_2 thresholds.

H_2-oxidizing process	Reactions	H_2 threshold (nM)	$\Delta G^{o\prime}$ (KJ mol/H_2)	Typical ambient $[H_2]$ (nmol/L)	Relevance for H_2 storage
Chromate Reduction	$\frac{1}{2}H_2 + \frac{1}{3}CrO_4^{2-} + \frac{5}{3}H^+ \rightarrow \frac{4}{3}H_2O$	< 0.1	NA	NA	low
Aerobic hydrogen oxidation (Knallgas)	$H_2 + \frac{1}{2}H_2 \rightarrow H_2$	0.051	− 237	NA	low
Iron (III) reduction	H_2 + ferric(oxy)hydroxides → ferrous iron + H_2O	< 0.11–0.8	− 228.3–182.5–114	0.2 0.2–1	Intermediate
Denitrification	$H_2 + \frac{2}{5}H^+ + \frac{2}{5}NO_3^{2-} \rightarrow \frac{1}{5}N_2 + \frac{6}{5}H_2O$	< 0.05–0.5	− 240.1–224	< 0.05	low
Hydrogenotrophic sulfate reduction	$4H_2 + SO_4^{2-} + H^+ \rightarrow HS^- + 4H_2O$	1–15	− 38–48 − 57	1–2	High
Halorespiration	H_2 + halogenated compounds → dehalogentated compounds + HCl	0.05–0.27 < 0.3 0.27–2	− 230– − 187	NA	low
Sulfur reduction	$H_2 + S \rightarrow HS^- + H^+$	2500	− 33.1	NA	Intermediate
Manganese (IV) reduction	$H_2 + MnO_2 \rightarrow Mn(OH)_2 + 2H_2O$	< 0.05	− 163	< 0.05	low
Hydrogenotrophic methanogenesis	$4H_2 + CO_2 \rightarrow CH_4 + 2H_2O$	0.4–95	− 34–43.9	5–10 7–13	High
Arsenate reduction	$H_2 + HAsO_4^{2-} + 2H^+ \rightarrow H_3AsO_3 + H_2O$	0.03–0.09	− 162.4	0.4–0.7	low
Ammonification	$4H_2 + 2H^+ + NO_3^- \rightarrow NH_4^+ + 3H_2O$	0.015–0.025	− 150	< 0.05	low
Homoacetogenesis	$4H_2 + 2CO_2 \rightarrow CH_3COOH + 2H_2$	328–3640	− 26–36.1	100 < 117–150	High
Fumarate reduction	H_2 + fumarate → succinate	0.015	− 86.2	NA	low

NA, Not available. Excluded are the reduction processes of cobalt, uranium, selenium, technetium, and vanadium, as they hold limited significance for H_2 storage.

From Thaysen, E. M., McMahon, S., Strobel, G. J., Butler, I. B., Ngwenya, B. T., Heinemann N., Wilkinson, M., Hassanpouryouzband, A., McDermott, C. I., & Edlmann, K. (2021). Estimating microbial growth and hydrogen consumption in hydrogen storage in porous media. Renewable and Sustainable Energy Reviews, 151, *https://doi.org/10.1016/j.rser.2021.111481 https://www.journals.elsevier.com/renewable-and-sustainable-energy-reviews.*

In the subsurface environment, oxygen and nitrate are limited, causing aerobic H_2 oxidation, denitrification, and ammonification to gain significance primarily when aquifer contamination arises, such as through drilling fluid intrusion (Hallbeck & Pedersen, 2008; Ranchou-Peyruse et al., 2017). Numerous DIRBs, along with a limited number of SSRM, have the capability to link H_2 oxidation with the reduction of various other trace metal oxides, such as MnO_4^{2-}/MnO_2, Co, CrO_4^{2-}, VO_4^-, SeO_4^{2-} and AsO_3^-. Hence, these compounds have minimal presence in the surroundings, resulting in their limited relevance to influencing H_2 storage (Jiang et al., 2018; Muyzer & Stams, 2008).

10.3.3 The impact of environmental factors on microbial growth

The rates of microbial growth and consumption of H_2 exhibit variability in response to fluctuations in nutrition supply and environmental factors (Gregory et al., 2019; Ver Eecke et al., 2013). Organisms can grow under conditions that are optimal for their growth. However, when exposed to conditions that deviate from the optimum, their growth rate may be slowed or they may enter a dormant state. This section focuses on the requirements for nutrients and geological factors such as temperature, salinity, pH, and pressure on the growth of the microbial.

An appealing environment is provided by porous media for bacteria to inhabit. The interplay of population growth, microbial communications, and passage within the porous media presents valuable prospects to enhance both industrial development and environmental practices. Fig. 10.5 illustrates a schematic representation of the specific processes that contribute to the deposition and separation of biofilms within the permeable channels.

10.3.3.1 Nutrients

The nutritional needs of microorganisms that oxidize H_2 are not well understood. Frequently, the examination of a restricted selection of individual strains from the distinct metabolic groups has been conducted, which is improbable to encompass the entirety of strains within those groups. In the following section, we provide a concise overview of the limited information that is currently available. In addition to water with adequate thermodynamic activity (as discussed in Section 10.3.3.4), hydrogenotrophs require H_2 as an electron acceptor and a carbon supply for cellular division. Furthermore, they require a range of macro and trace elements, along with diverse organic nutrients (Usher et al., 2015).

Nutrients have the potential to be absorbed from the solution or obtained directly from minerals, with the latter proving especially vital in oligotrophic environments. It is currently uncertain to what extent subsurface microbial communities rely on mineral weathering. However, mineral weathering leads to the liberation of crucial elements such as sulfur, carbon, iron, and phosphorus. Regarding soils, an analysis revealed that over half of the 1100 microbial strains exhibited the potential for mineral weathering, as demonstrated by their capacity to mineralize the biotite (Huang et al., 2014).

Methanogens and homoacetogens exhibit versatility in their growth, as they can adapt to both autotrophic and heterotrophic modes, and certain species can even adopt mixotrophic strategies. However, limited knowledge exists concerning the variations in nutrient necessities among functional groups and the diversity in nutrient demands within a single

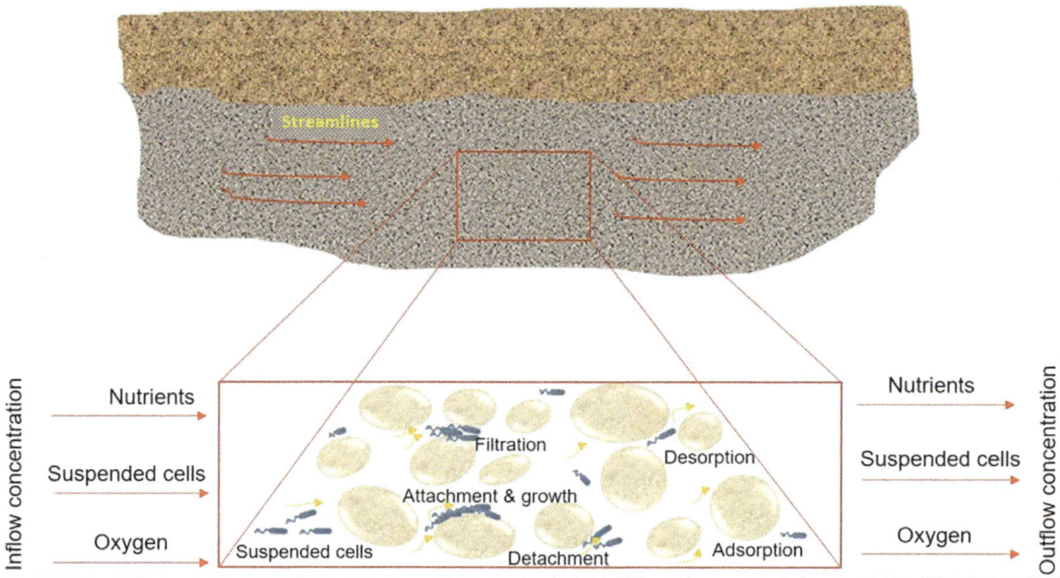

FIGURE 10.5 The phenomena driving the buildup and release of biofilms within the porous media. Source: *From Hassannayebi, N. (2019).* An assessment of underground hydrogen storage: transport, geochemistry, and bioactivity, *Montanuniversität. (Original work published 2019).*

functional group. While most SSRM generally exhibit heterotrophic growth, certain species are capable of autotrophic or mixotrophic growth with higher iron requirements (1.8×10^{-6} M) (Herbert et al., 1985). Methanogens belonging to the Methanosarcinae order achieve optimal growth at 29.6 mM Mg concentration, but their growth is restricted when the Mg concentration reaches 16.5 mM (Chen, Cheng, & Creamer, 2008). Under ideal circumstances, autotrophic organisms' growth rate might be constrained by the rate of H_2 and CO_2 transfer from gas to liquid. This was demonstrated in *Methanobacterium thermoautotrophicum* methanogens and *Desulfotomaculum* sp. sulfate reducers (Taylor & Pirt, 1977; Van Houten et al., 1997).

10.3.3.2 Temperature

Microorganisms are classified according to their temperature preference for growth (Thaysen et al., 2021). Temperature-dependent microbial growth can be categorized as follows:

- Psychrophiles—microorganisms that demonstrate the highest rate of growth at temperatures below 20°C.
- Psychotrophs—microorganisms capable of thriving at or above 20°C and can tolerate temperatures as low as 5°C.
- Mesophiles—microorganisms that grow within a temperature range of 20 to 45°C.
- Thermophiles—microorganisms that flourish at temperatures above 45°C–50°C.
- Hyperthermophiles—microorganisms that exhibit optimal growth at temperatures of 80°C or higher.

Elevated temperatures induce modifications in the energetic characteristics, such as vibrational modes, of biomolecules when immersed in an aqueous solvent. Additionally, high temperatures also impact the solubility or viscosity of the substrate and modify the ionization state of the aqueous medium (Jaenicke & Sterner, 2006). Thermophiles and hyperthermophiles face the issue of increased reaction rates at elevated temperatures. This situation suggests that abiotic reaction rates are sufficiently rapid, rendering no discernible advantage to the microbe in catalyzing the reaction (Amend & Shock, 2001). Consequently, it is believed that microbes adapted to high temperatures possess enzymes that exhibit accelerated rates of reaction. The majority of hydrogenotrophic methanogens that have been studied are mesophiles. However, it is worth noting that the ideal growth temperatures for methanogens can vary significantly, spanning a range of 30°C–40°C. However, Methanopyrus kandleri, a species of methanogen, has been seen to exhibit growth at temperatures as high as 122°C. Optimal growth temperatures for homoacetogens usually fall within the range of 20°C–30°C. However, thermophilic growth of *Moorella mulderi* and *Acetogenium kivui* occurs at temperatures of 60°C or higher (Schwab et al., 2022). SSRM generally exhibits optimal growth temperatures within the range of 20°C–30°C or 50°C–70°C. It is noteworthy that sulfur-reducing archaea tend to possess greater optimal growth temperatures compared to sulfur and SRB (Thaysen et al., 2021).

10.3.3.3 Pressure

Microbial cellular activity is often less constrained by the considerable pressures within the crustal pore spaces compared to the substantial temperatures. This is attributed partly to the relatively elevated osmotic pressure within the cytoplasm, especially among thermophiles and hyperthermophiles. There is currently no defined upper-pressure threshold for microbial existence. H_2 storage pressure thresholds span to 5–20 MPa or 1–50 MPa, and under such high pressures, thriving necessitates adjustments for maintaining homeostasis (Salvador-Castell et al., 2020). The growth of different mesophilic microorganisms accustomed to atmospheric pressure is restrained at 30–50 MPa. On the other hand, pressure impacts typically support the growth of hyperthermophiles. However, to sustain a liquid environment beyond 100°C, increased pressures are necessary. Increased pressure can raise the growth temperature maximum limit by 2°C–12°C compared to lower pressure ranges of 0.1–3 MPa (Schwab et al., 2022). The SSRM Desulfovibrio Indonesians exhibited a temperature-dependent pressure response while maintaining comparable growth rates at 45°C and a diminished growth rate at 30 MPa and 20°C compared to 0.1 MPa. Nevertheless, in empirical studies, pressure tolerance is not commonly encompassed in the description of a microorganism's environmental growth restrictions.

10.3.3.4 Salinity

The appropriate concentration range of salt for H_2 storage is between 0 and 5 M NaCl (Shi et al., 2020). Microorganisms are categorized based on their ability to tolerate varying levels of salt. Non-halophilic bacteria exhibit growth up to a concentration of 0.2 M NaCl. Slight halophiles, on the other hand, demonstrate growth within the range of 0.2–0.5 M NaCl. Moderate halophiles thrive in environments with NaCl concentrations ranging from 0.5 to 2.5 M. Lastly, extreme halophiles, which prefer hypersaline conditions, exhibit optimal growth in medium containing 2.5–5.2 M NaCl (Oren, 2013). Microorganisms residing in high salt concentrations experience osmotic stress, necessitating the maintenance of their intracellular environment to be at least isosmotic with the external environment. There are two primary techniques that can be employed to attain osmotic balance: the salt-in strategy and the exclusion of salt coupled with the manufacture and accumulation of organic "compatible" solutes (Oren, 2011). The in-depth analysis of various metabolic pathways is discussed somewhere else (Oren, 2011). It is frequently observed that the tolerance or necessity for salt is increased at elevated temperatures (Oren, 2013). However, there exist numerous instances of mesophilic halophiles that contradict this trend. The majority of hydrogenotrophic methanogens that have been cultured have a preference for salt concentrations of up to 0.77 M NaCl, which is about equivalent to the salinity of seawater (Thaysen et al., 2021). The vast majority of developed SSRM exhibit optimal growth at low salinity conditions, typically ranging from 0 to 0.4 M NaCl. Nevertheless, it is worth noting that a total of thirty-four SSRM (all of which are mesophiles) have been found to possess upper salinity limits for growth that are equal to or more than 1.7 M NaCl (Thaysen et al., 2021).

Dissolved ions in natural brines interact intricately, potentially creating challenges for brine habitability like high ionic strength, low water activity (a_w), and chaotropy (disruption of H_2 bonds in water molecules) either individually or combined. Many bacteria thrive at an a_w of approximately 0.98, similar to seawater, but only a limited number of species can flourish at an a_w of 0.96 or less (Payler et al., 2019). Chaotropic agents like $CaCl_2$, $MgCl_2$, LiBr, $FeCl_3$, and KI seem to have a greater influence than a_w in restricting microbial life, in contrast with examples of kosmotropic agents such as KCl, $MgSO_4$, NaCl, $FeSO_4$, and Na_2SO_4. However, the interactions and complexities among chaotropic or kosmotropic agents, aw, and other brine physicochemical characteristics remain largely uncharted.

10.3.3.5 pH

Underground microbial action is critically impacted by pH conditions and is known to influence the quality of the stored H_2 during UHS (Groenenberg et al., 2020). Microbes thrive between pH levels 0 and 11, with a peak population at pH 6–7 (Flesch et al., 2018). The pH of brine can influence microorganism growth through two routes: (1) directly influencing growth metabolic rate and (2) affecting redox reactions. Several methanogens, homoacetogens, and SSRM thrive at a pH range of 6.5–7.5; however, most SSRM and methanogens typically have growth limitations beyond a pH range of 4–9.5 (O'Flaherty et al., 1998). Eighteen SSRM, like *Desulfonatronovibrio hydrogenovorans* and *Desulfurispira natronophila*, seven methanogens, including *Methanocalculus natronophilus* and alkaliphilus, and six notable homoacetogenic forms like *Fuchsiella alkaliacetigena* and *Clostridium ultunense*, can thrive in highly alkaline environments above pH 10 (Thaysen et al., 2021).

10.3.3.6 Oxygen concentration

Organisms are sorted based on their capability to thrive in the presence of oxygen. The thriving capacity of aerobe microorganisms in atmospheric oxygen is in contrast to that of anaerobic microorganisms which can flourish in oxygen absence. Multi-cellular organisms that solely rely on atmospheric oxygen for their growth are classified as obligate aerobes. Consistent growth is demonstrated by aerotolerant anaerobes regardless of oxygen availability. Organisms known as facultative anaerobes can grow in oxygen absence; however, they experience improved growth in its presence. For the growth of microaerophiles, they necessitate 2%–10%. oxygen levels. Conversely, strictly anaerobic organisms, or obligate anaerobes, demonstrate a complete inability to withstand oxygen and experience fatality upon exposure to it (Hassannayebi, 2019).

10.3.3.7 Inhibitors

At 25°C, 10 MPa and pH 8.5, H_2S dissociates with 6.99 pK1, yielding over 99% bisulfite ion (HS^-). Exposure to both H_2S and HS^- detrimentally affects microbial proteins and coenzymes. While the exact cause of toxicity remains uncertain, it is widely agreed that H_2S can more easily pass through the microbial cell membranes compared to HS^-. Methanogens and SSRM growth are negatively impacted when H_2S concentrations exceed 3.8–4.0 mM, though the growth of SSRM is entirely halted at 5.0–6.3 mM H_2S, with some persisting metabolic activity (Choi & Rim, 1991; Kushkevych et al., 2019). Oxygen is harmful to numerous anaerobes, including homoacetogens and methanogens. Additionally, microorganisms such as Methanothermococcus hermolithotrophicus and Desulfotomaculum geothermicum of methanogen and SSRM strain, respectively, demonstrate toxicity when exposed to CO_2 pressure exceeding 1 bar. Homoacetogenesis is restrained by nitrate, while methanogenesis is impeded by ammonium and sulfate, with specific inhibitory concentrations varying across different environments (Schwab et al., 2022). However, under conditions of H_2 storage, the influence of sulfate in methanogenesis is not much, as studies have demonstrated the reversal of sulfate inhibition through H_2 addition.

10.4 Microbial growth and biofouling effects

Biofouling, also referred to as biological fouling, is often caused by microorganisms and algae, and triggers detrimental impacts on storage surfaces. This buildup not only leads to surface degradation but also gradually diminishes the overall structural integrity. The resulting issues encompass a wide spectrum, including mechanical blockages, the initiation of microbially influenced corrosion, the possibility of product contamination, hindrances to fire protection systems, and potential risks to public health. The scale discrepancies and processes are encapsulated in the summarized form of Fig. 10.6. As illustrated, it is imperative to appropriately examine the growth of biofilm inside the water layer and the flow of gas in porous media.

10.4.1 Microbial-induced corrosion

Microbially influenced corrosion (MIC) is a prevalent issue encountered by steel infrastructure in many environmental settings, including water, oil, and underground installations (Enning & Garrelfs, 2014). The intricate interaction between biotic and abiotic corrosion can cause steel structures to erode mostly in one place, resulting in equipment malfunction. The activity has a strong correlation with the development of microbial biofilms on the surfaces of steel. Biofilms are a prevalent manifestation of microorganisms in the surrounding environment (Donlan, 2002). A biofilm is comprised of cellular entities

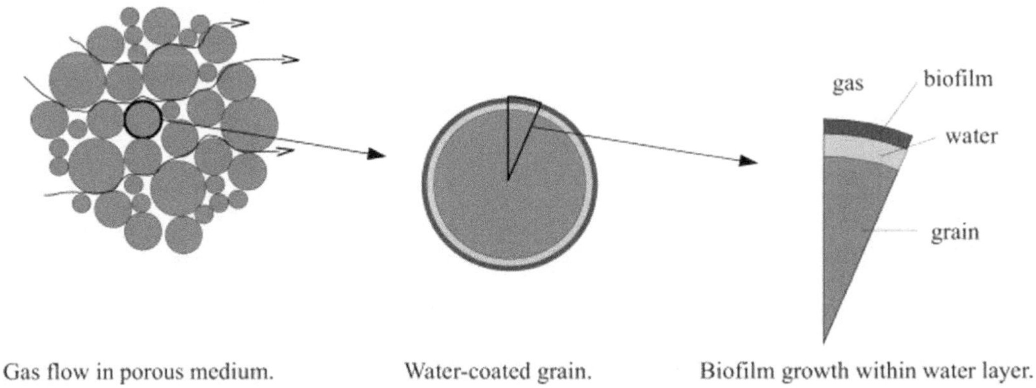

FIGURE 10.6 Visualization of mechanisms and presumptions: passage of gas through a permeable substance featuring particles enclosed by a slender water film, wherein the probability of microbial proliferation exists (Ebigbo, Golfier, & Quintard, 2013). Source: *With permission from Ebigbo, A., Golfier, F., & Quintard, M. (2013). A coupled, pore-scale model for methanogenic microbial activity in underground hydrogen storage.* Advances in Water Resources, 61, *74–85. https://doi.org/10.1016/j.advwatres.2013.09.004.*

enveloped by a matrix composed of exopolymeric components, such as proteins, sugars, nucleic acids, and occasionally minerals. The environment of biofilm offers a sheltered habitat for cells, wherein the presence of physical and chemical stressors is diminished. Corrosion frequently manifests beneath a biofilm due to notable alterations in redox and chemical variables resulting from restricted diffusion within the biofilm matrix.

Microbes that typically cause methanogens, biofilms, and corrosion issues include sulfate-reducing bacteria, methanogens, and acid-producing microbes. (Vigneron et al., 2018). The accurate identification of MIC is a significant problem due to its similarity to chemical corrosion and its variable expression in terms of corrosion rates and occurrences. To ascertain the presence and specific manifestation of MIC, it is most effective to analyze a grouping of factors, including microbial activity, corrosion properties, and corrosion products. An increase in the number of cells, specifically sulfate-reducing bacteria, detected in samples from either the liquid or solid phases might be considered as a suggestive factor for the occurrence of ongoing corrosion (Eckert & Skovhus, 2018; El-Shamy et al., 2009). H_2 has the capability to act as an electron donor within this particular process, promoting the growth and formation of microbial biofilms. Furthermore, it is possible for it to be produced as a secondary outcome of chemical reactions involving corrosion. (Atta et al., 2011). In addition, the corrosion rates can be heightened, and stress cracking is generated by the presence of H_2S that is formed through microbial activity (Bai et al., 2015).

10.4.2 Microbial-induced plugging

Microbial-induced plugging (MIP) refers to the phenomenon in which microorganisms cause obstruction or blockage of flow in many industrial systems. The obstruction or plugging of rock pores in formations or pipelines may arise from the presence of exopolymers, microbial biomass, or microbial-induced mineral precipitations. Microbial biomass plugging occurs because of the rapid expansion and propagation of cells and their accompanying biofilm structures within the confined spaces of pipelines and adjacent surfaces.

In the event of a porous structure, such as sediment or rock, a reduction in pore size will result in diminished permeability and injectivity (Kryachko, 2018). This phenomenon is frequently evident in the vicinity adjacent to the injection site, where the concentration of nutrients and cellular constituents tends to be the highest. Microbes can catalyze redox reactions, which in turn can initiate various mineral precipitations. Consequently, this process might result in clogging and a subsequent decrease in injectivity.

The process of sulfate reduction produces H_2S, which can react with ferrous iron in solution, leading to the precipitation of FeS minerals primarily. When dissolved iron is present in conjunction with either nitrate or low levels of oxygen, iron-oxidizing microorganisms induce the precipitation of ferric iron minerals such as ferrihydrite, goethite, magnetite, or other similar compounds. Microaerophilic iron-oxidizing microorganisms have a propensity for the development of substantial biofilms and mineral formations, frequently leading to the obstruction of systems due to the introduction of oxygen.

Microbial-induced carbonate precipitation (MICP) represents an additional avenue for possible clogging (Martin et al., 2013). Carbonate minerals are produced because of chemical transformations induced by a diverse range of metabolic processes. An instance of microbial-induced calcium carbonate precipitation can occur through the processes of nitrate and sulfate reduction if there is the degradation of an additional carbon source. This degradation results in the localized production of CO_2 and a subsequent rise in pH. The phenomenon under discussion is frequently associated with biofilms, as the restricted diffusion within the matrix creates favorable circumstances for localized precipitation. The identification of plugging occurrences can be accomplished by observing a noticeable reduction in injectivity and/or a concurrent surge in injection pressure. (Dopffel et al., 2018). Mineral precipitation and biomass plugging can both have significant and long-lasting effects, although in general, mineral precipitation is considered to be more trustworthy than biomass plugging.

10.4.3 Microbial-induced mineral dissolution

The presence of acid-producing microorganisms, specifically heterotrophic or acetogenic bacteria, in the formation fluid may indicate a pH drop. The aforementioned issue can be a challenge in the extraction and processing of some minerals. Consequently, the acids generated by these bacteria possess the capacity to induce the dissolution of readily soluble minerals like carbonates. This phenomenon has the potential to result in a substantial emission of CO_2 and bicarbonate ions. The procedure would be accompanied by a concurrent elevation in the concentration of dissolved CO_2 within the fluids that are being reproduced. Furthermore, the formations (either carbonate or sandstone) that have undergone dissolution can serve as a carbon source for bacteria, resulting in heightened growth and activity characterized by an augmentation in cell populations. In the event of a reservoir containing carbonate minerals, the dissolution of these minerals will result in alterations to the permeability and porosity characteristics. Consequently, this will induce modifications in fluid flow patterns and behavior inside the reservoir.

In the modern era, driven by the escalating need to combat global climate change and curtail greenhouse gas emissions, there has been a growing fascination with harnessing energy storage through pure H_2 or in conjunction with CO_2. This approach, employing power-to-gas technology, holds promise for generating green CH_4. In this context, it becomes imperative to consider the potential impact of gas mixture behavior and microbiological activity on the storage of H_2 (in the form of gas mixtures) within underground storage reservoirs. Thus, the occurrence of heightened microbial activity might result in the emission of excess heat because of exothermic metabolic processes. Studies have proposed that a potential sign of robust microbial activity in a reservoir might manifest as a minor elevation in temperature (Dopffel et al., 2021). However, such an occurrence has not been documented in the context of a H_2 storage facility.

10.5 Impacts on stored hydrogen quantity and quality

The diverse range of microbiological processes associated with molecular H_2 might lead to various consequences in the context of subsurface H_2 storage. Numerous industries, such as oil and gas or geothermal, frequently encounter substantial microbial activity, which is anticipated to be similarly observed during UHS operations (Gniese et al., 2014). This section provides an overview of several microbiological concerns, along with a selection of significant indicators for these processes. Microbial activity often gives rise to various impacts inside the system.

10.5.1 The change in the composition of a gas mixture

As previously mentioned, microorganisms can utilize molecular H_2 as part of their metabolic processes. Because of this consumption of H_2, the levels of H_2 may drop while the concentrations of other gases, such as CH_4 or H_2S, may rise. In addition, the use of H_2 by bacteria may also contribute to variations in the levels of CO_2, which may result in an increase or decrease in the concentration of this gas.The aforementioned factor may have a significant impact on the usefulness of the replicated H_2. Even low amounts of H_2S can have adverse consequences on the properties of gases, leading to negative impacts on the integrity of materials, safety, and human health. Consequently, the implementation of gas treatment becomes necessary. The concentration of H_2 in the injected gas mixture and the total gas volume can be decreased based on the field location's features, cell counts, and the presence or introduction of electron acceptors during operations. The anticipated H_2 loss is expected to remain constant or perhaps rise during the H_2 injection or production cycles. This is in contrast to the initial loss of H_2 produced by diffusion, which will be most significant in the initial cycle and decrease over time.

The simultaneous decrease of CO_2 within the mixture can serve as a sign that microorganisms are at work, specifically pointing towards methanogenesis or acetogenesis. The later phenomenon can be distinguished by an observed rise in CH_4 concentration within the resulting mixture as well as a notable alteration in the microbial community composition towards the domain Archaea. Specifically, this shift is characterized by an increase in the presence of several taxa, such as methanobacteriales, methanococcales, methanomicrobiales, methanosarcinales, and methanopyrales (Enzmann et al., 2018). Furthermore, the phenomenon of acetogenesis can lead to a reduction in H_2 levels and a concomitant increase in the concentration of acetate in the produced fluids. The aforementioned phenomenon is frequently found in the presence of specific bacterial taxa, namely Acetobacterium, Acetoanaerobium, Acetogenium, and Clostridium (Zabranska & Pokorna, 2018). The alteration in pH within a gas storage system can be effectively concealed by the robust buffering capacity of the environment and the presence of microorganisms that consume acetate.

10.5.2 Souring and the subsequent generation of hydrogen sulfide

The process of microbial sulfate reduction in an active state can result in the production of H_2S gas, which is known for its toxicity and corrosive properties. For a microbiological activity to take place, it is necessary for sulfate to be present in the water in the form of dissolved sulfate or through the existence of sulfidic minerals such as gypsum or anhydride (Wolicka & Borkowski, 2007). The existence of H_2 and readily accessible carbon sources, such as ethanol or organic acids, greatly enhances the reaction. In addition to the creation of H_2S, there would be a detectable decrease in the concentration of sulfates in the reproduced fluid. The examination of communities frequently uncovers a rise in Proteobacteria or microorganisms with names such as Desulfovibrio, Desulfobacterium, or Desulfotomaculum, which are associated with the reduction of sulfate, as well as sulfate-reducing Archaea. Sulfate reduction is a prevalent metabolic process observed in numerous microbial species across various environments (Oren, 2008). The process of sulfate reduction has high efficiency, as even small quantities of sulfate can result in substantial production of H_2S. If there are ferrous iron minerals or dissolved iron present, the H_2S may react with Fe^{2+} ions and form a precipitate. The identification of this process is characterized by a reduction in the quantities of dissolved iron and sulfide in the reconstituted liquids.

10.5.3 Potential hydrogen escape and consequences

Reports of gas leaks from subsurface gas storage sites are frequent, often stemming from operational malfunctions, cap rock diffusion, compromised well integrity, or seepage through geological fissures. The H_2 coefficient of diffusion surpasses that of CH_4, and over extended storage periods, there exists a potential for H_2 gas mixtures to potentially migrate to upper geological layers, reaching groundwater or the surface (Reitenbach et al., 2015). In addition to the evident hazard of H_2 release leading to explosions, the presence of H_2 can influence the microbial communities in soil and groundwater, impacting correlated nutrient sequences. A published study investigated the response of groundwater in the event of H_2 leakage at alkaline pH levels of 8.5–10 (Thaysen et al., 2023). A swift H_2 consumption was observed to be linked to acetate production, sulfate reduction, and an accompanying pH rise. Practically, this suggests that H_2 plumes would rapidly dissipate, and the resulting products could serve as valuable indicators for detecting leaks.

Alterations in groundwater microbiology and surface soil caused by H_2 could potentially result in lasting shifts in water and the chemistry of the soil, necessitating continuous monitoring. Additional, comprehensive studies on this significant subject are required to gain deeper insights into potential environmental impacts.

10.6 Risk assessment and monitoring

While the UHS presents an appealing prospect for energy solutions of the future, it is important to acknowledge that it also comes with inherent risks and uncertainties. Frequent reports have been made on the occurrence of gas leakages from subterranean gas storage sites (Evans & Chadwick, 2009). The occurrence of this phenomenon can be attributed to operational failures, compromised well integrity, diffusion through the cap rock, or leakage through geological fissures. The diffusion coefficient of H_2 is larger than that of CH_4. Consequently, during extended storage periods, H_2 gas mixtures or saturated liquids have the potential to migrate to higher geological strata, including groundwater or even the surface. Evaluating the enduring impact of microbial actions on an extensive scale is crucial, as these activities can lead to direct H_2 loss and contamination. This could significantly influence the efficiency of H_2 cycling and amplify the expenses associated with regasification processes.

In addition to the evident hazards of explosions resulting from the release of H_2, it is important to consider the impact of H_2 on microbial communities in soil and groundwater, as well as the associated nutrient cycles. Microbial actions impact the wetting of rock surfaces, leading to the entrapment and restriction of H_2 movement within porous materials. A scholarly publication has examined the response of groundwater to a H_2 leak under conditions of alkaline pH ranging from 8.5 to 10 (Berta et al., 2018). The study observed a rapid consumption of H_2, which was found to be correlated with sulfate reduction, acetate generation, and an increase in pH (Jagger, 1983). In practical terms, it may be inferred that H_2 plumes would exhibit rapid dissipation, rendering produced products potentially valuable indicators for the purpose of detecting leakage. Subsequent laboratory analyses revealed that under conditions of a more neutral groundwater pH of around 7, the process of acetogenesis was halted, but there was an observed increase in the generation of formate and CH_4. Additionally, the microbial processes exhibited inhibition over time or following alterations in pH. One plausible hypothesis for this phenomenon could be attributed to a constraint in the accessibility of trace elements within the groundwater over an extended period. This constraint may result in the persistence of a protracted H_2 plume within the groundwater system.

Alterations in the microbiological composition of groundwater and topsoil induced by the presence of H_2 have the potential to induce enduring modifications in the chemical properties of water and soil. These changes are essential for future UHS to assess and mitigate undesirable microbial actions. It is imperative to conduct a thorough and extended examination of soil and surface monitoring, as numerous environmental variables, such as seasonal changes or weather events, can obscure subtle alterations originating from subsurface gas leakage. To effectively manage microbial corrosion and souring, it is imperative to thoroughly analyze and maintain ongoing monitoring of the pertinent physical, geochemical, and microbial attributes of a reservoir. The initial findings of the ROSTOCK-H project have been released, presenting the results of tracer and specialized piezometer measurements conducted to identify possible H_2 plumes (Lafortune et al., 2020). Table 10.3 provides a compilation of significant parameters for conducting risk assessments. Microbial analysis methods serve as valuable tools for both prediction and monitoring, encompassing advanced molecular techniques and foundational culturing approaches.

10.7 Mitigation and control measures

The prevalent recognition of microbial actions underscores their significance as a primary consideration for the successful execution of UHS initiatives. Microbial actions can potentially jeopardize storage integrity through their influence on mineral dissolution and precipitation via various redox reactions. During the initial planning stage of the UHS site, it is crucial to consider not only the physical, geological, and economic factors but also the prospective microbial impacts, assessing and evaluating them. A summary of the crucial factors that can regulate the activities of distinct groups of microorganisms is presented in Table 10.4. The optimum conditions signify the points at which microorganism growth is most pronounced, while the critical ranges represent the upper limits within which growth remains feasible.

Broadly speaking, microbial actions exhibit heightened reactivity within environments characterized by diminished salinity (below 100,000 ppm), nearly neutral pH levels (ranging from 6.0 to 7.5), and low temperatures (below 40°C). Through deeper H_2 storage, raising the underground temperatures or elevating brine salinity with remarkably high concentrations of NaCl can efficiently mitigate microbial activities that lead to H_2 loss. Besides, when reservoirs encompass reactive minerals capable of initiating redox reactions, the ambient pH levels may rise to 9–12. Thus, it is anticipated that in highly alkaline environments, the extent of microbial activities affecting biotic mineral dissolution or precipitation and related storage integrity concerns is projected to be minimal. When dealing with elevated reservoir temperatures, the likelihood of microbial activity decreases. However, it is important to account for cooling effects stemming from prior or anticipated water or gas injections. Looking solely at the microorganism aspect, opting for H_2 storage in deeper formations characterized by elevated pH levels, salinity, and temperature appears advantageous for constraining subsurface bacterial activities.

Reservoir souring involves sulfide production by native or introduced sulfate-reducing microorganisms employing sulfate from brines or injected seawater. Countermeasures tested include sulfate removal from injection water, biocides usage, and different inhibitors such as molybdate, tungstate, or nitrate. The effectiveness of nitrate addition varies, with instances of activity control and, conversely, cases where it leads to heightened activity. Microbiological insights have a direct impact on material choices, the operational scheme, gas constituents, and monitoring approaches. Hence, incorporating these considerations into the design and construction phases of planning is crucial. Utilizing nucleic acid (DNA and RNA)-based techniques for analyzing the injected and reproduced gases, brine, and reservoir fluids should ideally involve multiple assessments over time. This strategy, which

TABLE 10.3 Crucial factors for conducting an assessment of microbial risks at an underground hydrogen storage site.

Parameter	Specimen	Applicable method	Otputs and risks
Mineral composition	Rock/ sediment	XRD, SEM-EDX, Microscopy, Raman	Minerals with a high carbonate and sulphate content can stimulate microbial activity.
Temperature	–	Down-hole measurement	Significant microbial activity between 20°C and 80°C
Gas analysis	Gas	GC	The reduction of CO_2/H_2 and presence of CH_4 signifies microbial activity.
Isotopic analysis of gas	Gas	Mass spectrometry	An isotope pattern can show if a geophysical or biological event is taking place.
Salinity	Water	EC, TDS	The observed diversity is less than 100 g/L.
Organic compounds (organic acids, fatty acids)		HPLC, LC-MS	Carbon sources for microbes trigger higher activity. The presence of acetogens leads to an elevation in acetate levels.
Sulfate		ICP, IC, Photometrical assays	The foundation for the process of sulfate reduction, which involves the loss of H_2, ultimately results in the creation of H_2S.
Nitrate			The fundamental mechanism underlying nitrate reduction is the loss of H_2.
Carbonate/CO_2			Microbes rely on carbon as a source of sustenance. Results in the production of either CH_4 or acetate.
Ferrous iron			Indicator for possible precipitation with H_2S.
pH		pH probe	The microbial activity exhibits a significant increase throughout the pH range of 3 to 8.
Microbial community	Water or sediment	Third generation-sequencing of DNA (presence) or RNA (activity)	The results of an initial community investigation indicate potential microbial concerns. Community shifts suggest metabolic activity.
Cell numbers for metabolism		q/d PCR, MPN, Microscopic cell count (only total cells)	Elevated cell counts are indicative of an increased likelihood of microbial activity. The phenomenon of increase through time signifies a progressive growth inside the system.

EC, Electrical conductivity; *TDS*, total dissolved solids; *IC*, ion chromatography; *ICP*, inductively coupled plasma mass spectrometry; *HPLC*, high-performance liquid chromatography; *LC-Ms*, liquid chromatography-mass spectrometry; *q/d PCR*, quantitative/digital polymerase chain reaction; *MPN*, most probable number; *XRD*, X-ray crystallography; *SEM*, scanning electron microscopy; *EDX*, energy dispersive X-ray; *GC*: gas chromatography.

With permission from Dopffel, N., Jansen, S., & Gerritse, J. (2021). Microbial side effects of underground hydrogen storage – Knowledge gaps, risks and opportunities for successful implementation. International Journal of Hydrogen Energy, 46*(12), 8594–8606. https://doi.org/10.1016/j.ijhydene.2020.12.058.*

involves tracking changes in water interaction, gas composition, microbial activity, cell numbers, and/or community composition, is vital for capturing the inherent variations in a natural context rather than being a one-time undertaking. In laboratory settings, incubations can

TABLE 10.4 The ideal and crucial pH, salinity, and temperature ranges impact microbial activities.

	pH		Salinity (ppm)		Temperature (°C)	
Microbial activity	**Optimum**	**Critical**	**Optimum**	**Critical**	**Optimum**	**Critical**
Methanogenesis	6.0–7.5	4.5–9.0	< 60,000	200,000	30–40	122
Acetogenesis	6.0–7.5	3.6–10.7	< 40,000	300,000	20–30	72
Sulfate reduction	6.0–7.5	0.8–11.5	< 100,000	240,000	20–30	113
Iron reduction	6.0–7.5	1.6–9.0	< 40,000	200,000	0–30	90

From Zeng, L., Sarmadivaleh, M., Saeedi, A., Chen, Y., Zhong, Z., & Xie, Q. (2023). Storage integrity during underground hydrogen storage in depleted gas reservoirs. Earth-Science Reviews, 247, *https://doi.org/10.1016/j.earscirev.2023.104625.*

be employed to assess the effectiveness of mitigation strategies, such as the effectiveness of biocides or specific inhibitors adjacent to distinct metabolic groups.

Microbial activity within reservoirs is a recognized phenomenon in industries like oil and gas, carbon capture and storage (CCS), and geothermal operations. Similar occurrences can be identified within the UHS systems, particularly in terms of the interaction amongst chemical environments, microbial action, and the possible significance of sulfate reducers. Also, carbonate precipitation through microbial activity can be explored as a possible technique to mitigate H_2 leakage risks, like what is done to prevent leakage of CO_2 to the surface in the CCS process. Thus, the essential discoveries and strategies for monitoring and control management learned from CCS can be utilized in the UHS.

10.8 Underground hydrogen storage case studies with microbial activity

In the era of renewable energy, the focus has shifted to the storage of H_2 on a large scale due to its cost-effectiveness and ability to scale up. Pure H_2 storage, in comparison, remains less frequent, with only a few storage sites having undergone testing. Ongoing operational activities for H_2 storage are primarily associated with salt caverns, whereas storage within aquifers and porous rock formations is at a field-test stage. Only a handful of UHS facilities are currently operational, for instance, in the United States at Moss Bluff, Spindletop, and Clemens Dome; the Hyunder, Vienna Basin Green H_2 storage, and MefHySto projects of the European Union; and at Teeside and HyStorPor of the United Kingdom (Schwab et al., 2022; Thaysen et al., 2023). Furthermore, there is a dearth of comprehensive information concerning microbial activity during UHS.

Insights gathered from the ongoing H_2 storage ventures within the industry reveal that the H_2 utilization rate by microbial activity varies. For example, 3% was observed in the Underground Sun Storage initiative in Austria (H_2 transformed into CH_4 via methanogenesis) and 45%–60% utilization of H_2 at the town gas storage unit of Lobodice, Czech Republic, with H_2 content ranging from 16% to 50% v/v (Zeng et al., 2023). Reports of microbial activities have also emerged from other UHS projects in countries like Germany, Argentina, and France (Table 10.5). A valuable lesson garnered across industries over the years is that preventing microbial activity is consistently more cost-effective than

TABLE 10.5 Microbial activity reported underground hydrogen storage field locations and pilot projects along with their accessible attributes.

Project location	Country	Year	Formation	Depth (m)	Pressure (bar)	Volume (m^3)	Gas composition
Beynes	France	1957–1972	Saline aquifer	430	–	3.3×10^8	H_2:16%–50%
HyChico	Argentina	2010–2018	Sandstone	815	26.5	750.00	H_2:10%
Ketzin	Germany	1964–2000		250–400	–	1.30×10^8	Town gas
Lobodice	Czech Rep	1989		400–500	–	–	H_2: 50%; CO_2: 12%; CH_4: 22%
Underground Sun Storage	Austria	2017		1200	78	115.44	H_2: 10%

From Dopffel, N., Jansen, S., & Gerritse, J. (2021). Microbial side effects of underground hydrogen storage – Knowledge gaps, risks and opportunities for successful implementation. International Journal of Hydrogen Energy, *46(12), 8594–8606. https://doi.org/10.1016/j.ijhydene.2020.12.058.*

subsequent treatment measures. Hence, implementing a clearly defined monitoring plan from the outset of UHS operations becomes imperative to prevent failures triggered by microbial activities.

10.9 Future perspectives

With the increasing number of field pilots for UHS, our understanding of the impacts of microbial activity within these storage sites continues to grow. However, numerous unresolved questions exist that are unique to UHS, where substantial distinctions from other industries are anticipated. One such crucial aspect is the requirement for additional microbiological data obtained from field tests. While microbiology is frequently highlighted as a significant risk factor at UHS, the microbiological components within the projects are often limited in scope or sometimes even nonexistent. The techniques employed for analyzing brine and reservoir fluids to monitor microbial growth are still relatively demanding in terms of labor and cost. These methods need to be adapted to account for the generally low cell numbers encountered in subsurface environments. Consequently, our understanding remains incomplete regarding whether other microbial metabolisms beyond the three main groups (methanogens, sulfate reducers, and acetogens) could potentially have a significant impact. The reservoir pressure maintenance by cushion gas (a base gas—probably CO_2, N_2, or CH_4) while H_2 is withdrawn is necessary. Still, the impact of high H_2 concentrations and partial pressures on the microbiome within the storage sites remains to be investigated. Further investigation into the composition and potential functions of native microbiomes within underground reservoirs is crucial for evaluating the potential opportunities and obstacles in the realm of future UHS. Subsequent efforts are necessary to tackle these issues, aiming to reduce limitations and mitigate the challenges related to storage integrity in the context of UHS within porous media.

10.10 Conclusion

Microbial presence and activity can be observed in multiple locations earmarked for UHS. Microbial activity is most probable in mature or depleted oil and gas reservoirs and aquifers, while it is less likely in salt caverns. Nonetheless, the precise activities will heavily rely on the unique local circumstances and remain feasible across all scenarios. Possible outcomes encompass microbial alteration of stored H_2, leading to the creation of unwanted by-products like H_2S, CH_4, and acids, as well as corrosion, blockages, and dissolution of reservoir materials. Various microorganisms, especially methanogens, sulfate reducers, and acetogens, facilitate these alterations. Additional insights gained from a wider array of field sites are necessary to establish comprehensive predictions regarding microbial risks. Countermeasures like biocides could potentially yield positive outcomes, but their application must be cautious, and their impact on H_2 consumption should be thoroughly assessed. Given that the microbial consequences of H_2 storage arise from the intricate interactions between geochemistry, microbiology, and physics, all of which are site-specific and predominantly unexplored, there is a pressing need for greater understanding. Through the approach of conducting research across laboratory and field scales and modeling studies by employing a multidisciplinary approach encompassing microbiologists, chemists, physicists, and applied scientists.

References

Amend, J. P., & Shock, E. L. (2001). Energetics of overall metabolic reactions of thermophilic and hyperthermophilic archaea and bacteria. *FEMS Microbiology Reviews, 25*(2), 175–243. Available from https://doi.org/10.1111/j.1574-6976.2001.tb00576.x.

Atta, N. F., Fekry, A. M., & Hassaneen, H. M. (2011). Corrosion inhibition, hydrogen evolution and antibacterial properties of newly synthesized organic inhibitors on 316L stainless steel alloy in acid medium. *International Journal of Hydrogen Energy, 36*(11), 6462–6471. Available from https://doi.org/10.1016/j.ijhydene.2011.02.134.

Bai, P., Zhao, H., Zheng, S., & Chen, C. (2015). Initiation and developmental stages of steel corrosion in wet H_2S environments. *Corrosion Science, 93*, 109–119. Available from https://doi.org/10.1016/j.corsci.2015.01.024, https://www.journals.elsevier.com/corrosion-science.

Bernardez, L. A., de Andrade Lima, L. R. P., de Jesus, E. B., Ramos, C. L. S., & Almeida, P. F. (2013). A kinetic study on bacterial sulfate reduction. *Bioprocess and Biosystems Engineering, 36*(12), 1861–1869. Available from https://doi.org/10.1007/s00449-013-0960-0.

Berta, M., Dethlefsen, F., Ebert, M., Schäfer, D., & Dahmke, A. (2018). Geochemical effects of millimolar hydrogen concentrations in groundwater: An experimental study in the context of subsurface hydrogen storage. *Environmental Science and Technology, 52*(8), 4937–4949. Available from https://doi.org/10.1021/acs.est.7b05467, http://pubs.acs.org/journal/esthag.

Caglayan, D. G., Weber, N., Heinrichs, H. U., Linßen, J., Robinius, M., Kukla, P. A., & Stolten, D. (2020). Technical potential of salt caverns for hydrogen storage in Europe. *International Journal of Hydrogen Energy, 45*(11), 6793–6805. Available from https://doi.org/10.1016/j.ijhydene.2019.12.161, http://www.journals.elsevier.com/international-journal-of-hydrogen-energy/.

Cavallaro, A. N., Martinez, M. E. G., Ostera, H., Panarello, H., & Cordero, R. R. (2005). Oilfield reservoir souring during waterflooding: A case study with low sulphate concentration in formation and injection waters. *Proceedings – SPE International Symposium on Oilfield Chemistry*, Society of Petroleum Engineers (SPE) Argentina, 131–142. https://doi.org/10.2523/92959-ms. https://www.onepetro.org/conferences/SPE/01OCS.

Chen, Y., Cheng, J. J., & Creamer, K. S. (2008). Inhibition of anaerobic digestion process: A review. *Bioresource Technology, 99*(10), 4044–4064. Available from https://doi.org/10.1016/j.biortech.2007.01.057.

Choi, E., & Rim, J. M. (1991). Competition and inhibition of sulfate reducers and methane producers in anaerobic treatment. *Water Science and Technology, 23*(7-9), 1259–1264. Available from https://doi.org/10.2166/wst.1991.0577.

Donlan, R. M. (2002). Biofilms: Microbial life on surfaces. *Emerging Infectious Diseases, 8*(9), 881–890. Available from https://doi.org/10.3201/eid0809.020063, http://www.cdc.gov/eid.

Dopffel, N., Jansen, S., & Gerritse, J. (2021). Microbial side effects of underground hydrogen storage – Knowledge gaps, risks and opportunities for successful implementation. *International Journal of Hydrogen Energy, 46*(12), 8594–8606. Available from https://doi.org/10.1016/j.ijhydene.2020.12.058, http://www.journals.elsevier.com/international-journal-of-hydrogen-energy/.

Dopffel, N., Kögler, F., Hartmann, H., Costea, P. I., Mahler, E., Herold, A., & Alkan, H. (2018). Microbial induced mineral precipitations caused by nitrate treatment for souring control during microbial enhanced oil recovery (MEOR). *International Biodeterioration and Biodegradation, 135*, 71–79. Available from https://doi.org/10.1016/j.ibiod.2018.09.004, http://www.elsevier.com/inca/publications/store/4/0/5/8/9/9.

Dutta, A., Dutta Gupta, S., Gupta, A., Sarkar, J., Roy, S., Mukherjee, A., & Sar, P. (2018). Exploration of deep terrestrial subsurface microbiome in Late Cretaceous Deccan traps and underlying Archean basement, India. *Scientific Reports, 8*(1). Available from https://doi.org/10.1038/s41598-018-35940-0, http://www.nature.com/srep/index.html.

Ebigbo, A., Golfier, F., & Quintard, M. (2013). A coupled, pore-scale model for methanogenic microbial activity in underground hydrogen storage. *Advances in Water Resources, 61*, 74–85. Available from https://doi.org/10.1016/j.advwatres.2013.09.004.

Eckert, R. B., & Skovhus, T. L. (2018). Advances in the application of molecular microbiological methods in the oil and gas industry and links to microbiologically influenced corrosion. *International Biodeterioration and Biodegradation, 126*, 169–176. Available from https://doi.org/10.1016/j.ibiod.2016.11.019, http://www.elsevier.com/inca/publications/store/4/0/5/8/9/9.

El-Shamy, A. M., Soror, T. Y., El-Dahan, H. A., Ghazy, E. A., & Eweas, A. F. (2009). Microbial corrosion inhibition of mild steel in salty water environment. *Materials Chemistry and Physics, 114*(1), 156–159. Available from https://doi.org/10.1016/j.matchemphys.2008.09.003.

Enning, D., & Garrelfs, J. (2014). Corrosion of iron by sulfate-reducing bacteria: New views of an old problem. *Applied and Environmental Microbiology, 80*(4), 1226–1236. Available from https://doi.org/10.1128/AEM.02848-13Germany, http://aem.asm.org/content/80/4/1226.full.pdf.

Enzmann, F., Mayer, F., Rother, M., & Holtmann, D. (2018). Methanogens: Biochemical background and biotechnological applications. *AMB Express, 8*(1). Available from https://doi.org/10.1186/s13568-017-0531-x, http://www.amb-express.com/.

Etiope, G., Schoell, M., & Hosgörmez, H. (2011). Abiotic methane flux from the *Chimaera seep* and *Tekirova ophiolites* (turkey): Understanding gas exhalation from low temperature serpentinization and implications for Mars. *Earth and Planetary Science Letters, 310*(1-2), 96–104. Available from https://doi.org/10.1016/j.epsl.2011.08.001.

Evans, D. J., & Chadwick, R. A. (2009). Underground gas storage: An introduction and UK perspective. *Geological Society, 313*(1), 1–11. Available from https://doi.org/10.1144/sp313.1.

Findlay, A. J. (2016). Microbial impact on polysulfide dynamics in the environment. *FEMS Microbiology Letters, 363*(11), 1–12. Available from https://doi.org/10.1093/femsle/fnw103, http://femsle.oxfordjournals.org/.

Flesch, S., Pudlo, D., Albrecht, D., Jacob, A., & Enzmann, F. (2018). Hydrogen underground storage—Petrographic and petrophysical variations in reservoir sandstones from laboratory experiments under simulated reservoir conditions. *International Journal of Hydrogen Energy, 43*(45), 20822–20835. Available from https://doi.org/10.1016/j.ijhydene.2018.09.112, http://www.journals.elsevier.com/international-journal-of-hydrogen-energy/.

Gniese, C., Bombach, P., Rakoczy, J., Hoth, N., Schlömann, M., Richnow, H. H., & Krüger, M. (2014). Relevance of deep-subsurface microbiology for underground gas storage and geothermal energy production. *Germany Advances in Biochemical Engineering/Biotechnology, 142*, 95–121. Available from https://doi.org/10.1007/10_2013_257, http://www.springer.com/series/10.

Gregory, S. P., Barnett, M. J., Field, L. P., & Milodowski, A. E. (2019). Subsurface microbial hydrogen cycling: Natural occurrence and implications for industry. *Microorganisms, 7*(2). Available from https://doi.org/10.3390/microorganisms7020053, https://www.mdpi.com/2076-2607/7/2/53/pdf.

Gregory, S. P., Maurice, L. D., West, J. M., & Gooddy, D. C. (2014). Microbial communities in UK aquifers: Current understanding and future research needs. *Quarterly Journal of Engineering Geology and Hydrogeology, 47* (2), 145–157. Available from https://doi.org/10.1144/qjegh2013-059.

Groenenberg, R. M., Juez-Larre, J., Machado, C., Wasch, L. J., Dijkstra, H. E., Wassing, B. B. T., Orlic, B., Brunner, L. G., van der Valk, K., Hajonides van der Meulen, T. C., & Kranenburg-Bruinsma, K. J. (2020). *Technoeconomic modelling of large-scale energy storage systems.*

Halim, A., Shapiro, A., Lantz, A. E., & Nielsen, S. M. (2014). Experimental study of bacterial penetration into chalk rock: Mechanisms and effect on permeability. *Transport in Porous Media, 101*(1), 1–15. Available from https://doi.org/10.1007/s11242-013-0227-x, http://www.wkap.nl/journalhome.htm/0169-3913.

Hallbeck, L., & Pedersen, K. (2008). Characterization of microbial processes in deep aquifers of the Fennoscandian Shield. *Applied Geochemistry, 23*(7), 1796–1819. Available from https://doi.org/10.1016/j.apgeochem.2008.02.012.

Hassannayebi, N. (2019). An assessment of underground hydrogen storage: Transport, geochemistry, and bioactivity. Montanuniversität. https://pure.unileoben.ac.at/en/publications/an-assessment-of-underground-hydrogen-storage-transport-geochemis.

Heinemann, N., Booth, M. G., Haszeldine, R. S., Wilkinson, M., Scafidi, J., & Edlmann, K. (2018). Hydrogen storage in porous geological formations – Onshore play opportunities in the midland valley (Scotland, UK). *International Journal of Hydrogen Energy, 43*(45), 20861–20874. Available from https://doi.org/10.1016/j.ijhydene.2018.09.149.

Hemme, C., & van Berk, W. (2018). Hydrogeochemical modeling to identify potential risks of underground hydrogen storage in depleted gas fields. *Applied Sciences (Switzerland), 8*(11). Available from https://doi.org/10.3390/app8112282, https://www.mdpi.com/2076-3417/8/11/2282/pdf.

Herbert, B. N., Gilbert, P. D., Stockdale, H., & Watkinson, R. J. (1985). *Factors controlling the activity of sulphate-reducing bacteria in reservoirs during water injection.* Dallas, TX, United States: Society of Petroleum Engineers of AIME. Available from https://www.osti.gov/biblio/5442684.

Hernsdorf, A. W., Amano, Y., Miyakawa, K., Ise, K., Suzuki, Y., Anantharaman, K., Probst, A., Burstein, D., Thomas, B. C., & Banfield, J. F. (2017). Potential for microbial H_2 and metal transformations associated with novel bacteria and archaea in deep terrestrial subsurface sediments. *ISME Journal, 11*(8), 1915–1929. Available from https://doi.org/10.1038/ismej.2017.39, http://www.nature.com/ismej/marketing/aims_scope.html.

Hoehler, T. M., & Jørgensen, B. B. (2013). Microbial life under extreme energy limitation. *Nature Reviews. Microbiology, 11*(2), 83–94. Available from https://doi.org/10.1038/nrmicro2939.

Huang, J., Sheng, X. F., Xi, J., He, L. Y., Huang, Z., Wang, Q., & Zhang, Z. D. (2014). Depth-related changes in community structure of culturable mineral weathering bacteria and in weathering patterns caused by them along two contrasting soil profiles. *Applied and Environmental Microbiology, 80*(1), 29–42. Available from https://doi.org/10.1128/AEM.02335-13China, http://aem.asm.org/content/80/1/29.full.pdf.

Jaenicke, R., & Sterner, R. (2006). *Life at high temperatures* (pp. 167–209). Springer Science and Business Media LLC. Available from 10.1007/0-387-30742-7_7.

Jagger, J. (1983). *Physiological effects of near-ultraviolet radiation on bacteria* (pp. 1–75). Springer Science and Business Media LLC. Available from 10.1007/978-1-4684-4505-3_1.

Jiang, Y., Zhang, B., He, C., Shi, J., Borthwick, A. G. L., & Huang, X. (2018). Synchronous microbial vanadium (V) reduction and denitrification in groundwater using hydrogen as the sole electron donor. *Water Research, 141*, 289–296. Available from https://doi.org/10.1016/j.watres.2018.05.033, http://www.elsevier.com/locate/watres.

Khajooie, S., Gaus, G., Dohrmann, A. B., Krüger, M., & Littke, R. (2023). Methanogenic conversion of hydrogen to methane in reservoir rocks: An experimental study of microbial activity in water-filled pore space. *International Journal of Hydrogen Energy.* Available from https://doi.org/10.1016/j.ijhydene.2023.07.065, http://www.journals.elsevier.com/international-journal-of-hydrogen-energy/.

Kryachko, Y. (2018). Novel approaches to microbial enhancement of oil recovery. *Journal of Biotechnology, 266*, 118–123. Available from https://doi.org/10.1016/j.jbiotec.2017.12.019, http://www.elsevier.com/locate/jbiotec.

Kushkevych, I., Dordević, D., & Vítězová, M. (2019). Toxicity of hydrogen sulfide toward sulfate-reducing bacteria *Desulfovibrio piger* Vib-7. *Archives of Microbiology, 201*(3), 389–397. Available from https://doi.org/10.1007/s00203-019-01625-z.

Lafortune, S., Gombert, P., Pokryszka, Z., Lacroix, E., de Donato, P., & Jozja, N. (2020). Monitoring scheme for the detection of hydrogen leakage from a deep underground storage. Part 1: On-site validation of an experimental protocol via the combined injection of helium and tracers into an aquifer. *Applied Sciences, 10*(17). Available from https://doi.org/10.3390/app10176058.

Liebscher, A., Wackerl, J., & Streibel, M. (2016). Geologic storage of hydrogen – Fundamentals, processing, and projects. *Hydrogen Science and Engineering: Materials, Processes, Systems and Technology, 2*. Available from https://doi.org/10.1002/9783527674268.ch26, http://onlinelibrary.wiley.com/book/10.1002/9783527674268.

Lovley, D. R. (1991). Dissimilatory Fe(III) and Mn(IV) reduction. *Microbiological Reviews, 55*(2), 259–287. Available from https://doi.org/10.1128/mr.55.2.259-287.1991.

Lovley, D. R., & Goodwin, S. (1988). Hydrogen concentrations as an indicator of the predominant terminal electron-accepting reactions in aquatic sediments. *Geochimica et Cosmochimica Acta, 52*(12), 2993–3003. Available from https://doi.org/10.1016/0016-7037(88)90163-9.

Magnabosco, C., Lin, L.-H., Dong, H., Bomberg, M., Ghiorse, W., Stan-Lotter, H., Pedersen, K., Kieft, T. L., van Heerden, E., & Onstott, T. C. (2018). The biomass and biodiversity of the continental subsurface. *Nature Geoscience, 11*(10), 707–717. Available from https://doi.org/10.1038/s41561-018-0221-6.

Martin, D., Dodds, K., Butler, I. B., & Ngwenya, B. T. (2013). Carbonate precipitation under pressure for bioengineering in the anaerobic subsurface via denitrification. *Environmental Science and Technology, 47*(15), 8692–8699. Available from https://doi.org/10.1021/es401270q.

Muyzer, G., & Stams, A. J. M. (2008). The ecology and biotechnology of sulphate-reducing bacteria. *Nature Reviews. Microbiology, 6*(6), 441–454. Available from https://doi.org/10.1038/nrmicro1892.

O'Flaherty, V., Mahony, T., O'Kennedy, R., & Colleran, E. (1998). Effect of pH on growth kinetics and sulphide toxicity thresholds of a range of methanogenic, syntrophic and sulphate-reducing bacteria. *Process Biochemistry, 33*(5), 555–569. Available from https://doi.org/10.1016/S0032-9592(98)00018-1.

Oren, A. (2008). Microbial life at high salt concentrations: phylogenetic and metabolic diversity. *Saline Systems, 4* (1). Available from https://doi.org/10.1186/1746-1448-4-2.

Oren, A. (2013). Life at high salt concentrations. *The prokaryotes: Prokaryotic communities and ecophysiology*. Israel: Springer-Verlag Berlin Heidelberg. Available from http://doi.org/10.1007/978-3-642-30123-0, 10.1007/978-3-642-30123-0_57.

Oren, A. (2011). Thermodynamic limits to microbial life at high salt concentrations. *Environmental Microbiology, 13* (8), 1908–1923. Available from https://doi.org/10.1111/j.1462-2920.2010.02365.x.

Panfilov, M. (2016). *Underground and pipeline hydrogen storage* (pp. 91–115). Elsevier BV. Available from 10.1016/b978-1-78242-362-1.00004-3.

Payler, S. J., Biddle, J. F., Lollar, B. S., Fox-Powell, M. G., Edwards, T., Ngwenya, B. T., Paling, S. M., & Cockell, C. S. (2019). An ionic limit to life in the deep subsurface. *Frontiers in Microbiology, 10*. Available from https://doi.org/10.3389/fmicb.2019.00426, https://www.frontiersin.org/journals/microbiology#.

Ranchou-Peyruse, M., Gasc, C., Guignard, M., Aüllo, T., Dequidt, D., Peyret, P., & Ranchou-Peyruse, A. (2017). The sequence capture by hybridization: A new approach for revealing the potential of mono-aromatic hydrocarbons bioattenuation in a deep oligotrophic aquifer. *Microbial Biotechnology, 10*(2), 469–479. Available from https://doi.org/10.1111/1751-7915.12426, http://onlinelibrary.wiley.com/journal/10.1111/(ISSN)1751-7915.

Reitenbach, V., Ganzer, L., Albrecht, D., & Hagemann, B. (2015). Influence of added hydrogen on underground gas storage: A review of key issues. *Environmental Earth Sciences, 73*(11), 6927–6937. Available from https://doi.org/10.1007/s12665-015-4176-2, http://www.springerlink.com/content/121380/.

Salvador-Castell, M., Oger, P., & Peters, J. (2020). *High-pressure adaptation of extremophiles and biotechnological applications. Physiological and biotechnological aspects of extremophiles* (pp. 105–122). France: Elsevier. Available from https://www.sciencedirect.com/book/9780128183229, 10.1016/B978-0-12-818322-9.00008-3.

Schwab, L., Popp, D., Nowack, G., Bombach, P., Vogt, C., & Richnow, H. H. (2022). Structural analysis of microbiomes from salt caverns used for underground gas storage. *International Journal of Hydrogen Energy, 47*(47), 20684–20694. Available from https://doi.org/10.1016/j.ijhydene.2022.04.170, http://www.journals.elsevier.com/international-journal-of-hydrogen-energy/.

Shi, Z., Jessen, K., & Tsotsis, T. T. (2020). Impacts of the subsurface storage of natural gas and hydrogen mixtures. *International Journal of Hydrogen Energy, 45*(15), 8757–8773. Available from https://doi.org/10.1016/j.ijhydene.2020.01.044, http://www.journals.elsevier.com/international-journal-of-hydrogen-energy/.

Šmigáň, P., Greksák, M., Kozánková, J., Buzek, F., Onderka, V., & Wolf, I. (1990). Methanogenic bacteria as a key factor involved in changes of town gas stored in an underground reservoir. *FEMS Microbiology Letters, 73*(3), 221–224. Available from https://doi.org/10.1016/0378-1097(90)90733-7.

Strobel, G., Hagemann, B., Huppertz, T. M., & Ganzer, L. (2020). Underground bio-methanation: Concept and potential. *Renewable and Sustainable Energy Reviews, 123*. Available from https://doi.org/10.1016/j.rser.2020.109747.

Taylor, G. T., & Pirt, S. J. (1977). Nutrition and factors limiting the growth of a methanogenic bacterium (*Methanobacterium thermoautotrophicum*). *Archives of Microbiology, 113*(1-2), 17–22. Available from https://doi.org/10.1007/BF00428574.

Thaysen, E. M., McMahon, S., Strobel, G. J., Butler, I. B., Ngwenya, B. T., Heinemann, N., Wilkinson, M., Hassanpouryouzband, A., McDermott, C. I., & Edlmann, K. (2021). Estimating microbial growth and hydrogen consumption in hydrogen storage in porous media. *Renewable and Sustainable Energy Reviews, 151*. Available from https://doi.org/10.1016/j.rser.2021.111481, https://www.journals.elsevier.com/renewable-and-sustainable-energy-reviews.

Thaysen, E. M., Armitage, T., Slabon, L., Hassanpouryouzband, A., & Edlmann, K. (2023). Microbial risk assessment for underground hydrogen storage in porous rocks. *Fuel, 352*. Available from https://doi.org/10.1016/j.fuel.2023.128852.

Thiyagarajan, S. R., Emadi, H., Hussain, A., Patange, P., & Watson, M. (2022). A comprehensive review of the mechanisms and efficiency of underground hydrogen storage. *Journal of Energy Storage, 51*. Available from https://doi.org/10.1016/j.est.2022.104490.

Usher, K. M., Kaksonen, A. H., Bouquet, D., Cheng, K. Y., Geste, Y., Chapman, P. G., & Johnston, C. D. (2015). The role of bacterial communities and carbon dioxide on the corrosion of steel. *Corrosion Science, 98*, 354–365. Available from https://doi.org/10.1016/j.corsci.2015.05.043.

Van Houten, R. T., Yun, S. Y., & Lettinga, G. (1997). Thermophilic sulphate and sulphite reduction in lab-scale gas-lift reactors using H_2 and CO_2 as energy and carbon source. *Biotechnology and Bioengineering, 55*(5), 807–814, https://doi.org/10.1002/(SICI)1097-0290(19970905)55:5 < 807::AID-BIT11% > 3.0.CO;2-8.

Varjani, S. J., & Gnansounou, E. (2017). Microbial dynamics in petroleum oilfields and their relationship with physiological properties of petroleum oil reservoirs. *Bioresource Technology, 245*, 1258–1265. Available from https://doi.org/10.1016/j.biortech.2017.08.028, http://www.elsevier.com/locate/biortech.

Ver Eecke, H. C., Akerman, N. H., Huber, J. A., Butterfield, D. A., & Holden, J. F. (2013). Growth kinetics and energetics of a deep-sea hyperthermophilic methanogen under varying environmental conditions. *Environmental Microbiology Reports, 5*(5), 665–671. Available from https://doi.org/10.1111/1758-2229.12065.

Vigneron, A., Alsop, E. B., Lomans, B. P., Kyrpides, N. C., Head, I. M., & Tsesmetzis, N. (2017). Succession in the petroleum reservoir microbiome through an oil field production lifecycle. *ISME Journal, 11*(9), 2141–2154. Available from https://doi.org/10.1038/ismej.2017.78, http://www.nature.com/ismej/marketing/aims_scope.html.

Vigneron, A., Head, I. M., & Tsesmetzis, N. (2018). Damage to offshore production facilities by corrosive microbial biofilms. *Applied Microbiology and Biotechnology, 102*(6), 2525–2533. Available from https://doi.org/10.1007/s00253-018-8808-9, link.springer.de/link/service/journals/00253/index.htm.

Wolicka, D., & Borkowski, A. (2007). The geomicrobiological role of sulphate-reducing bacteria in environments contaminated by petroleum products. *Geomicrobiology Journal, 24*(7-8), 599–607. Available from https://doi.org/10.1080/01490450701672117.

Yekta, A. E., Manceau, J.-C., Gaboreau, S., Pichavant, M., & Audigane, P. (2018). Determination of hydrogen–-water relative permeability and capillary pressure in sandstone: Application to underground hydrogen injection in sedimentary formations. *Transport in Porous Media, 122*(2), 333–356. Available from https://doi.org/10.1007/s11242-018-1004-7.

Zabranska, J., & Pokorna, D. (2018). Bioconversion of carbon dioxide to methane using hydrogen and hydrogenotrophic methanogens. *Biotechnology Advances, 36*(3), 707–720. Available from https://doi.org/10.1016/j.biotechadv.2017.12.003, http://www.elsevier.com/inca/publications/store/5/2/5/4/5/5/index.htt.

Zeng, L., Sarmadivaleh, M., Saeedi, A., Chen, Y., Zhong, Z., & Xie, Q. (2023). Storage integrity during underground hydrogen storage in depleted gas reservoirs. *Earth-Science Reviews, 247*. Available from https://doi.org/10.1016/j.earscirev.2023.104625.

Zivar, D., Kumar, S., & Foroozesh, J. (2021). Underground hydrogen storage: A comprehensive review. *International Journal of Hydrogen Energy, 46*(45), 23436–23462. Available from https://doi.org/10.1016/j.ijhydene.2020.08.138, http://www.journals.elsevier.com/international-journal-of-hydrogen-energy/.

CHAPTER

11

Experimental, modeling, and simulation approaches for underground hydrogen storage

Mohamed Gamal Rezk[1], *Omar Elkhatib*[2] *and Arshad Raza*[3]

[1]Center for Integrative Petroleum Research, College of Petroleum Engineering and Geosciences, King Fahd University of Petroleum and Minerals, Dhahran, Saudi Arabia [2]ChampionX USA Inc., Sugar Land, TX, United States [3]Department of Petroleum Engineering, King Fahd University of Petroleum and Minerals, Dhahran, Saudi Arabia

11.1 Introduction

The urgency of developing gigantic green energy resources, especially hydrogen gas, is encouraging the search for wide-scale storage options to ensure an uninterrupted energy supply during periods of peak energy consumption. The giga-ton capacity of underground storage sites, including saline aquifers and depleted oil and gas reservoirs, has the potential to meet the prospective needs of a sustainable hydrogen economy. The costs associated with the underground hydrogen storage (UHS) operations are still considered exorbitantly high. However, economizing such a process requires investing in promising technological advancements and increasing research efforts taking place around the world at present and in the foreseeable future (Tarkowski, 2019). Experimental, simulation, and pilot studies are spearheading the research efforts towards providing a state-of-the-art reference for the global application of UHS.

The H_2 geo-storage process is associated with various risk sources, including its potential underground losses, contamination, and injectivity problems. Fig. 11.1 shows various aspects involved in the hydrogen geological storage process, the successful completion of which requires comprehensively investigating a multitude of factors, including stress field changes, dynamics of multiphase flow, microbial interactions, and chemical disequilibria upon the introduction of hydrogen gas into subterranean storage sites. Various experimental methods and numerical models were employed in the literature for incorporating such

DOI: https://doi.org/10.1016/B978-0-443-24071-3.00012-1

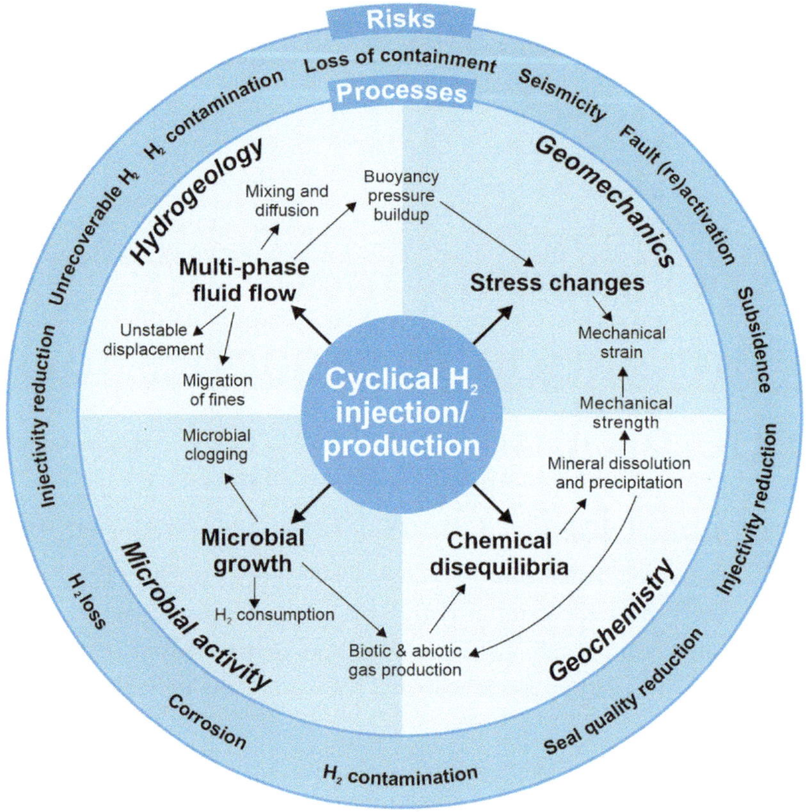

FIGURE 11.1 Factors controlling H_2 geological storage and sources of risks (Heinemann et al., 2021). Source: *From Heinemann, N., Alcalde, J., Miocic, J. M., Hangx, S. J. T., Kallmeyer, J., Ostertag-Henning, C., Hassanpouryouzband, A., Thaysen, E. M., Strobel, G. J., Schmidt-Hattenberger, C., Edlmann, K., Wilkinson, M., Bentham, M., Stuart Haszeldine, R., Carbonell, R., & Rudloff, A. (2021). Enabling large-scale hydrogen storage in porous media-the scientific challenges.* Energy and Environmental Science, 14*(2), 853–864. https://doi.org/10.1039/d0ee03536j.*

factors in order to optimize the hydrogen geological storage efficiency, and evaluate the formations best suited for its seamless, uninterrupted retrieval (Zivar et al., 2021).

The process of UHS includes the preinjection of a cushion gas that serves as a compression and expansion medium for efficient injectivity and deliverability of hydrogen gas. In addition, cushion gases reduce potential hydrogen losses into adjacent formations (Kanaani et al., 2022). However, the effect of cushion gas types on the storage capacity of hydrogen is an active area of research employing different experimental and modeling approaches.

Exposing rock minerals to hydrogen can impact the physicochemical properties of the porous medium. For instance, rock wettability is a potential factor that would be impacted by elongated exposure to hydrogen and its cushion gases. Considering the high diffusivity of the small hydrogen molecules, a higher threshold capillary pressure in the pore elements of the cap rock is necessary to reduce hydrogen seepage through its tight pores. In other words, a strong water-wet affinity is crucial for the success of UHS operations (Zivar et al., 2021).

Consequently, storage in depleted oil and gas reservoirs with mixed-to-oil wet conditions can pose additional challenges due to the sealing efficiency of oil-wet cap rocks. Probing these effects in the lab and utilizing simulation tools will facilitate the field application of UHS.

Investigating the abovementioned attributes of the UHS process has been attempted experimentally at high pressure and temperature (HPHT) conditions. Different experimental approaches have been employed, including pendant/sessile drop, microfluidic, and core flooding setups. In addition, pore scale and hydrodynamic modeling along with finite element and volume simulations, were attempted to utilize lab observations and incorporate them into representative UHS models. The next sections highlight the methodology employed in these experimental state-of-the-art and numerical simulations of hydrogen flow in porous media.

11.2 Experimental studies of underground hydrogen storage

11.2.1 Interfacial tension and wettability measurements

Interfacial tension (IFT) and rock wettability are key and crucial parameters that significantly impact the UHS process by dictating the fluid distribution in the formation, trapping efficiency, and storage capacity. The contact angle measurement is the most commonly used method to determine the wettability of a system. Largely, the sealing efficiency of candidate storage sites is controlled by the IFT and contact angle of the rock-fluid system as they shape the entry capillary pressure of the pore elements in the cap rock. IFT is experimentally measured using the pendant drop method, where a fluid droplet is suspended from a needle in a chamber filled with a different fluid, for example, gas. The IFT value is calculated by analyzing the suspended droplet profile at the experimental pressure and temperature. On the other hand, rock wettability can be experimentally measured by different methods, including the sessile drop and the tilted plate methods, both based on contact angle measurement. The sessile drop method involves placing a droplet of liquid on a solid surface and observing its shape. The contact angle formed by the droplet at the solid-liquid interface is indicative of the rock's wettability. The tilted plate method involves tilting a solid plate at various angles and observing its advancing and receding contact angles on the surface (Ali, Pan, et al., 2022; Iglauer et al., 2021). The angle at which the liquid begins to slide off the plate is related to the wettability of the surface. Fig. 11.2 shows a schematic diagram of the experimental apparatus that can be used to measure the contact angle and IFT between two fluids. The next sections discuss factors affecting IFT and rock wettability during the H_2 storage process.

11.2.1.1 Factor affecting H_2 interfacial tension at fluid–fluid interface

The interfacial tension of hydrogen at the interface separating two fluids is dependent on the system pressure, temperature, and salinity. In general, increasing the pressure of the gas–liquid system causes an increase in the gas density, and hence, the intermolecular interactions between the two fluids will be affected. Nevertheless, hydrogen has a relatively low density compared to other gases. Besides, the low solubility of hydrogen in reservoir brines

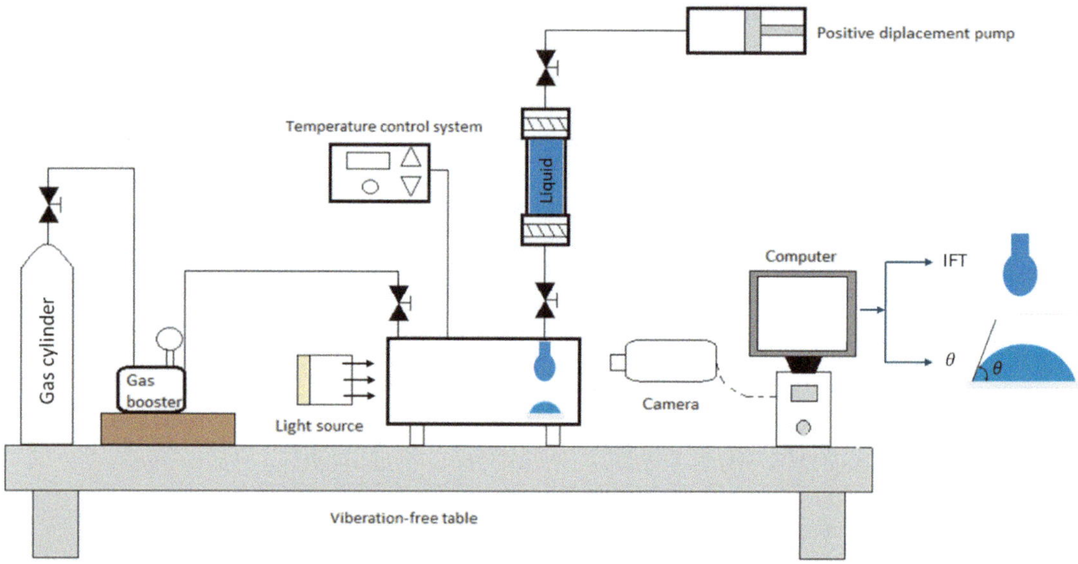

FIGURE 11.2 **Schematic of the experimental apparatus to measure the IFT and contact angle. In this system, the brine pendant drop or contact angle of a brine droplet surrounded by gas can be measured (Rezk and Foroozesh, 2019b).** Source: *Adapted from Rezk, M. G., & Foroozesh, J. (2019). Phase behavior and fluid interactions of a CO_2-light oil system at high pressures and temperatures.* Heliyon, 5(7). *https://doi.org/10.1016/j.heliyon.2019.e02057.*

promotes only an infinitesimal effect of pressure on H_2-reservoir fluid IFT (Al-Mukainah et al., 2022; Chow et al., 2018; Higgs et al., 2022). On the other side, reservoir temperature has a significant effect on the IFT between H_2 and reservoir fluids, which was reported to significantly decrease with increasing the system temperature (Hosseini et al., 2022a, 2022b). The higher thermal activity of the H_2 molecules at high temperatures can increase the fluid interactions at the interface, promoting higher IFT values. Fig. 11.3 illustrates the pressure and temperature effects on H_2-brine IFT. In addition, the H_2-brine IFT was found to increase with brine salinity. Increasing the brine salinity causes an increase in the brine density, which in turn induces a higher density difference between the brine and the hydrogen gas, resulting in the increase in the IFT. Fig. 11.4 shows the effect of brine salinity on H_2-brine IFT measured at different pressures.

11.2.1.2 Factor affecting rock wettability in H_2-fluid-rock system

Wettability is a rock-fluid property that is controlled by several factors, including pressure, temperature, brine salinity, rock surface roughness, the presence of organic acids, rock mineralogy, and surface contamination. Previous studies (Ali et al., 2021; Hosseini et al., 2022a, 2022b) experimentally measured the hydrogen wettability of both reservoir and cap rocks. Their results showed that the cap rocks are more water-wet in the presence of hydrogen, indicating higher entry capillary pressures and less risk of hydrogen leakage.

The effect of pressure on the H_2/brine/rock contact angle is not clearly defined in the literature. Previous studies reported an increase in the H_2/brine/rock contact angle on calcite,

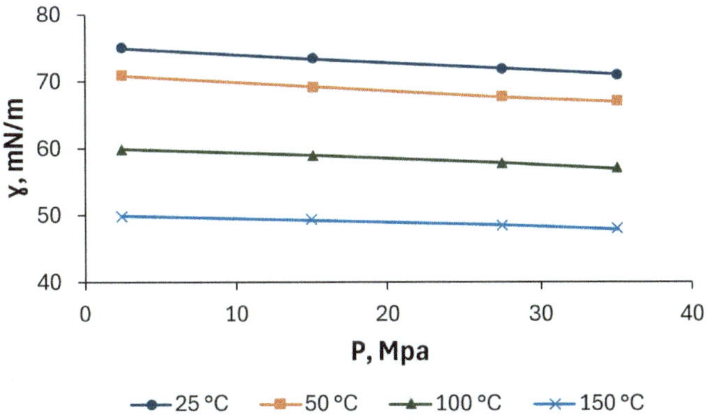

FIGURE 11.3 Pressure and temperature impact on H_2-brine IFT. Reservoir temperature had a significant effect on the IFT between H_2 and reservoir fluids while pressure had a minor impact on the H2-brine IFT. Source: *Adapted from Hosseini, M., Fahimpour, J., Ali, M., Keshavarz, A. & Iglauer, S. (2022). H_2 – brine interfacial tension as a function of salinity, temperature, and pressure; implications for hydrogen geo-storage.* Journal of Petroleum Science and Engineering, 213; *Gbadamosi, A. O. (2023). Underground hydrogen storage: A critical assessment of fluid-fluid and fluid-rock interactions.* Journal of Energy Storage, 72.

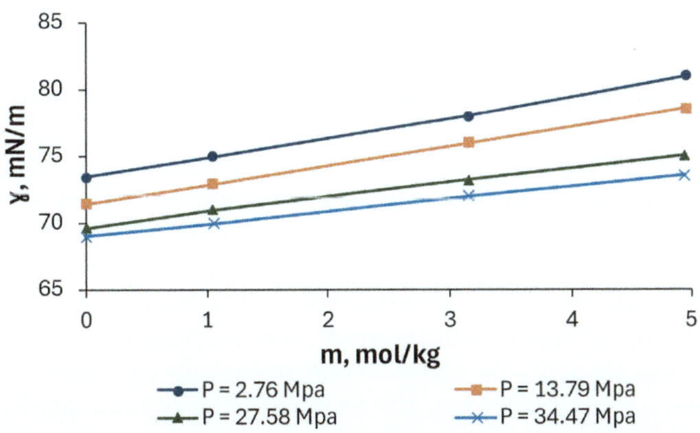

FIGURE 11.4 Effect of brine salinity on H_2-brine IFT measured at different pressures. Source: *Adapted from Hosseini, M., Fahimpour, J., Ali, M., Keshavarz, A. & Iglauer, S. (2022). H_2 – brine interfacial tension as a function of salinity, temperature, and pressure; implications for hydrogen geo-storage.* Journal of Petroleum Science and Engineering, 213; *Gbadamosi, A. O. (2023). Underground hydrogen storage: A critical assessment of fluid-fluid and fluid-rock interactions.* Journal of Energy Storage, 72.

quartz, and mica with pressure (Ali, Pan, et al., 2022; Hosseini et al., 2022a, 2022b). The increase in the contact angle was justified by the higher hydrogen density at higher pressures, which improved the intermolecular interactions in the hydrogen-brine system (Hosseini et al., 2022a, 2022b). Other studies reported that the contact angle of a H_2/brine/rock system is pressure-independent (Hashemi, Glerum, et al., 2021; Higgs et al., 2022). More studies and simulation work at the molecular level are required to further investigate the effect of pressure on contact angle. At high temperatures, the kinetic energy of hydrogen molecules increases,

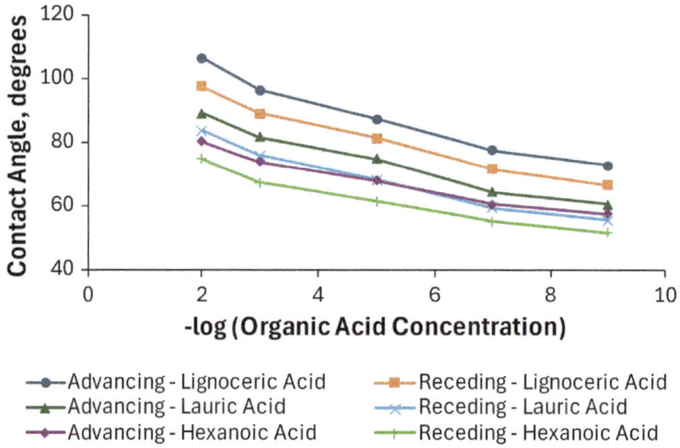

FIGURE 11.5 Impact of organic acid concentrations and structure on H_2/brine/mica wettability. H_2/brine/mica contact angle measured at various organic acid concentrations for different acid structures (Ali, Yekeen, et al., 2022; Gbadamosi, 2023). Source: *From Ali, M., Yekeen, N., Pal, N., Keshavarz, A., Iglauer, S., & Hoteit, H. (2022). Influence of organic molecules on wetting characteristics of mica/H₂/brine systems: Implications for hydrogen structural trapping capacities.* Journal of Colloid and Interface Science, *608, 1739–1749. https://doi.org/10.1016/j.jcis.2021.10.080; Gbadamosi, A. O. (2023). Underground hydrogen storage: A critical assessment of fluid-fluid and fluid-rock interactions.* Journal of Energy Storage, *72.*

which results in high diffusivity of the hydrogen gas and affects the H_2 gas molecular cohesive energy. Nevertheless, no clear relationship was reported in the literature for the impact of system temperature on H_2/brine/rock contact angle.

Based on the results of previous studies, the H_2/brine/rock contact angle was found to be insensitive to the brine salinity in sandstone reservoirs. However, the contact angle increased with brine salinity in H_2/brine/carbonate rocks. The reduction in rock hydrophilicity was attributed to the reduction in surface polarity due to the increase in brine salinity. The effect of different brine chemistries on H_2/brine/rock wettability still needs to be further investigated at representative reservoir conditions.

The presence of organic acids in hydrocarbon reservoirs can have a profound effect on their hydrogen storage kinetics. Hence, the impact of the organic acids on H_2/brine/rock wettability was investigated in many studies to provide a reliable representation of actual reservoir conditions. The results showed that the rock's hydrophilicity was reduced in the presence of organic acids. This was justified by the organic acid adsorption on the rock surface. The structure of different organic acids also had a noticeable impact on the wetting affinity of rock minerals. Fig. 11.5 shows the H_2/brine/mica contact angle at various organic acid concentrations for different acid structures.

11.2.2 Hydrogen solubility and diffusion in brine

Gas solubility and diffusivity can be measured by calculating the mass transfer between phases when gas is brought into contact with liquids. The gas solubility in brine is mainly controlled by the system pressure, temperature, and salinity. In general, brine salinity and system temperature have adverse effects on gas solubility in brine. While the dissolution of gases in liquids is enhanced at higher pressures, the solubility of hydrogen in brine is relatively low even at high pressures due to the low capability to form bonds between the two phases and the relative polar nature of water and the nonpolar feature of hydrogen (Gbadamosi, 2023). Although hydrogen has low solubility in brine, hydrogen molecular diffusion is high due to the small size of its molecules (Ghaedi et al., 2023). Depending on the capillary number during hydrogen flow in porous media, the mass transfer from the H_2 phase to the brine can take place through molecular diffusion or convection. For extremely low flow rates, the mass transfer process would be dominated by molecular diffusion, which is driven by a gas concentration gradient regulating hydrogen molecules movement from high concentration regions to low concentration regions. Molecular diffusion is defined by several types of equations. The most common relationship is Fick's first law:

$$J = -D\frac{\partial c}{\partial x} \tag{11.1}$$

where J is the H_2 diffusion flux measured in H_2 moles per unit area during a unit time interval, D is the diffusion coefficient in (m^2/sec), c is the concentration of hydrogen per unit volume, (m^3) and x is the length in meters.

Several experimental methods are available to measure gas solubility and diffusivity in liquids (Rezk et al., 2022). The gas solubility is calculated as the number of moles of gas dissolved in a certain mass of liquid. This can be obtained by monitoring the change in the pressure of a gas–liquid system under isothermal conditions or by taking samples during the experiment to be analyzed by gas chromatography (Rezk & Foroozesh, 2019a). Fig. 11.6 shows a schematic diagram of a hydrogen-brine solubility/diffusivity cell, while Fig. 11.7 shows the hydrogen solubility in water and brine with different salinities measured at two different temperatures and various pressures.

The experimental measurement of gas diffusion in liquids can be attained by direct and indirect methods (Rezk & Foroozesh, 2018). Direct methods depend on collecting fluid samples during the diffusion test to directly measure the fluid compositions, while indirect

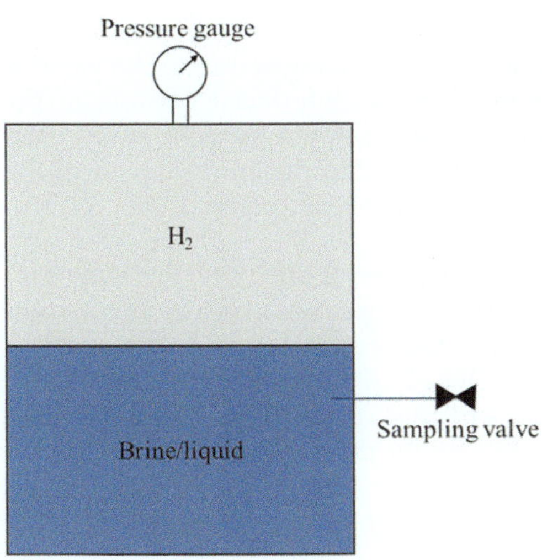

FIGURE 11.6 A schematic diagram of hydrogen solubility/diffusivity measurements cell.

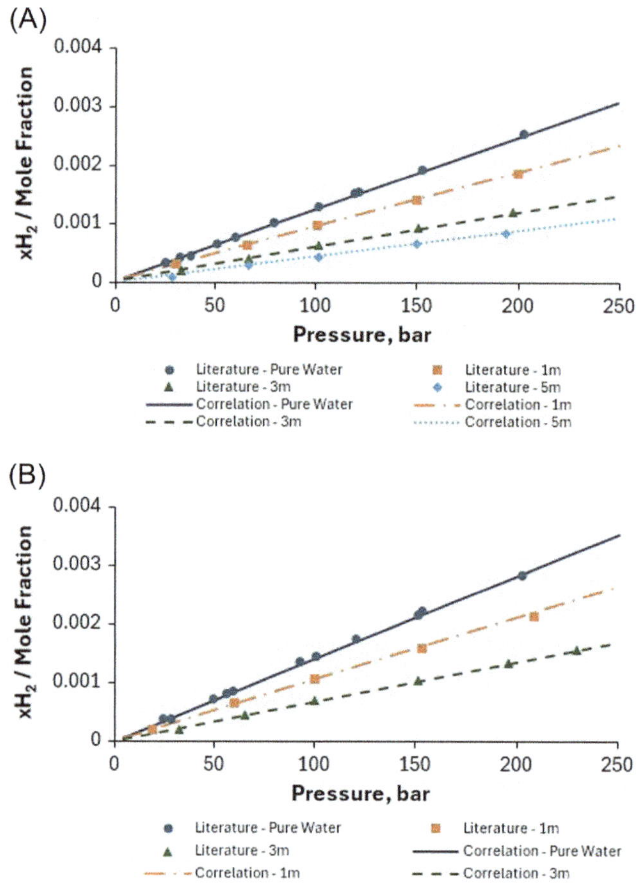

FIGURE 11.7 Hydrogen solubility in water and brine with different salinities measured at two different temperatures and various pressures (A) 323K and (B) 373K (Chabab et al., 2020). Source: *Adapted form Chabab, S., Théveneau, P., Coquelet, C., Corvisier, J., & Paricaud, P. (2020). Measurements and predictive models of high-pressure H_2 solubility in brine (H2O + NaCl) for underground hydrogen storage application.* International Journal of Hydrogen Energy, *45(56), 32206–32220. https://doi.org/10.1016/j.ijhydene.2020.08.192.*

methods are based on measuring a property of the system that changes due to the gas diffusion process. This property can be system pressure, volume, interfacial tension, or the position of the gas–liquid interface. The pressure decay test is the most commonly used gas diffusion measurement method. In a pressure decay test, the dynamic change in the pressure of gas in direct contact with a liquid is recorded. The system is sealed in a diffusion cell (see Fig. 11.6) under a constant temperature (Rezk et al., 2022). Due to the complexity and potential hazards of measuring H_2 solubility and diffusivity in laboratory settings, limited studies have reported the results of these properties at realistic reservoir conditions. The hydrogen diffusivity was measured in shale formation and compared to the CO_2 diffusivity in the same formation as shown in Fig. 11.8 at various pressures and a temperature of 333K. It is clear that the diffusion coefficient of hydrogen is higher than that of CO_2 by one order of magnitude when measured under the same conditions.

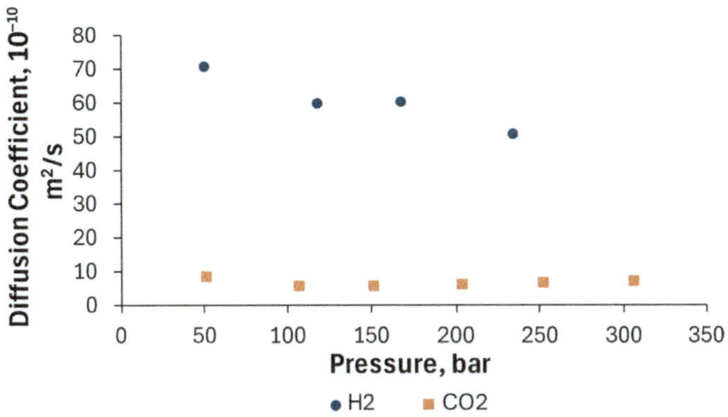

FIGURE 11.8 H_2 and CO_2 diffusion coefficients in shale formation as a function of pressure at 333K. Hydrogen diffusivity measured in shale formation and compared to the CO_2 diffusivity in the same formation at various pressures and a temperature of 333K (Samara et al., 2022). Source: *Adapted from Samara, H., Von Ostrowski, T., & Jaeger, P. (2022). geological storage of carbon dioxide and hydrogen in Jordanian shale formations. In:* Proceedings – SPE Annual Technical Conference and Exhibition *(Vols. 2022). Society of Petroleum Engineers (SPE). https://doi.org/10.2118/210202-Ms.*

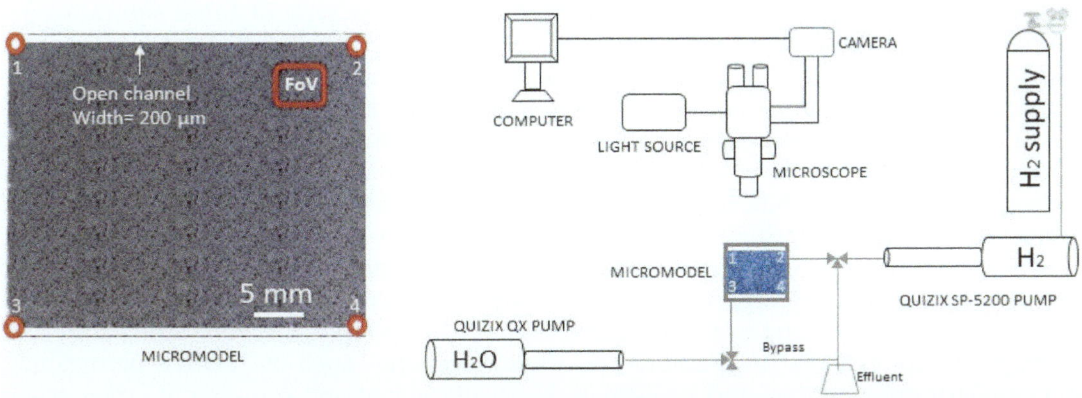

FIGURE 11.9 Micromodel setup used to study hydrogen displacement and trapping efficiencies. A micromodel setup with irregular pore pattern used to measure hydrogen displacement and trapping efficiencies. Source: *Adapted from Lysyy, M., Liu, N., Solstad, C. M., Fernø, M. A., & Ersland, G. (2023). Microfluidic hydrogen storage capacity and residual trapping during cyclic injections: Implications for underground storage.* International Journal of Hydrogen Energy, *48(80), 31294–31304. https://doi.org/10.1016/j.ijhydene.2023.04.253.*

11.2.3 Pore scale tests—microfluidics

Understanding hydrogen transport properties in porous media is essential for the appropriate selection of underground storage sites and for optimizing the hydrogen storage process. Utilizing micromodels provided direct observations of the multiphase flow properties. Fig. 11.9 shows a micromodel setup with an irregular pore pattern that represents a heterogeneous porous medium. The micromodel has four ports for fluid injection.

A microscope is used to observe the dynamic fluid flow, and a camera is used for recording. The setup is equipped with two pumps for hydrogen and brine injection.

In general, injected gases are trapped in a sealed, porous structure through four trapping mechanisms (Iglauer, 2011). These mechanisms are structural, mineral, solubility, and capillary trapping. During hydrogen geo-storage, capillary (residual) trapping is recognized as one of the main hydrogen loss mechanisms. Considering that underground hydrogen storage and recovery involve several cycles of hydrogen injection (drainage) and withdrawal (imbibition), gas residual trapping takes place during the imbibition process, where water partially displaces hydrogen from the pores. Previous studies used pore-scale imaging to investigate hydrogen residual trapping at different experimental conditions, including operating pressure, injection rate, and number of injection-withdrawal cycles (Lysyy et al., 2022; Medina et al., 2024). The impact of pore pressure on the initial hydrogen saturation was found to be insignificant. However, hydrogen recovery during brine imbibition was found to be dependent on the pore pressure. In other words, higher hydrogen residual saturations were obtained at higher pressures. Increasing the capillary number during brine imbibition increased the hydrogen recovery (Thaysen et al., 2023). In a recent study, the micromodel test results showed that similar hydrogen residual saturations were obtained for the tested five injection-withdrawal cycles (Lysyy et al., 2023). Additionally, the trapped hydrogen was reconnected by capillary forces after the brine imbibition process in the subsequent drainage process. The type of brine saturating the porous media and its salinity were found to control the hydrogen initial saturation value. Fig. 11.10 shows initial hydrogen saturation (S_{gi}) after drainage process in micromodels saturated with different salt types (KCl, NaCl, and KCl + NaCl) and brine concentrations (0.5, 2, and 4 M). Increasing the brine salinity resulted in lower hydrogen displacement efficiency. As can be observed, increasing salt concentration resulted in lower initial hydrogen saturations, while the highest S_{gi} was obtained in the KCl case. The observed difference in S_{gi} can be justified by the change in capillary number due to interfacial tension alteration, the change in mobility ratio due to the change in viscosity of brine, and wettability alteration (Medina et al., 2024). Although micromodels provide a qualitative and quantitative description of hydrogen flow and trapping at the pore scale, core-scale experiments are essential to measure rock-fluid properties such as relative permeability and capillary pressure data and investigate H_2 displacement and trapping at larger scales. The next section highlights core flooding experiments using H_2-brine systems.

11.2.4 Core flooding tests

Hydrogen has relatively low density and viscosity compared to reservoir brine, even at high pressures and temperatures. Hence, during hydrogen injection (hydrogen storage), viscous fingering and gravity segregation can reduce hydrogen displacement efficiency in saline aquifers, while during hydrogen reproduction (the imbibition process), brine will be displacing the less viscous and less dense hydrogen phase. Hence, depending on the rock properties and pore structures, pockets of unrecoverable hydrogen may exist in the reservoir. Therefore, a good understanding of hydrogen flow in porous media is essential to evaluating the performance of the UHS process. Relative permeability and capillary

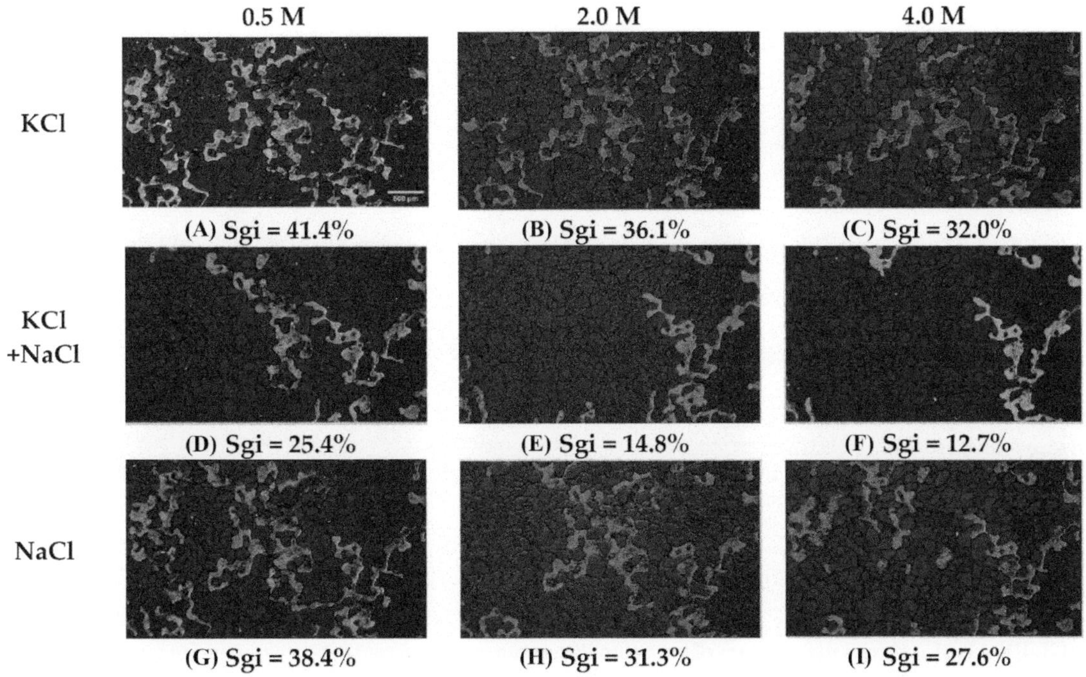

FIGURE 11.10 Initial H_2 saturation (S_{gi}) in light blue after gas drainage in porous media saturated with three different types of brine with different salinities. The tests were conducted at pressure = 30 bar and temperature = 20°C with H_2 injection rate = 10 mL/h. Source: *From Medina, O. E., Gallego, J. F., Moncayo-Riascos, I., Lysyy, M., Benjumea, P. N., Cortés, F. B. & Franco, C. A. (2024). Salinity influence on underground hydrogen storage: Insights from molecular dynamics and pore-scale analysis.* International Journal of Hydrogen Energy, *60, 959–975. https://doi.org/10.1016/j.ijhydene.2024.02.073.*

pressure are two key rock-fluid properties that describe the hydrogen-brine dynamic interactions in a specific rock. Capillary forces are responsible for the residual trapping of the injected hydrogen, that is, the nonwetting phase, during the imbibition process. Due to the high risk of running hydrogen experiments at high pressures and elevated temperatures, there are limited studies in the literature that have experimentally tested hydrogen injection in porous media at the core scale. Fig. 11.11 shows a typical schematic diagram of the core flooding system used for hydrogen tests. The main components of the setup include a positive displacement pump, a core holder, an X-ray source and detector, an LCR meter, accumulators, pressure gauges, and a back pressure regulator.

Relative permeability (K_r) can be measured directly by conducting steady- or unsteady-state core flooding tests. For the steady-state K_r method, the wetting and nonwetting phases are coinjected into a rock sample at various fractional flows. This method calculates K_r directly from Darcy's law. Nevertheless, steady-state conditions need to be achieved, and water saturation and the differential pressure along the core need to be measured for each fractional flow. On the other hand, the unsteady state method is based on injecting the nonwetting phase (gas) at a constant rate and recording the pressure drop. The measurement is

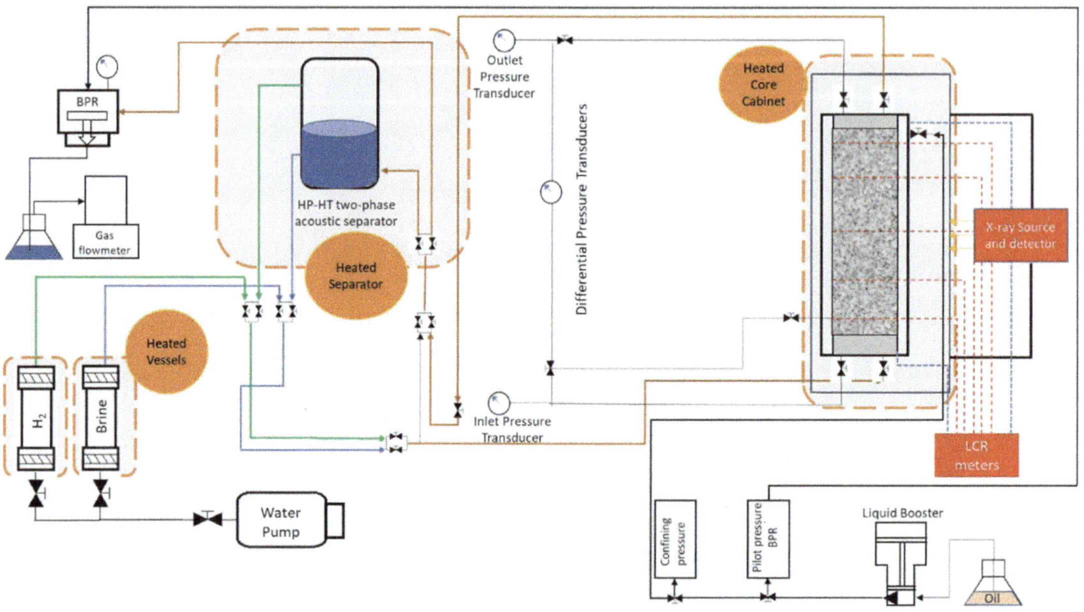

FIGURE 11.11 Schematic diagram of core flooding setup equipped with X-ray scanner and resistivity measurements. A typical schematic diagram of core flooding system used for hydrogen flooding tests (Adebayo et al., 2024). Source: *Adapted from Adebayo, A. R., Rezk, M. G., & Badmus, S. O. (2024). The effect of multiple cycles of surfactant-alternating-gas process on foam transient flow and propagation in a homogeneous sandstone.* SPE Journal, *29(2), 909–927. https://doi.org/10.2118/217999-PA.*

quicker compared to the steady state, but obtaining the K_r curves requires numerical modeling as only the end points of the K_r curves can be measured. H_2/brine capillary pressure curves can be obtained from mercury injection capillary pressure (MICP) measurement by converting the measured data from the Hg/air system to the H_2/brine system using the Young–Laplace equation. Additionally, the H_2/brine capillary pressure can be measured directly using the semipermeable disk method (Hassler & Brunner, 1945) or the steady-state technique (Pini et al., 2012). Fig. 11.12 shows steady-state K_r data and capillary pressure for the H_2/water system measured in a sandstone formation. It can be observed that hydrogen has low relative permeability values. Additionally, the low capillary pressure values shown in Fig. 11.12 indicate that hydrogen can easily invade large pore sizes. In summary, a very limited bulk of literature attempted to evaluate the H_2/brine drainage and imbibition relative permeability and capillary pressure behavior.

Previous studies investigated hydrogen displacement and trapping efficiencies in different types of lithologies using a core flooding system equipped with a medical X-ray CT scanner (Al-Yaseri et al., 2024; Boon & Hajibeygi, 2022). Gas drainage followed by water imbibition was conducted to quantify initial and residual gas trapping in porous media. Fig. 11.13 shows water saturation distributions during both drainage and imbibition processes using different fraction flows of hydrogen and water. The gravity segregation between the injected phases caused the H_2 to saturate the top sections of the core. High-density fingers were

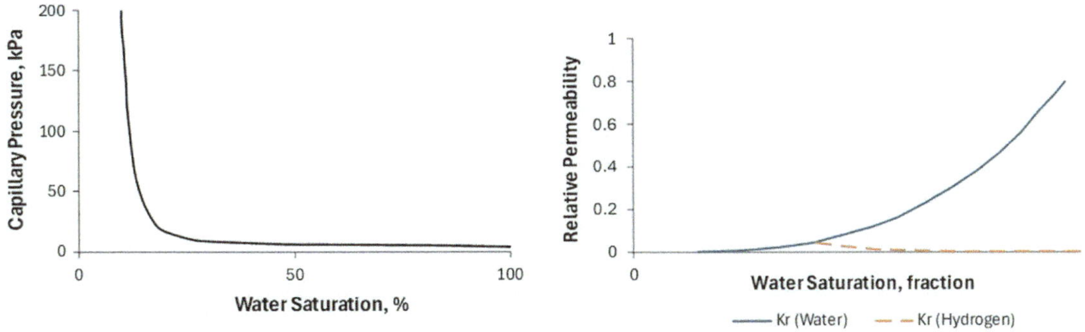

FIGURE 11.12 H_2/water capillary pressure and relative permeability curves in sandstone formation. Steady state K_r data and capillary pressure for H_2/water system measured in a sandstone formation (Zivar, Kumar and Foroozesh, 2021; Yekta et al., 2018). Source: *From Zivar, D., Kumar, S., & Foroozesh, J. (2021). Underground hydrogen storage: A comprehensive review.* International Journal of Hydrogen Energy, *46(45), 23436–23462. https://doi.org/10.1016/j.ijhydene.2020.08.138; Yekta, A. E., J. -C, Gaboreau, S., Pichavant, M., & Audigane,. (2018). Determination of hydrogen–water relative permeability and capillary pressure in sandstone: Application to underground hydrogen injection in sedimentary formations.* Transport in Porous Media, 122.

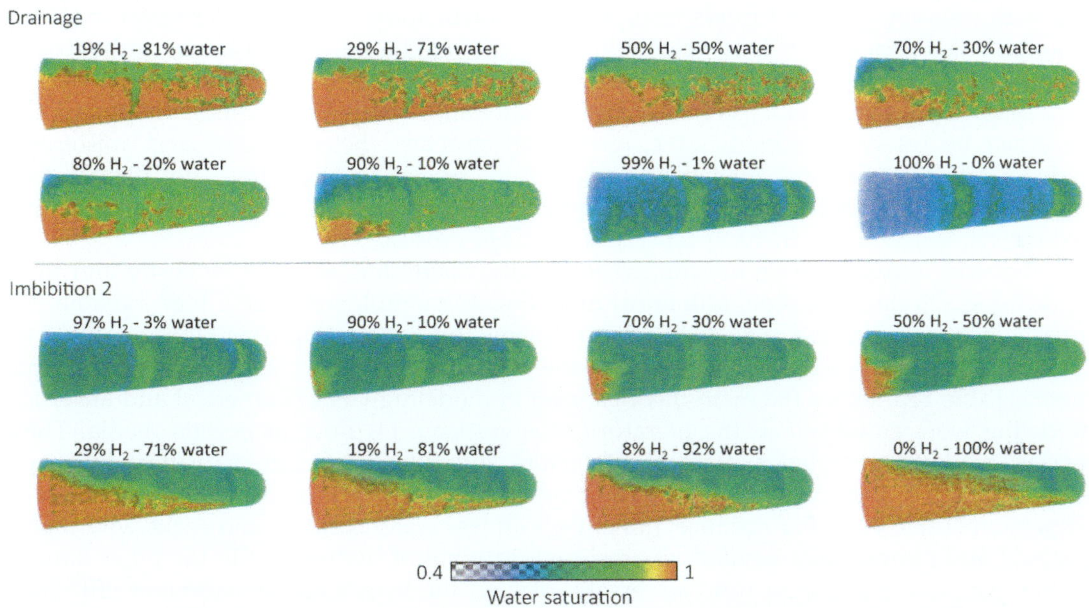

FIGURE 11.13 Water saturation distributions during drainage and imbibition processes, presented at various fractional flows. Source: *From Boon, M. & Hajibeygi, H. (2022). Experimental characterization of H_2/water multiphase flow in heterogeneous sandstone rock at the core scale relevant for underground hydrogen storage (UHS).* Scientific Reports, 12.

observed during the imbibition process, as shown in Fig. 11.13. Hydrogen residual trapping was found to be highly controlled by the rock wettability and pore structure. The reported residual saturations of hydrogen gas in sandstones at the end of the imbibition process varied from low values (e.g., 2% (Al-Yaseri et al., 2022)) to high values (e.g., 41% (Jha et al., 2021)). It is worth mentioning that residual gas saturation is a function of the maximum initial gas saturation achieved at the end of the gas drainage process; that is, higher initial gas saturation resulted in higher saturation of residual gas. Nevertheless, it is required to maximize the initial gas saturation while keeping the lowest residual gas trapped during the hydrogen recovery process. In other words, hydrogen is recommended to be temporarily stored in formations with low residual trapping efficiency (S_{gr}/S_{gi}) and high H_2 displacement efficiency.

11.3 Modeling and numerical simulations of underground hydrogen storage

The performance of UHS is controlled by hydrogeological, geochemical, geo-mechanical, biochemical, and mineralogical processes. Comprehensive studies of these processes help in understanding the H_2-rock-fluid interactions, H_2 storage capacity, H_2 losses, and efficiency of H_2 injection and withdrawal cycles. Geochemical modeling evaluates the reactivity of hydrogen with reservoir rock and fluids using defined kinetic rates. The geo-mechanical modeling mainly investigates rock failure potential during the injection and withdrawal of hydrogen. Most of the previous studies, nevertheless, focused on the hydrodynamic modeling of UHS by assessing properties such as relative permeability and hysteresis. Additionally, the UHS is governed by reservoir properties (i.e., geological structure, heterogeneity, and anisotropy) and operating conditions such as injection rate, operating pressure, and temperature. Fig. 11.14 shows various parameters and phenomena studied in modeling hydrodynamic, geo-mechanical, and geochemical aspects of the UHS process.

Laboratory measurements are limited to specific conditions and can be costly and time-consuming. On the other side, numerical modeling and simulation have a high capability of investigating various and realistic field conditions and conducting sensitivity analyses for UHS. Various mathematical models describe hydrogen transport and storage in porous media. Table 11.1 shows the main theories used in modeling UHS. Numerical and analytical modeling were employed in the literature to investigate H_2 flow in porous media. These models have incorporated a plethora of factors and flow attributes that affect hydrogen injection and recovery, including operating conditions, rock-fluid properties, diffusion, and dispersion (Table 11.2). For instance, minerals with less water wetness and rocks with higher porosity and permeability yielded less residual trapping of hydrogen. On the other hand, a higher pore-to-throat aspect ratio adversely affected the hydrogen displacement efficiency, as observed by pore-scale modeling (Hashemi, Blunt, et al., 2021). Besides, heterogeneity in the storage formation negatively impacted the storage and withdrawal efficiency of hydrogen. Finite element modeling portrayed the diffusion and dispersion of hydrogen into aquifers, which entailed significant gravity override and fingering effects compared to its seamless and controlled injection into gas reservoirs. The hydrogen injection rate was also analyzed by the finite volume method, resulting in a more unstable displacement at higher rates due to more pronounced viscous forces (Hagemann et al., 2015).

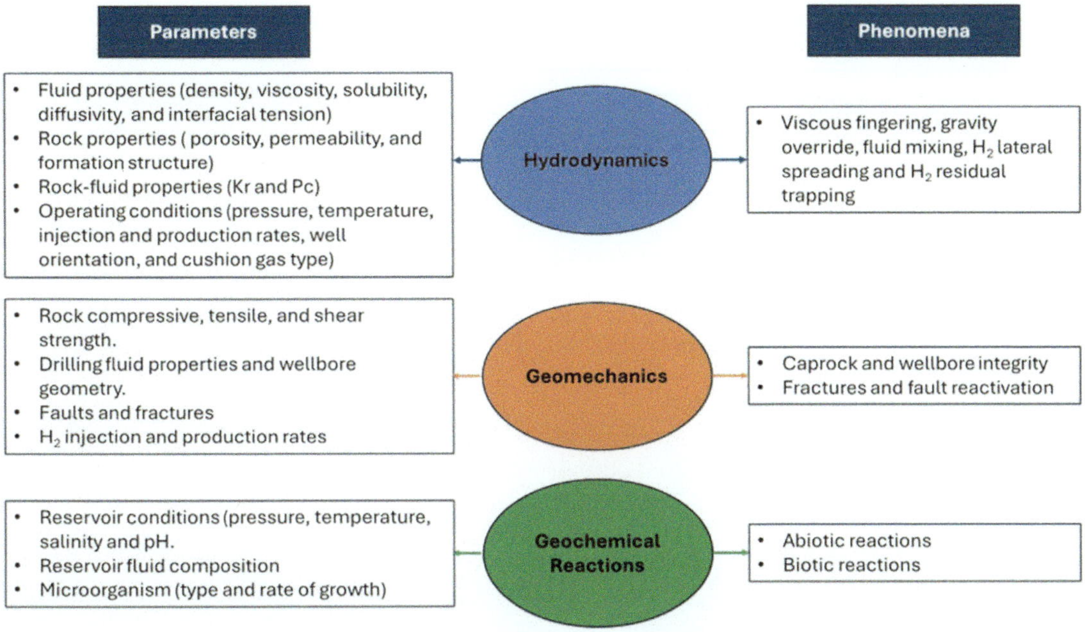

FIGURE 11.14 Various parameters and phenomena studied in modeling hydrodynamic, geomechanical, and geochemical aspects of UHS process.

The type of cushion gas has a marked effect on hydrogen storage and recovery efficiency. The use of denser supercritical CO_2, for example, managed to increase the hydrogen storage capacity while providing good control over its lateral spreading away from the wellbore. Yet, considerable fingering was recorded in finite difference models, as shown in Fig. 11.15, which highlights the potential losses during subsequent hydrogen recovery cycles. The use of lighter gases like methane and nitrogen as a cushion for hydrogen injection negatively impacted its storage capacity. Besides, nitrogen showed a similar fingering tendency, while methane significantly minimized hydrogen fingering (Saeed & Jadhawar, 2024).

Other factors, including pressure, temperature, and geochemical reactions, produced an insignificant impact on UHS (Table 11.2). The minimal solubility of hydrogen in water is the underlying reason for this behavior. Geomechanically, aquifer reservoirs with sealing cap rocks are more stable compared to shallow salt caverns. Nonetheless, the hydrogen losses were observed to be strictly limited in salt caverns, while higher residual hydrogen trapping was induced in aquifer storage sites.

11.3.1 Molecular dynamics simulation

Molecular modeling serves as a powerful tool for investigating binding affinities and providing detailed molecular-level analysis across a wide range of conditions where experimental methods may not be feasible. For instance, atomistic modeling has been applied in molecular equilibrium simulations (Ms) to obtain the capacity of adsorption on

TABLE 11.1 Main mathematical equations used in numerical simulation and modeling underground hydrogen storage (Zivar, Kumar and Foroozesh, 2021).

Theory	Mathematical equation	Nomenclature
Mass balance	$\varnothing\frac{\partial(\rho_g c^k{}_g S_g + \rho_w c^k{}_w S_w)}{\partial t} - \nabla.\left(\rho_g c^k{}_g v_w + J^k{}_w + \rho_w c^k{}_w v_g + J^k{}_g\right) = q^k$	$\varnothing$: porosity, ρ: molar density, c: mole fraction, q: source or sink term, S: saturation, v: fluid velocity, J: dispersive/diffusive flux. The subscripts g is for gas phase, w is for water phase and k stands for chemical components.
Darcy's law	$v_i = \lambda_i.(\nabla P_i - \rho_i g)$	v is velocity. λ, P and ρ are the phase mobility, pressure, and density, respectively. g is the gravitational acceleration. i stands for gas or water phase.
Stefan-Maxwell equation	$j^k{}_g = -\left(\sum_{j=1}^{n} \frac{c^i{}_g}{\rho_g D^{ij}{}_{diff,g}}\right)^{-1} \nabla c^k{}_g - \rho_g D^k{}_{disp,g} \nabla c^k{}_g$	$D^k{}_{disp,g}$ is mechanical dispersion coefficient of component k in water. $D^{ij}{}_{diff,g}$ is the effective binary diffusion coefficient between components i and j
Microbial population dynamics	$\frac{\partial n_m}{\partial t} = S_w \psi_m{}^{growth}.n_m - \psi_m{}^{decay}.n_m + \nabla.(D_m \nabla n_m)$	n is the number of microorganisms, $\psi_m{}^{growth}$ is the microbial growth function, D_m is the microbial diffusion coefficient, $\psi_m{}^{decay}$ is the decay function.
H_2 storage amount	$\frac{V_G{}^H}{V_R} = \left(\frac{Y_H \varnothing(1 - S_{wi})}{Z}\right)\left(\frac{P}{P_o}\right)\left(\frac{T_o}{T}\right)$	$V_G{}^H$ is total capacity of hydrogen, Y_H is volume fraction of hydrogen in the gas, S_{wi} is the irreducible water saturation, P and T are pressure and temperature, respectively. $\varnothing$ is rock porosity, and o stands for standard conditions.
Fraction of hydrogen loss due todissolution and diffusion through caprock or aquifer after a time (L_d)	$L_d = \frac{2S_H\sqrt{(D_e \varnothing_d t)/\pi}}{\left(\frac{Y_H \varnothing(1 - S_{wi})}{Z}\right)\left(\frac{P}{P_o}\right)\left(\frac{T_o}{T}\right)H}$ where $D_e = (D_H \varnothing_d \delta)/\tau$	S_H is dissolved hydrogen concentration in the reservoir at T and P, D_e (effective diffusivity), $\varnothing_d$ is porosity of caprock or aquifer, D_H is diffusivity of hydrogen in water, δ is pores constriction factor, and τ is the tortuosity.

From Zivar, D., Kumar, S., & Foroozesh, J. (2021). Underground hydrogen storage: A comprehensive review. International Journal of Hydrogen Energy, *46(45), 23436–23462. https://doi.org/10.1016/j.ijhydene.2020.08.138.*

TABLE 11.2 Investigated parameters/properties of underground hydrogen storage using numerical simulation and modeling.

Investigated process/property	Method/software	Main findings	References
Wettability Rock properties	Pore scale modeling	A less water-wet system resulted in lower H_2 trapping. The highest porosity and permeability sample had the lowest H_2 residual trapping. The smallest pores and restricted connectivity sample had the highest H_2 residual trapping.	Hashemi, Blunt, et al. (2021)
Hydrodynamics	DuMux code -finite volume method	Low injection rates resulted in a uniform displacement (gravitational forces are dominant). High injection rates resulted in unstable displacement (viscous forces are dominant).	Hagemann et al. (2015)
Diffusion Dispersion Hydrodynamics	COMSOL Multiphysics (finite element) and DuMux software (finite volume)	Diffusion and mechanical dispersion amplified gas components mixing. Gravity override and viscous fingering appeared during H_2 injection in saline aquifers. While both phenomena had a minor role in gas reservoirs.	Feldmann et al. (2016)
Type of cushion gas	CMG-GEM compositional reservoir simulator (finite difference)	CO_2 as cushion gas resulted in the highest H_2 storage capacity due to its relatively high density. Lighter cushion gases, such as N_2 and CH_4, had higher recovery efficiency by reducing unfavorable phenomena such as gravity overriding and viscous fingering.	Saeed and Jadhawar (2024)
Heterogeneity Hysteresis Hydrogen solubility and diffusivity in water	CMG-GEM compositional reservoir simulator (finite difference)	Formation heterogeneity highly affected the total volume of produced H_2 (lower H_2 recovery in case of heterogeneous formation compared to homogeneous aquifer). Relative permeability resulted in lower total H_2 stored and produced volumes due to gas entrapment. Hydrogen dissolution and diffusion in water have a noticeable impact on UHS when included with hysteresis.	Jadhawar and Saeed (2024)
Effect of pressure, temperature, mineral composition on H_2 losses due to geochemical reactions.	PHREEQC	Pressure and temperature had a negligible effect on H_2 losses due to low solubility in brine. From geochemical perspective, calcite-free reservoir is better for UHS.	Bo et al. (1999)
Geochemical effects on gas storage and rock properties.	PHREEQC	Negligible H_2 losses due to geochemical reactions.	Saeed et al. (2023)

(Continued)

TABLE 11.2 (Continued)

Investigated process/property	Method/software	Main findings	References
		With using CO_2 as a cushion gas, rock porosity and permeability may increase due to carbonate dissolution.	
Hydraulic and geomechanical impacts on UHS.	3D fluid flow coupled with a geomechanical model	78% of injected H_2 can be extracted. During H_2 storage, formation rock and caprock still can maintain a geomechanical stability.	Bai and Tahmasebi (2022)
Salt caverns integrity	Hydromechanical modeling	Deep caverns are more subjected to mechanical stability issues compared to shallow caverns. Limited H_2 loss was observed.	Coarita-Tintaya et al. (2023)

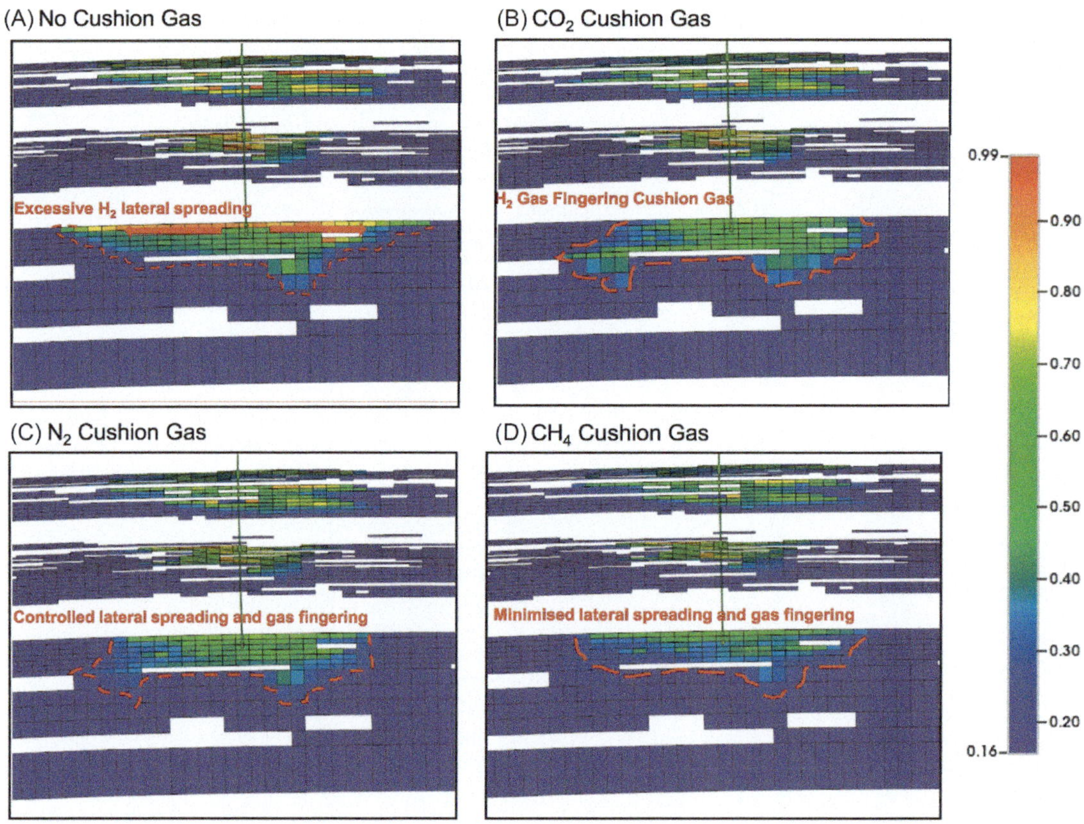

FIGURE 11.15 H_2 saturation distribution after injection-production cycles simulated without cushion gas and with different types of cushion gases: (A) no-cushion gas, (B) CO_2 cushion gas, (C) N_2 cushion gas, and (D) CH_4 cushion gas. Source: *From Saeed, M. & Jadhawar, P. (2024). Optimizing underground hydrogen storage in aquifers: The impact of cushion gas type.* International Journal of Hydrogen Energy, 52, *1537–1549. https://doi.org/10.1016/j.ijhydene.2023.08.352.*

solid structures (adsorbents) and the diffusion behavior of gases. Besides, molecular dynamics (MD) simulation has been applied to the hydrogen and reservoir fluids and rocks in UHS systems. More specifically, hydrogen diffusion in underground formations and solubility in brine have been studied using MD simulation. Additionally, MD was employed to study the behavior of sealing materials in the presence of hydrogen at various pressures and temperatures, which is essential for selecting relevant materials that can maintain integrity during hydrogen storage. Hydrogen diffusion was studied in three clay types. The MD simulation results showed lower hydrogen diffusion in clay minerals compared to bulk water. The hydrogen diffusion coefficient was estimated to increase with larger pore sizes (up to 2 nm), while at pores greater than 2 nm, the diffusion coefficient remains constant (Ghasemi et al., 2022). Hydrogen solubility in brine under various conditions, that is, pressure, temperature, and salinity, was studied at the molecular level as well (Zhang et al., 2023). The modeling results are consistent with the laboratory-measured parameters. Temperature was found to have a nonlinear impact on hydrogen solubility in brine.

The hydrogen adsorption behavior in kerogen was studied at different pressures and temperatures (Raza et al., 2022). Before conducting molecular simulations on kerogen systems, the methodology was validated through a molecular loading approach, wherein loading hydrogen into an empty cell yields a density range consistent with experimental data (from NIST) for hydrogen, as presented in Fig. 11.16. Thus, a validated modeling approach coupled with appropriate force field assignment forms the foundation for molecular simulations, enabling the testing of structures under high-pressure and high-temperature conditions. The adsorption process was simulated utilizing the Monte Carlo (MC) technique, employing the grand-canonical Monte Carlo (GCMC) approach. Following the simulation, the total average number of gas molecules adsorbed was determined based on the chemical potential. For a more detailed understanding of the GCMC simulation procedure, additional information can be found in the referenced literature (Zhao et al., 2017).

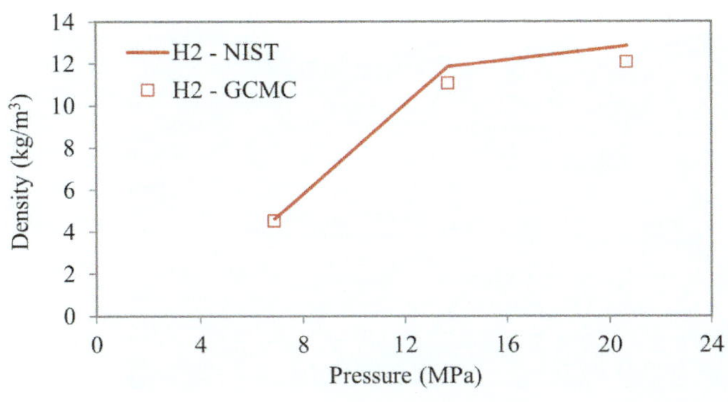

FIGURE 11.16 Molecular loading approach: density from predicted versus experimental data (from NIST) for hydrogen. Source: *From Raza, A., Mahmoud, M., Alafnan, S., Arif, M., & Glatz, G. (2022). H_2, CO_2, and CH_4 adsorption potential of kerogen as a function of pressure, temperature, and maturity.* International Journal of Molecular Sciences, 23 *(21). https://doi.org/10.3390/ijms232112767.*

Hydrogen adsorption in an organic-rich formation was investigated, focusing on over-mature shale. A nanopore layer model representing type II-D kerogen was constructed as illustrated in Fig. 11.17. The study encompassed a broad pressure range, spanning from 2.75 to 20 MPa, for investigating H_2 adsorption on the kerogen structure. Fig. 11.18 illustrates a direct relationship between pressure and the adsorption of H_2. Additionally, it was noted that the adsorption behavior tended to plateau beyond 20 MPa pressure. Regarding H_2 adsorption, there is currently no effort directed towards examining kerogen to be compared with the findings of this study. In conclusion, the studied factors had a remarkable impact on the thermodynamics of adsorption, wherein the bonding between the adsorbate and adsorbent heavily relies on gas density.

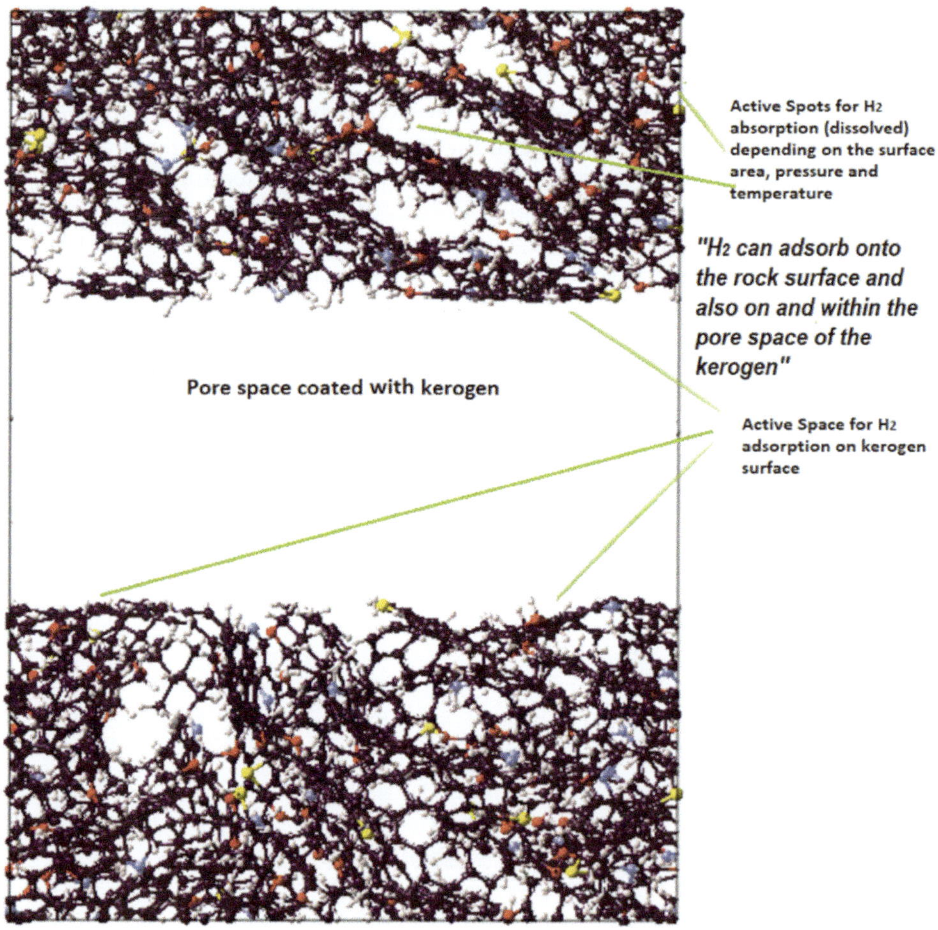

FIGURE 11.17 Illustration of kerogen shale structure with nanopore to study adoption on the surface. Source: *From Raza, A., Mahmoud, M., Alafnan, S., Arif, M., & Glatz, G. (2022). H_2, CO_2, and CH_4 adsorption potential of kerogen as a function of pressure, temperature, and maturity.* International Journal of Molecular Sciences, *23(21). https://doi.org/10.3390/ijms232112767.*

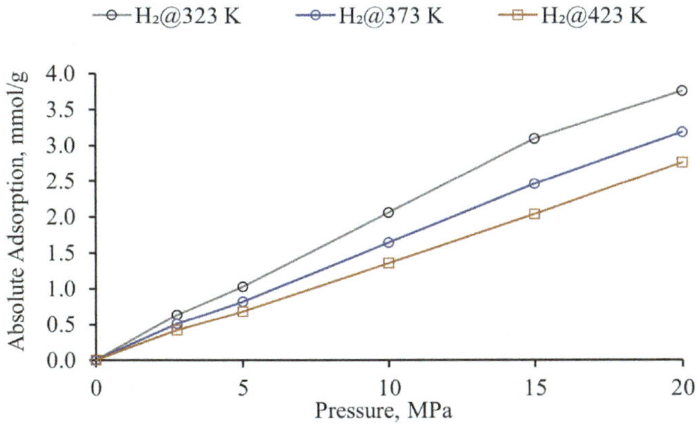

FIGURE 11.18 Pressure and adsorption relationship of H_2 with kerogen shale at various temperatures. Source: *From Raza, A., Mahmoud, M., Alafnan, S., Arif, M., & Glatz, G. (2022). H_2, CO_2, and CH_4 adsorption potential of kerogen as a function of pressure, temperature, and maturity.* International Journal of Molecular Sciences, *23(21). https://doi.org/10.3390/ijms232112767.*

TABLE 11.3 Isotherm models used to describe gas adsorption.

Model name	Equation	Nomenclature
Freundlich isotherm model	$q_e = K_F C^{\frac{1}{n_F}}{}_e$	K_F is Freundlich constant $(mg/g)(mg/L)^{-\frac{1}{n}}$, n_f is dimensionless factor representing surface heterogeneity.
Toth isotherm model	$q_e = q_T \frac{K_T C_e}{(1+(K_T C_e)^t)^{\frac{1}{t}}}$	q_T and K_T refer to the Toth maximum adsorption capacity $(mmol/g)$ and Toth constant (bar^{-1}) respectively, while t characterizes the heterogeneity of the system and generally takes the value of ≤ 1. The system is considered heterogeneous when $t < 1$, otherwise the Langmuir model is recovered when $t = 1$
Langmuir isotherm model	$V = V_L \frac{P}{P_L + P}$	where V is the adsorbed volume (scf/ton), V_L is maximum sorption capacity, P is the absolute pressure, and P_L is the Langmuir pressure.

11.3.2 Adsorption isotherm models

Adsorbed gas is usually expressed and calculated by several isothermal equations and has been extensively used to describe the adsorbate's interaction with adsorbents at equilibrium to provide input to numerical simulators. These isotherm models correspond to their underlying theories and assumptions. Different models reveal distinct pieces of information, such as the basis of surface properties and the adsorption relationship. There are several isotherm models, such as the Langmuir and Freundlich-Toth models, etc. Table 11.3 presents some of these models.

11.4 Underground hydrogen storage projects

Several significant underground storage initiatives for hydrogen have been implemented globally (Panfilov, 2016; Zivar et al., 2021). One example is the Kiel project in Germany, which commenced hydrogen storage in a salt cavern in 1971, reaching a capacity of 32,000 m^3 at 8–10 MPa, although it now serves as a natural gas storage facility. In Teesside, Yorkshire,

England, three salt caverns store approximately 1 million/m^3 of pure hydrogen (with a composition of 95% H_2 and 3%–4% CO_2) at a depth of 400 m and a pressure of 5 MPa. The Clemens project in Texas has been storing 95% hydrogen in the Clemmons salt dome at a depth of 1000 m since the 1980s, directly supplying the Old Ocean refinery. In Texas, the Praxair Moss Bluff project has catered to the petrochemical demands of Texas and Louisiana by storing hydrogen in salt caverns for several years. Similarly, the Beynes project in France utilized a saline aquifer to store natural gas with a 50% hydrogen composition at a depth of 430 m, boasting a capacity of 385 million/m^3 from 1956 to 1972. In Germany, a manufactured gas with a 62% hydrogen content was stored in an aquifer at depths ranging from 200 to 250 m within the Ketzin reservoir. The Lobodice project in the Czech Republic involved storing town gas, comprising 50% hydrogen and 25% CH_4, in an aquifer at a depth of 430 m. Argentina initiated the Hychico project in 2015, combining UHS with natural gas storage (at 50% hydrogen content) at depths of 600–800 m, under a pressure of 1 MPa and a temperature of 323K. UHS endeavors in Austria have explored the use of depleted gas media for injecting natural gas containing 10% hydrogen at a depth of 1000 m, a pressure of 7.8 MPa, and a temperature of 313.15K. Notably, three UHS projects worldwide, spanning the USA and the UK, are focused on storing 95% pure hydrogen.

11.5 Conclusion

In summary, estimates of the hydrogen storage capacity and efficiency of hydrogen injection-withdrawal cycles in various types of formations are highly uncertain due to a lack of experimental data and reliable mathematical models. The most challenging aspect of understanding the UHS process is accurately measuring hydrogen-brine-rock interactions at reliable reservoir conditions and using various formation types. The impact of reservoir conditions, for example, pressure, temperature, brine salinity, and rock composition, on rock wettability, hydrogen-brine interfacial tension, and hydrogen solubility and diffusivity in brine needs further investigation. The performance of UHS projects is mainly evaluated by hydrogen storage capacity during gas injection and hydrogen losses during gas storage and withdrawal. Hydrogen displacement efficiency governs the storage capacity of hydrogen in porous media. While the gas trapping mechanisms, mainly hydrogen structural and residual trapping and hydrogen diffusion, quantify hydrogen losses. Investigating these processes at different operating conditions and reservoir types will lead to the successful implementation of UHS. For modeling and simulation of UHS, integration of multiscale phenomena, including hydrodynamics, geochemistry, and geomechanics, is essential to obtain representative results. Additionally, upscaling from micro and core scales to a field scale while capturing the real physics that occurs during the UHS process is necessary. Hence, it is essential to develop methodologies that can adapt and scale laboratory results to align with complex field heterogeneities.

References

Adebayo, A. R., Rezk, M. G., & Badmus, S. O. (2024). The effect of multiple cycles of surfactant-alternating-gas process on foam transient flow and propagation in a homogeneous sandstone. *SSPE Journal*, *29*(2), 909–927. Available from https://doi.org/10.2118/217999-PA, https://watermark.silverchair.com/spe-217999-pa.pdf?.

Ali, M., Pan, B., Yekeen, N., Al-Anssari, S., Al-Anazi, A., Keshavarz, A., Iglauer, S., & Hoteit, H. (2022). Assessment of wettability and rock-fluid interfacial tension of caprock: Implications for hydrogen and carbon dioxide geo-storage. *International Journal of Hydrogen Energy, 47*(30), 14104–14120. Available from https://doi.org/10.1016/j.ijhydene.2022.02.149, http://www.journals.elsevier.com/international-journal-of-hydrogen-energy/.

Ali, M., Yekeen, N., Pal, N., Keshavarz, A., Iglauer, S., & Hoteit, H. (2021). Influence of pressure, temperature and organic surface concentration on hydrogen wettability of caprock; implications for hydrogen geo-storage. *Energy Reports, 7*, 5988–5996. Available from https://doi.org/10.1016/j.egyr.2021.09.016, http://www.journals.elsevier.com/energy-reports/.

Ali, M., Yekeen, N., Pal, N., Keshavarz, A., Iglauer, S., & Hoteit, H. (2022). Influence of organic molecules on wetting characteristics of mica/H_2/brine systems: Implications for hydrogen structural trapping capacities. *Journal of Colloid and Interface Science, 608*, 1739–1749. Available from https://doi.org/10.1016/j.jcis.2021.10.080, http://www.elsevier.com/inca/publications/store/6/2/2/8/6/1/index.htt.

Al-Mukainah, H., Al-Yaseri, A., Yekeen, N., Hamad, J. A., & Mahmoud, M. (2022). Wettability of shale–brine–H_2 system and H_2-brine interfacial tension for assessment of the sealing capacities of shale formations during underground hydrogen storage. *Energy Reports, 8*, 8830–8843.

Al-Yaseri, A., Esteban, L., Giwelli, A., Sarout, J., Lebedev, M., & Sarmadivaleh, M. (2022). Initial and residual trapping of hydrogen and nitrogen in Fontainebleau sandstone using nuclear magnetic resonance core flooding. *International Journal of Hydrogen Energy, 47*(53), 22482–22494. Available from https://doi.org/10.1016/j.ijhydene.2022.05.059, http://www.journals.elsevier.com/international-journal-of-hydrogen-energy/.

Al-Yaseri, A., Yekeen, N., Al-Mukainah, H., Sarmadivaleh, M., & Lebedev, M. (2024). Snap-off effects and high hydrogen residual trapping: Implications for underground hydrogen storage in sandstone aquifer. *Energy and Fuels, 38*(4), 2983–2991. Available from https://doi.org/10.1021/acs.energyfuels.3c04261, http://pubs.acs.org/journal/enfuem.

Bai, T., & Tahmasebi, P. (2022). Coupled hydro-mechanical analysis of seasonal underground hydrogen storage in a saline aquifer. *Journal of Energy Storage, 50*, 104308. Available from https://doi.org/10.1016/j.est.2022.104308.

Bo, Z., Zeng, L., Chen, Y., & Xie, Q. (1999). Geochemical reactions-induced hydrogen loss during underground hydrogen storage in sandstone reservoirs. *International Journal of Hydrogen Energy, 46*, 19998–20009. Available from https://doi.org/10.1016/j.ijhydene.2021.03.116.

Boon, M., & Hajibeygi, H. (2022). Experimental characterization of H_2/water multiphase flow in heterogeneous sandstone rock at the core scale relevant for underground hydrogen storage (UHS). *Scientific Reports, 12*, 14604. Available from https://doi.org/10.1038/s41598-022-18759-8.

Chabab, S., Théveneau, P., Coquelet, C., Corvisier, J., & Paricaud, P. (2020). Measurements and predictive models of high-pressure H_2 solubility in brine ($H_2O + NaCl$) for underground hydrogen storage application. *International Journal of Hydrogen Energy, 45*(56), 32206–32220. Available from https://doi.org/10.1016/j.ijhydene.2020.08.192, http://www.journals.elsevier.com/international-journal-of-hydrogen-energy/.

Chow, Y. T. F., Maitland, G. C., & Trusler, J. P. M. (2018). Interfacial tensions of ($H_2O + H_2$) and ($H_2O + CO_2 + H_2$) systems at temperatures of (298–448)K and pressures up to 45 MPa. *Fluid Phase Equilibria, 475*, 37–44. Available from https://doi.org/10.1016/j.fluid.2018.07.022, https://www.journals.elsevier.com/fluid-phase-equilibria.

Coarita-Tintaya, E.-D., Golfier, F., Grgic, D., Souley, M., & Cheng, L. (2023). Hydromechanical modelling of salt caverns subjected to cyclic hydrogen injection and withdrawal. *Computers and Geotechnics, 162*, 105690. Available from https://doi.org/10.1016/j.compgeo.2023.105690.

Feldmann, F., Hagemann, B., Ganzer, L., & Panfilov, M. (2016). Numerical simulation of hydrodynamic and gas mixing processes in underground hydrogen storages. *Environmental Earth Sciences, 75*(16). Available from https://doi.org/10.1007/s12665-016-5948-z.

Gbadamosi, A. O. (2023). Underground hydrogen storage: A critical assessment of fluid-fluid and fluid-rock interactions. *Journal of Energy Storage, 72*, 108473. Available from https://doi.org/10.1016/j.est.2023.108473.

Ghaedi, M., Andersen, P. Ø., & Gholami, R. (2023). Hydrogen diffusion into caprock: A semi-analytical solution and a hydrogen loss criterion. *Journal of Energy Storage, 64*, 107134. Available from https://doi.org/10.1016/j.est.2023.107134.

Ghasemi, M., Omrani, S., Mahmoodpour, S., & Zhou, T. (2022). Molecular dynamics simulation of hydrogen diffusion in water-saturated clay minerals; implications for underground hydrogen storage (UHS). *International*

Journal of Hydrogen Energy, 47(59), 24871–24885. Available from https://doi.org/10.1016/j.ijhydene.2022.05.246, http://www.journals.elsevier.com/international-journal-of-hydrogen-energy/.

Hagemann, B., Rasoulzadeh, M., Panfilov, M., Ganzer, L., & Reitenbach, V. (2015). Mathematical modeling of unstable transport in underground hydrogen storage. *Environmental Earth Sciences, 73*(11), 6891–6898. Available from https://doi.org/10.1007/s12665-015-4414-7.

Hashemi., Blunt, M., & Hajibeygi, H. (2021). Pore-scale modelling and sensitivity analyses of hydrogen-brine multiphase flow in geological porous media. *Scientific Reports, 11*, 8348. Available from https://doi.org/10.1038/s41598-021-87490-7.

Hashemi., Glerum, W., Farajzadeh, R., & Hajibeygi, H. (2021). Contact angle measurement for hydrogen/brine/sandstone system using captive-bubble method relevant for underground hydrogen storage. *Advances in Water Resources, 154*, 103964. Available from https://doi.org/10.1016/j.advwatres.2021.103964.

Hassler, G. L., & Brunner, E. (1945). Measurement of capillary pressures in small core samples. *Transactions of the AIME, 160*(01), 114–123. Available from https://doi.org/10.2118/945114-g.

Heinemann, N., Alcalde, J., Miocic, J. M., Hangx, S. J. T., Kallmeyer, J., Ostertag-Henning, C., Hassanpouryouzband, A., Thaysen, E. M., Strobel, G. J., Schmidt-Hattenberger, C., Edlmann, K., Wilkinson, M., Bentham, M., Stuart Haszeldine, R., Carbonell, R., & Rudloff, A. (2021). Enabling large-scale hydrogen storage in porous media-the scientific challenges. *Energy and Environmental Science, 14*(2), 853–864. Available from https://doi.org/10.1039/d0ee03536j, http://pubs.rsc.org/en/journals/journal/ee.

Higgs, S., Da Wang, Y., Sun, C., Ennis-King, J., Jackson, S. J., Armstrong, R. T., & Mostaghimi, P. (2022). In-situ hydrogen wettability characterisation for underground hydrogen storage. *International Journal of Hydrogen Energy, 47*(26), 13062–13075. Available from https://doi.org/10.1016/j.ijhydene.2022.02.022, http://www.journals.elsevier.com/international-journal-of-hydrogen-energy/.

Hosseini, M., Fahimpour, J., Ali, M., Keshavarz, A., & Iglauer, S. (2022a). Hydrogen wettability of carbonate formations: Implications for hydrogen geo-storage. *Journal of Colloid and Interface Science, 614*, 256–266. Available from https://doi.org/10.1016/j.jcis.2022.01.068, http://www.elsevier.com/inca/publications/store/6/2/2/8/6/1/index.htt.

Hosseini, M., Fahimpour, J., Ali, M., Keshavarz, A., & Iglauer, S. (2022b). H_2 – brine interfacial tension as a function of salinity, temperature, and pressure; implications for hydrogen geo-storage. *Journal of Petroleum Science and Engineering, 213*. Available from https://doi.org/10.1016/j.petrol.2022.110441.

Iglauer, S. (2011). *Dissolution trapping of carbon dioxide in reservoir formation brine – A carbon storage mechanism*. InTech. Available from https://doi.org/10.5772/20206.

Iglauer, S., Ali, M., & Keshavarz, A. (2021). Hydrogen wettability of sandstone reservoirs: Implications for hydrogen geo-storage. *Geophysical Research Letters, 48*(3). Available from https://doi.org/10.1029/2020GL090814, http://agupubs.onlinelibrary.wiley.com/hub/journal/10.1002/(ISSN)1944-8007/.

Jadhawar, P., & Saeed, M. (2024). Mechanistic evaluation of the reservoir engineering performance for the underground hydrogen storage in a deep North Sea aquifer. *International Journal of Hydrogen Energy, 50*, 558–574. Available from https://doi.org/10.1016/j.ijhydene.2023.07.272, https://www.sciencedirect.com/science/journal/03603199.

Jha, N. K., Al-Yaseri, A., Ghasemi, M., Al-Bayati, D., Lebedev, M., & Sarmadivaleh, M. (2021). Pore scale investigation of hydrogen injection in sandstone via X-ray micro-tomography. *International Journal of Hydrogen Energy, 46*(70), 34822–34829. Available from https://doi.org/10.1016/j.ijhydene.2021.08.042, http://www.journals.elsevier.com/international-journal-of-hydrogen-energy/.

Kanaani, M., Sedaee, B., & Asadian-Pakfar, M. (2022). Role of cushion gas on underground hydrogen storage in depleted oil reservoirs. *Journal of Energy Storage, 45*, 103783. Available from https://doi.org/10.1016/j.est.2021.103783, https://www.sciencedirect.com/science/article/pii/S2352152X21014560.

Lysyy, M., Ersland, G., & Fernø, M. (2022). Pore-scale dynamics for underground porous media hydrogen storage. *Advances in Water Resources, 163*, 104167. Available from https://doi.org/10.1016/j.advwatres.2022.104167.

Lysyy, M., Liu, N., Solstad, C. M., Fernø, M. A., & Ersland, G. (2023). Microfluidic hydrogen storage capacity and residual trapping during cyclic injections: Implications for underground storage. *International Journal of Hydrogen Energy, 48*(80), 31294–31304. Available from https://doi.org/10.1016/j.ijhydene.2023.04.253, http://www.journals.elsevier.com/international-journal-of-hydrogen-energy/.

Medina, O. E., Gallego, J. F., Moncayo-Riascos, I., Lysyy, M., Benjumea, P. N., Cortés, F. B., & Franco, C. A. (2024). Salinity influence on underground hydrogen storage: Insights from molecular dynamics and pore-scale analysis. *International Journal of Hydrogen Energy, 60*, 959–975. Available from https://doi.org/10.1016/j.ijhydene.2024.02.073, https://www.sciencedirect.com/science/journal/03603199.

Panfilov, M. (2016). *Underground and pipeline hydrogen storage* (pp. 91–115). Elsevier BV. Available from 10.1016/b978-1-78242-362-1.00004-3.

Pini, R., Krevor, S. C. M., & Benson, S. M. (2012). Capillary pressure and heterogeneity for the CO_2/water system in sandstone rocks at reservoir conditions. *Advances in Water Resources, 38*, 48–59. Available from https://doi.org/10.1016/j.advwatres.2011.12.007.

Raza, A., Mahmoud, M., Alafnan, S., Arif, M., & Glatz, G. (2022). H_2, CO_2, and CH_4 adsorption potential of kerogen as a function of pressure, temperature, and maturity. *International Journal of Molecular Sciences, 23*(21). Available from https://doi.org/10.3390/ijms232112767.

Rezk, M. G., & Foroozesh, J. (2018). Determination of mass transfer parameters and swelling factor of CO_2-oil systems at high pressures. *International Journal of Heat and Mass Transfer, 126*, 380–390. Available from https://doi.org/10.1016/j.ijheatmasstransfer.2018.05.043, http://www.journals.elsevier.com/international-journal-of-heat-and-mass-transfer/.

Rezk, M. G., & Foroozesh, J. (2019a). Effect of CO_2 mass transfer on rate of oil properties changes: Application to CO_2-EOR projects. *Journal of Petroleum Science and Engineering, 180*, 298–309. Available from https://doi.org/10.1016/j.petrol.2019.05.053, http://www.sciencedirect.com/science/journal/09204105/81.

Rezk, M. G., & Foroozesh, J. (2019b). Phase behavior and fluid interactions of a CO_2-light oil system at high pressures and temperatures. *Heliyon, 5*(7). Available from https://doi.org/10.1016/j.heliyon.2019.e02057, http://www.journals.elsevier.com/heliyon/.

Rezk, M. G., Foroozesh, J., Abdulrahman, A., & Gholinezhad, J. (2022). CO_2 diffusion and dispersion in porous media: Review of advances in experimental measurements and mathematical models. *Energy Fuels, 36*, 133–155. Available from https://doi.org/10.1021/acs.energyfuels.1c03552.

Saeed, M., & Jadhawar, P. (2024). Optimizing underground hydrogen storage in aquifers: The impact of cushion gas type. *International Journal of Hydrogen Energy, 52*, 1537–1549. Available from https://doi.org/10.1016/j.ijhydene.2023.08.352, https://www.sciencedirect.com/science/journal/03603199.

Saeed, M., Jadhawar, P., & Bagala, S. (2023). Geochemical effects on storage gases and reservoir rock during underground hydrogen storage: A depleted north sea oil reservoir case study. *Hydrogen (Switzerland), 4*(2), 323–337. Available from https://doi.org/10.3390/hydrogen4020023, https://www.mdpi.com/journal/hydrogen.

Samara, H., Von Ostrowski, T., & Jaeger, P. (2022). Geological storage of carbon dioxide and hydrogen in Jordanian shale formations. *Proceedings – SPE annual technical conference and exhibition*. https://www.onepetro.org/conferences/SPE/97SPE 10.2118/210202-MS.

Tarkowski, R. (2019). Underground hydrogen storage: Characteristics and prospects. *Renewable and Sustainable Energy Reviews, 105*, 86–94. Available from https://doi.org/10.1016/j.rser.2019.01.051, https://www.sciencedirect.com/science/article/pii/S1364032119300528.

Thaysen, E. M., Butler, I. B., Hassanpouryouzband, A., Freitas, D., Alvarez-Borges, F., Krevor, S., Heinemann, N., Atwood, R., & Edlmann, K. (2023). Pore-scale imaging of hydrogen displacement and trapping in porous media. *International Journal of Hydrogen Energy, 48*(8), 3091–3106. Available from https://doi.org/10.1016/j.ijhydene.2022.10.153, http://www.journals.elsevier.com/international-journal-of-hydrogen-energy/.

Yekta, A. E., Manceau, J.-C., Gaboreau, S., Pichavant, M., & Audigane, P. (2018). Determination of hydrogen--water relative permeability and capillary pressure in sandstone: Application to underground hydrogen injection in sedimentary formations. *Transport in Porous Media, 122*, 333–356. Available from https://doi.org/10.1007/s11242-018-1004-7.

Zhang, J., Clennell, M. B., Sagotra, A., & Pascual, R. (2023). Molecular dynamics simulation and machine learning for predicting hydrogen solubility in water: Effects of temperature, pressure, finite system size and choice of molecular force fields. *Chemical Physics, 564*, 111725. Available from https://doi.org/10.1016/j.chemphys.2022.111725.

Zhao, T., Li, X., Zhao, H., & Li, M. (2017). Molecular simulation of adsorption and thermodynamic properties on type II kerogen: Influence of maturity and moisture content. *Fuel, 190*, 198–207. Available from https://doi.org/10.1016/j.fuel.2016.11.027, http://www.journals.elsevier.com/fuel/.

Zivar, D., Kumar, S., & Foroozesh, J. (2021). Underground hydrogen storage: A comprehensive review. *International Journal of Hydrogen Energy, 46*(45), 23436–23462. Available from https://doi.org/10.1016/j.ijhydene.2020.08.138, http://www.journals.elsevier.com/international-journal-of-hydrogen-energy/.

CHAPTER

12

Government policies, guidelines, initiatives, and supports for underground hydrogen storage

Ramesh Guduru (Kumar)[1], *Ricurve Patel*[1], *Robin Singh*[2] and *Rakesh Kumar Vij*[3]

[1]Department of Mechanical Engineering, School of Technology, Pandit Deendayal Energy University, Gandhinagar, Gujarat, India [2]Department of Chemistry, School of Energy Technology, Pandit Deendayal Energy University, Gandhinagar, Gujarat, India [3]Department of Petroleum Engineering, School of Energy Technology, Pandit Deendayal Energy University, Gandhinagar, Gujarat, India

12.1 Introduction

The world is currently facing the challenges of global warming and the energy transition as the excessive greenhouse gas (GHG) emissions are causing climate change. While the Russia–Ukraine conflict has disrupted European energy supplies, in response, "Net-Zero" policies have been introduced to facilitate the shift from fossil fuels to clean energy sources, ensuring more reliable energy supplies (Kuzemko et al., 2022). With global energy demand expected to double by 2050, there is a growing focus on increasing energy production from renewables to meet international climate protection agreements and improve quality of life (Guduru et al., 2024a, 2024b; Kuzemko et al., 2022). However, existing renewable sources like solar and wind energy are weather-dependent, leading to energy fluctuations and mismatches between supply and demand. Their intermittent nature has prompted global attention, and the challenge now is to find safe, secure, and efficient ways to store this energy with a lower carbon footprint. In 2020, the United Nations Climate Change Conference in Paris resulted in the Paris Agreement, in which 196 countries committed to limiting global temperature rise to 2.0°C, preferably 1.5°C (Guduru et al., 2024c). However, as of 2022, Earth's average temperature has already

Subsurface Hydrogen Energy Storage
DOI: https://doi.org/10.1016/B978-0-443-24071-3.00013-3

increased by about 1.1°C. To achieve the 1.5°C target, the Paris Agreement calls for a 45% reduction in global emissions by 2030, reaching net zero by 2050. Currently, efforts are falling short, with projections indicating a 10–11% increase in emissions by 2030. A UN report highlights that 75% of emissions come from 20 entities, and half of all emissions originate from just seven: China, the United States, India, the European Union, Indonesia, Russia, and Brazil. Addressing these major emitters is crucial for achieving global climate goals. Governments should prioritize hydrogen development due to its role in achieving climate goals (Kumar, Regassa Jufar, et al., 2023). Hydrogen is a clean energy source crucial for decarbonizing challenging sectors like heavy industry and transportation. Policies supporting hydrogen foster international collaboration, create economic opportunities, and drive technological innovation. Investing in hydrogen infrastructure reduces reliance on fossil fuels and aligns with global efforts to combat climate change, especially given the significant contribution of major emitters like China, United States, and others (Solomon & Banerjee, 2006). Hydrogen is considered a key energy carrier for the future, with a high specific energy capacity and clean combustion properties, making it an appealing green energy option. Converting surplus renewable energy into hydrogen during periods of low demand and storing it for use when demand is high could provide a green energy solution. However, there are significant obstacles to overcome for this hydrogen revolution, especially concerning its storage. Due to their very low density, surface-based facilities are impractical and expensive for large-scale storage. Instead, underground storage in geological formations like aquifers, depleted hydrocarbon reservoirs, and salt caverns offers ample storage volume and the potential for economically storing hydrogen at high pressure, resulting in high energy densities. In the context of hydrogen storage, underground hydrogen storage (UHS) is seen as a vital component in the current energy mix. In the global pursuit of sustainable energy solutions and the mitigation of climate change, hydrogen has emerged as a key player in the transition to cleaner energy systems. Governments around the world are formulating and implementing hydrogen policies to harness its potential as a versatile and clean energy carrier. These policies encompass a spectrum of strategies, ranging from research and development initiatives to incentives for private sector investment (Kovač et al., 2021). As nations grapple with the imperative to reduce GHG emissions and achieve carbon neutrality, hydrogen has gained prominence as a solution to address sectors that are challenging to decarbonize through electrification alone (Aglietta et al., 2021). For example, the US hydrogen industry is receiving substantial government support through key policies such as tax credits for clean hydrogen production, incentives for advanced energy projects and carbon capture, funding for regional clean hydrogen hubs, and increased loan guarantees (Iida & Sakata, 2019). These measures aim to boost hydrogen production, reduce emissions, and drive the transition to a more sustainable energy landscape. Another expiable includes the political agreement in Denmark designates Energinet and Evida for hydrogen transmission and distribution, marking a significant step toward developing the country's hydrogen economy. The focus on green hydrogen from renewable sources aligns with the broader goals of Denmark's and Europe's energy systems. Energinet's CEO highlights the pivotal role green hydrogen is expected to play in the near future. Mitsubishi Power Americas and Magnum Development are launching the world's largest UHS project in Delta, Utah, with a capacity of 300 GWh (Oshiro et al., 2017). The facility, supported by a $504.4 million loan guarantee from the US

Department of Energy, will store hydrogen from an 840 MW hydrogen-capable power plant. Initially, it will operate on a 30% green hydrogen blend, increasing to 100% by 2045. WSP USA will oversee the construction of two caverns, part of the Advanced Clean Energy Storage Hub, aiming to produce up to 100 metric tons of green hydrogen daily. The project capitalizes on Utah's business-friendly environment and unique geological features. Understanding this type of policy adopted by various countries provides valuable insight into the global effort to harness hydrogen's capabilities and achieve a more sustainable energy future (Erbach & Jensen, 2021). These policies show a shared vision to explore hydrogen's potential across industries and applications, from transportation to industrial processes, and to establish a foundation for a cleaner and more resilient energy system. Fig. 12.1 shows the countries working toward sustainable hydrogen economic approaches. Government support for hydrogen technology and the worldwide transition toward hydrogen-based energy solutions are two examples of governments playing a crucial role in the battle against climate change and the achievement of sustainable energy goals. Many nations have started implemented plans and regulations to encourage the development and acceptance of hydrogen technology on a worldwide scale (Navaid et al., 2023). Notably, the European Clean Hydrogen Alliance, formed in July 2020, is a collaborative effort uniting various stakeholders to facilitate the extensive deployment of clean hydrogen technologies by 2030 (Tarkowski & Uliasz-Misiak, 2022). This alliance spans renewable and low-carbon

FIGURE 12.1 Major countries working toward sustainable hydrogen economical approaches.

hydrogen production, industrial demand, mobility, and distribution infrastructure. In January 2021, the UAE Ministry of Energy and Infrastructure, along with key government-backed entities in Abu Dhabi, launched a "Hydrogen Alliance." This initiative aims to produce both blue and green hydrogen by utilizing natural gas and renewable energy. The ambitious plan includes international marketing of hydrogen and self-sufficiency in gas production for the UAE. Significant investments, such as ADNOC's $122 billion commitment, underscore the nation's dedication to hydrogen development. Canada and Germany are forging a significant partnership in the hydrogen sector. The Canada-Germany Hydrogen Alliance, solidified through a Joint Declaration of Intent, demonstrates a shared commitment to collaborate on exporting clean Canadian hydrogen to Germany by 2025. This alliance focuses on harmonizing policies, supporting secure supply chains, establishing a transatlantic supply corridor, and promoting investment in hydrogen projects. This joint effort aligns with Canada's Hydrogen Strategy, positioning the country as a leading global supplier of clean hydrogen (Kumar, Regassa Jufar, et al., 2023). India, through the IH2A (India Hydrogen Alliance), is actively pursuing the development of a green hydrogen hub in Kochi. Proposals include a 150 MW electrolyzer plant and comprehensive infrastructure. IH2A envisions 25 national green hydrogen projects and five H_2 hubs by 2025, emphasizing the need for a national development plan. The formation of the energy transition coalition "India H_2 Alliance" further reinforces a collaborative approach by global players to build a robust hydrogen economy in India. IH2A suggests initiatives like a National Electrolyzer Manufacturing Mission to promote domestic manufacturing and aims for a substantial investment pool by 2030, directed toward national hydrogen projects (Global Hydrogen Review, 2022).

- Many governments are allocating substantial funding for hydrogen R&D to accelerate technological advancements. These investments cover various aspects, including hydrogen production, storage, transportation, and utilization. The European Union has pledged to invest billions of euros in hydrogen-related research through programs such as Horizon Europe.
- Most of the hydrogen which was used in 2021 came from fossil fuels, resulting in over 900 million metric tons of CO_2 emissions. About 62% of this hydrogen came from natural gas, and 18% was a byproduct of refining processes. China led in coal-based hydrogen production, making 19% of the total. Very little hydrogen, less than 1%, was from oil, and only 0.7% was low-emission, mainly using carbon capture with fossil fuels. Only 35,000 metric tons came from clean methods like water electrolysis, showing a 20% increase from the previous year (Bartlett & Krupnick, 2020). Many countries are trying to make cleaner hydrogen, like Germany, which aims for 10 GW of green hydrogen by 2030 to cut carbon emissions (OSTI.GOV, 1980).
- Building a strong hydrogen infrastructure is essential for more people to use hydrogen technologies. The international energy agency (IEA) has revised down its global demand growth forecast for the fourth quarter of 2023 by almost 400 kb/d, with Europe making up more than half of the decline. Future projections indicate that the growth in oil demand will slow down in 2024 and reach 930,000 barrels per day (b/d). Several factors have contributed to this slowdown, including the energy transition, stricter efficiency regulations, and the increasing prevalence of electric vehicles on the road.

According to IEA projections, global oil consumption growth is expected to halve to 1.1 million b/d in 2024 as major economies continue to develop below trend and c ovid-related distortions start to lessen.

- Hydrogen is crucial for cleaning and improving oil, and in 2021, the demand for it went up to nearly 40 million tons because more diesel was needed, as shown in Fig. 12.2. In 2020, the demand went down to 38 million tons because of the pandemic. But in 2022, it's expected to go even higher, maybe over 41 million tons, mostly because of more oil processing and rules to reduce sulfur emissions, even though this creates a lot of carbon emissions from fossil fuels.
- Governments are spending money to make hydrogen refueling stations, pipelines, and networks. For example, as a major renewable energy source, green hydrogen has set major ambitions in Japan. The nation wants the price of hydrogen to drop to 20 yen/m (Kumar, Regassa Jufar, et al., 2023) by 2050 and 30 yen/m (Kumar, Regassa Jufar, et al., 2023) by 2030. In order to expedite efforts to use hydrogen to transition to a low-carbon economy, the government has also announced a revised objective to increase the supply of hydrogen to 12 million tons annually by 2040. Over the next 15 years, the plan calls for investing 15 trillion yen ($107.5 billion). The four objectives of the 2023 hydrogen strategy are to: manufacture hydrogen using renewable energy, switch to low-carbon hydrogen, expand the availability of hydrogen and ammonia in Japan, and create an international hydrogen energy supply chain. The plan also seeks to build a supply chain in Japan and cut expenses. Japan wants 200 hydrogen refueling stations to be working by 2025 (Lackey et al., 2023).
- Governments are encouraging people to use hydrogen as a clean fuel in different areas like cars, buses, and trucks and in making things like steel and chemicals. For example, In India, The Indian Railways Organization for Alternate Fuels (IROAF) is working on a number of initiatives that will involve hydrogen. Among the important projects are, hydrogen fuel cell train by 2024, India will see the launch of its first hydrogen-powered

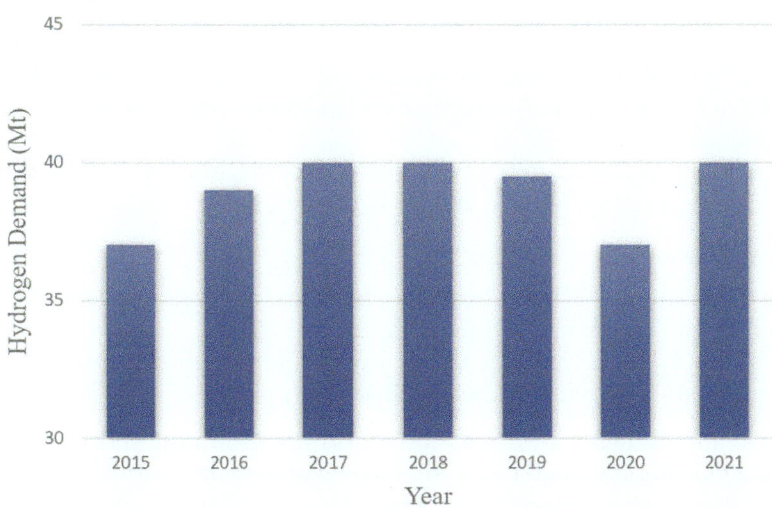

FIGURE 12.2 Hydrogen demand in refining.

train, which will travel on classic narrow-gauge routes run by the Kalka-Shimla Railway, the Darjeeling Himalayan Railway, the Nilgiri Mountain Railway, the Matheran Hill Railway, and other companies. The railroads intend to operate 35 hydrogen trains on eight historic routes. Project for hydrogen-fueled buses and cars in Leh and Delhi: For the Leh and Delhi lines, IROAF is also developing a project for hydrogen fuel-powered buses and cars. 10 H_2 fuel cell buses and electric vehicles: The Jind-Sonipat section of the Northern Railways 3 is to be funded by IROAF as a trial project for retrofitting hydrogen fuel cells on existing diesel electrical multiple rakes with ground infrastructure. These programs seek to lessen the railway system's carbon footprint while promoting environmentally friendly and sustainable transportation in India. Also, South Korea intends to have 6.2 million hydrogen-powered vehicles on its roads by 2040, according to a number of sources. The nation has set high goals for hydrogen as a major renewable energy source, with 5.3 million fuel cell electric cars (FCEVs) on the road by 2050 and a 40% decrease in emissions by 2030, all of which are supported by at least 2000 hydrogen filling stations.

- Governments are making rules like tax breaks, financial help, and laws to support the use of hydrogen more. They are also deciding how much hydrogen should be part of the energy we use (Lackey et al., 2023). Various policies are being put in place by governments all over the world to encourage the use of hydrogen as a sustainable and clean energy source. These policies include financial aid, tax advantages, and legislation to encourage the production and use of hydrogen. By 2050, Japan, for instance, wants to use 20% hydrogen in its energy mix. Several important programs and policies in various nations include:

Japan: By 2030, the nation wants to have consumed three million tons of hydrogen annually. The main objectives of Japan's new hydrogen strategy are to increase supply, generate demand, switch to low-carbon hydrogen, produce hydrogen with renewable energy, and set up a supply chain within the country. Additionally, the government is creating international hydrogen energy supply chains and passing new laws to provide financial assistance to businesses and build necessary infrastructure (Oshiro et al., 2017).

- United States and Japan: As part of their Competitiveness and Innovation Partnership (CoRE), the two nations have decided to work together on the development of clean energy, particularly hydrogen.
- Australia and Brunei are actively involved in demonstration projects, and by 2030, Japan intends to purchase roughly 300,000 tons of hydrogen from these nations.
- European nations: A number of European governments have released their hydrogen policies, emphasizing the development of global supply chains for hydrogen energy, demand creation, and supply chain expansion (Erbach & Jensen, 2021).
- China and other Asian nations: To encourage the use of hydrogen in their energy mix and lower emissions, several nations have also created hydrogen policies.

These programs show how governments all across the world are becoming more interested in and committed to supporting hydrogen as a vital renewable energy source. Some countries want to sell clean hydrogen to places with high demand. This can help their

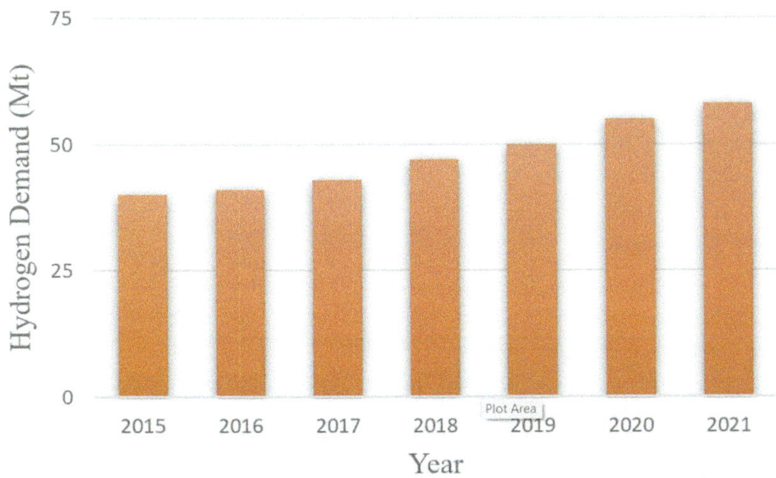

FIGURE 12.3 Hydrogen demand in industry.

economies grow and make the world cleaner. For instance, Australia is planning to become a major hydrogen exporter to Asian markets.

As shown in Fig. 12.3, in 2021, hydrogen was utilized in emerging sectors, accounting for about 40 kt (0.04% of global hydrogen use), primarily in road transport, with a notable 60% increase in hydrogen-powered truck use in China, while China, the United States, and the Middle East led in hydrogen consumption, with Europe and India also seeing increased usage due to economic recovery and industrial sectors' growth.

Hydrogen is a clean energy source, but it has some tricky aspects when it comes to storing it. It's highly flammable and takes up a lot of space in gaseous phase for the energy it holds. The usual ways to store it above the ground, like putting it in high-pressure tanks or super-cold containers, aren't great for storing a lot of hydrogen for a long time. That's where UHS comes in as a revolutionary technology (Navaid et al., 2023). Comparing hydrogen storage options, using underground spaces like salt caverns and depleted gas fields offers lots of space and uses existing infrastructure. But it depends on having the right geological formations, and it can be expensive. The global UHS market is anticipated to reach 5.05 BCM by 2030, growing at a rate of 10.7% from 2023 to 2030 (Steinberg et al., 2023). Key contributors include major companies like Air Products and Chemicals, Air Liquide, Linde Plc, Engie, Uniper SE, and others. This storage method is deemed essential due to its safety advantages, eliminating the risks associated with hydrogen's flammability. Salt caverns, with their structural strength derived from existing reserves, dominate the market. In 2022, Europe held a significant market share of over 53.88%, driven by lower costs and the enhanced safety offered by salt caverns. The market's growth is fueled by increased hydrogen production and the absence of viable alternatives for large-scale storage. Ongoing research and development efforts in regions like North America, the Middle East and North Africa, and the Asia Pacific are further propelling market expansion. Notable collaborations, such as Storengy and HyPSTER's initiative to convert a salt cavern facility into UHS in Etrez, France, highlight the industry's commitment to advancing hydrogen storage technologies for a sustainable energy future (Marfaing, n.d.). On the other hand, methods like compressed gas, liquid hydrogen, chemical hydrogen, and metal hydride storage are more flexible for different needs. Each has its own efficiency and technical challenges. One example of underground storage is the Teesside facility in the UK. It has been storing hydrogen for ammonia and methanol production since 1972. It can hold 0.2 million m^3 of hydrogen and has around 25 GWh of stored energy at high pressure. Similar storage is also happening along the Gulf Coast in Texas (Kumar, Regassa Jufar, et al., 2023).

12.2 Policies, guidelines, and initiatives for hydrogen technology

The successful integration of hydrogen technology into the global energy landscape requires a comprehensive framework that considers the diverse needs and capabilities of industrialized, emerging, and impoverished nations. A future with sustainable and inclusive hydrogen can be achieved through cooperation, information sharing, and focused activities. Within the ever-evolving field of sustainable energy solutions, hydrogen technology has emerged as a significant player in the global quest for more efficient and cleaner alternatives. As governments all over the world come to understand the transformative potential of hydrogen, it is essential for them to develop and implement comprehensive policies to successfully integrate hydrogen (Merkeley et al., 2010). Since it enables the implementation of sophisticated strategies that are tailored to the diverse requirements and capacities of countries at different stages of development, the classification of these policies becomes crucial. This section explores an organized framework that categorizes nations into developed, developing, and underdeveloped groups based on their levels of economic and social development. Specific laws, regulations, and programs are examined within each region to gain a comprehensive understanding of how different countries are navigating the complex field of hydrogen technology. Every group requires a different set of policies, ranging from industrial powerhouses that lead the way in cutting-edge research to emerging economies seeking sustainable alternatives and developing countries that embrace innovative solutions for basic infrastructure. This classification emphasizes the significance of global collaborative efforts while also acknowledging the variety of opportunities and challenges. We will explore the principles that drive creativity, the regulations that ensure security and consistency, and the programs that foster global cooperation. There is no one-size-fits-all solution to the transition toward a hydrogen-powered future. Instead, shaping an inclusive and sustainable global energy landscape requires understanding the customized strategies employed by various countries. By 2030, the following wealthy nations are anticipated to be among the front-runners in green hydrogen projects:

1. United States of America
2. Spain
3. Canada
4. Chile
5. Egypt
6. Germany

7. India
8. Brazil
9. Morocco
10. Australia

By the end of 2030, these nations are expected to lead the way in green hydrogen production and utilization.

12.2.1 Developed countries

In hydrogen technology, developed nations lead the way with a strong commitment to advanced solutions for cleaner energy. Their policies go beyond borders, working together for a more sustainable global future.

12.2.1.1 *United States of America*

The United States is working toward a greener energy future by introducing policies and guidelines to encourage the use of cleaner hydrogen technology. These include regulations for low-carbon fuels and requirements for eco-friendly transportation. The goal is to motivate industries to adopt cleaner hydrogen solutions and promote a more environmentally conscious approach. Furthermore, the U.S. government is actively championing the integration of hydrogen-powered transportation alternatives (Global Hydrogen Review, 2022). Specifically, there is a notable emphasis on supporting the deployment of hydrogen-powered buses and the utilization of cleaner ferries. These initiatives aim to revolutionize public transportation by introducing vehicles with significantly reduced emissions or, in some cases, zero emissions altogether. By steering toward the incorporation of hydrogen as a primary fuel source, the United States is demonstrating its commitment to technological innovation and sustainable practices, solidifying its status as a developed and environmentally responsible nation (Bartlett & Krupnick, 2020). In recent years, the United States has shown a strong commitment to advancing hydrogen projects, notably through the Bipartisan Infrastructure Law, which provides substantial funding for creating hydrogen centers, developing hydrogen infrastructure, and advancing hydrogen technology. The U.S. Department of Energy has played a crucial role, offering a $504 million loan guarantee for a significant hydrogen storage project and committing $1.04 billion to a project converting methane into hydrogen. Additionally, the government gained authority through the Defense Production Act to invest in critical energy technologies, such as hydrogen production. As of 2021, the U.S. had approximately 257 specialized hydrogen production facilities, 25 hydrogen pipelines spanning around 1600 miles collectively, and four UHS facilities either in use or under development (Jensterle et al., n.d.). About half of the produced hydrogen is used onsite at nearby facilities, while the remainder is sold to end users through contracts and transported via pipeline or truck. Notably, clean hydrogen production in the U.S. is currently limited. Between June 2021 and August 2022, there were announcements of approximately 375 new projects dedicated to clean hydrogen research, demonstration, and deployment. Looking forward, the Biden administration has set ambitious targets to significantly boost clean hydrogen production in the U.S., aiming for 10 million metric tons annually by 2030, 20 million metric tons annually by 2040, and an even more ambitious

50 million metric tons annually by 2050 (Caglayan et al., 2020). Looking ahead, there are plans for new UHS facilities in the United States, such as Advanced Clean Energy Storage, scheduled for 2025 and developed by Mitsubishi Power and Magnum Development, with each cavern capable of storing up to 150 GWh and currently in the feasibility study phase. Additionally, there's the Hydrogen City-South Texas project, planned for 2026 by Green Hydrogen International and Energy Estate, which is in its conceptual stage, starting with two caverns initially and with the potential for up to 50 caverns in the future (Ozawa et al., 2018). Financially, the U.S. has allocated significant resources to hydrogen projects, including \$8 billion for clean hydrogen centers, \$1 billion for a program promoting affordable hydrogen production, and \$0.5 billion for hydrogen technology development and recycling. Companies producing low-emission hydrogen can benefit from tax breaks of up to \$3/kg of hydrogen, thanks to the Inflation Reduction Act. Furthermore, there's a substantial \$40 billion available for loan guarantees to support low-emission hydrogen projects, exemplified by a significant hydrogen storage project (Borowski & Karlikowska, 2023). The Bipartisan Infrastructure Law also allocates \$1 billion for hydrogen research and development and \$0.5 billion for clean hydrogen technology (Steinberg et al., 2023). The U.S. is also working on a rule called the Clean Hydrogen Standard, which will tell companies how clean their hydrogen should be. They sent some changes they want to make to a worldwide rule about hydrogen and fuel cell cars to the United Nations. These changes will make the rules better and include bigger vehicles that use hydrogen, showing that the United States is really serious about making rules for hydrogen, not just in the U.S. but around the world. Rising equipment and financial costs are presenting challenges to various projects, potentially undermining the positive impact of government support for their deployment. Several initiatives have found it necessary to revise their initial cost estimates upward by up to 50%. Complicating matters further, inflationary pressures coincide with recent declines in natural gas prices, especially in Europe, and disruptions in the supply chain have affected project timelines (Lebrouhi et al., 2022). As a result, the initially announced government funding may now cover a smaller number of projects than anticipated, given the increased investment needed to bridge the cost gap between low-emission hydrogen and hydrogen derived from unabated fossil fuels. In the European Union, there have been important projects to study how we can store hydrogen underground. One of these projects, called H2STORE and supported by the Fuel Cells and Hydrogen Joint Undertaking, aimed to show that storing hydrogen in salt caverns is a good idea (Tarkowski & Uliasz-Misiak, 2022). They wanted to make sure the technology was safe and worked well for storing a lot of hydrogen underground. At the same time, there was another project called HyUnder in the European Union (Simon et al., 2015). This project, part of the Horizon 2020 program by the European Commission, looked into using empty gas fields and underground water reservoirs (aquifers) to store hydrogen. They wanted to figure out if it was technically and economically practical, which could make these underground storage methods more widespread. Moreover, these initiatives highlight the commitment of developed countries in Europe to explore and implement innovative solutions for sustainable energy storage (Merkeley et al., 2010). Another project called STORE&GO, which also got funding from Horizon 2020, looked at lots of different ways to make hydrogen from renewable sources and how to move it around and store it, even underground. They wanted to make it easier to use hydrogen with the existing natural gas system (Walsh et al., 2021). There was also the H2RIZON

project, supported by the European Commission, and it focused on figuring out if we could put hydrogen in empty oil and gas fields safely. They wanted to solve any technical problems and make guidelines for doing this kind of underground storage in these old reservoirs. Apart from these separate projects, the European Union is really interested in turning the North Sea into a special place for renewable energy projects, like making hydrogen and storing it (Lycourghiotis, 2022). It was concluded that using the ground to store hydrogen in this area can help reduce pollution from our energy, and it's a big part of their plan to make things more environmentally friendly.

12.2.1.2 Germany

Green hydrogen production will be the primary focus of implementing the National Hydrogen Strategy. Promoting investment in hydrogen technology through favorable feed-in tariffs and regulatory support Establishing industry-specific guidelines for the safe and compatible use of hydrogen working together to develop a unified regulatory framework within the European Union and neighboring countries (Shin, 2022). Leading research projects in collaboration with Japan to advance hydrogen storage and transportation technologies (Jensterle et al., n.d.). With its National Hydrogen Strategy, Germany aims to become a global leader in hydrogen technologies and drastically cut carbon emissions. In order to meet roughly 26%–35% of Germany's anticipated hydrogen consumption in 2030, the German government has increased its national objective for green hydrogen production by 2030 from 5 to 10 GW. This means that a million tons of green hydrogen would be produced yearly. Germany is also creating an import plan for hydrogen and its derivatives, since the country is predicted to import between 50% and 70% of the 95–130 TWh of hydrogen that it would require in 2030. The 2020 strategy's initial target of 10 GW must be doubled by 2030 in order to meet the hydrogen demand for domestic electrolyzers. During the hydrogen market's ramp-up phase, the policy places a strong emphasis on supporting low-carbon hydrogen, including blue hydrogen made from fossil fuels with carbon capture and storage (CCS). By 2030, Germany hopes to take the lead in the production of hydrogen-related devices including fuel cells and electrolyzers. Germany has made a strong commitment to promoting the use of hydrogen as a crucial component of its energy transition and as a tool to considerably decrease carbon emissions, as evidenced by these aims and measures (German National Hydrogen Council, T. Statement on Germany's Updated National Hydrogen Strategy of 24th, 2023).

12.2.1.3 Japan

Japan is actively pursuing policies and initiatives to advance hydrogen technology as part of its strategy to become a global leader in the hydrogen economy while achieving sustainability and energy goals. Two key initiatives exemplify these efforts. First, Japan's strategic roadmap for hydrogen and fuel cells is a comprehensive policy document outlining the government's vision and strategy for hydrogen. It encompasses various aspects of hydrogen technology, including production, transportation, and storage. The roadmap establishes ambitious targets, such as achieving carbon neutrality by 2050, with hydrogen playing a pivotal role in decarbonizing the energy sector (Cheng et al., 2023). It also emphasizes the expansion of hydrogen production capacity, particularly focusing on the development of green hydrogen from renewable sources. Infrastructure development is a key component, with plans to build a robust hydrogen infrastructure, including pipelines, filling stations, and storage facilities. The importance of research and development (R&D) is highlighted to advance hydrogen technologies, enhance efficiency, and reduce costs, with government funding allocated to support R&D initiatives (Caglayan et al., 2020). Additionally, Japan actively engages in international collaborations to promote hydrogen technology and create a global hydrogen supply chain (Lee & Woo, 2020). Second, the Kushiro hydrogen production and supply base, located in Hokkaido, is a significant initiative that includes UHS facilities (Chu et al., 2022). This project involves the utilization of salt caverns for UHS and serves as a demonstration project to validate the feasibility and safety of storing hydrogen in such facilities. The base employs various methods, including renewable energy sources like wind power and thermal power generation, coupled with CCS technology to reduce emissions during hydrogen production (Lee & Kim, 2021). The stored hydrogen can be supplied to industrial users and the transportation sector, showcasing Japan's commitment to advancing hydrogen production and storage technologies. This project acts as a model for future developments in the country, emphasizing the importance of innovation in the hydrogen sector (Stangarone, 2021).

- **Policies**: Supporting a civilization centered on hydrogen and providing a clear path for technological advancement. Creating financial incentives to encourage private sector investment in hydrogen infrastructure (Global Hydrogen Review, 2022).
- **Guidelines**: Taking the lead in establishing global safety and compatibility standards for hydrogen. Working together with international partners to ensure compatibility.
- **Initiatives**: Collaborating on cooperative research efforts with Australia to enhance the efficiency of hydrogen production.

12.2.1.4 South Korea

South Korea has developed a comprehensive hydrogen roadmap to support its hydrogen industry and integrate hydrogen into its energy transition plan. In 2020, South Korea introduced its Green New Deal, placing a significant emphasis on hydrogen as a pivotal component in the country's efforts to combat climate change. The government has set ambitious targets, aiming to produce 6.2 million tons of hydrogen by 2040, as outlined in the Hydrogen Economy Roadmap. This commitment involves substantial investments in hydrogen production, with a focus on green hydrogen generated from renewable sources like wind and solar power. South Korea aims to develop the capacity to produce 5.26 million tons of green hydrogen by 2040 (Medisetty et al., 2020). To support the growth of hydrogen technology, South Korea is actively expanding its hydrogen infrastructure, including refueling stations and storage facilities, crucial for the adoption of fuel cell vehicles (FCVs) and other hydrogen-powered vehicles (Muhammed et al., 2022). The country is also exploring innovative storage solutions like UHS in caverns or salt caverns to ensure a stable supply and manage production and demand fluctuations (Nouni et al., 2021). Additionally, South Korea promotes research and development in hydrogen technologies, collaborates internationally, offers incentives and subsidies for hydrogen adoption, and works on creating a robust regulatory framework to ensure the safe production, transportation, and storage of hydrogen. These combined efforts underscore South Korea's commitment to the development and deployment of hydrogen technology in the fight against climate change.

The nation's programs and policies consist of the following:

- South Korea aims to establish a global supply chain and develop the necessary infrastructure to significantly expand its hydrogen industry. The government expects that these actions will have a significant impact on the economy.
- The country plans to build 70 liquid hydrogen fueling stations and introduce 30,000 hydrogen-powered commercial vehicles by 2030. Furthermore, by 2040, South Korea aims to have 6.2 million fuel cell electric cars on its highways. Hydrogen in power generation: By 2036, South Korea aims to generate 7.1% of clean power through the use of hydrogen and ammonia in power generation.
- International cooperation: To accelerate the adoption of hydrogen-powered vehicles, the nation is collaborating with the US.
- Safety guidelines and regional framework: South Korea is developing comprehensive safety guidelines for the use of hydrogen in various industries. Additionally, it is collaborating with ASEAN countries to establish a regional framework for hydrogen.
- Targets for hydrogen production: South Korea's "First Basic Plan for the Implementation of the Hydrogen Economy" aims to increase the nation's hydrogen output to 28 million tons by 2050, with a short-term target of 3.9 million tons by 2030.
- Challenges and realities: South Korea has faced difficulties in meeting its ambitious goals for hydrogen-powered cars and refueling facilities. The nation is striving to address the realities and challenges of transitioning to a hydrogen economy, and the government has revised its objectives. These programs and regulations demonstrate South Korea's commitment to developing its hydrogen industry and integrating hydrogen into its transportation and energy sectors.

12.2.1.4.1 Policies and guidelines

- Including hydrogen in the country's energy transition plan.
- Supplying cash and incentives to promote the development, research, and commercialization of projects related to hydrogen (Shin, 2022).
- Guidelines for creating comprehensive safety standards for the utilization of hydrogen across various industries.
- Working together to create a regional framework for hydrogen with the ASEAN nations (Shin, 2022).
- Initiatives: Cooperating with the US to accelerate the adoption of hydrogen-powered cars.

12.2.1.5 Spain

The aim of the Spanish government's 2020 hydrogen roadmap is to help the country achieve climate neutrality and develop a hydrogen economy based entirely on renewable energy sources. Recent developments in Spain's green hydrogen sector include the following:

- Updated goals: By 2030, Spain aims to have 11 GW of electrolyzer capacity, with the objective of obtaining 74% of the hydrogen used in industry from renewable sources by that time (Lombardi, 2023).

- Green hydrogen plant: Enagás, a Spanish energy company, is developing the largest hydrogen plant in Spain with an estimated investment of USD 309 million.
- Investment: The ambitious goals outlined in the Spanish Hydrogen Roadmap will require an estimated billion between 2020 and 2030. The project consists of a 32 MW electrolyzer powered by a 150 MW photovoltaic plant, with the goal of producing 12 tons of green hydrogen per day.
- Collaboration between Masdar and VERBUND: A partnership has been established between VERBUND Green Hydrogen GmbH, a Spanish company focused on developing green hydrogen plants, and Masdar, a renewable energy company. To achieve net-zero targets, Spain and Central Europe could decarbonize their hard-to-abate sectors by harnessing the potential of green hydrogen.
- Achieving goals: Spain exceeded the 2.5% target set out in the national energy and climate plan for 2020 by accounting for 20% of the renewable hydrogen projects announced globally in 2022, as reported by the Spanish Ministry of Ecological Transition.

These advancements demonstrate Spain's commitment to expanding the production of green hydrogen and positioning itself as a key player in the renewable hydrogen market.

12.2.1.6 Australia

The coordinated initiatives highlight Australia's proactive approach to tackling the obstacles associated with the energy transition and advancing a sustainable future. Australia is leveraging its abundant renewable resources to strategically position itself as a major player in the global hydrogen industry. With a variety of initiatives and ambitious targets, the country is actively promoting the growth of a strong hydrogen sector.

12.2.1.6.1 Production of green hydrogen

Australia is producing green hydrogen by utilizing its abundant solar, wind, and hydroelectric power resources. The goal of this clean energy initiative is to establish Australia as a prominent supplier of green hydrogen to the global market while also meeting the country's domestic energy needs (Panchenko et al., 2023).

12.2.1.6.2 Investment in renewable infrastructure

To ensure affordable power generation, transmission, and storage, the Australian government is making substantial investments in renewable energy infrastructure. This pledge demonstrates the country's commitment to creating an environment conducive to the efficient and sustainable production of hydrogen.

12.2.1.6.3 Goal for achieving affordable hydrogen costs

Australia aims to produce hydrogen for less than 2 AUD/kg by 2030. By reducing the cost of hydrogen compared to traditional fossil fuel options, this pricing target aims to promote widespread adoption.

12.2.1.6.4 Emphasis on blue and green hydrogen

The nation's hydrogen plan prioritizes the production of green and blue hydrogen. While blue hydrogen is produced using natural gas combined with CCS, green hydrogen is created using renewable resources. Australia can diversify its hydrogen production methods by employing this dual approach.

12.2.1.6.5 Origin guarantee program

The Australian government is implementing a guarantee of origin program to safeguard the environmental integrity of hydrogen production. To ensure that the hydrogen produced meets zero-emission criteria, this regulatory framework may require hydrogen producers to demonstrate the emissions intensity of their operations.

12.2.1.6.6 Development of projects and regional centers

Australia is home to over 15 projects that have reached the final investment decision (FID) stage, and there are currently 80 additional renewable hydrogen projects under development. Furthermore, the government has announced funding for the establishment of regional hydrogen centers in strategic areas, including Pilbara, Kwinana, Hunter, Bell Bay, Gladstone, and Port Bonython. These centers serve as focal points for boosting local economic development and generating hydrogen.

12.2.1.6.7 Ambitions for hydrogen export

Australia is actively pursuing the goal of becoming the world's leading exporter of hydrogen. Significant investments are being made in the development of the green and blue hydrogen sectors. Gladstone is being recognized as a crucial location for the production and export of green hydrogen because of its abundant renewable energy sources.

12.2.1.6.8 Government investment

The Australian government has shown its commitment to the growth of the hydrogen industry by investing over $500 million. This funding is essential for conducting research, developing infrastructure, and executing projects. Australia is showing a strong commitment to leading the global hydrogen economy through a diverse approach that includes substantial investments, regulatory frameworks, regional hubs, and technological development (Walsh et al., 2021).

12.2.1.7 Other countries

Recent developments in green hydrogen production in the United Kingdom, France, Sweden, Norway, Denmark, Switzerland, the Netherlands, Finland, New Zealand, Singapore, Belgium, Austria, Ireland, and Iceland include the following (Lycourghiotis, 2022):

- Policies and regulations: These countries have implemented various policies and regulations to promote the development of green hydrogen, including tax incentives, financial aid, and legislation to encourage its use. For example, the UK has implemented a hydrogen strategy and allocated £240 million for the development of a hydrogen economy.

- Targets for 2040 and 2070: The European Union has set ambitious goals for green hydrogen production, aiming to produce 40 TWhs of renewable hydrogen by 2030 and 400 TWh by 2040. The UK has also set a target to produce hydrogen at a cost of less than £2/kg by 2030.
- Green hydrogen production: Countries like France and Sweden are investing in facilities for green hydrogen production to reduce their dependence on fossil fuels and meet their climate targets.
- International collaboration: The Netherlands and Germany are working together on the development of green hydrogen, aiming to establish a hydrogen production plant in the Port of Antwerp*38*.
- Investment: The UK government has allocated £240 million for the development of a hydrogen economy, aiming to establish a hydrogen supply chain and support the growth of a hydrogen-based economy.
- Research and development: Countries like Norway are investing in research and development to reduce the cost of producing green hydrogen and enhance its competitiveness compared to fossil fuels.

These developments demonstrate the growing interest and commitment of these countries to promoting green hydrogen as a vital renewable energy source.

12.2.2 Developing countries

Several developing countries have adopted green hydrogen policies to encourage the use of renewable energy and decrease carbon emissions. For instance, the World Bank has been working on a report to improve the comprehension of the opportunities and challenges of green hydrogen in developing countries. The report emphasizes potential use cases, strategic value, technology risks, and implementation challenges. The International Renewable Energy Agency emphasized the importance of expanding sustainability standards for hydrogen beyond carbon emissions. This involves addressing other environmental impacts and social aspects, as well as establishing suitable governance structures. The IEA reported that more than 30 countries have developed or begun preparing national hydrogen strategies in the last 5 years. This trend has been influenced by the Paris climate goals, Russia's conflict in Ukraine, and the increasing gas prices. Regions such as Latin America and the Caribbean have the potential to expand green hydrogen production, with countries like Brazil, Uruguay, and Paraguay having over 85% renewable power in their energy mix. Chile is working on a $150 million project to promote the development of clean hydrogen, supporting the establishment of a green hydrogen market from pilot to industrial scale. These policies and initiatives demonstrate the growing interest and commitment of developing countries in green hydrogen as a crucial element in their transition to a low-carbon economy. In developing countries, hydrogen technology plays a crucial role in shaping sustainable energy futures. These nations recognize the potential of hydrogen as a versatile and clean energy carrier, and their policies reflect a commitment to harnessing its benefits. Despite facing unique challenges, these countries are actively implementing initiatives and strategies to integrate hydrogen into their energy landscapes (Shin, 2022).

12.2.2.1 China

Electrolyzer manufacturers have revealed their ambitious expansion initiatives, highlighting the availability of approximately 14 GW of manufacturing capacity, with half of it located in China (Gupta et al., 2024). In 2022, the estimated electrolyzer production is slightly over 1 GW. Looking ahead, manufacturers have articulated plans for significant growth, aiming for a manufacturing capacity of 155 GW/year by 2030 (Caglayan et al., 2020). However, it's noteworthy that only 8% of this capacity has reached the FID stage. The realization of these manufacturers' ambitious goals is contingent upon a robust demand for electrolyzers, a factor currently characterized by a high degree of uncertainty. This uncertainty has already translated into delays in the execution of expansion plans, with some projects temporarily put on hold. It underscores the nuanced landscape and challenges of aligning manufacturing capacity expansion with the dynamic and evolving demand for electrolysis technology. China has implemented various policies and initiatives to promote the development of green hydrogen and to support its role in decarbonizing hard-to-electrify sectors. Some of the key policies and targets include:

- Hydrogen industry development plan (2021–2035): The plan aims to annually produce 100,000 to 200,000 tons of renewable-based hydrogen and to have 50,000 hydrogen fuel-cell vehicles on the road by 2025.
- Action plan for carbon dioxide peaking before 2030: This plan emphasizes the importance of hydrogen in industries such as steel, petrochemicals, and transportation, as well as technologies for renewable hydrogen production.
- Hydrogen strategy by 2035: This strategy positions hydrogen as a critical element of China's future energy system and a key enabler for achieving a low-carbon energy transition in the country.
- 14th 5-year plan (2021–2025): The plan incorporates hydrogen as a critical element in China's energy transition and aims to boost the production and utilization of green hydrogen.
- Green hydrogen in China: According to the reports, China has the potential to install 100 GW or more of capacity by 2030, with a focus on utilization in the chemical, steel, and heavy-duty transportation sectors (Cheng et al., 2023).

These policies and plans demonstrate China's commitment to promoting green hydrogen as a crucial component in its transition to a low-carbon economy.

12.2.2.2 India

In India, hydrogen technology is actively advancing, involving various production methods such as steam methane reforming and electrolysis, along with pilot projects by organizations like the Indian Oil Corporation. Additionally, pilot programs for hydrogen fuel cell vehicles have begun in cities like Delhi and Bengaluru, exploring various storage methods, including underground storage (Ciotta et al., 2023). Upcoming trends include a shift toward green hydrogen production, the development of hydrogen corridors, ongoing investments in research and development, expanded international collaborations, the establishment of comprehensive regulatory frameworks, and the formulation of growth targets for the hydrogen economy, including the number of refueling stations and green hydrogen production capacity. India's hydrogen industry was ready for substantial expansion, driven by a promise to make green hydrogen using renewable energy to cut down on carbon emissions. There were expectations of upcoming announcements about green hydrogen projects and how much they could produce, showing India's dedication to eco-friendly hydrogen. Additionally, plans for hydrogen corridors, which are like special roads for hydrogen transportation, were in the works to make it easier for hydrogen vehicles to be used. At the same time, India was investing in research and development to make hydrogen production and storage better, and they were considering setting aside money specifically for these efforts to become leaders in hydrogen technology (Chantre et al., 2022). India has been actively promoting its green hydrogen policies and initiatives to support its renewable energy goals. Some of the notable advancements and initiatives include:

- The national hydrogen mission (NHM) has been approved by India with a target to produce at least 5 million tons of green hydrogen by 2030. The government will provide incentives for the production of green hydrogen and electrolyzers.
- Green hydrogen definition and certification: India has defined green hydrogen and green ammonia as products produced by the electrolysis of water using renewable energy sources. The country is in the process of establishing the definition and standards for green hydrogen certification, which are aligned with international standards.
- India has approved a carbon credit trading scheme, which aims to incentivize entities to adopt green hydrogen and renewable energy sources.
- Transportation and vehicles: India has approved a policy to promote the use of renewable energy in all economic sectors, including the widespread adoption of green hydrogen-powered vehicles. The government is investigating the use of hydrogen-powered vehicles as a potential alternative to traditional internal combustion engine vehicles.
- State-level initiatives: States like Gujarat, Rajasthan, Uttar Pradesh, Odisha, and Tamil Nadu are formulating policies to establish green hydrogen hubs.

India's commitment to promoting the production and use of green hydrogen is evident through these policies and initiatives, which align with its renewable energy goals and transition to a low-carbon economy. India has implemented various tax incentives and schemes to encourage green hydrogen projects. Some of the main incentives and programs include:

- Green hydrogen incentives: India plans to provide incentives to green hydrogen fuel producers, amounting to at least 10% of their costs, as part of a $2 billion scheme set to commence before the end of 2023. The government will provide incentives of at least 30 ₹/kg for the production of green hydrogen (Gupta et al., 2024).
- Transmission fee waiver: India has extended a waiver of transmission fees for renewable power, with the aim of reducing the cost of green hydrogen. The waiver will make more green hydrogen manufacturing projects eligible for the 25-year waiver previously available for projects set up before July 2025.
- National green hydrogen mission: India's National Green Hydrogen Mission aims to produce a minimum of 5 million tons of green hydrogen by 2030. The mission will

facilitate export opportunities through supportive policies and strategic partnerships. Additionally, MNRE will develop a suitable regulatory framework for the certification of green hydrogen and its derivatives.

These incentives and schemes aim to promote the implementation of green hydrogen projects in India, reduce reliance on fossil fuels, and position the country as a leader in green hydrogen technology and market. India is planning to work more with other countries in the hydrogen field, aiming to share technology and knowledge. They were expected to announce new partnerships with international groups. Additionally, India was likely to create rules to make sure hydrogen is used safely and efficiently, which would be important for everyone involved. Lastly, India had goals for its hydrogen industry, like having more hydrogen refueling stations, making more eco-friendly hydrogen, and getting more people to use hydrogen-powered vehicles, which would help shape the future of hydrogen in India (Cogho et al., 2023).

12.2.2.3 Brazil

Brazil has been making significant strides in developing policies and regulations for green hydrogen. The lower house in Brazil's legislature has approved a crucial bill defining low-carbon and renewable hydrogen and enabling 10-year subsidies for the green hydrogen industry. Additionally, discussions on hydrogen as a low-carbon energy source are growing in Brazil, resulting in several business opportunities in Brazilian states, with various federal bills and state decrees and laws being put in place to regulate and encourage the use of sustainable and renewable hydrogen. The Brazilian Senate's Environment Committee has also approved a draft bill regulating the use of hydrogen as an energy source, including the production of green hydrogen. These developments demonstrate Brazil's commitment to establishing a legal framework and regulatory stability to attract investment in the green hydrogen sector. Brazil's policies and regulations on green hydrogen encompass several key provisions aimed at promoting the use of renewable hydrogen and establishing a legal framework for its production and consumption. Some of the key provisions include:

- State and federal legislation: It also outlines the responsibilities of the Brazilian national agency for oil, natural gas, and biofuels (ANP) in regulating the hydrogen sector, including production, import, export, and storage.
- Regulatory stability and legal certainty: The approved bill aims to establish the legal certainty and regulatory stability necessary to attract investment in the low-carbon hydrogen sector (Chantre et al., 2022).
- ANP's duties and the national hydrogen program: The Brazilian ANP will be responsible for regulating companies operating in export processing zones in Brazil and overseeing the production and use of hydrogen. The bill also outlines the national hydrogen program.
- Subsidies and funding: The approved legislation provides for 10-year subsidies for the green hydrogen industry. These subsidies will be financed through federal government budget allocations, contributions from the public and private sectors, loans, and a percentage of profits that exceed a certain threshold achieved by local development agencies (Ciotta et al., 2023).

These provisions demonstrate Brazil's commitment to promoting the development and use of green hydrogen through clear regulatory frameworks, financial incentives, and the establishment of low-carbon and renewable hydrogen standards.

12.2.2.4 South Africa

South Africa has made significant progress in developing its policy and strategy for green hydrogen (Cogho et al., 2023). Here are some key points from the sources provided:

- South Africa has identified three green hydrogen hubs in the eastern part of the country, and a feasibility study was released in October 2021.
- Hydrogen society roadmap: The country has unveiled a hydrogen society roadmap with the aim of producing 500,000 tons of green hydrogen annually by 2030. The government also aims to install 10 GW of electrolysis capacity in the Northern Cape region by 2030 and 15 GW by 2040 (AbouSeada & Hatem, 2022).
- Economic impact: South Africa's plans for green hydrogen are expected to create at least 20,000 new jobs by 2030, and the country aims to produce green hydrogen for \$1.60/kg by 2030.
- International partnerships: South Africa has partnered with the governments of the Netherlands and Denmark to secure USD 1 billion for the development of green hydrogen projects in the country.
- Export market: The country aims to develop an export market for South African green hydrogen and ammonia, with a focus on reducing carbon emissions in the power generation, transport, and heavy industry sectors.
- Cost and production targets: South Africa has unveiled plans to produce over five million tons of green hydrogen annually by 2040, with an estimated cost of \$230 billion.

These developments demonstrate South Africa's commitment to becoming a significant player in the green hydrogen economy, emphasizing domestic production, job creation, and international partnerships.

12.2.2.5 Turkey

Turkey has made significant progress in developing green hydrogen as a key component of its energy strategy. The country's hydrogen strategy emphasizes the development of green hydrogen energy, which is essential for improving energy security, minimizing environmental impact, and fulfilling the country's energy requirements. The South Marmara Development Agency (GMKA) plans to build a sustainable energy production facility in Bandırma that can produce 500 tons of green hydrogen annually. Turkey's strategic investments and policy adjustments could increase its green hydrogen production to 3.4 million tons annually by 2050.Turkey has announced its national energy plan and hydrogen strategy, with a focus on green hydrogen playing a significant role in the country's efforts to achieve net-zero emissions (Hilali et al., 2024). The Regulation on Rules and Procedures Regarding Increasing Energy Efficiency in Transportation aims to promote energy efficiency by utilizing hydrogen. Key aspects of Turkey's hydrogen policy include:

- The hydrogen technologies strategy and roadmap focuses on the development of green hydrogen energy, which is essential for enhancing energy security, minimizing environmental impact, and fulfilling the country's energy requirements.
- National energy plan: The plan aims to reduce Turkey's energy consumption, which was 147.2 million tons of oil equivalent in 2020, and to increase the installed capacity to 70 GW by 2053. The electrolysis of water using renewable energy sources will be a crucial factor in achieving the country's net-zero emission target.
- Hydrogen production targets: Turkey aims to increase the proportion of hydrogen blended into natural gas to 12% by 2030 and 30% by the end of 2053. The cost of hydrogen production per kilogram is expected to decrease to $2.4 by 2035 (Yilmaz et al., 2024).
- Electrolyzer capacity: Under the new plan, the installed electrolyzer capacity will reach 2 GW in 2030, 5 GW in 2035, and 70 GW by 2053.

While Turkey has made progress in developing a national hydrogen strategy, it is essential to address the lack of specific legislation defining "gas" and ensure a stable and affordable supply of green hydrogen.

12.2.2.6 Chile

Chile has been actively developing its policies and regulations for green hydrogen. The Chilean government has outlined a regulatory roadmap, which includes the implementation of a series of regulations such as the hydrogen facilities general regulation, the multi-fuel service stations regulation, and the technical, construction, and security requirements for GH2-powered vehicles regulation. The government is amending its regulations on Law DL 2.224 (Article 3), which established the Ministry of Energy, to include standards and safety regulations for green hydrogen. However, there are currently no specific laws regulating green hydrogen in Chile, although a bill is being processed in the Chilean Congress. The Chile green hydrogen facility aims to promote investment in green hydrogen projects in Chile to support sustainable, resilient, and inclusive economic development. This initiative accelerates the country's progress toward green growth and energy transition, in line with its commitment to achieve carbon neutrality by 2050. The World Bank has been assisting Chile in developing regulations and establishing a conducive environment for green hydrogen since 2020 (Garrido et al., 2024).

The report concludes with recommendations for developing safety standards and regulations for the use of hydrogen in mining. It also includes a set of recommended regulatory assessments for renewable energy quotas, including an analysis of the value chain for green hydrogen.

12.2.2.7 United Arab Emirates

The United Arab Emirates (UAE) has implemented various incentives to encourage the production of green hydrogen, aiming to become one of the leading global producers of green hydrogen. Some of the main incentives and initiatives include:

- Investment in green hydrogen: The UAE is investing in green hydrogen, which is produced using renewable energy sources such as solar and wind power. This method ensures that the hydrogen is produced with minimal carbon emissions (Gandhi et al., 2022).

- The UAE has approved a national hydrogen strategy that includes plans to develop at least two production hubs by 2031, with the intention of establishing three more by 2050. The strategy also aims to establish a green hydrogen research and development center by 2031.
- Public spending and energy strategy: The UAE's new national energy strategy pledges to increase public investment in all forms of energy, with clean energy accounting for 30% of the total mix by 2031. The strategy aims to enhance the UAE's status as a producer and exporter of low-emission hydrogen. It involves establishing hydrogen refueling stations and a national research and development center.
- Export goals: By 2050, the UAE aims to export between 4.8 and 9.6 million tons per year of green hydrogen and its derivatives, including ammonia, synthetic fuels, and green steel. The strategy encompasses various types of incentives aimed at promoting the growth of the hydrogen production industry. This includes cost support to reduce the levelized cost of hydrogen production projects in the UAE, as well as revenue support guarantees.
- Production Capacity Targets: The UAE aims to achieve a production capacity of 1.4 million tons per annum (mtpa) of low-carbon hydrogen by 2031, 7.5 mtpa by 2040, and 15 mtpa by 2050.

These incentives and targets demonstrate the UAE's strong commitment to and investment in the production of green hydrogen, positioning the country as a significant player in the global market for green hydrogen.

12.2.2.8 *Other developing nations*

Colombia, Morocco, and Mexico have been developing policies and regulations for green hydrogen. Here is a summary of their respective policies and rules in Colombia. Senate Bill 371 of 2020 aims to promote the production and use of green hydrogen across various industries. Green hydrogen pilot projects are expected to commence in 2022. By 2050, Colombia aims to substantially increase the production of green hydrogen from renewable sources. The country's commitment to reducing GHG emissions and leading the way in energy transition may lead to the enactment of a green hydrogen law in the future. Mexico has a high potential for renewable energy and a well-distributed energy infrastructure, which could facilitate the development of green hydrogen. The existing energy policy in the country and the difficult investment environment for renewable projects in Mexico are obstacles to the advancement of green hydrogen. The lack of regulations aimed at achieving the goals of the Paris Agreement is one of the main barriers to the development of green hydrogen in Mexico (Obregón et al., 2023). The Energy Sectoral Program 2020–2024 outlines the "rescue and promotion" of the energy sector, with a focus on hydrogen. In summary, Colombia is in the process of developing a regulatory roadmap to establish safety and technology standards for the use of hydrogen in new applications. Mexico faces barriers in its energy policy and lacks regulations, but it has a high potential for developing green hydrogen using renewable sources.

12.2.3 Least developed countries

Many nations, especially some least developed countries (LDCs), have expressed interest in integrating green hydrogen into their energy strategies, as the concept of "green hydrogen" and its related policies have gained international prominence. In an era characterized by a global demand for innovative and sustainable energy solutions, countries with diverse economic landscapes are demonstrating unprecedented interest in green hydrogen. Remarkably, even the LDCs are now actively involved in the effort to incorporate green hydrogen into their energy plans. This increased interest is evidence of our shared commitment to addressing climate change, improving energy security, and promoting economic growth. As the LDCs embrace the potential of green hydrogen, they are advancing their economies by harnessing a wealth of renewable resources, seizing international market opportunities, and pursuing sustainable development goals to establish a cleaner and more resilient energy future. This paradigm shift places LDCs at the center of the rapidly evolving global energy innovation landscape, while also reflecting a commitment to environmental stewardship. The following are examples of the least developed nations that have been exploring or expressing interest in green hydrogen policies:

12.2.3.1 Bangladesh

Bangladesh has been developing policies and initiatives for green hydrogen, recognizing its potential to achieve energy independence and reduce the country's expenditure on energy imports. Some significant developments and initiatives include:

- Bangladesh hydrogen and fuel cell association (BHFCA): The BHFCA is a dedicated organization that aims to promote the advancement of green hydrogen in Bangladesh. It has partnered with the Bangladesh Atomic Energy Commission to create a plan for producing green hydrogen.
- Pilot Plant: Bangladesh has established a pilot plant at the Bangladesh Council of Scientific and Industrial Research (BCSIR) to produce green hydrogen using waste and biomass as feedstock. The pilot plant is part of the government's broader initiative to diversify the country's energy sources, as it is currently heavily reliant on gas and coal.
- Research and Training: The BCSIR has commenced training programs for scientists and is conducting research on hydrogen production, storage, and quality control. The project aims to accelerate energy efficiency and assist the government in achieving its target of reducing GHG emissions.
- Future Prospects: Salam, the project director and senior principal engineer at BCSIR, is optimistic that hydrogen fuel can be made available for industrial and vehicular use by 2030. The project aims to conduct research on quality control related to hydrogen production, storage, and supply, as well as infrastructure development (Hossain et al., 2023).

While specific policies and regulations for green hydrogen are not yet in place, these initiatives demonstrate Bangladesh's commitment to developing a sustainable energy future and harnessing the potential of green hydrogen for energy independence and reduced emissions.

12.2.3.2 *Bhutan*

Bhutan's green hydrogen policy is centered on the National Green Hydrogen Roadmap, which aims to accelerate the implementation of green hydrogen projects in the country. The roadmap will include strategies to support the deployment of green hydrogen infrastructure in accordance with relevant laws, regulations, government commitments, incentives, and targets of the Royal Government of Bhutan. The roadmap will also outline priority economic sectors for the utilization of hydrogen, the economic impact of the hydrogen initiatives, a process roadmap for stakeholder participation in the hydrogen sector, national milestones and targets for hydrogen sector growth, and a strategy for financing hydrogen projects and market access. Bhutan's Sustainable Hydropower Development Policy, 2020, emphasizes the development of green hydrogen in conjunction with hydropower. The country's Department of Energy is developing a roadmap and has engaged technical and economic hydrogen experts to conduct a market study on green hydrogen in Bhutan. The study identified potential domestic and export markets by projecting hydrogen production prices, considering existing transportation infrastructure, and evaluating potential end-use markets within and outside of Bhutan. The policy and regulatory recommendations for advancing green hydrogen development will be proposed based on the findings of the market study.

12.2.3.3 *Cambodia*

Cambodia has attracted significant foreign direct investment interest in its energy sector, especially in green hydrogen projects. For instance, a $9.4 billion green hydrogen project has been announced in the Tsau Khaeb National Park, and there are plans to produce green hydrogen with a total investment estimated at $2.29 billion. This indicates a growing interest and momentum in green hydrogen initiatives in Cambodia. Additionally, although there is no specific information available on Cambodia's green hydrogen policy, the European Union has established regulations for green hydrogen, highlighting the importance of producing renewable hydrogen in locations where ample renewable energy is accessible. The global context may influence Cambodia's approach to green hydrogen policy and regulations (Handayani et al., 2023).

12.2.3.4 *Challenges faced by least developed countries*

It is important to remember that the specific challenges and reasons for delays may vary significantly among LDCs. To tackle these issues, it is often necessary to combine international cooperation, financial assistance, and targeted capacity-building initiatives. The challenges faced by LDCs in implementing green hydrogen policies include:

- Elevated costs of production: The high production costs of green hydrogen present a significant obstacle to its widespread implementation. This cost challenge must be addressed to make green hydrogen more economically viable.
- Lack of existing infrastructure: The lack of established infrastructure for green hydrogen production and distribution poses a challenge to its implementation. Developing the essential infrastructure necessitates substantial investment and planning.

- Policy barriers and a lack of demand for green hydrogen can hinder its full potential contribution to the industrial sector. Industrial policies need to address these barriers and stimulate the demand for green hydrogen.
- Geopolitical implications and import dependency: Countries that depend on imports for green hydrogen face challenging geopolitical implications that must be addressed. Securing a stable and affordable supply of green hydrogen, especially through imports, is a major concern.
- Integration into the broader energy system: As green hydrogen becomes more widely adopted, policies should consider its integration into the broader energy system. Civil society and industry must collaborate to maximize the benefits of green hydrogen.

In summary, the challenges faced by LDCs in implementing green hydrogen policies include high production costs, lack of existing infrastructure, policy barriers, geopolitical implications, and the need for integration into the broader energy system. Addressing these challenges is crucial for the successful implementation of green hydrogen policies in the LDCs.

12.3 Policies and regulations with a global perspective with focus on underground hydrogen storage

Worldwide, caverns and aquifers have been the most frequently found places where hydrogen (H_2) is effectively stored in underground geological formations, as previously mentioned. But none of the hydrogen storage techniques described above for storing 100% pure H_2 are documented in the literature. The successful storage of H_2 (containing 95% H_2 and 3%−4% CO_2) by a British company in salt caverns located 400 m beneath the surface of Teesside, UK, is noteworthy. ConocoPhillips, which operates salt caverns on the Texas US Gulf Coast, has also achieved success in storing hydrogen at a depth of 1000 m (AbouSeada & Hatem, 2022). The storage consists of 95% hydrogen and 3%−4% carbon dioxide (Hilali et al., 2024). Engineers and scientists have welcomed the growing interest in hydrogen (H_2) storage and its conversion into other energy sources, emphasizing the significance of innovation and advancement. The ongoing global effort to develop a carbon-free hydrogen economy is the main reason for this increased interest. This significant change in focus reflects everyone's shared commitment to shifting energy paradigms. A proactive step toward realizing an environmentally friendly energy future is represented by this revitalized exploration. Using research from studies conducted by Chantre et al. (2022) and Yilmaz et al (2024), it has been thoroughly documented that hydrogen (H_2) storage facilities have been successfully implemented worldwide. The Teesside site in the United Kingdom consists of three separate caverns, each with a volume of 70,000 m^3 and a depth of 400 m. This facility operates at a pressure level of 45 bar, resulting in an impressive 95% H_2 content. Similar to this, a cavern with a substantial volume of 580,000 m^3 is utilized for storing H_2 at the Clemens site in the United States (Garrido et al., 2024). The cavern is situated at a depth of 1000 m. With a pressure range of 70−137 bars, this cavern produces hydrogen with a volume composition of 95%. The Moss Bluff location, another successful endeavor in the US, is situated at a depth of 1200 m and has a cavern volume of 566,000 m^3. A composition of 95% H_2 is possible under operating conditions ranging from 55 to 152 bars. In Germany, at the Kiel facility, which has a cavern capacity of 32,000 m^3, operational pressures of 100 bars result in an H_2 concentration of 60% (Gandhi et al., 2022). The Lobodice facility in the Czech Republic operates at depths ranging from 400 to 500 m. In this facility, the concentration of H_2 is 54% at a pressure of 90 bars and a temperature of 34°C. Similar to this, the Beynes facility in France, which is located at a depth of 430 m, stores hydrogen in a space with a volume of 3.3×10^8 m^3 (Rojas Obregón et al., 2023).

The building maintains a hydrogen (H_2) concentration of 50%. Although specific volume and pressure information was not provided, the Ketzin location in Germany maintains an H_2 content of 62% and ranges in depth from 200 to 250 m (Hossain et al., 2023). These productive hydrogen storage projects, which utilize various geological formations, operational settings, and hydrogen compositions, showcase the global endeavors to develop effective and efficient solutions for hydrogen storage while ensuring the integrity of the original data.

12.4 Considerations for infrastructure development and regulation for underground hydrogen storage

UHS development and regulation entails a number of essential factors to enable its safe and successful deployment. Geological appropriateness, adapting existing infrastructure, safety precautions, legal frameworks, and public acceptability all play important roles in the development of UHS projects. By addressing these concerns, UHS can provide a dependable method of balancing hydrogen supply and demand, improving energy security, and easing the transition to a sustainable energy future. Fig. 12.4 shows the different factors affecting UHS.

- Comprehensive geological assessments are necessary when selecting suitable geological formations, such as salt caverns, depleted gas fields, or aquifers, to assure security, confinement, and enough storage capacity (Tarkowski & Uliasz-Misiak, 2022).
- Existing facilities, such as salt caverns or exhausted natural gas reservoirs, may be retrofitted for hydrogen storage at a lower cost and faster than creating brand-new infrastructure from scratch (Kumar, Honorio, et al., 2023).
- In order to get permits and licenses for UHS, regulatory bodies require proof of compliance with safety, environmental, and land-use laws (Al-Shafi et al., 2023).
- In-depth risk assessments, monitoring procedures, and efficient mitigation techniques are required in order to prevent hydrogen leaks, subsurface contamination, and unintentional discharges (Zivar et al., 2021).
- To avoid leaks and structural failures, it is crucial to maintain the containment integrity of storage formations, which calls for regular integrity checks and maintenance procedures (Thiyagarajan et al., 2022).
- The key to handling the various patterns of hydrogen supply and demand successfully is the development of flexible systems capable of both quick cycling to suit short-term demand changes and leisurely cycling for prolonged storage durations (Jafari Raad et al., 2022).

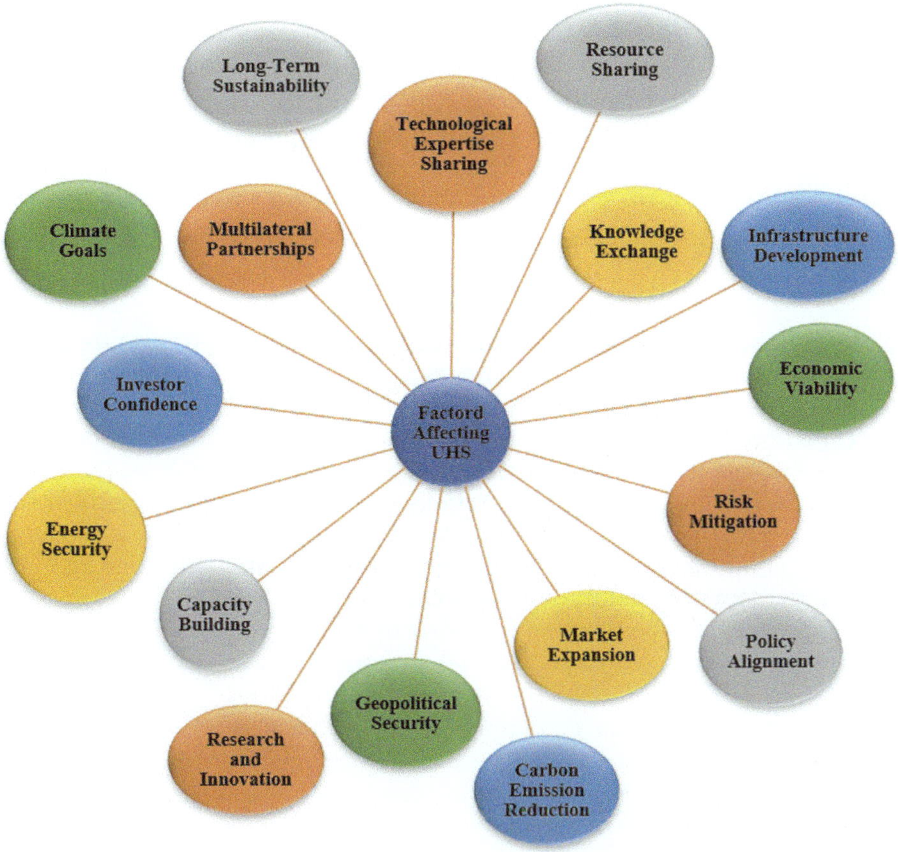

FIGURE 12.4 Schematic showing the different factors affecting underground hydrogen storage.

- Establishing monitoring systems to monitor hydrogen levels, pressure, and possible leaks, together with frequent reporting to regulatory authorities, ensures that compliance criteria are met on a continuous basis (Qin et al., 2023).

12.5 Global collaborations on underground hydrogen storage

12.5.1 Public/ecological considerations and popular acceptability

The global push for UHS faces significant public and legal challenges, exemplified by the debate in Canada surrounding the Impact Assessment Act (IAA), formerly known as Bill C-69. This legislation enables the Canadian government to evaluate the environmental and social impacts of major projects, including those related to UHS (Laidlaw, 2018). Public opinion on UHS is polarized, with stakeholders such as the Indian Resource Council expressing concerns about inadequate consultations and the perceived federal veto against exploration and production activities. Indigenous groups argue that their rights were not sufficiently considered during the development of the bill. Conversely, environmental and legal organizations support the IAA, viewing it as a vital tool to address climate change and protect the environment. This legal battle underscores the complexities and divergent perspectives surrounding environmental legislation, particularly in the context of emerging technologies like UHS. The outcome of the Supreme Court case will likely influence the trajectory of UHS projects in Canada, setting a precedent for how public and ecological considerations are navigated globally (Barbier & Agnoletti, 2023).

As shown in Fig. 12.5, the five factors exerting influence on global collaborations are clearly illustrated.

1. Environmental impact: One of the primary concerns regarding UHS is its potential environmental impact. The extraction and storage of hydrogen may involve processes that release GHGs or other pollutants, depending on the hydrogen production method. Public acceptance of UHS technologies may be influenced by how well these technologies address and mitigate environmental impacts (Zhao et al., 2023).
2. Safety concerns: The public may be apprehensive about the safety aspects of storing large volumes of hydrogen underground. Accidents or leaks could raise safety concerns among nearby communities. Demonstrating robust safety measures and risk assessments is essential to gaining public trust (Pan et al., 2021).
3. Land use and asthetics: The physical infrastructure required for UHS, such as storage facilities and pipelines, may impact local landscapes and land use. Public acceptance

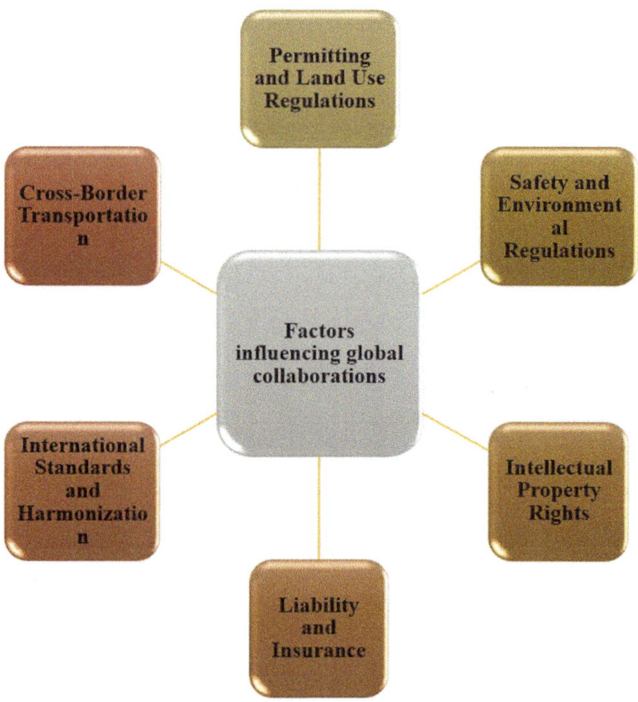

FIGURE 12.5 Factors influencing major collaborations across the globe.

may depend on how well such projects integrate with existing surroundings and minimize visual and environmental disruptions (Haghi et al., 2018).

4. Community engagement: Involving local communities in the decision-making process and considering their concerns and preferences can positively influence public acceptance of UHS technologies. Transparency and open dialog are essential to building trust and addressing community needs (Graetz, 2009).
5. Awareness and education: Public acceptance may be influenced by the level of awareness and understanding of hydrogen technologies and their potential benefits. Educating the public about the role of UHS in supporting renewable energy integration and decarbonization can foster positive attitudes (Iordache et al., 2014).

In conclusion, public acceptance of UHS and hydrogen technologies depends on addressing ecological concerns, demonstrating safety measures, and engaging with local communities. Legal frameworks and regulatory compliance are critical for obtaining necessary permits and ensuring that UHS projects adhere to safety and environmental standards. Harmonization of international standards can facilitate global collaboration while also addressing issues related to intellectual property and technology transfer (Jafari Raad et al., 2022). By considering public and ecological factors and navigating legal complexities, the successful implementation of UHS technologies can contribute to a more sustainable and decarbonized energy future (Qin et al., 2023).

12.5.2 Underground hydrogen storage collaborations across the world

Global collaborations in UHS are of paramount importance, particularly for densely populated and high-GHG emitting countries like China, India, and the USA (Laidlaw, 2018). According to data from the Global Carbon Atlas, China is the largest emitter of CO_2 globally, contributing approximately 28% of the world's total emissions, while India follows closely as the third-largest emitter, at around 7% (Barbier & Agnoletti, 2023). These statistics underscore the critical need for effective UHS solutions in these nations to curb their significant carbon footprints. While cutting-edge technologies often originate in smaller European nations, the widespread implementation required by larger countries underscores the critical need for technology transfer. Collaborative efforts should prioritize the adaptation of UHS technologies to align with the specific challenges and infrastructure of more populous nations, such as China and India. In fostering collaborative work, international research initiatives are key. Partnerships between developed nations like Germany, a leader in hydrogen technology, and developing nations can accelerate the development of efficient and cost-effective UHS technologies. Platforms for the exchange of knowledge, best practices, and lessons learned should be established to facilitate a more streamlined development and deployment process (Tarkowski, 2019). Monetary support is a fundamental aspect of successful UHS collaborations. According to the IEA, global hydrogen investment is expected to reach $300 billion by 2030 (Olemedia, n.d.). This financial commitment is crucial for the success of UHS projects, with developed nations contributing significantly to funds dedicated to clean energy projects, particularly in developing countries like those in Southeast Asia and Africa (Pflugmann & De Blasio, 2020). Bilateral and multilateral agreements can also play a crucial role, incorporating technology transfer

FIGURE 12.6 Major underground hydrogen storage collaborations across the world.

clauses and commitments to jointly address climate change through hydrogen storage solutions.

The European Union's commitment to hydrogen as part of its Green Deal strategy serves as an example of a regional bloc actively supporting the development and deployment of clean energy technologies (Haghi et al., 2018). Capacity building is essential to ensuring the sustainable implementation of UHS technologies. Training programs for local professionals in developing countries can transfer the necessary skills and knowledge, empowering the local workforce. This is especially relevant for countries like Brazil and South Africa, which are emerging as key players in the global transition to cleaner energy. Beyond the technological realm, collaborative efforts should extend to environmental diplomacy. Advocacy for the importance of UHS projects in international forums and diplomatic endeavors to garner support for such initiatives are vital components. Recognizing the global impact of addressing climate change, developed nations can play a diplomatic role in ensuring the success of UHS projects on a worldwide scale. Fig. 12.6 shows the major UHS collaborations across the world (Lankof & Tarkowski, 2023).

12.6 Future prospects and potentials of underground hydrogen storage technologies

1. Growing importance: UHS technologies are expected to play an increasingly critical role in the development of a sustainable and decarbonized energy system. As renewable

energy sources like solar and wind become more prevalent, UHS can act as a vital enabler for storing excess energy and ensuring a stable supply of hydrogen when renewables are not available.

2. Integration with renewable energy: UHS technologies can facilitate the integration of renewable energy into the grid by offering a way to store excess electricity as hydrogen during periods of low demand. This stored hydrogen can be converted back to electricity or used as a clean fuel for various applications, such as transportation and industry.
3. Support for grid balancing: UHS can contribute to grid stability by providing a buffer for intermittent renewable energy sources. It can help balance supply and demand and reduce the need for curtailment of renewable energy during times of oversupply.
4. Decarbonization in various sectors: Hydrogen produced and stored using UHS can be utilized as a clean energy carrier in sectors such as industry, transportation, and heating, thus supporting efforts to reduce GHG emissions in these sectors (Bai et al., 2014).

12.6.1 New trends and integration opportunities for underground hydrogen storage

1. Power-to-gas: The integration of UHS with power-to-gas technologies presents a promising trend. Excess electricity from renewable sources can be used to produce hydrogen through electrolysis, which is then stored underground, creating a link between the electricity and gas sectors (Aslannezhad et al., 2023).
2. Sector coupling: UHS can be integrated with various sectors, allowing for the utilization of stored hydrogen in power generation, mobility, heating, and industrial processes, creating an interconnected and versatile energy system.
3. Hydrogen hubs: Developing hydrogen hubs with UHS facilities can facilitate regional and cross-border hydrogen trading, promote international collaboration, and leverage the strengths of different countries in the hydrogen value chain (Mori et al., 2023).

12.6.2 Suggestions to improve underground hydrogen storage implementation

1. Research and development: Continued investment in research and development is crucial for advancing UHS technologies, optimizing storage capacities, enhancing safety measures, and reducing costs (Bo et al., 2023).
2. Standardization: Establishing international standards for UHS technologies can create a more consistent and predictable regulatory environment, fostering global cooperation and investment (Bahrami et al., 2023).
3. Infrastructure development: Governments and private sectors should invest in building the necessary infrastructure, including pipelines, to transport hydrogen from UHS facilities to end-users (Yazdi et al., 2023).

12.6.3 Government actions, legislation, and support for underground hydrogen storage

Incentives and funding: Governments can provide financial incentives, grants, and subsidies to support the development and implementation of UHS technologies, encouraging private investment and accelerating deployment. Whereas enacting supportive policies and regulations can create a favorable business environment for UHS projects and ensure compliance with safety and environmental standards. Governments should encourage international collaboration on UHS technologies through partnerships, knowledge exchange, and joint research initiatives. In conclusion, UHS technologies hold significant promise for the future energy landscape, enabling the integration of renewable energy, promoting sector coupling, and contributing to decarbonization efforts (Miłek et al., 2022). To maximize its potential, continued research, standardization, and infrastructure development are essential. Government support through incentives, policies, and international collaboration will be crucial in unlocking the full benefits of UHS for a sustainable and low-carbon energy future. This chapter has revealed a space where creativity and government collide and where aspiration and action intersect. The adoption of hydrogen technology and ultra-high-strength (UHS) materials is being driven by governments, who are not only observers but also agents of change. The synergy of national strategies, cooperative alliances, and legal frameworks has the potential to reshape industries, decrease carbon footprints, and propel society toward a sustainable and advanced future. We find ourselves at a crossroads of possibilities as this journey progresses, where the decisions we make today will have an impact on future generations.

12.7 Conclusions

The convergence of forward-thinking policies and innovative technologies is crucial for the pursuit of a sustainable future. The chapter thoroughly examines the intricate network of government regulations, directives, programs, and support for two revolutionary components: hydrogen technology and UHS materials. These locations emerge as beacons, guiding us toward a brighter tomorrow as the world grapples with environmental challenges and the need for technological progress. On a global scale, there has been a growing interest in the revival of hydrogen as a versatile and emission-free energy carrier. The potential to reduce carbon emissions and diversify energy portfolios has led nations to adopt proactive measures. Infrastructure development, investment incentives, and national hydrogen plans all indicate a coordinated effort to establish the "hydrogen economy." The collaborative platforms, such as the Hydrogen Council and the European Clean Hydrogen Alliance, demonstrate the determination of all those involved to dismantle barriers and promote sustainable solutions. Parallel to this, the importance of UHS materials has increased across industries (Graetz, 2009). UHS materials have revolutionized the regulation of design, efficiency, and environmental impact in various industries, from aeronautical marvels to automotive advancements. Recognizing the importance of these projects, governments are directing industrial innovation initiatives, establishing regulatory frameworks, and cultivating a skilled workforce. These programs will enable UHS materials to shape a future in which lighter, more durable constructions coexist harmoniously with sustainability objectives. These efforts have a complex background, shaped by public opinion and ecological concerns. A global audience, concerned about the health of the Earth, is drawn to the transition toward sustainable energy and innovative materials. Governments pay attention to these voices, ensuring that the community is involved and that decisions are made in a transparent manner. This is done in order to develop long-term solutions that align with societal ideals. Governments are laying the groundwork for a safe and successful deployment by navigating the legal maze. The development of hydrogen technology and UHS materials is limited by safety regulations and intellectual property laws, both of which aim to encourage innovation and protect the welfare of society.

References

AbouSeada, N., & Hatem, T. M. (2022). Climate action: Prospects of green hydrogen in Africa. *Energy Reports, 8*, 3873–3890. Available from https://doi.org/10.1016/j.egyr.2022.02.225, http://www.journals.elsevier.com/energy-reports/.

Aglietta, M., Bai, G., & Macaire, C. (2021). The 14th five-year plan in the new era of China's reform – Asian Integration, Belt and Road Initiative and Safeguarding Multilateralism.

Al-Shafi, M., Massarweh, O., Abushaikha, A. S., & Bicer, Y. (2023). A review on underground gas storage systems: Natural gas, hydrogen and carbon sequestration. *Energy Reports*, *9*, 6251–6266. Available from http://www.journals.elsevier.com/energy-reports/, 10.1016/j.egyr.2023.05.236.

Aslannezhad, M., Ali, M., Kalantariasl, A., Sayyafzadeh, M., You, Z., Iglauer, S., & Keshavarz, A. (2023). A review of hydrogen/rock/brine interaction: Implications for Hydrogen Geo-storage. *Progress in Energy and Combustion Science, 95*. Available from https://doi.org/10.1016/j.pecs.2022.101066, https://www.journals.elsevier.com/progress-in-energy-and-combustion-science.

Bahrami, M., Izadi Amiri, E., Zivar, D., Ayatollahi, S., & Mahani, H. (2023). Challenges in the simulation of underground hydrogen storage: A review of relative permeability and hysteresis in hydrogen-water system. *Journal of Energy Storage, 73*. Available from https://doi.org/10.1016/j.est.2023.108886, http://www.journals.elsevier.com/journal-of-energy-storage/.

Bai, M., Song, K., Sun, Y., He, M., Li, Y., & Sun, J. (2014). An overview of hydrogen underground storage technology and prospects in China. *Journal of Petroleum Science and Engineering, 124*, 132–136. Available from https://doi.org/10.1016/j.petrol.2014.09.037, http://www.sciencedirect.com/science/journal/09204105/81.

Barbier, L., & Agnoletti, M. F. (2023). Are French people ready to accept hydrogen underground storage? An answer through the distance from object model. *International Journal of Hydrogen Energy, 48*(52), 19792–19802. Available from https://doi.org/10.1016/j.ijhydene.2023.02.077, http://www.journals.elsevier.com/international-journal-of-hydrogen-energy/.

Bartlett, J., & Krupnick, A. (2020). Decarbonized hydrogen in the US power and industrial sectors: Identifying and incentivizing opportunities to lower emissions about the authors.

Bo, Z., Boon, M., Hajibeygi, H., & Hurter, S. (2023). Impact of experimentally measured relative permeability hysteresis on reservoir-scale performance of underground hydrogen storage (UHS). *International Journal of Hydrogen Energy, 48*(36), 13527–13542. Available from https://doi.org/10.1016/j.ijhydene.2022.12.270, http://www.journals.elsevier.com/international-journal-of-hydrogen-energy/.

Borowski, P. F., & Karlikowska, B. (2023). Clean Hydrogen is a challenge for enterprises in the era of low-emission and zero-emission economy. *Energies, 16*(3). Available from https://doi.org/10.3390/en16031171, http://www.mdpi.com/journal/energies/.

Caglayan, D. G., Weber, N., Heinrichs, H. U., Linßen, J., Robinius, M., Kukla, P. A., & Stolten, D. (2020). Technical potential of salt caverns for hydrogen storage in Europe. *International Journal of Hydrogen Energy, 45*(11), 6793–6805. Available from https://doi.org/10.1016/j.ijhydene.2019.12.161, http://www.journals.elsevier.com/international-journal-of-hydrogen-energy/.

Chantre, C., Andrade Eliziário, S., Pradelle, F., Católico, A. C., Branquinho Das Dores, A. M., Torres Serra, E., Campello Tucunduva, R., Botelho Pimenta Cantarino, V., & Leal Braga, S. (2022). Hydrogen economy development in Brazil: An analysis of stakeholders' perception. *Sustainable Production and Consumption, 34*, 26–41. Available from https://doi.org/10.1016/j.spc.2022.08.028, http://www.journals.elsevier.com/sustainable-production-and-consumption/.

Cheng, G., Luo, E., Zhao, Y., Yang, Y., Chen, B., Cai, Y., Wang, X., & Dong, C. (2023). Analysis and prediction of green hydrogen production potential by photovoltaic-powered water electrolysis using machine learning in China. *Energy, 284*. Available from https://doi.org/10.1016/j.energy.2023.129302, https://www.journals.elsevier.com/energy.

Chu, K. H., Lim, J., Mang, J. S., & Hwang, M. H. (2022). Evaluation of strategic directions for supply and demand of green hydrogen in South Korea. *International Journal of Hydrogen Energy, 47*(3), 1409–1424. Available from https://doi.org/10.1016/j.ijhydene.2021.10.107, http://www.journals.elsevier.com/international-journal-of-hydrogen-energy/.

Ciotta, M., Tassinari, C., Larizatti Zacharias, L. G., van der Zwaan, B., & Peyerl, D. (2023). Hydrogen storage in depleted offshore gas fields in Brazil: Potential and implications for energy security. *International Journal of Hydrogen Energy, 48*(100), 39967–39980. Available from https://doi.org/10.1016/j.ijhydene.2023.08.209, https://www.sciencedirect.com/science/journal/03603199.

Cogho, E., Beukes, J. P., van Zyl, P. G., Vakkari, V., Laakso, L., Josipovic, M., & Kulmala, M. (2023). Concentration contextualisation, temporal patterns and sources of hydrogen sulphide at a site on the South African Highveld. *Atmospheric Environment, 315*. Available from https://doi.org/10.1016/j.atmosenv.2023.120140.

Erbach G. and Jensen L. 2021. EU hydrogen policy. Hydrogen as an energy carrier for a climate-neutral economy, Briefing. Towards climate neutrality, EPRS, February 2021.

Gandhi, K., Apostoleris, H., & Sgouridis, S. (2022). Catching the hydrogen train: economics-driven green hydrogen adoption potential in the United Arab Emirates. *International Journal of Hydrogen Energy, 47*(53), 22285–22301. Available from https://doi.org/10.1016/j.ijhydene.2022.05.055.

Garrido, R. A., Manrique, R., Fredes, J., Rodriguez, P., Rodríguez, A., Serafini, D., Mena, M., Masip, Y., & Díaz, I. (2024). Evaluating hydrogen production from grape pomace gasification: Unveiling the potential for Chile's wine industry and its solid waste recovery as energy source. *Renewable Energy, 223*. Available from https://doi.org/10.1016/j.renene.2024.119953.

German National Hydrogen Council, T. Statement on Germany's Updated National Hydrogen Strategy of 24th. (2023).

Global Hydrogen Review. (2022). International Energy Agency.

Graetz, J. (2009). New approaches to hydrogen storage. *Chemical Society Reviews, 38*(1), 73–82. Available from https://doi.org/10.1039/b718842k.

Guduru, R. K., Singh, R., & Vij, R. K. (2024a). Unlocking the Opportunities: Green Hydrogen from Renewable Energy Sources. In P. Singh, A. K. Agarwal, A. Thakur, & R. K. Sinha (Eds.), *Challenges and Opportunities in Green Hydrogen Production. Energy, Environment, and Sustainability*. Singapore: Springer. Available from https://doi.org/10.1007/978-981-97-1339-4_7.

Guduru, R. K., Singh, R., Vij, R. K., & Tiwari, A. K. (2024b). Hydrogen Production From Biomass. In R. K. Arya, G. D. Verros, O. P. Verma, & C. M. Hussain (Eds.), *From Waste to Wealth*. Singapore: Springer. Available from https://doi.org/10.1007/978-981-99-7552-5_15.

Guduru, R. K., Singh, R., & Vij, R. K. (2024c). *Desalination of seawater and related technologies: a step towards achieving sustainable development goals, . Water, The Environment, and the Sustainable Development Goals,* (1st, pp. 305–325). (1, pp. 305–325). Elsevier. Available from https://doi.org/10.1016/B978-0-443-15354-9.00011-6.

Gupta, S., Kumar, R., & Kumar, A. (2024). Green hydrogen in India: Prioritization of its potential and viable renewable source. *International Journal of Hydrogen Energy, 50*, 226–238. Available from https://doi.org/10.1016/j.ijhydene.2023.08.166.

Haghi, E., Raahemifar, K., & Fowler, M. (2018). Investigating the effect of renewable energy incentives and hydrogen storage on advantages of stakeholders in a microgrid. *Energy Policy, 113*, 206–222. Available from https://doi.org/10.1016/j.enpol.2017.10.045, http://www.journals.elsevier.com/energy-policy/.

Handayani, K., Overland, I., Suryadi, B., & Vakulchuk, R. (2023). Integrating 100% renewable energy into electricity systems: A net-zero analysis for Cambodia, Laos, and Myanmar. *Energy Reports, 10*, 4849–4869. Available from https://doi.org/10.1016/j.egyr.2023.11.005.

Hilali, İ., Işıker, Y., & Ulker, N. (2024). The hydrogen perspective for Türkiye, which is on the Asia-Europe energy transition route. Can Türkiye become hydrogen hub? *International Journal of Hydrogen Energy*. Available from https://doi.org/10.1016/j.ijhydene.2023.12.247.

Hossain, S., Saha, C. K., Ismail, M., Reza, T., Kabir, K. B., & Kirtania, K. (2023). Prospects and challenges of renewable hydrogen generation in Bangladesh. *International Journal of Hydrogen Energy, 48*(54), 20588–20612. Available from https://doi.org/10.1016/j.ijhydene.2023.03.059, http://www.journals.elsevier.com/international-journal-of-hydrogen-energy/.

Iida, S., & Sakata, K. (2019). Hydrogen technologies and developments in Japan. *Clean Energy, 3*(2), 105–113. Available from https://doi.org/10.1093/ce/zkz003, https://academic.oup.com/ce/pages/About.

Iordache, I., Schitea, D., Gheorghe, A. V., & Iordache, M. (2014). Hydrogen underground storage in Romania, potential directions of development, stakeholders and general aspects. *International Journal of Hydrogen Energy, 39*(21), 11071–11081. Available from https://doi.org/10.1016/j.ijhydene.2014.05.067, http://www.journals.elsevier.com/international-journal-of-hydrogen-energy/.

Jafari Raad, S. M., Leonenko, Y., & Hassanzadeh, H. (2022). Hydrogen storage in saline aquifers: Opportunities and challenges. *Renewable and Sustainable Energy Reviews, 168*. Available from https://doi.org/10.1016/j.rser.2022.112846, https://www.journals.elsevier.com/renewable-and-sustainable-energy-reviews.

Jensterle, M., Narita, J., Piria, R., Samadi, S., Prantner, M., Crone, K., et al. STUDY the role of clean hydrogen in the future energy systems of Japan and Germany.

Kovač, A., Paranos, M., & Marciuš, D. (2021). Hydrogen in energy transition: A review. *International Journal of Hydrogen Energy, 46*(16), 10016–10035. Available from https://doi.org/10.1016/j.ijhydene.2020.11.256, http://www.journals.elsevier.com/international-journal-of-hydrogen-energy/.

Kumar, K. R., Honorio, H., Chandra, D., Lesueur, M., & Hajibeygi, H. (2023). Comprehensive review of geomechanics of underground hydrogen storage in depleted reservoirs and salt caverns. *Journal of Energy Storage, 73*. Available from https://doi.org/10.1016/j.est.2023.108912.

Kumar, S., Regassa Jufar, S., Kumar, S., Foroozesh, J., Kumari, S., & Bera, A. (2023). Underground hydrogen storage and its roadmap and feasibility in India toward Net-Zero target for global decarbonization. *Fuel, 350*. Available from https://doi.org/10.1016/j.fuel.2023.128849, http://www.journals.elsevier.com/fuel/.

Kuzemko, C., Blondeel, M., Dupont, C., & Brisbois, M. C. (2022). Russia's war on Ukraine, European energy policy responses & implications for sustainable transformations. *Energy Research and Social Science, 93*. Available from https://doi.org/10.1016/j.erss.2022.102842, http://www.journals.elsevier.com/energy-research-and-social-science/.

Lackey, G., Freeman, G. M., Buscheck, T. A., Haeri, F., White, J. A., Huerta, N., & Goodman, A. (2023). Characterizing hydrogen storage potential in U.S. underground gas storage facilities. *Geophysical Research Letters, 50*(3). Available from https://doi.org/10.1029/2022GL101420, http://agupubs.onlinelibrary.wiley.com/hub/journal/10.1002/(ISSN)1944-8007/.

Laidlaw, D. (2018). Bill C-69, the impact assessment act, and indigenous process considerations.

Lankof, L., & Tarkowski, R. (2023). GIS-based analysis of rock salt deposits' suitability for underground hydrogen storage. *International Journal of Hydrogen Energy, 48*(71), 27748–27765. Available from https://doi.org/10.1016/j.ijhydene.2023.03.415, http://www.journals.elsevier.com/international-journal-of-hydrogen-energy/.

Lebrouhi, B. E., Djoupo, J. J., Lamrani, B., Benabdelaziz, K., & Kousksou, T. (2022). Global hydrogen development – A technological and geopolitical overview. *International Journal of Hydrogen Energy, 47*(11), 7016–7048. Available from https://doi.org/10.1016/j.ijhydene.2021.12.076, http://www.journals.elsevier.com/international-journal-of-hydrogen-energy/.

Lee, D., & Kim, K. (2021). Research and development investment and collaboration framework for the hydrogen economy in South Korea. *Sustainability (Switzerland), 13*(19). Available from https://doi.org/10.3390/su131910686, https://www.mdpi.com/2071-1050/13/19/10686/pdf.

Lee, J. H., & Woo, J. (2020). Green new deal policy of south korea: Policy innovation for a sustainability transition. *Sustainability (Switzerland), 12*(23), 1–17. Available from https://doi.org/10.3390/su122310191, https://www.mdpi.com/2071-1050/12/23/10191/pdf.

Lombardi, P. (2023). Spain raises bet on green hydrogen, biogas in draft plan.

Lycourghiotis, S. (2022). Trends in renewable energy: An overview. *Global Nest Journal, 24*(3), 505–525. Available from https://doi.org/10.30955/gnj.004286, http://journal.gnest.org/taxonomy/term/all.

Marfaing, O., Réalités Industrielles-Novembre 2022 © Annales Des Mines A Major Asset in the French Strategy: The Important Project of Common European Interest (PIIEC/IPCEI) on Hydrogen.

Medisetty, V. M., Kumar, R., Ahmadi, M. H., Vo, D. V. N., Ochoa, A. A. V., & Solanki, R. (2020). Overview on the current status of hydrogen energy research and development in India. *Chemical Engineering and Technology, 43*(4), 613–624. Available from https://doi.org/10.1002/ceat.201900496, http://onlinelibrary.wiley.com/journal/10.1002/(ISSN)1521-4125.

Merkeley, T., Abou-Khalil, R., Mira Hettesova., Christensen, V., Jull, M., & Tiauriman, Y. (2010). *Roadmap 2050: A practical Guide to a Prosperous, Low-carbon Europe*. OMA.

Miłek, D., Nowak, P., & Latosińska, J. (2022). The development of renewable energy sources in the European Union in the light of the european green deal. *Energies, 15*(15). Available from https://doi.org/10.3390/en15155576.

Mori, M., Žvar Baškovič, U., Stropnik, R., Lotrič, A., Katrašnik, T., Šipec, R., Lipar, J., Lesar, Ž., & Drobnič, B. (2023). Green energy hubs for the military that can also support the civilian mobility sector with green hydrogen. *International Journal of Hydrogen Energy* ((). Available from https://doi.org/10.1016/j.ijhydene.2023.05.026, http://www.journals.elsevier.com/international-journal-of-hydrogen-energy/.

Muhammed, N. S., Haq, B., Al Shehri, D., Al-Ahmed, A., Rahman, M. M., & Zaman, E. (2022). A review on underground hydrogen storage: Insight into geological sites, influencing factors and future outlook. *Energy Reports, 8*, 461–499. Available from https://doi.org/10.1016/j.egyr.2021.12.002, http://www.journals.elsevier.com/energy-reports/.

Navaid, H. B., Emadi, H., & Watson, M. (2023). A comprehensive literature review on the challenges associated with underground hydrogen storage. *International Journal of Hydrogen Energy, 48*(28), 10603–10635. Available from https://doi.org/10.1016/j.ijhydene.2022.11.225, http://www.journals.elsevier.com/international-journal-of-hydrogen-energy/.

Nouni, M. R., Jha, P., Sarkhel, R., Banerjee, C., Tripathi, A. K., & Manna, J. (2021). Alternative fuels for decarbonisation of road transport sector in India: Options, present status, opportunities, and challenges. *Fuel, 305*. Available from https://doi.org/10.1016/j.fuel.2021.121583, http://www.journals.elsevier.com/fuel/.

Obregón, A. R., De Kerpel Ramírez, A., Giarola, S., & Kerdan, I. G. (2023). The role of hydrogen in a decarbonised future transport sector: A case study of Mexico. *Energy Reports, 9*, 26–31. Available from https://doi.org/10.1016/j.egyr.2023.09.143.

Olemedia. Global hydrogen review 2022. ESCAPED_CHARACTER_LESS_THAN, http://www.iea.org/t&c/.

Oshiro, K., Kainuma, M., & Masui, T. (2017). Implications of Japan's 2030 target for long-term low emission pathways. *Energy Policy, 110*, 581–587. Available from https://doi.org/10.1016/j.enpol.2017.09.003.

OSTI.GOV. (1980). *Design and development of Stirling Engines for stationary power generation applications in the 500 to 3000 hp range. Subtask 1A report: state-of-the-art conceptual design*. United States. Available from https://doi.org/10.2172/5426798.

Ozawa, A., Kudoh, Y., Murata, A., Honda, T., Saita, I., & Takagi, H. (2018). Hydrogen in low-carbon energy systems in Japan by 2050: The uncertainties of technology development and implementation. *International Journal of Hydrogen Energy, 43*(39), 18083–18094. Available from https://doi.org/10.1016/j.ijhydene.2018.08.098.

Pan, B., Yin, X., Ju, Y., & Iglauer, S. (2021). Underground hydrogen storage: Influencing parameters and future outlook. *Advances in Colloid and Interface Science, 294*. Available from https://doi.org/10.1016/j.cis.2021.102473, https://www.journals.elsevier.com/advances-in-colloid-and-interface-science.

Panchenko, V. A., Daus, Y. V., Kovalev, A. A., Yudaev, I. V., & Litti, Y. V. (2023). Prospects for the production of green hydrogen: Review of countries with high potential. *International Journal of Hydrogen Energy, 48*(12), 4551–4571. Available from https://doi.org/10.1016/j.ijhydene.2022.10.084, http://www.journals.elsevier.com/international-journal-of-hydrogen-energy/.

Pflugmann, F., & De Blasio, N. (2020). The geopolitics of renewable hydrogen in low-carbon energy markets. *Geopolitics, History, and International Relations, 12*(1), 9–44. Available from https://doi.org/10.22381/GHIR12120201, http://www.addletonacademicpublishers.com/geopolitics-history-and-international-relations.

Qin, B., Li, H., Wang, Z., Jiang, Y., Lu, D., Du, X., & Qian, Q. (2023). New framework of low-carbon city development of China: Underground space based integrated energy systems. *Underground Space*. Available from https://doi.org/10.1016/j.undsp.2023.06.008.

Shin, J. E. (2022). Hydrogen technology development and policy status by value Chain in South Korea. *Energies, 15*(23). Available from https://doi.org/10.3390/en15238983, http://www.mdpi.com/journal/energies/.

Simon, J., Ferriz, A. M., & Correas, L. C. (2015). HyUnder – Hydrogen underground storage at large scale: Case study Spain. *Energy Procedia, 73*, 136–144. Available from https://doi.org/10.1016/j.egypro.2015.07.661, http://www.sciencedirect.com/science/journal/18766102.

Solomon, B. D., & Banerjee, A. (2006). A global survey of hydrogen energy research, development and policy. *Energy Policy, 34*(7), 781–792. Available from https://doi.org/10.1016/j.enpol.2004.08.007, http://www.journals.elsevier.com/energy-policy/.

Stangarone, T. (2021). South Korean efforts to transition to a hydrogen economy. *Clean Technologies and Environmental Policy, 23*(2), 509–516. Available from https://doi.org/10.1007/s10098-020-01936-6, https://link.springer.com/journal/10098.

Steinberg, D.C., Brown, M., Wiser, R., Donohoo-Vallett, P., Gagnon, P., Hamilton, A., Mowers, M., Murphy, C., Prasana, A. 2023. Evaluating Impacts of the Inflation Reduction Act and Bipartisan Infrastructure Law on the

U.S. Power System. Golden, CO: National Renewable Energy Laboratory. NREL/TP-6A20-85242. https://www.nrel.gov/docs/fy23osti/85242.pdf.

Tarkowski, R. (2019). Underground hydrogen storage: Characteristics and prospects. *Renewable and Sustainable Energy Reviews, 105*, 86–94. Available from https://doi.org/10.1016/j.rser.2019.01.051, https://www.journals.elsevier.com/renewable-and-sustainable-energy-reviews.

Tarkowski, R., & Uliasz-Misiak, B. (2022). Towards underground hydrogen storage: A review of barriers. *Renewable and Sustainable Energy Reviews, 162*. Available from https://doi.org/10.1016/j.rser.2022.112451, https://www.journals.elsevier.com/renewable-and-sustainable-energy-reviews.

Thiyagarajan, S. R., Emadi, H., Hussain, A., Patange, P., & Watson, M. (2022). A comprehensive review of the mechanisms and efficiency of underground hydrogen storage. *Journal of Energy Storage, 51*. Available from https://doi.org/10.1016/j.est.2022.104490, http://www.journals.elsevier.com/journal-of-energy-storage/.

Walsh, S. D. C., Easton, L., Weng, Z., Wang, C., Moloney, J., & Feitz, A. (2021). Evaluating the economic fairways for hydrogen production in Australia. *International Journal of Hydrogen Energy, 46*(73), 35985–35996. Available from https://doi.org/10.1016/j.ijhydene.2021.08.142, http://www.journals.elsevier.com/international-journal-of-hydrogen-energy/.

Yazdi, M., Zarei, E., Pirbalouti, R. G., & Li, H. (2023). A comprehensive resilience assessment framework for hydrogen energy infrastructure development. *International Journal of Hydrogen Energy*. Available from https://doi.org/10.1016/j.ijhydene.2023.06.271, http://www.journals.elsevier.com/international-journal-of-hydrogen-energy/.

Yilmaz, G., Sadvakasova, A. K., Kossalbayev, B. D., Bauenova, M. O., Zharmukhamedov, S. K., Ziyayeva, G. K., Zaletova, D. E., Alharby, H. F., & Allakhverdiev, S. I. (2024). Hydrogen energy development in Turkey: Challenges and opportunities. *International Journal of Hydrogen Energy, 52*, 1304–1311. Available from https://doi.org/10.1016/j.ijhydene.2023.11.230, https://www.sciencedirect.com/science/journal/03603199.

Zhao, Q., Wang, Y., & Chen, C. (2023). Numerical simulation of the impact of different cushion gases on underground hydrogen storage in aquifers based on an experimentally-benchmarked equation-of-state. *International Journal of Hydrogen Energy*. Available from https://doi.org/10.1016/j.ijhydene.2023.07.262, http://www.journals.elsevier.com/international-journal-of-hydrogen-energy/.

Zivar, D., Kumar, S., & Foroozesh, J. (2021). Underground hydrogen storage: A comprehensive review. *International Journal of Hydrogen Energy, 46*(45), 23436–23462. Available from https://doi.org/10.1016/j.ijhydene.2020.08.138, http://www.journals.elsevier.com/international-journal-of-hydrogen-energy/.

CHAPTER

13

Underground hydrogen storage case studies: current status, ongoing activities, and global potential

Anooja Sara Mathew, Soubhik Mondal, Sandeep D. Kulkarni and Vishnu Chandrasekharan Nair

Deysarkar Centre of Excellence in Petroleum Engineering, IIT Kharagpur, Kharagpur, India

13.1 Introduction

Energy is one of the basic needs of everyone. From the primitive period itself energy is obtained from different sources and has been transformed from one form to another. With the continuous growth of population, economies expand, living standards rise and energy demand also increases. Among the different sources of energy, fossils fuels have traditionally held a dominant role, providing around 80% or more of the world's energy (Hematpur et al., 2023; Iordache et al., 2014). According to the CO_2 emission report 2023 by the international energy agency (IEA), worldwide energy-related carbon-dioxide (CO_2)emissions have increased by 1.1%, reaching a new high of 37.4 billion tons (Gt) in 2023, with a rise of 410 million tons (Mt) (CO_2 Emissions in 2023, 2023). Also, the increase of 1.5°C in the average atmospheric temperature above the preindustrial levels has triggered a need for the shift from fossil fuels to low-carbon energy alternatives. This transition is mainly driven by the acknowledgment that approximately 74% of the total global greenhouse gas emissions result from CO_2 emissions generated by the combustion of fossil fuel (Mouli-Castillo et al., 2021). To effectively manage the growing energy demand and facilitate the transition toward a net-zero carbon economy, it is imperative to adopt a comprehensive strategy that encompasses the supply and demand aspects of the energy equation. Renewable energy sources, including solar, wind, and hydropower, play a pivotal role in decarbonization endeavors from the supply perspective. According to the world energy outlook, renewable sources have the capacity to fulfill 20%–30% of the world's energy demands by 2040 and achieve full coverage, meeting 100% of global energy

Subsurface Hydrogen Energy Storage
DOI: https://doi.org/10.1016/B978-0-443-24071-3.00014-5

requirements, by 2050 (Gernaat et al., 2021; Tackie-Otoo & Haq, 2024). In this scenario, hydrogen has evolved as a low carbon energy carrier alternative as well as storage means of renewables. It is recognized as one of the four pillars of global energy transformation as chartered by IEA. Hydrogen production can be achieved from renewable energy sources through electrolysis of water without the emission of CO_2. This will result in the sustainable, long-term production of decarbonized hydrogen (Mouli-Castillo et al., 2021). Fig. 13.1 gives the idea of hydrogen economy which can make our energy sector green with hydrogen geo-storage. The 26th Conference of the Parties (COP26) summit highlighted the pivotal role of hydrogen in the global vision for 2050, with nearly 30 countries unveiling their hydrogen roadmaps (Global Hydrogen Review, 2022). The Indian Government, aiming for energy self-sufficiency by 2047 and net zero emissions by 2070, has prioritized the hydrogen economy through the national hydrogen mission launched in the 2021 budget (Energy Transition Advisory Committee ETAC, 2023). The United States has implemented the US National Clean Hydrogen Strategy and Roadmap, modifying its approach to align with recent incentives and financing opportunities for carbon capture, utilization, and storage, renewable energy, and sustainable aviation fuel production. In addition to this, countries like Germany, Belgium, Korea and Japan have updated their existing hydrogen policies. Along with this, several countries, especially in Europe, are currently in the process of reviewing and modifying their strategy (Global Hydrogen Review, 2023).

Hydrogen, which remains as the most widely seen element in the universe is colorless and odorless and it boasts a high energy content per weight standing at 120 MJ/kg (Amirthan & Perera, 2022; Tackie-Otoo & Haq, 2024), which is around three times the

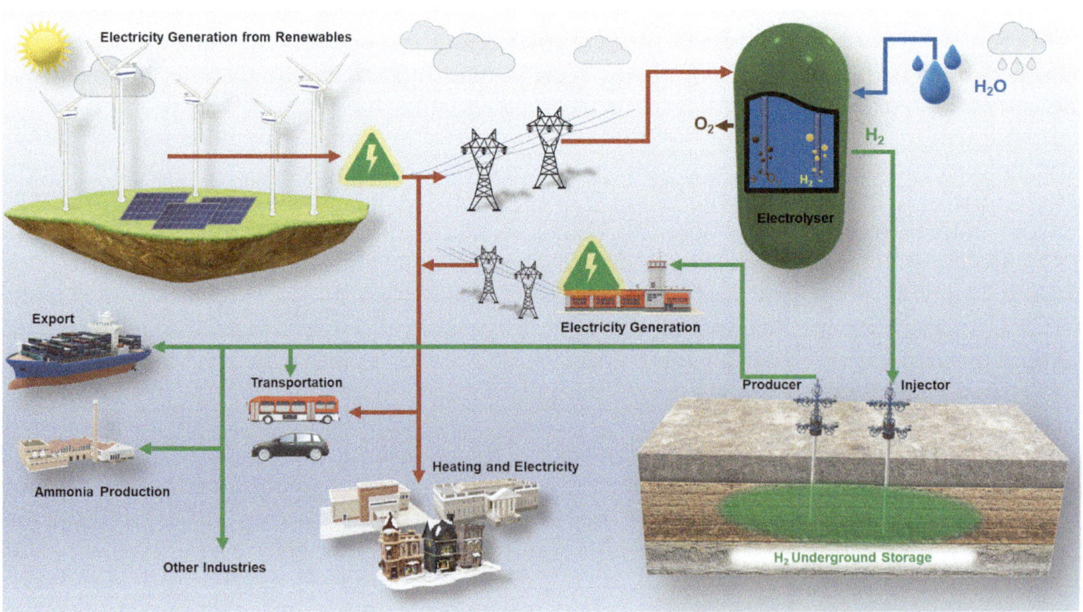

FIGURE 13.1 Schematic representation of hydrogen economy with hydrogen geo-storage.

energy content possessed by gasoline. But on a volume basis, the reverse happens as the energy content per volume of hydrogen is around 8 MJ/L whereas that of gasoline is 32 MJ/L. Given its substantial energy potential, hydrogen has the capability to substitute approximately 60% of natural gas usage in nonindustrial applications (Navaid et al., 2023). Hydrogen is known in different color codes such as blue, green, gray, yellow, white, turquoise, black and brown depending on the different processes through which it is produced (Hematpur et al., 2023). Given hydrogen's lower density and viscosity compared to methane (CH_4) and carbon dioxide (CO_2), storage challenges arise, necessitating large volumes and higher pressure (Muhammed et al., 2023). Hydrogen has various storage options, including compressed gas, liquid form, adsorption onto materials with extensive surface area, absorption into interstitial sites of host metals, chemical bonding, and reactive metal oxidation, such as lithium, magnesium, sodium, aluminum, zinc, and others with water (Hematpur et al., 2023).

All the available surface storage methods of hydrogen storage possess the challenge of high cost and limited volume for storage. The volume requirement of 1 kg (500 moles) of hydrogen in gaseous state under different pressure and 25°C temperature along with its volume in liquid and solid states at atmospheric pressure is shown in Fig. 13.2. The diagram provides a clearer depiction of the issues that are associated with surface hydrogen storage systems. Hydrogen has the capability to be stored in substantial volumes (10^8 Nm^3)

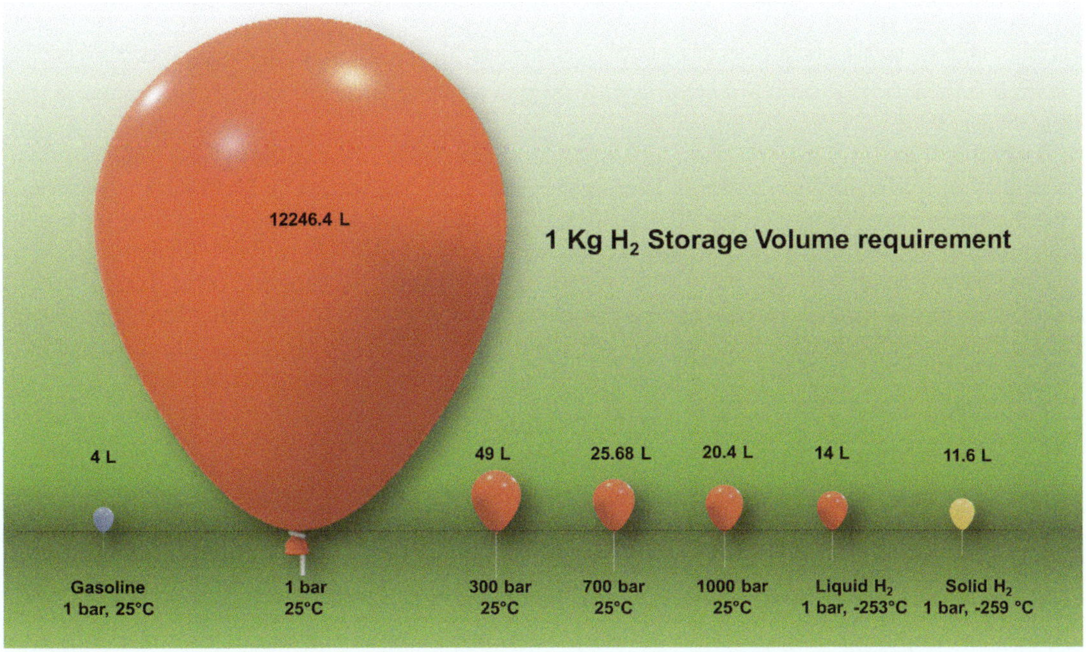

FIGURE 13.2 Hydrogen storage volume requirement (Hsiung, 2001; Hydrogen as an Alternative Fuel, 2024; H2 Hydrogen Tools, 2016). Source: *Data from Hsiung, S. (2001). Energy Technology Training Center, College of the Desert Hydrogen Fuel Cell Engines and Related Technologies; Hydrogen as an Alternative Fuel. (2024). 1991 4 18. https://afdc.energy.gov/fuels/hydrogen-basics; Hydrogen Tools. (2016).*

for extended periods, ranging from months to years, within underground formations (Zivar et al., 2021), thereby generating thousands of megawatts of power (Higgs et al., 2022). The underground storage of hydrogen also referred to as subsurface hydrogen storage and geo-storage of hydrogen acquire the principles of underground gas storage (UGS) which has been in practice since ages. The major identified subsurface hydrogen storage sites are salt caverns, rock caverns, saline aquifers and depleted hydrocarbon reservoirs (Amirthan & Perera, 2022). When hydrogen is injected as compressed gas into these storage sites, it is trapped there by the gas-trapping mechanisms such as structural or stratigraphic trapping, residual or capillary trapping, solubility or dissolution trapping and mineral trapping (Muhammed et al., 2023) From experimental investigations, it has been found that structural trapping and residual trapping are the primary trapping mechanism for hydrogen retention underground (Muhammed et al., 2023). Among the different subsurface storage options, choosing depleted hydrocarbon reservoirs is the most favorable option for large-scale underground hydrogen storage (UHS) due to their recognized geological structure, significant compactness, intact source rock, and the presence of both surface and underground facilities. Oil and gas reservoirs, extensively researched as subsurface geological porous media, are well-suited for the advancement of UHS. Specifically, depleted gas reservoirs offer advantages stemming from substantial experience, simplifying their exploration, management, and operation. With well-defined geological parameters and stable caprock conditions, these reservoirs prove highly suitable for UHS (Muhammed et al., 2023; Navaid et al., 2023). Currently, governments of 41 countries have established a hydrogen strategy, with early adopters revising their initial plans to elevate their ambitions (Global Hydrogen Review, 2023). Most of the developed and developing nations have proactively initiated research on the transition from storing natural gas to storing hydrogen in underground geological formations (Navaid et al., 2023).

Several research initiatives and pilot projects worldwide are exploring the feasibility and scalability of hydrogen geo storage. HyStock Phase II in Netherlands, Hychico in Argentina, Underground Sun Storage (USS) 2030 in Austria, HyCAVmobil (hydrogen cavern for mobility) in Germany, HySecure in United Kingdom and H_2CAST (H_2 cavern storage transition) in Germany are a few among those projects (Hydrogen Production and Infrastructure Projects Database, 2023). In this chapter, we examine the principles, challenges, and recent advancements in hydrogen geo-storage, drawing insights from relevant studies and ongoing projects. Along with this, some case studies which practiced UHS such as salt cavern storage in Teesside, United Kingdom, depleted gas reservoir storage in Molasse basin, Austria (USS) and the upcoming USS 2030 in Austria are discussed in detail. By evaluating the technical, economic, and environmental aspects of this emerging storage method, we aim to contribute to the understanding and deployment of hydrogen geo-storage as a key enabler of sustainable energy systems.

13.2 Geological characterization of site

Subsurface storage of natural gas has already been in practice since ages. Hence, the same method is also applicable for hydrogen which can be used as a season independent energy carrier. Geological site selection is one of the critical aspects of hydrogen geo-storage, which involves storing hydrogen in subsurface porous geological formations like salt caverns, aquifers and depleted oil and gas reservoirs. For the success and safety of UHS projects, the choice of a suitable geological site is very essential. The selection criteria for potential UHS sites can be drawn from the previous experiences of UGS, carbon capture and storage, and nuclear waste disposal (Amirthan & Perera, 2023; Navaid et al., 2023).

The utilization of depleted hydrocarbon reservoirs is highly suitable for UGS owing to their clearly defined geological structures, robust caprock tightness and integrity, and the presence of the required surface and subsurface infrastructure (Zivar et al., 2021). Depleted reservoirs hold potential to be the most economical large-scale storage options for hydrogen. Their investment and operational costs are lower, and prevailing infrastructure can be repurposed for hydrogen storage with proper modifications. The cushion gas requirement in this case is also lower due to the presence of remaining hydrocarbons. Also, the Austrian USS pilot project have effectively demonstrated blending 10% hydrogen in underground natural gas storage sites (Amirthan & Perera, 2022).

While some pilot projects such as HyStorPor (hydrogen storage in porous media), sun storage, and hychico predominantly focused on porous mediums, particularly depleted reservoirs for hydrogen storage, previous research has indicated practical experiences with caverns and aquifers (Muhammed et al., 2023). In regions lacking depleted reservoirs, aquifers become viable options for UHS. Aquifers, typically consisting of porous rocks like sandstone situated deep underground, store water. To match the storage efficiency of depleted oil and gas reservoirs, aquifers must possess similar geological properties including high capacity, permeability and porosity (Navaid et al., 2023). Aquifers in projects such as Beynes in France, Ketzin in Germany and Lobodice in Czech Republic have showcased the underground storage of hydrogen in conjunction with other gases like carbon dioxide, methane, carbon monoxide and nitrogen (Muhammed et al., 2023). Economically, aquifers lack the cost advantages of depleted hydrocarbon reservoirs and salt caverns. Despite facing similar challenges in UHS, aquifers require substantially higher infrastructural investments (Navaid et al., 2023).

Salt caverns have already proven to be one of the suitable candidates for the storage of gases underground. Notably, a salt cavern in Teesside, United Kingdom, reported a mixture of 95% hydrogen and 3%–4% of carbon dioxide. Similarly, locations like the Gulf Coast of Texas and the town gas project of Kiel in Germany have reported high hydrogen concentrations (Muhammed et al., 2023). According to the research conducted by Matos et al. (2019), it is suggested that cavern storage should be conducted within a depth range of 200–2000 m. These caverns should be strategically located in areas where the hydraulic pressure surpasses the intended storage pressure. Additional suggested measurements consist of a minimum height of 300 m and a diameter of 70 m. The construction cost becomes high when the cavern is situated at lower depths and the compressor costs are low in this case. Conversely, at greater depths, volumetric requirements for caverns are lower, and compressor costs become dominant (Amirthan & Perera, 2022; Matos et al., 2019).

In addition to these underground storage options, rock caverns and coal seams are also under use and consideration. The suitability of coal seams for hydrogen storage was suggested on the basis of the swelling behavior exhibited by them in the presence of hydrogen. However, research indicates that the capacity of coal seams for storing hydrogen is ten times less than that observed for natural gkas. Abandoned mines with large underground openings can be used for hydrogen storage, but the problem with them is that they are not intended for the cyclic storage process. While they can exclude costs related to mining, significant expenses are incurred to make the storage impermeable. Salt caverns, when available, are the favored UHS option due to low associated leaks, enabling pure hydrogen storage, and speedy storage cycles, despite higher initial capital investment (Amirthan & Perera, 2022).

The economic feasibility of UHS projects depends on various capital costs encompassing site selection, gas separation, danger evaluation, bacterial activity inquiry, injection and withdrawal durations, infrastructural requirements, and labor expenditures. Depleted reservoirs emerge as economically attractive UHS candidates due to minimal site characterization costs. For them, cost savings of 4.7% and 23.6% were estimated, on average, compared to aquifer and salt cavern storage, respectively (Lord et al., 2014; Navaid et al., 2023). Among the total cost required for converting depleted reservoirs into hydrogen storage sites, the cost for cushion gas is the highest. This can be minimized by the selection of a site with the presence of some remaining gas. The levelized cost per kilogram of hydrogen for different storage options was found to be 1.61 USD for salt caverns, 1.23 USD for depleted hydrocarbon reservoirs, and 2.77 USD for saline aquifers (Lord et al., 2014; Muhammed et al., 2023).

13.3 Physiochemical interactions and assessment

The physiochemical interactions involved in hydrogen storage are critical considerations in developing efficient and safe storage solutions for this promising energy carrier. Hydrogen, with its unique properties, presents both challenges and opportunities in geo-storage applications. Understanding how hydrogen interacts with geological formations, as well as its physiochemical behavior under varying storage conditions, is essential for designing effective storage systems. Physiochemical interactions influence storage capacity, efficiency, and safety, shaping the design and operation of hydrogen geo-storage facilities. This section briefly describes the physiochemical interactions associated with UHS in terms of hydrogen's density, viscosity, solubility, diffusivity, and adsorption.

13.3.1 Density and viscosity

As mentioned earlier, hydrogen is characterized with low volumetric density and viscosity. With formation depth, pressure and temperature increases drastically due to the geothermal and hydrostatic gradients. Density of hydrogen is found to increase with pressure and temperature whereas viscosity is only slightly affected by pressure and temperature (Pan et al., 2021). Also, in comparison to carbon dioxide and methane, the density of hydrogen is significantly lower than the density of water under same physico-thermal circumstances (Navaid et al., 2023). The low viscosity of hydrogen over methane and carbon dioxide makes its retention challenging during UHS (Muhammed et al., 2023). It is also found that, the viscosity of hydrogen is approximately one or two magnitudes smaller than that of water at the same thermo physical conditions (Pan et al., 2021). The effect of its low viscosity and density can result in more viscous fingering which will result in early

H_2 surface leakage, poor conformance, low hydrogen sweep efficiency and other problems (Carden & Paterson, 1979; Muhammed et al., 2023; Pan et al., 2021). Due to its low sweep efficiency, hydrogen is able to sweep only a small amount of water which reduces the available storage space for hydrogen (Kumar et al., 2023).

13.3.2 Hydrogen solubility

Hydrogen solubility during underground storage has been considered to be of paramount importance. Solubility is the extent to which a material can dissolve in a solvent, resulting in the formation of a solution, primarily in the case of water (Jahanbakhsh et al., 2024). Hydrogen has low solubility in its liquid and gaseous states (near to critical point) in pure water (Jahanbakhsh et al., 2024; Muhammed et al., 2022). Despite being subjected to high pressure, the solubility of hydrogen is significantly lower in comparison to methane, which is 15 times more than carbon dioxide (although it may vary depending on the pressure of the reservoir). At the temperature of 298K and pressure of 1 bar, hydrogen solubility in water is 0.00001411 mol fraction (Navaid et al., 2023). At high pressure and temperature also, hydrogen solubility in pure water is around 0.00103 by mole fraction at 323K and 79 bar (Navaid et al., 2023; Pan et al., 2021). Since, the pore spaces in depleted hydrocarbon reservoirs and aquifers contain water, it is essential to study the solubility characteristic of hydrogen under reservoir conditions in the view of UHS (Hematpur et al., 2023).

13.3.3 Hydrogen diffusivity

Being the lightest and smallest of the molecules ever known to exist, hydrogen has high diffusivity or penetrability in the reservoir during its storage (Jahanbakhsh et al., 2024). The study conducted by Hemme and van Berk (2018) revealed the effective diffusion coefficients of H_2 and CH_4 in pure water to be 5.13×10^{-9} m^2/s and 1.85×10^{-9} m^2/s respectively, at temperature of 25°C. Also, it was found that hydrogen diffusivity increased with an increase in temperature and a decrease in pressure (Hemme & van Berk, 2018; Pan et al., 2021; Perera, 2023). The influence of temperature can be directly associated with alterations in molecular momentum, Brownian motion, and intermolecular forces that are generated by temperature (Bird et al., n.d.; Perera, 2023). The increase of pressure causes a reduction in the molecular distance of water, hence restricting the movement of hydrogen through the pore fluid (Perera, 2023).

13.3.4 Hydrogen adsorption

Though studied very little, hydrogen's adsorption and desorption hold a crucial role in the context of geo-storage. From the experimental studies involving adsorption of hydrogen, it was found that hydrogen adsorption increased with pressure and decreased with temperature (Abdulelah et al., 2023; Didier et al., 2012; Iglauer, Abid, et al., 2021; Pan et al., 2021; Perera, 2023). The evaluation of hydrogen physisorption on earth minerals was undertaken in a recent work by Abdulelah et al. (2023). From this study it was observed that the quantity of hydrogen adsorbed was in the order of illite > kerogen > kaolinite > quartz > calcite (Abdulelah et al., 2023). Thus, it is known from this mineral study that in case of reservoirs rock system, sandstone reservoirs exhibit high hydrogen adsorption capacity than carbonate reservoirs. More experimental studies in hydrogen adsorption under reservoir conditions is expected in future.

13.4 Multiphase flow characteristics

The multiphase flow behavior of hydrogen in porous media holds significant importance mainly in the storage of hydrogen in aquifers and depleted hydrocarbon reservoirs. Here we need to consider and characterize the parameters such as wettability, interfacial tension (IFT), capillary pressure (P_c), relative permeability and hysteresis. This section contains a brief note on the influence of these parameters in hydrogen geo storage.

Wettability is the ability of a fluid to remain in contact with a solid surface when another immiscible fluid is present. It plays a crucial role in selecting sites for UGS and hydrogen storage (Navaid et al., 2023). In laboratory methods, wettability is quantified by the measurement of contact angle (Aghaei et al., 2023). Wettability of hydrogen/reservoir rock system depends on the type of rock, pore fluid, properties of hydrogen stored (Perera, 2023) and the reservoir conditions. Experimental studies conducted by Hosseini et al. (2022) and Aghaei et al. (2023) to examine the wettability of hydrogen over carbonate rocks have found that the system remained strongly water wet under standard conditions where as it shifted to intermediate wet under higher physio-thermal conditions (Aghaei et al., 2023; Hosseini et al., 2022). The study results of Esfandyari et al. (2022)on the hydrogen rock mineral wettability have also shown that quartz, anhydrite, calcite, gypsum, dolomite, basalt, granite, and shale remained in a strongly water-wet condition even with pressure varying from 10 to 100 bar and temperature varying from 20°C to 80°C (Esfandyari et al., 2022). This indicates the higher storage capacities of reservoir with these minerals for hydrogen storage. However, in a recent study carried out by Iglauer, Abid, et al. (2021), it was found that the inclusion of stearic acid on pure quartz at elevated pressure and temperature circumstances resulted in an increased hydrogen wettability (Iglauer, Ali, et al., 2021).

The IFT between fluids can be measured directly in the laboratory using capillary rise method and pendant drop method (Perera, 2023). The study performed by Esfandyari et al. (2022) to investigate IFT of hydrogen/distilled water system and hydrogen/formation brine system showed an increase in IFT with increase in temperature and salinity. However, a sharp decrease in IFT was seen with increase in pressure for both the cases of distilled water and formation brine (Esfandyari et al., 2022). According to Pan et al. (2021) the hydrogen water interfacial is almost independent of pressure but it decreases strongly with increase in temperature (Pan et al., 2021). The IFT between hydrogen and water found by Yekta, Manceau, et al. (2018) using an indirect method by fitting H_2−water−sand stone core flood data with associated mercury injection capillary pressure data was slightly less than that of directly measured values (Yekta, Manceau, et al., 2018). The measurement of IFT in the context of UHS is highly important as it determines the fluid spreading behavior and the fluid rock interaction (Perera, 2023). Also, according to Esfandyari et al. (2022), the effect of hydrogen on capillary pressure is assumed to be dependent on H_2−water IFT (Esfandyari et al., 2022).

The concept of capillary pressure (P_c) refers to the difference in pressure between the wetting and nonwetting phases, which gives rise to an interaction force existing between the fluids and the surfaces of the rock grains (Perera, 2023) and the concept of threshold capillary pressure, also known as entry capillary pressure, refers to the pressure at which hydrogen is able to permeate the largest accessible pore within a rock. For a particular pore radius, the two parameters that control the threshold pressure are IFT and the contact angle (Zivar et al., 2021). Before injecting hydrogen, the pore space is filled with formation water. Therefore, the injected hydrogen must displace the existing brine. The volume of displaced brine (and consequently the injected hydrogen) and the pressure needed for this process are determined by capillary pressure, P_c. Hence, capillary pressure serves as a crucial parameter in evaluating the viability of UHS projects (Pan et al., 2021).

The ease with which a fluid flows, is quantified in terms of relative permeability as a function of saturation. The amount of hydrogen that is trapped in the porous space by the capillary force is indicated by the end points of relative permeability curve. This trapped hydrogen cannot be recovered and leads to the permanent gas loss and low efficiency (Hematpur et al., 2023). According to the findings of Yekta, Manceau, et al. (2018), the relative permeability and capillary pressure curves demonstrate that hydrogen exhibits a comparatively low level of relative permeability. Additionally, the capillary pressure curve reveals that the minimum pressure required for hydrogen to penetrate the largest pore is relatively low (Yekta, Manceau, et al., 2018). The relative permeability curves of hydrogen exhibit a hysteresis effect as a result of its entrapment inside the reservoir, causing variations in its mobility between the drainage and imbibition processes (Zivar et al., 2021).

However, there exists a scarcity of experimental data to quantify the capillary pressure and relative permeability of the hydrogen—brine system. A straightforward method to evaluate multiphase flow for hydrogen storage is to utilize data from literatures related to natural gas (methane) reservoirs, assuming similarity in properties (Hassannayebi et al., 2019). This assumption is supported by the previous observation that in the presence of water, hydrogen remains as the nonwetting phase similar to natural gas (Hematpur et al., 2023).

13.5 Geochemical and mineralogical interactions and assessment

Geochemical aspects involve the study of the chemical interactions between hydrogen and the porous geological formations when hydrogen is injected into it. Hydrogen is a highly reactive element and the reactivity of hydrogen in the domain of underground storage depends upon the conditions of temperature, pressure, salinity, pH and the various minerals present in the porous media (Al-Yaseri, Amao, et al., 2023). At reservoir conditions, hydrogen gets dissolved in the formation water present and can react with the sensitive rock minerals. These reactions can cause mineral precipitation or dissolution which can modify the porosity and permeability conditions of reservoir rocks. It can also lead to the change in composition of stored hydrogen by the generation of CO_2, CH_4, and hydrogen sulfide (H_2S) (Van Gessel et al., 2023). The major challenges associated with the geochemical reactions in UHS are the loss of hydrogen due to its conversion into other unwanted chemical compounds and the alteration of reservoir rock properties (Amirthan & Perera, 2023). While considering the four main geo-storage options viz depleted reservoirs, saline aquifers, salt caverns, and rock caverns, the depleted reservoirs and aquifers have found to be more prone to reactions and salt caverns are considered to have low reactivity to chemical reactions (Al-Yaseri, Amao, et al., 2023).

Major minerals that are susceptible to chemical reactions underground are carbonate minerals like calcite, magnesite, dolomite, siderite and dawsonite, sulfate minerals such as gypsum, anhydrite, barite, and anglesite. In carbonate minerals, CO_3^{2-} gets reduced to CH_4 in the presence of hydrogen with water and OH^- as by-products. Sulfate ions SO_4^{2-} get reduced in the presence of hydrogen resulting in the generation of H_2S. The carbonate minerals if present in the rock matrix may help to mitigate H_2S formation. In the presence of hydrogen, reductive dissolution can occur in case of ferric ion associated minerals consisting of oxides where Fe^{3+} gets partially reduced to Fe^{2+}. In case of clays, alteration in structure can occur when hydrogen reduces the structural Fe^{3+} (Van Gessel et al., 2023). Experimental studies conducted by Al-Yaseri, Al-Mukainah, et al. (2023), Zeng et al. (2022) have found out that in the presence of hydrogen, carbonate minerals undergo precipitation and dissolution (Al-Yaseri, Al-Mukainah, et al., 2023; Al-Yaseri, Amao, et al., 2023; Zeng et al., 2022). The study conducted by Truche et al. (2013) have experimentally found out the conversion of pyrite to pyrrhotite on a clay rich sandstone under temperature varying from 90°C to 250°C and pressures varying from 0.3 to 3 MPa (Truche et al., 2013). Reitenbach et al. (2015) also reported that the abiotic pyrite reduction has resulted in the production of hydrogen sulfide in the town gas storage site at Beynes, France (Hematpur et al., 2023; Reitenbach et al., 2015). Hassannayebi et al. (2019) conducted a geochemical study to examine the interaction of minerals in a gas-brine system in the Molasse Basin, Austria, in order to evaluate the loss of H_2. The research utilized an equilibrium methodology to estimate the long-term impacts and deployed kinetic modeling techniques to examine interactions across a standard storage cycle. Results indicated that the reduction of pyrite into pyrrhotite have increased reservoir pH and hydrogen loss (Hassannayebi et al., 2019; Muhammed et al., 2023).

Among the different minerals present on rock, calcite and pyrite have found to be more reactive in the case of UHS. Hence the absence of these minerals is preferred for the long-term storage of hydrogen in geological media. These abiotic reactions will result in the reduction of overall storage integrity as well as loss of hydrogen during subsurface storage. Also, the effect of long-term storage should be considered separately. The study conducted by Yekta et al. (2018) has determined that hydrogen primarily exhibits inert gas behavior in depleted gas reservoirs mainly composed of sandstone. This finding supports the feasibility of storing hydrogen in these reservoirs (Yekta et al., 2018).

13.6 Microbial interactions

In addition to abiotic reactions leading to hydrogen loss in underground storage sites, there are biotic reactions triggered by microorganisms found in subsurface conditions. As shown in Fig. 13.3, the major reactions that occur during UHS in the presence of microbes are methanogenesis, acetogenesis, sulfate and iron reduction reactions. In methanogenesis,

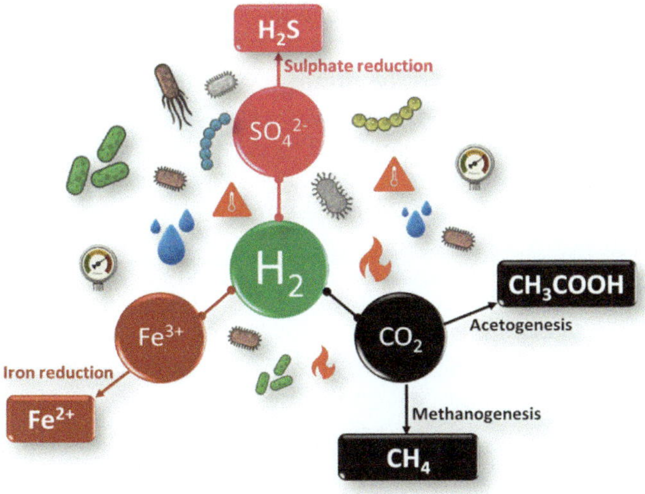

FIGURE 13.3 Major microbial (biotic) reactions occurring in hydrogen geo-storage.

the methanogens convert hydrogen into methane and leads to the loss of hydrogen. Acetogenesis results in the loss of hydrogen due to the production of acetate by acetogens. Sulfate reduction leads to the formation of hydrogen sulfide by the sulfate reducing bacteria where as in case of iron reduction, Fe^{3+} gets reduced to Fe^{2+} in the presence of hydrogen (Heinemann et al., 2021). All these reactions will pave way to hydrogen loss as well as issues such as blockage and corrosion (Dopffel et al., 2021). Microbial activities are influenced by certain variables such as mineral composition and the existence of specific bacteria. The activities take place within a temperature range of 15°C–121°C (Colman et al., 2017; Lassin et al., 2011).

Thaysen et al. (2021) investigated microbial proliferation and hydrogen utilization for storage of hydrogen in porous media. Their findings revealed that those reservoirs having salinity and temperature of low values exhibited increased rates of bacterial growth and subsequent consumption of hydrogen. They observed losses of 3% attributed to methanogens and 2%–4% attributed to sulfate reducers. Moreover, they suggested that these phenomena might escalate with recurrent storage cycles, potentially replenishing nutrients via mineral weathering, influx of water, and decomposition of microbial cells. It was also indicated from their research that under storage conditions characterized by high temperatures, such as in deep reservoirs, microbial activity could be suppressed. However, operational challenges and higher costs are likely to accompany with increased depths (Jahanbakhsh et al., 2024; Thaysen et al., 2021). The microbes present in these subsurface conditions are thought to be originated from natural transportation or sedimentation processes or might have introduced by human activities such as drilling, pumping or mining. Even though microbial activity is considered to be detrimental in the case of hydrogen storage, two pilot projects which are the underground sun conversion project of Austria and the Hychico project of Argentina aims to produce green methane underground by injecting hydrogen and carbon dioxide into depleted gas fields (Van Gessel et al., 2023).

13.7 Case studies

On giving due consideration to the UHS projects worldwide, it is seen that salt caverns have been utilized for storage of high purity hydrogen since 1970 from the Teesside facility in UK, later from 1980 in US at the US Gulf Coast and at the Chevron Phillips Clemens Terminal in Texas. Similarly, Moss Bluff, Praxair in US, Yakshunovskoe in Russia, Spindletop in US, Kiel, Germany have also embraced salt cavern hydrogen storage. Nevertheless, there are currently no known examples of high purity hydrogen storage initiatives in aquifers and depleted reservoirs throughout the globe. Ketzin in Germany, Beynes in France, Labodice in Czech, and Kasimovskoieimovskoie in Russia are the known projects of hydrogen storage in aquifers as mixture of gases whereas Hychico in Argentina, German hydrogen to store (H_2STORE), USS in Austria are the underground storage projects involving hydrogen gas mixtures in depleted reservoirs. Another project named USS 2030 is expected to the first project to store 100% hydrogen in depleted hydrocarbon reservoir (GaffneyCline, 2022).

13.7.1 Underground hydrogen storage in salt caverns

In Texas, there are two significantly larger caverns, as noted by Crotogino et al. (2010). One of these caverns, located at the ConocoPhillips Clemens Terminal, has been storing hydrogen since 1980. Situated approximately 850 m underground, this cylindrical cavern has a diameter of 49 m and a height of 300 m, offering a usable hydrogen capacity of 30 MMm^3, equivalent to 2520 metric tons at a pressure of 150 bar. This storage facility is directly linked to the Old Ocean refinery and contains gas that is 95% hydrogen, providing an energy capacity of 892 GWh. Additionally, in Texas, Air Liquide, a major hydrogen producer in North America, utilizes a large salt cavern along the Gulf Coast for storing 95% hydrogen. This cavern which is situated at a depth of 850–1400 m has a total capacity of 600,000 m^3. The Moss Bluff, Praxair cavern in US which is operational since 2007 has an energy capacity 80 GWh with an overall capacity of 566,000 m^3 (Crotogino et al., 2010). The German salt cavern storage Kiel, is situated at a depth of around 1335 m with a total capacity of 32,000 m^3. It is operational since 1971, storing 62% of hydrogen under 80–100 bar pressure. This cavern is currently in use for methane storage. There still exists a scarcity of projects that store 100% hydrogen even in salt caverns. The "Hydrogen Pilot Storage for large Ecosystem Replication" (HyPSTER) project in France employs a salt cavern located in Etrez for the purpose of storing pure hydrogen. Similarly, the "H_2 Cavern Storage Transition" (H_2CAST) project in Germany involves the storage of 100% hydrogen within a salt cavern situated in Etzel. Additionally, the HyStock project in the Netherlands utilizes a recently created salt cavern in Zuidwending to store 100% hydrogen (Raza et al., 2024). Fig. 13.4 shows the updated status (till October 31, 2023) of worldwide UHS projects in salt caverns as per the data base from IEA. The Teesside salt cavern storage of UK is discussed in detail under the following section.

13.7.1.1 Teesside project, Yorkshire, UK

Teesside storage of hydrogen is a salt cavern storage operated by Sabic since 1972. It is a bed formation with storage volume of 210,000 m^3 with a mean cavern depth of 365 m

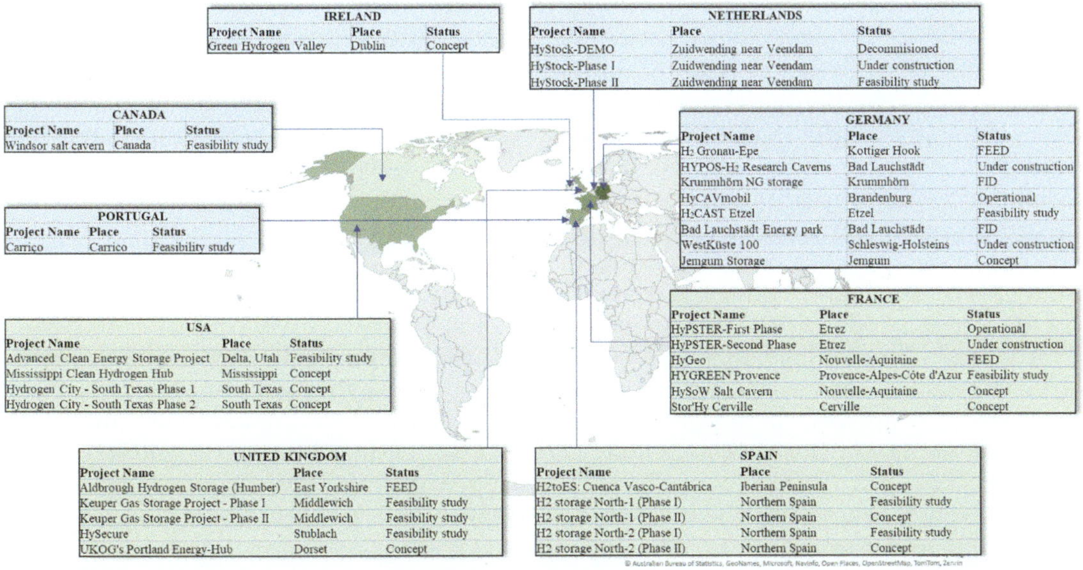

FIGURE 13.4 Current status of underground hydrogen storage in salt caverns worldwide. Map lines delineate study areas and do not necessarily depict accepted national boundaries. Source: *Data from Hydrogen Production and Infrastructure Projects Database. (2024). 2023 10 2024/03.*

(Bérest et al., 2021). The three prior brine caverns located in the Saltholme Brinefield, which is part of the larger Teesside salt field, store around one million cubic meters of hydrogen (95% Hydrogen and 3%–4% carbon dioxide) at a pressure of 50 bar (Thiyagarajan et al., 2022; Williams et al., 2022). These caverns utilize Permian salt beds that exhibit a little eastward dip, which are located inside the Boulby Halite formation. These salt beds are situated at relatively lower depths, ranging from 350 to 450 m. These caverns are distinguished by their comparatively small dimensions and display a planar and elliptical configuration. Each cavern has the capacity to store approximately 70,000 m^3 of hydrogen, intended for utilization in various industrial processes (GaffneyCline, 2022; Williams et al., 2022). The limited thickness of the halite imposes constraints on cavern size which results in elliptical caverns with heights ranging from 15 to 40 m and diameters around 70 m (Bérest et al., 2021; Williams et al., 2022). This hydrogen is later used by the surrounding industrial units to facilitate the synthesis of ammonia and methanol (Tarkowski, 2019). It is noteworthy that the facility's estimated energy storage capacity for the working gas is 30 GWh (GaffneyCline, 2022; Stolten & Emonts, 2016).

Regarding the Teesside representative cavern, it functions as a wet storage facility, operating through brine displacement, maintaining a constant and relatively low pressure within the caverns (Atkins, 2018; Bérest et al., 2021; GaffneyCline, 2022). This constant internal pressure equivalent to the halmostatic pressure corresponds to the full brine head. An examination of the distribution of the minor principal stress σ_3 above the cavern's roof reveals that, as the storage pressure remains equal to the corresponding hydrostatic full brine head, σ_3 is compressive. Thus, under the current structural conditions of the caverns, no tensile stresses are anticipated to develop (Atkins, 2018).

Due to the constant pressure operation, there is no requirement for cushion gas, and the occupied gas corresponds to the total volume of hydrogen gas, which sums to about 0.76×10^6 kg of hydrogen or 8.9×10^6 m^3 (standard cubic meters) of hydrogen (Stolten & Emonts, 2016). The Boulby Halite deposit is not recognized as a significant gas storage site in the study conducted by Williams et al. (2022) due to uncertainties associated with rapid fluctuations in lithology and lateral thickness. Nevertheless, it could potentially provide localized opportunities for storage of hydrogen. The utilization of salt caverns for brine production in the Saltholme, Billingham, Wilton and Greatham brine fields indicates a restricted potential for the development of a significant number of new multi-cavern storage initiatives, designed specifically for hydrogen storage. Other than this, there exists a potential for the construction of new caverns, while certain preexisting caverns could be modified to accommodate additional capacity for hydrogen storage in particular cases. The utilization of former caverns for hydrogen storage in Teesside showcases the prospect of repurposing existing caverns for comparatively small-scale hydrogen storage in the region (Williams et al., 2022). Significant geochemical activity and microbial activity that causes loss of hydrogen during subsurface storage are not reported from this site (Van Gessel et al., 2023). Sabic Petro has been overseeing the operation of this underground storage facility since its commissioning in approximately 1972 (Stolten & Emonts, 2016).

Although small caverns, like those found in Teesside, can serve local purposes, their restricted capacity may not be adequate for a nationwide hydrogen storage and distribution system. The Boulby Halite, located in the Teesside formation, has been employed for many storage applications, with a thickness of up to 45 m and depths varying from 274 to 366 m. The Teesside region has utilized former salt caverns and purpose-built caverns, specifically at Saltholme and Wilton, for the purpose of storing natural gas, nitrogen and various other liquid hydrocarbons (Williams et al., 2022).

13.7.2 Underground hydrogen storage in aquifers

Significant insights into hydrogen storage in saline aquifers were gathered from the storage of town gas in aquifers from 1950s to the 1970s. Town gas, derived from coal gasification, comprises approximately 50%–60% hydrogen, 30% methane, and 20% carbon dioxide and carbon monoxide resulting from coal oxidation by oxygen and steam. The storage of town gas in aquifers was practiced in various locations such as Beynes in France, the Lobodice in Czech Republic, and Bad Lauchstaedt and Engelborstel, in Germany. Despite decades of commercial operation, no containment failures were reported at these town gas storage sites. However, variations in the composition of the gas stored is believed to have occurred due to biological and geo-chemical reactions within the storage sites, as indicated by studies by Buzek et al. (1994), Kruck et al. (2013), and Panfilov (2015) Buzek et al., 1994; Kruck et al., 2013; Panfilov (2015). One of the notable changes in them is the generation of hydrogen sulfide which is probably due to the abiotic pyrite reduction and the generation of methane from methanogenesis (Miocic et al., 2022). Fig. 13.5 shows the current status of aquifer hydrogen storage projects worldwide. Some of the practiced projects of UHS in aquifers are discussed in detailed here.

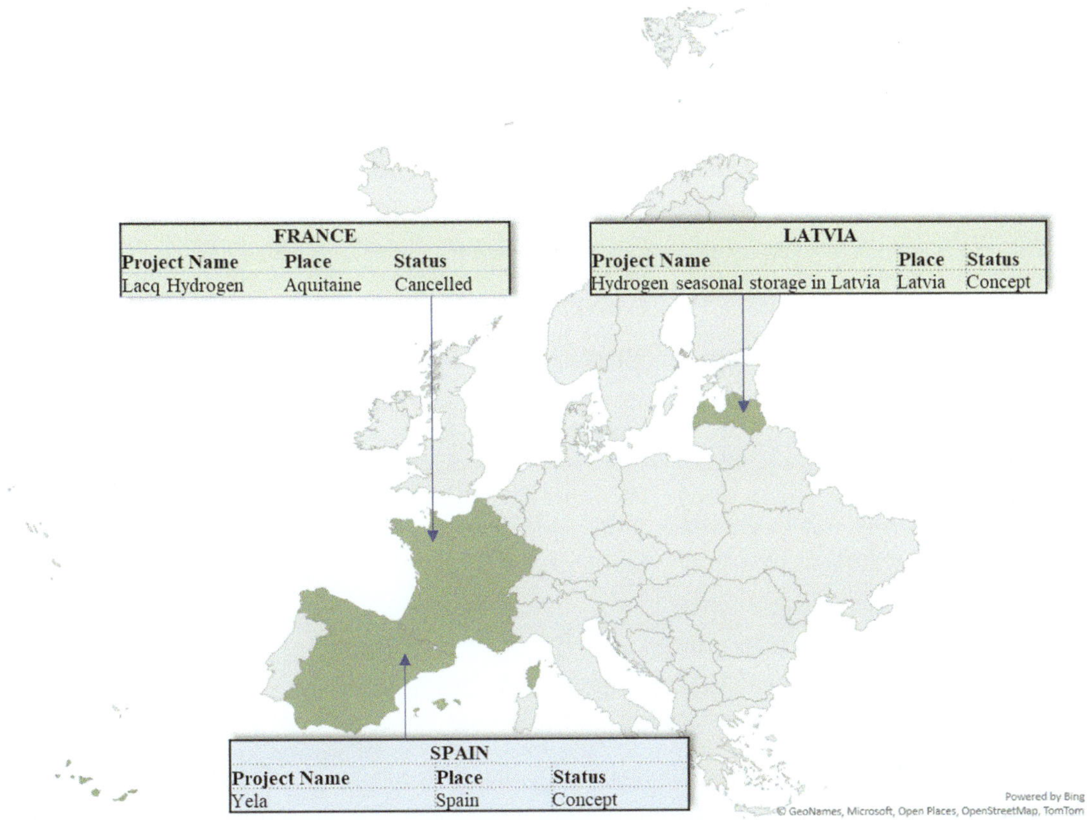

FIGURE 13.5 Current status of underground hydrogen storage in aquifers worldwide. Map lines delineate study areas and do not necessarily depict accepted national boundaries. Source: *Data from Hydrogen Production and Infrastructure Projects Database. (2024). 2023 10 2024/03.*

13.7.2.1 Beynes (France)

From 1956 to 1972, hydrogen was stored in an aquifer near the French city of Beynes, which is located at a depth of 366 m (1200 ft) (Foh et al., 1979; Jahanbakhsh et al., 2024). The hydrogen composition of this manufactured gas , which was produced from the steel and coking coal in eastern France was around 50%. The purpose of this storage was to regulate variations in gas production and demand. With the 10 m thickness and 3–5 Darcy permeability, the aquifer is composed of unconsolidated sandstone. The aquifer has a storage capacity of 360 million m^3, out of a total capacity of around 500 million m^3. Intensive microbial activity and consecutive gas transformation was observed from this site (Jahanbakhsh et al., 2024; Panfilov, 2015). After a year in storage, the gas that was collected contained trace amounts of iron and nickel carbonyls. The gas had to be desulfurized, dried, and oxygenated before the carbonyls could be scrubbed off (Foh et al., 1979; Jahanbakhsh et al., 2024). During the initial 18 years of operation, no hydrogen losses and safety issues were reported from this site. However, the upper stratum of this site was transformed to store natural gas in 1975. Later, the lower reservoir was also used for storing natural gas at a depth of 740 m with a capacity of 800 million standard m^3 capacity (Panfilov, 2015).

13.7.2.2 Ketzin (Germany)

In this project, which is situated 40 km west of Berlin in Ketzin, Germany, involves the storage of hydrogen-rich gas in an aquifer at a depth of 200–250 m, namely in sandstone (Jahanbakhsh et al., 2024; Panfilov, 2015). From 1964 to 1985, the project's monitoring system revealed that gas compositions evolved throughout the storage period. Carbon monoxide levels were found to be lower, but hydrogen, methane, and carbon dioxide levels were found to be higher. Additionally, there was a noticeable rise in temperature of 30°C–40°C (Jahanbakhsh et al., 2024; Strobel et al., 2020). In its lower sand stone reservoir, the first European onshore pilot site for carbon dioxide storage has been established in at a depth of 630 m since 2008. Three wells were drilled for this purpose. However, the injection of carbon dioxide ended by 2013 and based on the recommendation of German GeoResearch Center (GFZ) Potsdam, this geological structure has the potential to serve as a pilot project for further hydrogen storage (Panfilov, 2015).

13.7.2.3 Lobodice (Czech Republic)

The underground storage sites at Lobodice in Czech Republic, functioned to store the surplus coke gas from Ostrava. This project involved the storage of town gas, which consisted of a mixture of 25% methane and 50% hydrogen, in an aquifer. The reservoir is situated at a depth of 430 m, operating under a pressure of 90 bar and a temperature of 43°C. This study has demonstrated various phenomena, including structural trapping, dispersion, partial hydrogen leakage, and significant alterations in gas composition, as contributing to gas losses (De Ville De Goyet et al., 2018; Jahanbakhsh et al., 2024). The first in situ data on the uncommon behavior of hydrogen in underground storage was published from this site in 1990 (Buzek et al., 1994; Panfilov, 2015; Smigan et al., 1990). The initial injection mixture comprised 55% hydrogen, 20% carbon dioxide + carbon monoxide, and 20% methane. Following several months of injection and storage, at the commencement of the "production" phase of the cycle, the retrieved gas contained 37% hydrogen, 12% carbon dioxide + carbon monoxide, and 40% methane (Panfilov, 2010). From the data, the changes in the composition of gas after one injection cycle was clear. Specifically, there was a twofold increase in methane content, accompanied by a notable decrease in hydrogen and acid gases (Panfilov, 2015). When Smigan et al. (1990) analyzed the samples of fluid from the storage site, the presence of methanogenic bacteria was detected with a density of 103–104 cells/mL. The cultivation experiments later conducted proved the potential of this bacteria to convert hydrogen and carbon dioxide into methane (Smigan et al., 1990; Strobel et al., 2020). Additionally, a significant reduction in reservoir pressure was observed. The decrease is thought to be a result of in situ methane creation, whereby water was generated as a secondary outcome, resulting in the compression of gas in specific regions and its expansion in other locations (Panfilov, 2015). In 1991, the plant was fully converted to store natural gas (Jahanbakhsh et al., 2024).

As per the present records, the Beynes, Lobodice and the German projects Hahnlein and Eschenfelden are repurposed as natural gas storage sites whereas Ketzin and Engelborstel projects of Germany are decommissioned (Miocic et al., 2022).

13.7.3 Underground hydrogen storage in depleted reservoirs

Even though UHS potential of depleted hydrocarbon reservoirs is constantly studied and attaining massive attention, only two pilot projects have been conducted till date on storing and recovering hydrogen. Those are the Hychico project of Argentina and USS project of Austria which stored 10% of hydrogen produced from renewable energy into sandstone reservoirs (Miocic et al., 2022). The present status of worldwide hydrogen storage projects in depleted gas reservoirs are shown in Fig. 13.6. This is based upon the data from IEA's "Hydrogen Production and Infrastructure Projects Database."

13.7.3.1 Hychico (Argentina)

Hychico project was an hydrogen energy project started by the Argentinean company Hychico S a few years before (Strobel et al., 2020). Wind farms, hydrogen manufacturing, and subterranean storage are all part of this project. In the first stage, they built a 6.3 MW wind park and its 120 Nm^3/h hydrogen production facility which are situated close to a depleted gas reserve in Patagonian Plateau (Perez et al., 2016). In 2010, they began to study

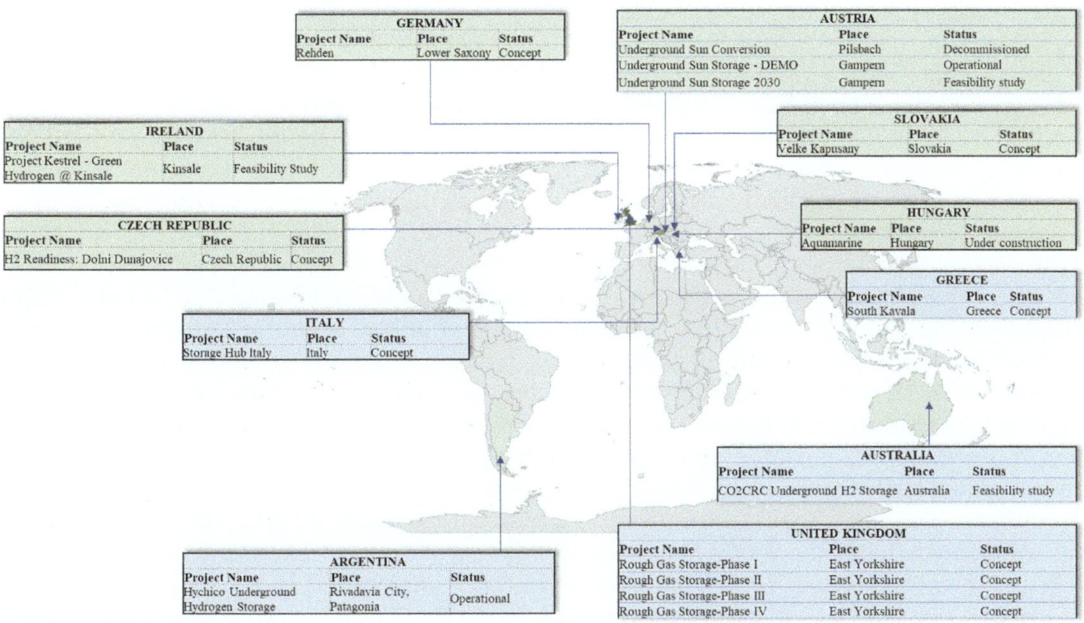

FIGURE 13.6 Current status of underground hydrogen storage in depleted gas fields worldwide. Map lines delineate study areas and do not necessarily depict accepted national boundaries. Source: *Data from Hydrogen Production and Infrastructure Projects Database. (2024). 2023 10 2024/03.*

the potential of storing green hydrogen in this depleted reservoir. Initially the reservoir sealing was verified by injecting natural gas (Jahanbakhsh et al., 2024; Perez et al., 2016). The original pressure of this 2.5 m thick reservoir was 26.5 bar. It had a porosity of 25%, absolute permeability of 500 mD and a water saturation of 55%. The ground temperature was around 55°C (Perez et al., 2016). Green hydrogen is transported to an injection well via a 2.3 km pipeline. Later, field tests were conducted by operating various cycles of hydrogen–natural gas storage in a reservoir depth of 815 m. The hydrogen percentage was then reached 10% during the test and each storage cycle was followed by laboratory analysis for the change in gas composition and reservoir properties. Some hydrogen conversion to methane was observed from this site. The next stage of this study is expected to explore the possibility of underground microbial methanation. For this, they have planned laboratory experiments, biological characterization and field tests in collaboration with the French Geological Survey (BRGM). The first results reported were favoring the microbial methanation process (Perez et al., 2016; Strobel et al., 2020).

The Austrian USS project and the upcoming USS 2030 which aims at storing 100% hydrogen are discussed in detail in the following sections.

13.7.3.2 Underground sun storage

The USS is an experimental project which started in the mid 2013 for the subsurface storage of hydrogen. There remained always a question of how to store renewable energy to maximize its contribution to the energy economy and to net zero goals. In this context, the energy storage company renewables and gas (RAG), Austria came up with this small-scale project as part of the "power to gas" technology for storing renewable energy in geological media in the form of hydrogen. This project which concluded in 2017, aimed to investigate the feasibility of storing renewable energy in repurposed natural gas reserves for extended periods and during different seasons. This goal was successfully achieved by storing gas mixtures containing hydrogen up to 10% concentration. It demonstrated the feasibility of storing solar energy during the summer and utilizing it during winter by injecting it back into the grid when demand increases (Underground Sun Storage, 2024).

In the progression of the USS project, the "Lehen-002" former gas field (Strobel et al., 2020) was selected as the suitable target site. This field was one among the smallest depleted gas reserves situated within the Molasse Basin located in Upper Austria. The Molasse Basin, situated in the region between Linz and Salzburg, is widely recognized as a prominent area in Austria for the production of natural gas. Thus far, a total of 40 gas fields have been located within this basin. A significant number of these regions are located within the Oligocene–Miocene Puchkirchen and Hall formations, which are distinguished by the presence of deep water sandstone and conglomerates (Hassannayebi et al., 2019). The field has unique characteristics, including homogeneity, effective sealing by shale layers, and absence of aquifer connections, making it a promising location for a trial project. The formation layer has an estimated thickness of 800 m, a saturation of 22% irreducible water, a porosity of 22% and a temperature of 40°C (Hematpur et al., 2023; Pichler, 2019).

The main objective of this project was to illustrate that certain gas storage reservoirs can withstand a hydrogen content of up to 10% (Underground Sun Storage, 2024). They aimed at storing the solar energy at summer to make it available at winter in the form of hydrogen. The project comprised of two phases. Phase I involved foundational inquiries with the objective of confirming the viability of a field test. Specifically, it was necessary to demonstrate that the presence of hydrogen admixtures at a 10% concentration does not compromise the integrity of the storage facility. In Phase II, activities encompassed the design, authorization, construction, and functioning of the field test facility. The planning and execution were guided by the geological conditions in the foothills of the Alps (Molasse zone) and factored in the technical specifications aligned with commercial UGS facilities operated by RAG Austria. The injection of a blend of hydrogen and natural gas finally commenced in April 2016 totaling 1.22 million standard cubic meters over a period of 3 months until early July. Subsequently, a 3-month shut-in phase was implemented to monitor pressure and temperature dynamics, as well as the variations in gas composition. Gas reproduction then occurred until mid-January 2017. Throughout these stages, continuous pressure and temperature measurements were undertaken. (Bauer & Pichler, 2017).

The investigators conducted an analysis of the entire cycle encompassing hydrogen injection, storage, and its recovery. In the concluding stage, the analysis of volume balance revealed that an estimated 82% of the hydrogen injected could be successfully recovered. The remaining hydrogen underwent diffusion, dissolution, or chemical conversion through microbial activities. (GaffneyCline, 2022; Jahanbakhsh et al., 2024). The information also discusses the potential for hydrogen leakage due to its high diffusivity and possibility of migration through fractures and faults in the caprock. Numerical simulation studies suggest hydrogen losses by diffusion through the caprock in the range of 2%–6%, and the final report of the project (2017) indicates a whole loss of 18%, including losses from diffusion and solubility. Over the 4-month shut-in period and the subsequent 3-month withdrawal phase, changes in gas composition were noted. The hydrogen concentration reduced to 7%, and the concentration of carbon dioxide changed from 0.2% to 0.06% (Austria, 2020). The slight variations in gas composition may have arisen due to several reasons, such as diffusion, leakage of injected gas, the blending of injected gas with the residual reservoir gas, and complex biological and geochemical interactions involving gas, brine, and rock. Also, the microorganism's on-site isotopic analysis affirmed that the transformation of carbon dioxide and hydrogen into methane primarily occurred by means of the catalytic activity of methanogens (Strobel et al., 2020; Wu et al., 2023). From the identified microbial consortium shift, it was determined that 3% of the hydrogen injected was converted to methane by methanogens (GaffneyCline, 2022). This shows that methanogens were able to survive even in extremely low CO_2 levels. Examination of fluids obtained from the reservoir during extraction indicated acetate production peaking at 100 mg/L and a decrease in sulfate levels from 20 to 0 mg/L (Thiyagarajan et al., 2022). Hassannayebi et al. (2019) conducted a modeling case study to explore these reactions considering the USS project. Their research developed a systematic approach to evaluate the interactions between gas, brine, and minerals. They began by studying the geo-chemical system using equilibrium batch modeling. The study utilized the Geochemist's Workbench as a geochemical modeling tool, integrating the internal thermodynamic database of Lawrence Livermore National Laboratory (LLNL) (Hassannayebi et al., 2019).

The other findings of this pilot project as reported by Pichler (2019) includes that hydrogen can be stored up to a partial pressure of 10 bar in these field without any negative effect on the reservoir and facilities. The testing pipe installed at the facility containing samples of all the installed materials such as cement, steel grades and elastomers had no degradation or corrosion above the industrial standard when exposed to the gas during the entire project cycle. Also, there were no significant change in the reservoir temperature and loss of hydrogen due to geochemical reactions (Pichler, 2019). The modeling study results as conducted by Hassannayebi et al. (2019) also supports the absence of geochemical activity during the storage period of this project (Hassannayebi et al., 2019). Addition to this the presence of hydrogen sulfide was not detected in this reservoir at any time which indicated that the sulfate reducing bacteria present in the reservoir had no role of converting hydrogen. This reservoir was known to have the presence of hydrogen even 2 years after the storage which indicates its low reactivity with the reservoir elements (Pichler, 2019).

This study aims to explore the possibility of making use of existing geological porous medium as a means of storing renewable energy, specifically hydrogen. The presence of hydrogen did not adversely affect the structural integrity of the cap rock and reservoir (Bauer & Pichler, 2017; Jahanbakhsh et al., 2024). This project's success in demonstrating the feasibility of storing hydrogen in subsurface gas storage facilities, coupled with its implications for seasonal energy storage, marks a significant advancement in the utilization of existing infrastructure for renewable energy integration (Underground Sun Storage, 2024). The results of RAG's research suggest that the enormous storage capacity of natural gas facilities can be utilized as hydrogen stores in the future energy system. These facilities could serve as large-scale, seasonal storage solutions for renewable energy. This concept aligns with the goal of transitioning toward a more sustainable and flexible energy landscape (Austria, 2023).

13.7.3.3 Underground sun storage 2030 (Austria)

Since there is no pilot projects of hydrogen geo-storage in depleted hydrocarbon reservoirs operating with 100% hydrogen, the first of that kind is anticipated in 2030, operated by RAG Austria (Amirthan & Perera, 2023). The primary objective of the USS 2030 initiative is to establish a secure, multi-seasonal, and extensive storage system for renewable energy, specifically hydrogen, within subterranean gas reserves. This interdisciplinary research project, unique in the world, involves converting solar energy into green hydrogen through electrolysis, followed by storing it in former gas reservoirs. The project is being carried out in a former gas reservoir located in the municipality of Gampern, Upper Austria, under realistic conditions, under the leadership of RAG Austria AG (Rag Austria, 2024).

The USS 2030 project has established as the first hydrogen storage facility in the world which stores it in an underground porous reservoir, setting new international standards (Austria, 2023; Muhammed et al., 2023). This demonstration facility is expected to convert 4.2 million kWh (4.2 GWh) of electricity generated during summer into hydrogen to meet the energy needs of winter, thereby ensuring a steady supply of renewable energy (Austria, 2023).

The research includes building a custom research facility, developing processing technologies, demonstrating future energy scenarios, and conducting techno-economic analyses. The USS 2030 project is positioned as a fundamental step toward achieving large-scale storage of solar energy in underground gas reservoirs in the form of hydrogen (Muhammed et al., 2023). Operational since April 2023, the facility aims to address the challenge of storing summer surplus energy (equivalent to about 1000 photovoltaic systems on family homes) and making it available during winter in the form of electricity and heat. The project covers the entire value chain, focusing on the seamless integration of generation, conversion, storage, and future utilization pathways for green hydrogen. This underground storage facility is positioned to transform the surplus solar power generated by around 1000 family homes into hydrogen and store it for seasonal use, thereby promoting sustainable energy practices (Austria, 2023).

Building on predecessor projects, USS 2030 investigates the storage of hydrogen with a potential increase in content up to 100%, surpassing the previous limit of 20%. The project form an integral component of the WIVA P&G energy showcase region, so making a significant contribution toward the advancement of a sustainable hydrogen economy (Bauer & Pichler, 2017). As per the recent reports from Gampern, hydrogen injection was successfully launched in this site which can be considered as an important milestone of the project (Hydrogen Injection Underway Gampern, 2023).

These pilot projects hold a crucial role in understanding the technical, economic, and environmental aspects of UHS, providing valuable insights for the development and implementation of large-scale UHS facilities (GaffneyCline, 2022). Numerous storage operators are contemplating extensive examinations involving various mixtures of hydrogen and methane, including the potential storage of pure hydrogen in a series of field tests (Hematpur et al., 2023). The case studies discussed above are summarized in Table 13.1.

TABLE 13.1 Summary of discussed case studies.

Project	Storage type	Major findings and limitations
Teesside (United Kingdom)	Salt cavern	Operational project storing pure hydrogen gas (95% hydrogen and 3%–4% carbon dioxide). No significant loss of hydrogen due to microbial and geochemical activity. Capacity is less. Some of the former caverns are utilized for storing other fluids now.
Beynes (France)	Aquifer	The gas mixture contained 50% hydrogen. Microbial activity and change in gas composition observed. Currently storing natural gas only.
Ketzin (Germany)	Aquifer	Stored hydrogen rich gas which changed its composition throughout the period of storage. 30°C–40°C rise in temperature was noticed from the site. Site is decommissioned now.
Lobodice (Czech Republic)	Aquifer	Stored town gas mixture contained only 50% hydrogen. Significant production of methane was observed from the site. Reduction in pressure was observed. Site is repurposed for natural gas storage.
Hychico (Argentina)	Depleted gas field	Hydrogen concentration was only 10%. Methane conversion of hydrogen was observed.
Underground sun storage (Austria)	Depleted gas field	Stored hydrogen gas up to 10% concentration along with natural gas out of which 82% of hydrogen was recovered after storage cycle. No significant geochemical reactions occurred. 3% hydrogen got converted to methane. Hydrogen losses due to diffusion and solubility were reported.
Underground sun storage 2030 (Austria)	Depleted gas field	Aims to store 100% hydrogen.

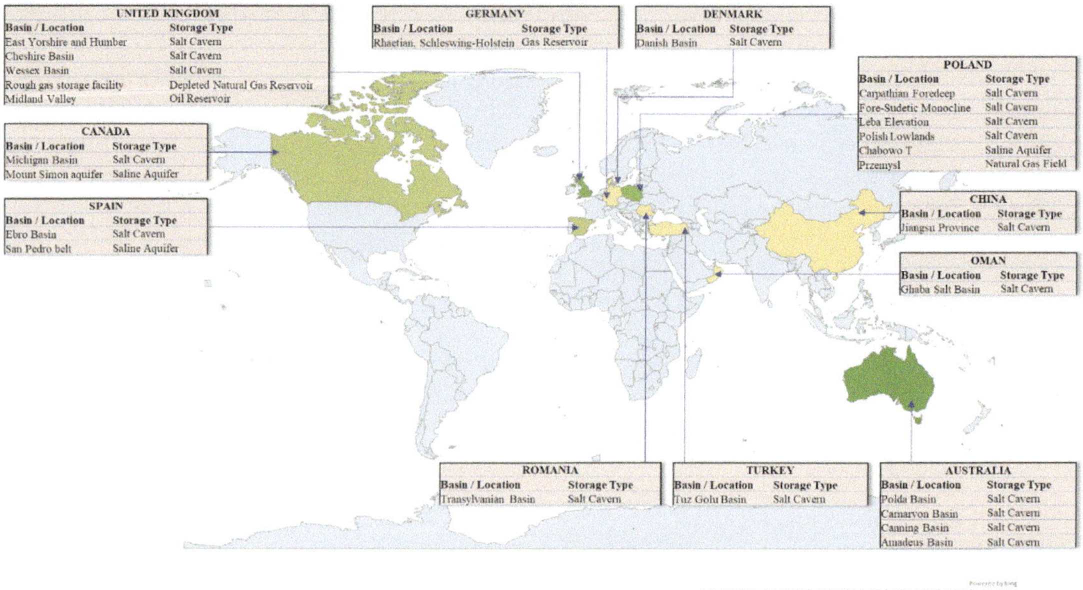

FIGURE 13.7 Potential sites for underground hydrogen storage worldwide. Map lines delineate study areas and do not necessarily depict accepted national boundaries. Source: *Data from Tackie-Otoo, B. N., & Haq, M. B. (2024). A comprehensive review on geo-storage of* H_2 *in salt caverns: Prospect and research advances.* Fuel, 356, *doi: 10.1016/j.fuel.2023.129609; Zivar, D., Kumar, S., & Foroozesh, J. (2021). Underground hydrogen storage: A comprehensive review.* International Journal of Hydrogen Energy, 46*(45), 23436–23462. Available from: http://www.journals.elsevier.com/international-journal-of-hydrogen-energy/, https://doi.org/10.1016/j.ijhydene.2020.08.138.*

13.8 Worldwide possibilities of underground hydrogen storage

Apart from the ongoing and upcoming hydrogen geo-storage projects discussed above, there are certain sites which are identified as the possible sites to store hydrogen gas. These sites are evaluated based on many criteria including storage capacity, geological depth, mineralogy, chemical stability, cost, and proximity to renewable energy producing facilities. In order to achieve a successful cyclic storage operation of hydrogen in porous geological media, it is necessary to consider all of these elements and feasibility studies should be conducted to determine whether a basin or location is suitable for safe storage of hydrogen (Zivar et al., 2021). Fig. 13.7 shows the worldwide potential sites for UHS.

13.9 Conclusion and future perspectives

Hydrogen has marked its role in the energy sector and will continue to increase its demand in the upcoming period of time as a carbon free carrier of energy. With increased demand, increased storage options are also required. Research has identified subsurface porous sites as the viable and cost-effective option for hydrogen storage. The investigation of this storage technique is currently underway as a prospective resolution to mitigate the intermittent and fluctuating nature of sustainable energy sources such as solar and wind power (Jahanbakhsh et al., 2024). This system enables the storage of high pressure, maintains strict safety protocols to safeguard against external factors, operates at significant depths beneath the surface and mitigates environmental consequences. It also enables extended operational durations, decreases investment and storage expenses, and offers a substantial storage and buffer capacity to address energy supply requirements during periods of scarcity (Amirthan & Perera, 2022).

The application of data from UGS to UHS operations is of crucial significance due to the developing stage of the field and the limited experimental experience. Consequently, the implementation of robust risk assessment strategies is imperative to safeguard the successful operation of subsurface hydrogen storage projects (Navaid et al., 2023). Therefore, many studies in the field of UHS highly depend on comparing it to UGS and drawing lessons from geological carbon dioxide storage projects. This is due to the scarcity of established large-scale UHS projects that may serve as a point of reference. Restricted measurements, information, and data are a consequence of the limited availability of practical experience. In this particular setting, numerous observations and assumptions are taken from either laboratory experiments or numerical models, hence presenting distinct restrictions (Jahanbakhsh et al., 2024). The properties of hydrogen and its behavior in reservoir conditions is very much different from methane and CO_2 which are commonly used for comparison. Hence subsurface hydrogen storage is a domain that requires to be studied separately. One of the major hinderances for the detailed experimental study as seen from the literatures is the safety concerns associated with handling hydrogen under reservoir conditions (high pressure and high temperature). There are only a limited number of hydrogen laboratories, primarily situated in Europe, that have released experimental findings regarding hydrogen under subsurface conditions. To encourage broader global experimentation with hydrogen, it is imperative to establish publicly accessible laboratory protocols and benchmarking data platforms. Moreover, there is a notable scarcity of publicly available information concerning the injection and recovery of hydrogen in genuine subsurface environments (Van Gessel et al., 2023).

From the pilot studies, microbial activity and hydrogen conversion was seen in aquifers and depleted reservoirs with limited geochemical activity and hydrogen loss. However, most of the sites considered for pilot studies on hydrogen storage has been either converted for natural gas storage or are decommissioned. Unlike CO_2 storage, one critical aspect of UHS in porous media is the attainment of an efficient recovery rate of hydrogen during extraction while ensuring minimal recovery of residual reservoir fluid. This consideration is particularly important when contemplating the use of an abandoned hydrocarbon reservoir for hydrogen storage. The extraction of hydrogen should not be hindered by low recovery rates or excessive separation procedures (Bechara et al., 2023). Necessity of more studies involving the behavior of hydrogen in reservoir conditions with the presence of residual oil saturations also comes from this. As seen from the literature, solubility of hydrogen is low whereas its diffusivity is higher in reservoir conditions. Hence, diffusive losses are a critical parameter to be studied. Also, there exists a scarcity of studies in the domain of hydrogen adsorption/desorption in subsurface storage. Examining rock-fluid interactions through wettability studies can offer valuable perceptions into depleted reservoir conditions, especially concerning trapping mechanisms. Current studies on rock/

brine/hydrogen systems have explored the impact of wettability using various rock minerals. This research should focus on understanding the effects of wettability changes and identifying common trapping mechanisms for immobilizing hydrogen (Muhammed et al., 2023). Also, to predict the efficiency of storage and recovery of hydrogen from UHS sites, more data in the domains of reactions and interactions of hydrogen under reservoir conditions are required. The aspects such as importance of cushion gas, geo-mechanical aspects over cyclic injection and withdrawal, well integrity, hydrogen embrittlement, hydrogen leakage etc. which are not included in this chapter are also significant in the background of subsurface hydrogen storage.

As the globe transitions toward a sustainable energy future, the need for effectual and scalable energy storage technologies becomes increasingly paramount. Hydrogen, widely regarded as the future fuel, provides a versatile energy carrier and has the capacity to store significant amounts of energy for long periods of time. Furthermore, the utilization of subsurface hydrogen storage presents itself as a potentially viable approach to tackle the difficulties linked to the intermittent nature and seasonal misalignment of renewable energy sources. Even though studies have shown that subsurface hydrogen storage have huge potential, there is lack of pilot projects addressing the same. Various pilot projects and research initiatives are underway to address technical, economic, and regulatory challenges associated with hydrogen geo-storage and its widespread implementation may take time. The technical readiness level of cavern storage is found to be at a higher level than porous reservoir storage which exhibits huge variations in geological characteristics. Hence more pilot and demonstration projects have to be established in them. Also, the monitoring of hydrogen in the subsurface conditions is still in its early stage (Van Gessel et al., 2023).

Addressing the challenges and exploring the future scope of UHS requires a multifaceted approach involving technological innovation, increased project implementation, data accumulation, regulatory development, and seamless integration with evolving energy systems.

References

Abdulelah, H., Keshavarz, A., Hoteit, H., Abid, H., Goudeli, E., Ennis-King, J., & Iglauer, S. (2023). Hydrogen physisorption in earth-minerals: Insights for hydrogen subsurface storage. *Journal of Energy Storage, 66*. Available from https://doi.org/10.1016/j.est.2023.107440.

Aghaei, H., Al-Yaseri, A., Toorajipour, A., Shahsavani, B., Yekeen, N., & Edlmann, K. (2023). Host-rock and caprock wettability during hydrogen drainage: Implications of hydrogen subsurface storage. *Fuel, 351*. Available from https://doi.org/10.1016/j.fuel.2023.129048.

Al-Yaseri, A., Al-Mukainah, H., & Yekeen, N. (2023). Experimental insights into limestone-hydrogen interactions and the resultant effects on underground hydrogen storage. *Fuel, 344*. Available from https://doi.org/10.1016/j.fuel.2023.128000.

Al-Yaseri, A., Amao, A., & Fatah, A. (2023). Experimental investigation of shale/hydrogen geochemical interactions. *Fuel, 346*. Available from https://doi.org/10.1016/j.fuel.2023.128272.

Amirthan, T., & Perera, M. S. A. (2022). The role of storage systems in hydrogen economy: A review. *Journal of Natural Gas Science and Engineering, 108*. Available from https://doi.org/10.1016/j.jngse.2022.104843.

Amirthan, T., & Perera, M. S. A. (2023). Underground hydrogen storage in Australia: A review on the feasibility of geological sites. *International Journal of Hydrogen Energy, 48*(11), 4300–4328. Available from https://doi.org/10.1016/j.ijhydene.2022.10.218, http://www.journals.elsevier.com/international-journal-of-hydrogen-energy/.

Atkins. (2018). Salt cavern appraisal for hydrogen and gas storage.

Austria. (2020). VERBUND GesmbH, AG, T LEOBEN UNIVERSITÄT für Bodenkultur Wien, J.K. der, Universität, Underground Sun Linz, Storage.

Austria, A. G. (2023). Gas renewables, we are opening doors to energy future, RAG Austria AG.

Bauer, S., & Pichler, M. (2017). *Underground sun storage*, 1–63.

Bechara, E., Gamadi, T., Watson, M., Mahamadene, A., Emadibaladehi, H., Hussain, A., & Ispas, I. (2023). Effect of hydrogen interaction with crude oil on the extraction of hydrogen if stored in depleted shale reservoirs: An experimental study analyzing the outcome of cyclic hydrogen injection into oil-saturated core samples. Proceedings – SPE International Symposium on Oilfield Chemistry, Society of Petroleum Engineers (SPE), United States. Available from https://doi.org/10.2118/213851-M, https://www.onepetro.org/conferences/SPE/01OCS, 9781613998748.

Bird, R. B., Stewart, W. E., & Lightfoot, E. N. Transport phenomena (2nd ed.).

Buzek, F., Onderka, V., Vancura, P., & Wolf, I. (1994). Carbon isotope study of methane production in a town gas storage reservoir. *Fuel*, *73*(5), 747–752. Available from https://doi.org/10.1016/0016-2361(94)90019-1.

Bérest, P., Brouard, B., Hévin, G., & Réveillère, A. (2021). 55th U.S. Rock Mechanics/Geomechanics Symposium 2021 9781713839125 433–445 American Rock Mechanics Association (ARMA) France Tightness of salt caverns used for hydrogen storage 3.

Carden, P. O., & Paterson, L. (1979). Physical, chemical and energy aspects of underground hydrogen storage. *International Journal of Hydrogen Energy*, *4*(6), 559–569. Available from https://doi.org/10.1016/0360-3199(79)90083-1.

CO_2 Emissions in 2023. (2023).

Colman, D. R., Poudel, S., Stamps, B. W., Boyd, E. S., & Spear, J. R. (2017). The deep, hot biosphere: Twenty-five years of retrospection. *Proceedings of the National Academy of Sciences of the United States of America*, *114*(27), 6895–6903. Available from https://doi.org/10.1073/pnas.1701266114, http://www.pnas.org/content/114/27/6895.full.pdf.

Crotogino, F., Donadei, S., Bünger, U., & Landinger, H. (2010). 18th World Hydrogen Energy Conference 2010, WHEC 2010, Proceedings 9783893366545 37-45 Forschungszentrum Julich GmbH Germany Large-Scale Hydrogen Underground Storage for Securing Future Energy Supplies 4.

De Ville De Goyet, B., Giouse, H., & Dequidt, D. (2018). International Gas Union World Gas Conference Papers 9781510885837 1534-1546 International Gas Union Belgium Guidance: Injection of hydrogen/natural gas admixtures in underground gas storages facilities (UGS) http://members.igu.org/.

Didier, M., Leone, L., Greneche, J. M., Giffaut, E., & Charlet, L. (2012). Adsorption of hydrogen gas and redox processes in clays. *Environmental Science and Technology*, *46*(6), 3574–3579. Available from https://doi.org/10.1021/es204583h.

Dopffel, N., Jansen, S., & Gerritse, J. (2021). Microbial side effects of underground hydrogen storage – Knowledge gaps, risks and opportunities for successful implementation. *International Journal of Hydrogen Energy*, *46*(12), 8594–8606. Available from https://doi.org/10.1016/j.ijhydene.2020.12.058, http://www.journals.elsevier.com/international-journal-of-hydrogen-energy/.

Energy Transition Advisory Committee (ETAC). (2023). Ministry of Petroleum & Natural Gas, Government of India. The Green Shift – The low carbon transition of India's. *Oil & Gas secto*.

Esfandyari, H., Sarmadivaleh, M., Esmaeilzadeh, F., Ali, M., Iglauer, S., & Keshavarz, A. (2022). Experimental evaluation of rock mineralogy on hydrogen-wettability: Implications for hydrogen geo-storage. *Journal of Energy Storage*, *52*. Available from https://doi.org/10.1016/j.est.2022.104866, http://www.journals.elsevier.com/journal-of-energy-storage/.

Foh, S., Novil, M., Rockar, E., & Randolph, P. (1979). Underground hydrogen storage. *Final Report*. Available from https://doi.org/10.2172/6536941.

GaffneyCline. (2022). Gaffney cline underground hydrogen storage gaffneycline_underground_hydrogen_storage_article.

Gernaat, D. E. H. J., de Boer, H. S., Daioglou, V., Yalew, S. G., Müller, C., & van Vuuren, D. P. (2021). Climate change impacts on renewable energy supply. *Nature Climate Change*, *11*(2), 119–125. Available from https://doi.org/10.1038/s41558-020-00949-9, http://www.nature.com/nclimate/index.html.

Global Hydrogen Review. (2022). OECD. Available from https://doi.org/10.1787/a15b8442-en.

Global Hydrogen Review. (2023). OECD. Available from https://doi.org/10.1787/cb2635f6-en.

H2 Hydrogen Tools, Hydrogen Analysis Resource Center, 2016. Hydrogen Compressibility at different temperatures and pressures. Pacific Northwest National Laboratory. Available from https://h2tools.org/hyarc/hydrogen-data/hydrogen-compressibility-different-temperatures-and-pressures.

Hassannayebi, N., Azizmohammadi, S., De Lucia, M., & Ott, H. (2019). Underground hydrogen storage: Application of geochemical modelling in a case study in the Molasse Basin, Upper Austria. *Environmental Earth Sciences*, *78*(5). Available from https://doi.org/10.1007/s12665-019-8184-5, http://www.springerlink.com/content/121380/.

Heinemann, N., Alcalde, J., Miocic, J. M., Hangx, S. J. T., Kallmeyer, J., Ostertag-Henning, C., Hassanpouryouzband, A., Thaysen, E. M., Strobel, G. J., Schmidt-Hattenberger, C., Edlmann, K., Wilkinson, M., Bentham, M., Haszeldine,

R. S., Carbonell, R., & Rudloff, A. (2021). Enabling large-scale hydrogen storage in porous media – The scientific challenges. *Energy & Environmental Science, 14*(2), 853–864. Available from https://doi.org/10.1039/d0ee03536j.

Hematpur, H., Abdollahi, R., Rostami, S., Haghighi, M., & Blunt, M. J. (2023). Review of underground hydrogen storage: Concepts and challenges. *Advances in Geo-Energy Research, 7*(2), 111–131. Available from https://doi.org/10.46690/ager.2023.02.05, https://www.yandy-ager.com/index.php/ager/article/view/667.

Hemme, C., & van Berk, W. (2018). Hydrogeochemical modeling to identify potential risks of underground hydrogen storage in depleted gas fields. *Applied Sciences (Switzerland), 8*(11). Available from https://doi.org/10.3390/app8112282, https://www.mdpi.com/2076-3417/8/11/2282/pdf.

Higgs, S., Da Wang, Y., Sun, C., Ennis-King, J., Jackson, S. J., Armstrong, R. T., & Mostaghimi, P. (2022). In-situ hydrogen wettability characterisation for underground hydrogen storage. *International Journal of Hydrogen Energy, 47*(26), 13062–13075. Available from https://doi.org/10.1016/j.ijhydene.2022.02.022, http://www.journals.elsevier.com/international-journal-of-hydrogen-energy/.

Hosseini, M., Fahimpour, J., Ali, M., Keshavarz, A., & Iglauer, S. (2022). Hydrogen wettability of carbonate formations: Implications for hydrogen geo-storage. *Journal of Colloid and Interface Science, 614*, 256–266. Available from https://doi.org/10.1016/j.jcis.2022.01.068, http://www.elsevier.com/inca/publications/store/6/2/2/8/6/1/index.htt.

Hsiung, S. (2001). Energy Technology Training Center, College of the Desert Hydrogen Fuel Cell Engines and Related Technologies.

Hydrogen as an Alternative Fuel. (2024). 1991, 4, 18. https://afdc.energy.gov/fuels/hydrogen-basics.

Hydrogen Injection Underway Gampern. (2023).

Hydrogen Production and Infrastructure Projects Database. (2023). 2024, 10 2024/03.

Iglauer, S., Abid, H., Al-Yaseri, A., & Keshavarz, A. (2021). Hydrogen adsorption on sub-bituminous coal: Implications for hydrogen geo-storage. *Geophysical Research Letters, 48*(10). Available from https://doi.org/10.1029/2021GL092976, http://agupubs.onlinelibrary.wiley.com/hub/journal/10.1002/(ISSN)1944-8007/.

Iglauer, S., Ali, M., & Keshavarz, A. (2021). Hydrogen wettability of sandstone reservoirs: Implications for hydrogen geo-storage. *Geophysical Research Letters, 48*(3). Available from https://doi.org/10.1029/2020GL090814, http://agupubs.onlinelibrary.wiley.com/hub/journal/10.1002/(ISSN)1944-8007/.

Iordache, I., Schitea, D., Gheorghe, A. V., & Iordache, M. (2014). Hydrogen underground storage in Romania, potential directions of development, stakeholders and general aspects. *International Journal of Hydrogen Energy, 39*(21), 11071–11081. Available from https://doi.org/10.1016/j.ijhydene.2014.05.067, http://www.journals.elsevier.com/international-journal-of-hydrogen-energy/.

Jahanbakhsh, A., Louis Potapov-Crighton, A., Mosallanezhad, A., Tohidi Kaloorazi, N., & Maroto-Valer, M. M. (2024). Underground hydrogen storage: A UK perspective. *Renewable and Sustainable Energy Reviews, 189*. Available from https://doi.org/10.1016/j.rser.2023.114001, https://www.journals.elsevier.com/renewable-and-sustainable-energy-reviews.

Kruck, O., Crotogino, F., Prelicz, R., Rudolph, T. (2013). Assessment of the potential, the actors and relevant business cases for large scale and seasonal storage of renewable electricity by hydrogen underground storage in Europe.

Kumar, S., Regassa Jufar, S., Kumar, S., Foroozesh, J., Kumari, S., & Bera, A. (2023). Underground hydrogen storage and its roadmap and feasibility in India toward net-zero target for global decarbonization. *Fuel, 350*. Available from https://doi.org/10.1016/j.fuel.2023.128849, http://www.journals.elsevier.com/fuel/.

Lassin, A., Dymitrowska, M., & Azaroual, M. (2011). Hydrogen solubility in pore water of partially saturated argillites: Application to Callovo-Oxfordian clayrock in the context of a nuclear waste geological disposal. *Physics and Chemistry of the Earth, Parts A/B/C, 36*(17-18), 1721–1728. Available from https://doi.org/10.1016/j.pce.2011.07.092.

Lord, A. S., Kobos, P. H., & Borns, D. J. (2014). Geologic storage of hydrogen: Scaling up to meet city transportation demands. *International Journal of Hydrogen Energy, 39*(28), 15570–15582. Available from https://doi.org/10.1016/j.ijhydene.2014.07.121, http://www.journals.elsevier.com/international-journal-of-hydrogen-energy/.

Matos, C. R., Carneiro, J. F., & Silva, P. P. (2019). Overview of large-scale underground energy storage technologies for integration of renewable energies and criteria for reservoir identification. *Journal of Energy Storage, 21*, 241–258. Available from https://doi.org/10.1016/j.est.2018.11.023, http://www.journals.elsevier.com/journal-of-energy-storage/.

Miocic, J., Heinemann, N., Edlmann, K., Scafidi, J., Molaei, F., & Alcalde, J. (2022). Underground hydrogen storage: A review. *Geological Society, London, Special Publications, 528*(1), 73–86. Available from https://doi.org/10.1144/sp528-2022-88.

Mouli-Castillo, J., Heinemann, N., & Edlmann, K. (2021). Mapping geological hydrogen storage capacity and regional heating demands: An applied UK case study. *Applied Energy, 283*. Available from https://doi.org/10.1016/j.apenergy.2020.116348, https://www.journals.elsevier.com/applied-energy.

Muhammed, N. S., Haq, B., Al Shehri, D., Al-Ahmed, A., Rahman, M. M., & Zaman, E. (2022). A review on underground hydrogen storage: Insight into geological sites, influencing factors and future outlook. *Energy Reports, 8*, 461–499. Available from https://doi.org/10.1016/j.egyr.2021.12.002, http://www.journals.elsevier.com/energy-reports/.

Muhammed, N. S., Haq, M. B., Al Shehri, D. A., Al-Ahmed, A., Rahman, M. M., Zaman, E., & Iglauer, S. (2023). Hydrogen storage in depleted gas reservoirs: A comprehensive review. *Fuel, 337*. Available from https://doi.org/10.1016/j.fuel.2022.127032, http://www.journals.elsevier.com/fuel/.

Navaid, H. B., Emadi, H., & Watson, M. (2023). A comprehensive literature review on the challenges associated with underground hydrogen storage. *International Journal of Hydrogen Energy, 48*(28), 10603–10635. Available from https://doi.org/10.1016/j.ijhydene.2022.11.225, http://www.journals.elsevier.com/international-journal-of-hydrogen-energy/.

Pan, B., Yin, X., Ju, Y., & Iglauer, S. (2021). Underground hydrogen storage: Influencing parameters and future outlook. *Advances in Colloid and Interface Science, 294*. Available from https://doi.org/10.1016/j.cis.2021.102473.

Panfilov, M. (2010). Underground storage of hydrogen: in situ self-organisation and methane generation. *Transport in Porous Media, 85*(3), 841–865. Available from https://doi.org/10.1007/s11242-010-9595-7.

Panfilov, M. (2015). Underground and pipeline hydrogen storage. *Compendium of hydrogen energy: hydrogen storage, distribution and infrastructure* (Volume 2, pp. 91–115). Elsevier. Available from https://www.sciencedirect.com/book/9781782423621, 10.1016/B978-1-78242-362-1.00004-3.

Perera, M. S. A. (2023). A review of underground hydrogen storage in depleted gas reservoirs: Insights into various rock-fluid interaction mechanisms and their impact on the process integrity. *Fuel, 334*. Available from https://doi.org/10.1016/j.fuel.2022.126677.

Perez, A., Pérez, E., Dupraz, S., & Bolcich, J. (2016). 21st World Hydrogen Energy Conference Patagonia Wind – Hydrogen Project: Underground Storage and Methanation.

Pichler, M. (2019). EAGE/DGMK joint workshop on underground storage of hydrogen DGMK 03 underground sun storage results and outlook.

Rag Austria, A. G. (2024). *Underground sun storage, 2030*.

Raza, A., Mahmoud, M., Arif, M., & Alafnan, S. (2024). Underground hydrogen storage prospects in the Kingdom of Saudi Arabia. *Fuel, 357*. Available from https://doi.org/10.1016/j.fuel.2023.129665, http://www.journals.elsevier.com/fuel/.

Reitenbach, V., Ganzer, L., Albrecht, D., & Hagemann, B. (2015). Influence of added hydrogen on underground gas storage: A review of key issues. *Environmental Earth Sciences, 73*(11), 6927–6937. Available from https://doi.org/10.1007/s12665-015-4176-2, http://www.springerlink.com/content/121380/.

Smigan, M., Greksak, J., Kozankova, F., Buzek, V., & Onderka, I. (1990). Wolf, methanogenic bacteria as a key factor involved in changes of town gas stored in an underground reservoir. *FEMS Microbiology Letters, 73*, 90733–90737. Available from https://doi.org/10.1016/0378-1097(90).

Stolten, D., Emonts, B. (2016). Handbook of hydrogen storage efficient carbon capture for coal power plants transition to renewable energy systems hydrogen storage technologies hydrogen and fuel cells.

Strobel, G., Hagemann, B., Huppertz, T. M., & Ganzer, L. (2020). Underground bio-methanation: Concept and potential. *Renewable and Sustainable Energy Reviews, 123*. Available from https://doi.org/10.1016/j.rser.2020.109747, https://www.journals.elsevier.com/renewable-and-sustainable-energy-reviews.

Tackie-Otoo, B. N., & Haq, M. B. (2024). A comprehensive review on geo-storage of H_2 in salt caverns: Prospect and research advances. *Fuel, 356*. Available from https://doi.org/10.1016/j.fuel.2023.129609.

Tarkowski, R. (2019). Underground hydrogen storage: Characteristics and prospects. *Renewable and Sustainable Energy Reviews, 105*, 86–94. Available from https://doi.org/10.1016/j.rser.2019.01.051, https://www.journals.elsevier.com/renewable-and-sustainable-energy-reviews.

Thaysen, E. M., McMahon, S., Strobel, G. J., Butler, I. B., Ngwenya, B. T., Heinemann, N., Wilkinson, M., Hassanpouryouzband, A., McDermott, C. I., & Edlmann, K. (2021). Estimating microbial growth and hydrogen consumption in hydrogen storage in porous media. *Renewable and Sustainable Energy Reviews, 151*. Available from https://doi.org/10.1016/j.rser.2021.111481, https://www.journals.elsevier.com/renewable-and-sustainable-energy-reviews.

Thiyagarajan, S. R., Emadi, H., Hussain, A., Patange, P., & Watson, M. (2022). A comprehensive review of the mechanisms and efficiency of underground hydrogen storage. *Journal of Energy Storage, 51*. Available from https://doi.org/10.1016/j.est.2022.104490, http://www.journals.elsevier.com/journal-of-energy-storage/.

Truche, L., Jodin-Caumon, M. C., Lerouge, C., Berger, G., Mosser-Ruck, R., Giffaut, E., & Michau, N. (2013). Sulphide mineral reactions in clay-rich rock induced by high hydrogen pressure. Application to disturbed or natural settings up to 250°C and 30 bar. *Chemical Geology, 351*, 217–228. Available from https://doi.org/10.1016/j.chemgeo.2013.05.025.

Underground Sun Storage. (2024).

Van Gessel, S., Hajibeygi, H., Yuste Fernández, C., & Fournier, C. (2023). Hydrogen TCP-task 42 underground hydrogen storage technology monitor report.

Williams, J. D. O., Williamson, J. P., Parkes, D., Evans, D. J., Kirk, K. L., Sunny, N., Hough, E., Vosper, H., & Akhurst, M. C. (2022). Does the United Kingdom have sufficient geological storage capacity to support a hydrogen economy? Estimating the salt cavern storage potential of bedded halite formations. *Journal of Energy Storage, 53*. Available from https://doi.org/10.1016/j.est.2022.105109, http://www.journals.elsevier.com/journal-of-energy-storage/.

Wu, L., Hou, Z., Xie, Y., Luo, Z., Huang, L., Wu, X., Luo, J., Fang, Y., Chen, Q., Sun, W., Lüddeke, C. T., & Yang, L. (2023). Carbon capture, circular utilization, and sequestration (CCCUS): A multifunctional technology coupling underground biomethanation with geothermal energy production. *Journal of Cleaner Production, 426*. Available from https://doi.org/10.1016/j.jclepro.2023.139225.

Yekta, A. E., Manceau, J. C., Gaboreau, S., Pichavant, M., & Audigane, P. (2018). Determination of hydrogen--water relative permeability and capillary pressure in sandstone: Application to underground hydrogen injection in sedimentary formations. *Transport in Porous Media, 122*(2), 333–356. Available from https://doi.org/10.1007/s11242-018-1004-7, http://www.wkap.nl/journalhome.htm/0169-3913.

Yekta, A. E., Pichavant, M., & Audigane, P. (2018). Evaluation of geochemical reactivity of hydrogen in sandstone: Application to geological storage. *Applied Geochemistry, 95*, 182–194. Available from https://doi.org/10.1016/j.apgeochem.2018.05.021, http://www.journals.elsevier.com/applied-geochemistry.

Zeng, L., Keshavarz, A., Xie, Q., & Iglauer, S. (2022). Hydrogen storage in Majiagou carbonate reservoir in China: Geochemical modelling on carbonate dissolution and hydrogen loss. *International Journal of Hydrogen Energy, 47*(59), 24861–24870. Available from https://doi.org/10.1016/j.ijhydene.2022.05.247, http://www.journals.elsevier.com/international-journal-of-hydrogen-energy/.

Zivar, D., Kumar, S., & Foroozesh, J. (2021). Underground hydrogen storage: A comprehensive review. *International Journal of Hydrogen Energy, 46*(45), 23436–23462. Available from https://doi.org/10.1016/j.ijhydene.2020.08.138, http://www.journals.elsevier.com/international-journal-of-hydrogen-energy/.

CHAPTER

14

Prospects and challenges of underground hydrogen storage

Sunil Kumar[1] *and Achinta Bera*[2]

[1]RINOLYST, SciGlyph Exploration Private Limited, Patna, Bihar, India [2]Department of Petroleum Engineering, School of Energy Technology, Pandit Deendayal Energy University, Gandhinagar, Gujarat, India

14.1 Introduction

Transitioning to sustainable energy represents a pivotal step in addressing climate change and reducing greenhouse gas emissions (Capurso, 2022; Zhang, 2024). Hydrogen, hailed as a versatile and clean energy carrier, has emerged as a frontrunner in this global transition. However, realizing its full potential hinges significantly on developing efficient storage solutions, especially to meet the demands of large-scale energy storage. This necessity underscores the importance of exploring both the prospects and challenges of underground hydrogen storage (UHS). Geological hydrogen storage offers a groundbreaking approach to addressing the intermittent nature of renewable energy sources and the challenges associated with seasonal energy demands. By utilizing subsurface formations such as depleted oil and gas reservoirs, salt caverns, or aquifers, hydrogen can be safely stored during periods of high renewable energy generation and subsequently released during periods of high energy demand (Amid et al., 2016; Caglayan, 2020; Sambo, 2022; Tarkowski & Uliasz-Misiak, 2022). This unique storage solution provides a reliable and flexible means of meeting energy needs while significantly reducing reliance on carbon-intensive power generation (Amid et al., 2016; Elberry et al., 2021; Gabrielli, 2020). This not only ensures grid stability but also enhances energy reliability during peak demand periods. The advantages of UHS are multifaceted, making it an attractive option for energy storage solutions. Its ability to offer large-scale capacity, long-term stability, and minimal environmental impact aligns seamlessly with the objectives of sustainable energy infrastructure (Olabi, 2021; Tarkowski & Uliasz-Misiak, 2022). Among the emerging technologies, the understanding of geological hydrogen storage for seasonal uses stands out as a promising technology.

Subsurface Hydrogen Energy Storage
DOI: https://doi.org/10.1016/B978-0-443-24071-3.00015-7

Despite these evident advantages, UHS presents a myriad of challenges that require careful consideration and innovative solutions. Selecting suitable geological sites demands an in-depth understanding of reservoir properties, geological barriers, and proximity to energy sources (Kharel & Shabani, 2018; Matos et al., 2019; Sambo, 2022). Technical complexities, such as optimizing hydrogen storage mechanisms in porous media and implementing efficient operational strategies, pose significant hurdles. Economic feasibility and regulatory compliance further complicate the landscape, highlighting the need for meticulous planning and execution. Safety concerns hold paramount importance in UHS, given the inherent properties of hydrogen as a highly flammable gas. Ensuring the integrity and reliability of mechanical components, along with stringent safety protocols, is imperative to mitigate risks effectively (Epelle, 2022; Kumar, 2023; Wallace, 2021). Furthermore, ongoing research and technological advancements at UHS are essential to addressing these challenges comprehensively.

Several studies on UHS demonstrate that it is still in the early stages of development requiring extensive research on the geological, technical, and economic challenges associated with hydrogen's characteristics (Du, 2024; Osman, 2022; Raad et al., 2022; Scafidi, 2021). It is important to express interest in understanding the research advancements made in corresponding physical phenomena involved with storing hydrogen in geological media (Chai, 2023; Liebscher et al., 2016; Williams, 2022). An economic outlook as well as a roadmap streamlining future research should be taken into consideration as well.

This chapter aims to explore the potential of UHS for storing excess renewable energy during peak periods and utilizing it during low-demand seasons. Later on, it discusses the current targets and challenges of UHS, including geological, technical, and economic aspects, and reviews recent advancements in relevant physical phenomena. Moreover, it covers technical considerations in selecting geological formations, developing infrastructure, and addressing safety protocols. Additionally, this chapter examines various applications of UHS in power generation, transportation, and industry, along with optimizing storage materials and efficiency. Insights and recommendations on UHS's role in decarbonization efforts will be provided, considering geological, technical, economic, and regulatory dimensions. By offering actionable insights and fostering innovation, it aims to accelerate the global transition towards a sustainable energy future with hydrogen storage solutions.

14.2 Prospects of underground hydrogen storage

The hydrogen economy can play a significant role not only in the global energy mix but also in mitigating the impacts of global warming. Addressing this dual challenge requires a collective effort—politicians, business leaders, technologists, and consumers—to take careful and meaningful action. Global H_2 production is around 70 million tons per annum (MTPA). And it is mainly produced from natural gas and coal; an Energy Information Administration (EIA) report shows that 7% of natural gas is dedicated to the production of H_2. It is predominately used for the production of value-added products such as methanol, ammonia, gas-to-liquids (GTL) fuels, and in refining and steel production. Hence, H_2 is a major intermediate in industry. But only around 6% of the produced H_2 is for pure H_2 demand, which indicates that it has a limited role in the global energy mix, as

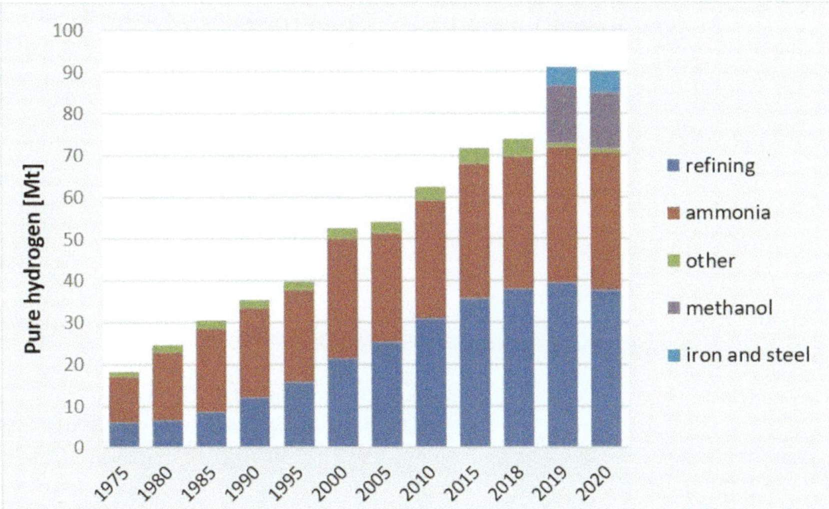

FIGURE 14.1 Global demand for pure hydrogen from 1975 to 2020.

shown in Fig. 14.1. Although H_2 is key for the production of value-added products, the challenge with hydrogen is that traditionally it is produced through fuel-based schemes such as methane reforming, coal, and biomass gasification, all of which lead to carbon emissions and other greenhouse gases.

With the necessary shift of today's global markets toward a sustainable future, H_2 can play a key role when it comes to decarbonizing the energy, industrial, and transportation sectors. The energy outlook scenarios for 2050 are focused on meeting the Paris Agreement's targets of keeping the earth's temperature rise below 1.5°C–2°C. And those scenarios have H_2 playing a bigger role in the world's energy mix. It is projected that 10 × the current production capacity of H_2, nearly 600 MTPA, is necessary to meet the carbon neutral targets for 2050, as shown in Fig. 14.2 (Hydrogen Council, 2017).

UHS is all about storing produced hydrogen using any method or surplus energy in the form of hydrogen underground to be used when needed. This method is gaining traction due to its safety advantages and alignment with global energy and emission goals, as seen in projects like H2STORE, HYSTOREPORT, HyUnder, InSpEE, and HyINTEGER (Epelle, 2022). These efforts underscore the growing interest and investment in exploring geological hydrogen storage. While UHS benefits from lessons learned in natural gas and CO_2 storage, hydrogen's unique properties pose specific challenges. For instance, hydrogen's low density and viscosity mean it moves differently and leaks more easily than natural gas, which calls for careful storage design. Geological formations like depleted reservoirs, aquifers, and artificial cavities are promising storage sites due to their depth and tightness (Epelle, 2022). However, extracting hydrogen from these sites involves managing water production, dealing with contaminants, and ensuring cost-effective operations. Comparing UHS with natural gas and CO_2 storage highlights various factors like storage capacity, leak risks, environmental impacts, and operational costs. Each method has its pros and

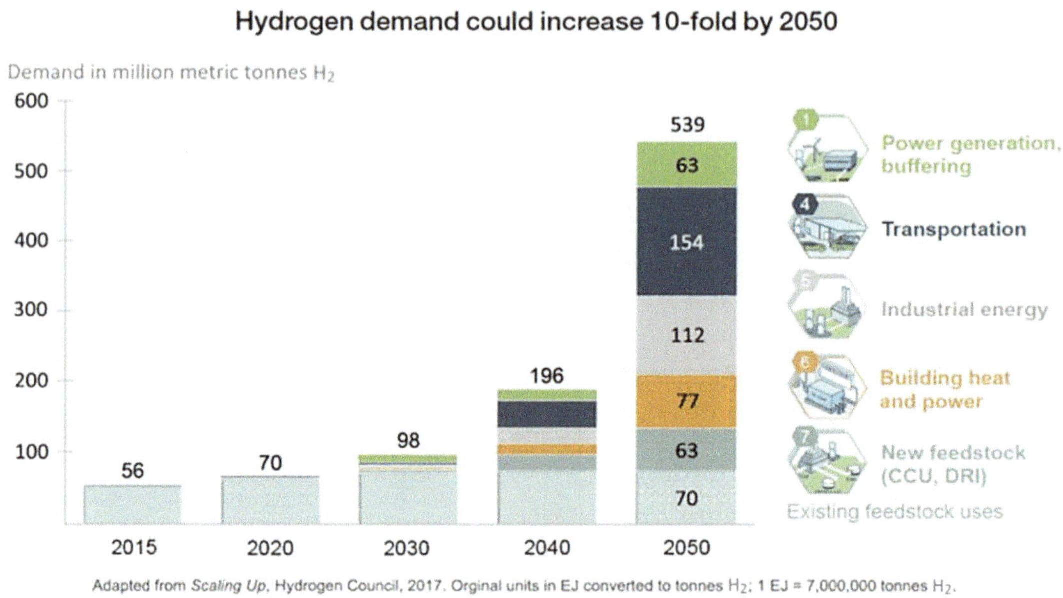

FIGURE 14.2 Global hydrogen demand projection up to 2050 (Hydrogen Council, 2017).

cons, emphasizing the need for site-specific assessments and tailored strategies (Epelle, 2022). Collaboration among stakeholders and ongoing research efforts will be crucial for refining storage techniques, enhancing efficiency, and addressing environmental concerns. Despite challenges, ongoing advancements in technology, regulatory frameworks, and market conditions bode well for the future of UHS. It's about finding the right balance between technical feasibility, economic viability, and environmental sustainability to make UHS a key player in our energy landscape.

14.3 Challenges for underground hydrogen storage

Addressing the challenges of developing UHS systems requires a holistic approach encompassing technical, regulatory, economic, and environmental considerations. From a technical perspective, ensuring optimal fluid dynamics in porous media, managing geochemical interactions to prevent reservoir damage, mitigating microbial impacts, and understanding geomechanical stresses are crucial for ensuring the long-term integrity and operational safety of storage facilities (Epelle, 2022; Heinemann, 2021). Compliance with regulatory frameworks and evolving policies necessitates seamless integration with existing infrastructure, the establishment of robust safety protocols, and alignment with emerging standards to ensure reliability and compatibility across the hydrogen value chain. Managing costs and optimizing efficiency entails addressing production costs through technological advancements, improving storage efficiency, and maximizing energy utilization while upholding stringent purity standards. International collaboration and standardization efforts play a key role in promoting knowledge sharing,

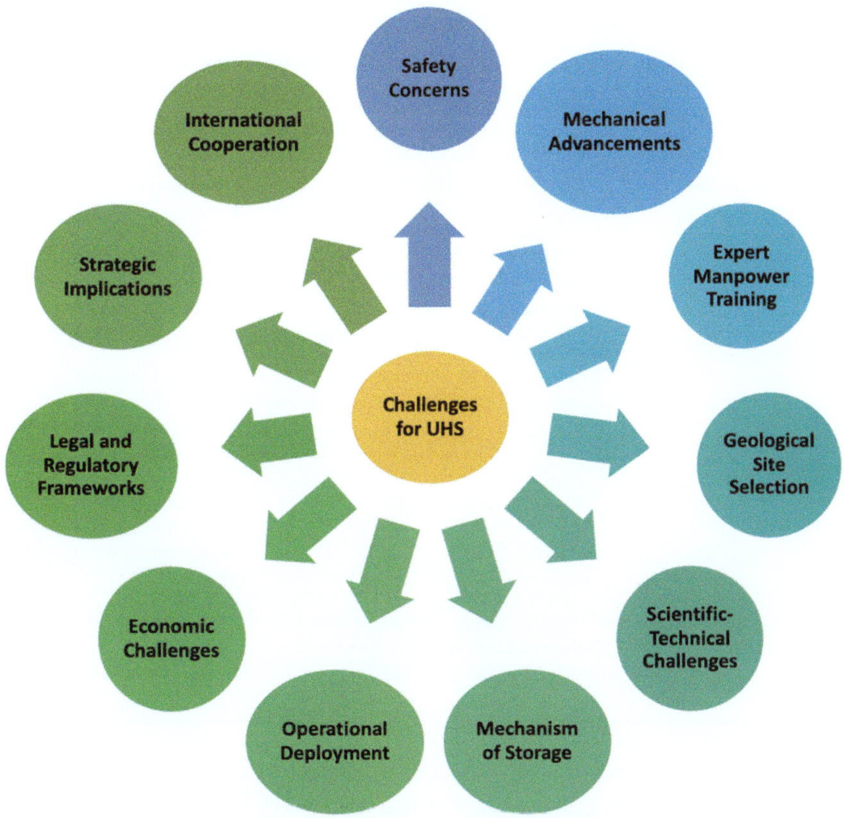

FIGURE 14.3 Different challenges associated with underground hydrogen storage.

developing global standards, and facilitating technology transfer to expedite the widespread adoption of hydrogen technologies worldwide (Kumar, 2023; Wallace, 2021). Strategic long-term planning and investment initiatives must prioritize sustainable infrastructure development, foster viable business models, and emphasize environmental sustainability through comprehensive assessments, carbon footprint reductions, and ecosystem preservation strategies. These multifaceted challenges underscore the necessity of a collaborative and integrated approach to unlocking the potential of UHS to meet future energy demands sustainably. Fig. 14.3 depicts different challenges for subsurface hydrogen storage.

14.3.1 Geological site selection

Reservoir properties, geological barriers, and proximity to energy sources are important factors for geological site selection at UHS. Through a critical examination of these considerations, the aim is to provide insights into optimizing site selection processes and advancing the development of UHS infrastructure. The selection of suitable geological formations for UHS poses both challenges and prospects that warrant careful consideration. One of the primary factors in site selection is reservoir properties, including porosity,

TABLE 14.1 Challenges, strategies, and recommendations for geological site selection (Hagemann et al., 2016; Ozarslan, 2012).

Aspect	Challenges and risks	Feasible strategies	Recommendations
Reservoir properties	Contribute to hydrogen storage, deliverability, and risks.	Optimal rock formations with high porosity and permeability, impermeable caprocks.	Prioritize formations with suitable porosity, permeability, and robust caprock integrity.
Geological barriers	Impermeable caprocks, fault lines, seismic activity.	Thorough geological assessments, suitable caprock selection, continuous gas migration monitoring.	Rigorous assessments of caprock tightness and hydraulic integrity to prevent hydrogen losses and surface leakage.
Proximity to energy sources	Economic feasibility, reduced transportation costs.	Strategic site placement near energy and pipeline infrastructures.	Opt for sites close to hydrogen sources and pipelines for enhanced operational efficiency and economic feasibility.
Advanced survey techniques	Enhanced geological insights.	Utilize seismic imaging, electromagnetic surveys, gravity mapping.	Leverage advanced techniques for accurate reservoir and barrier assessments, refine site selection criteria.
Caverns and reservoir types	Limited availability, untested aspects in aquifers.	Prioritize caverns for accessibility and resilience, rigorous testing and monitoring for aquifers	Select suitable sites based on accessibility, resilience, and tested capabilities of different reservoir types.

permeability, and storage capacity. Identifying formations with optimal reservoir characteristics is essential to ensuring efficient storage and retrieval of hydrogen (Harati, 2024; Lewandowska-Śmierzchalska et al., 2018; Muhammed, 2022). Geological barriers also play a crucial role in site selection, as they can affect the integrity and security of the storage reservoir. Factors such as the presence of impermeable cap rocks, fault lines, and seismic activity must be thoroughly assessed to mitigate risks and ensure the long-term stability of the storage site (Amirthan & Perera, 2023; Calabrese, 2024; Schultz, 2023).

Proximity to energy sources is another key consideration, as it directly impacts the feasibility and economics of UHS projects. Situating storage facilities close to renewable energy generation sites, such as wind or solar farms, can enhance operational efficiency and reduce transportation costs associated with hydrogen production and delivery (Ershadnia, 2023; Sambo, 2022). In the pursuit of optimal site selection, the prospect of utilizing advanced geological survey techniques holds promise for enhancing decision-making processes. Technologies such as seismic imaging, electromagnetic surveys, and gravity mapping offer valuable insights into subsurface geology, enabling more accurate assessments of reservoir characteristics and geological barriers (Tarkowski & Uliasz-Misiak, 2022).

Thus, the geological site selection process for UHS encompasses various critical factors, each posing distinct challenges and offering unique prospects. These factors (summarized in Table 14.1) include reservoir properties, geological barriers, proximity to energy sources, advanced survey techniques, and specific types of reservoirs like caverns, depleted hydrocarbon reservoirs, and aquifers (Heinemann et al., 2018; Sainz-Garcia et al., 2017;

Tarkowski and Czapowski, 2018). By leveraging these advanced techniques, stakeholders can make informed decisions regarding site selection, thereby maximizing the efficiency and effectiveness of UHS projects. Furthermore, ongoing advancements in geospatial analysis and data modeling provide opportunities to refine site selection criteria and identify previously overlooked geological formations with potential for hydrogen storage.

14.3.2 Scientific and technical challenges at hydrogen-rock interface

UHS presents a myriad of scientific and technical challenges at the hydrogen-rock interface, including flow dynamics, physiochemical, geochemical, and microbial activities, that must be addressed to ensure the safe and efficient operation of storage facilities. One of the primary challenges lies in reservoir characterization, which involves understanding the complex geological and petrophysical properties of subsurface formations (Aftab et al., 2022; Kumar, 2023). Characterizing reservoirs accurately is essential for predicting storage capacity, flow dynamics, and the behavior of hydrogen within porous media. Fluid flow dynamics present challenges such as lateral spreading, fingering, and gas rising, which can lead to unstable displacement and uncontrolled hydrogen leakage beyond geological traps. Employing strategies that leverage reduced dynamic effects, such as gravitational forces, can stabilize displacement, inhibit spreading and fingering, and optimize injection rates. The behavior of hydrogen in porous media poses another significant challenge, as it is influenced by factors such as adsorption, diffusion, and capillary forces. Understanding these phenomena is crucial for optimizing injection and withdrawal processes, minimizing storage losses, and ensuring the long-term integrity of storage reservoirs (Raza et al., 2022; Zivar et al., 2021). To address these challenges, the prospect of employing advanced modeling and simulation techniques holds promise for enhancing our understanding of UHS operations. Techniques like controlled injection rates, chemical use, and advanced modeling play key roles in mitigating challenges like viscous fingering and gravity segregation phenomena. Concurrently, maintaining cap rock integrity through precise pressure management is essential. Numerical models can simulate complex fluid flow phenomena and predict the behavior of hydrogen under varying reservoir conditions. Addressing flow dynamics is crucial for optimizing storage efficiency and reducing undesirable flow patterns. By integrating data from geological surveys, laboratory experiments, and field observations, these models can provide valuable insights into reservoir behavior and aid in the optimization of storage operations. Furthermore, advancements in computational fluid dynamics (CFD) and reservoir simulation software enable engineers to simulate the injection, migration, and extraction of hydrogen with unprecedented accuracy. These tools facilitate the design of efficient injection and withdrawal strategies, the assessment of storage performance, and the evaluation of potential risks and uncertainties.

Furthermore, the interplay between hydrogen and geological materials can lead to reactions affecting storage stability and efficiency. These reactions include the dissolution or precipitation of minerals due to geochemical interactions, impacting porosity and permeability within the reservoir. Understanding these processes is crucial for selecting suitable storage sites and designing effective containment strategies. Additionally, the potential for hydrogen generation during storage operations, especially through biotic processes like hydrogenotrophic sulfate reduction, adds another layer of complexity that necessitates comprehensive monitoring and management strategies. The movement of hydrogen within water-saturated reservoir pores is influenced by a range of physiochemical phenomena, such as surface properties, wetting behaviors, and capillarity effects, all of which impact hydrogen dissolution dynamics. These intricate interactions underscore the need for a nuanced approach to managing storage integrity, especially concerning potential hydrogen loss due to reactions with minerals like carbonates. For example, the presence of CO_2, SO_4^{2-}, Fe^{3+}, and clay minerals within reservoir environments introduces additional complexities. CO_2 presence can lead to methanogenic and acetogenic reactions, causing hydrogen conversion into CH_4 and CH_3COO^-, necessitating meticulous monitoring and control of environmental conditions. Similarly, sulfate reduction reactions facilitated by SO_4^{2-}, convert hydrogen into H_2S and iron reduction convert hydrogen into Fe^{2+}, highlighting the need for targeted mitigation strategies (Kumar, 2023; Zivar et al., 2021). Clay minerals, ubiquitous in reservoir formations, can induce precipitation and dissolution reactions, alter porosity and permeability, and contribute to hydrogen trapping, underscoring the importance of continuous monitoring and proactive management strategies. Moreover, monitoring microbial growth to curtail potential hydrogen consumption becomes paramount, considering microbial activities such as methanogenesis and sulfate reduction in the subsurface environment. Collaboration across disciplines—geology, microbiology, geochemistry, and engineering—is pivotal for developing innovative solutions and ensuring the long-term sustainability of UHS as a viable energy storage option. Table 14.2 illustrates the challenges and risks, feasible strategies, and recommendations for UHS aspects.

14.3.3 Strategic storage mechanism

Understanding the intricacies of hydrogen storage mechanisms in geological formations is paramount for optimizing storage operations and ensuring the efficiency of underground facilities (Kumar, 2023; Luo et al., 2024; Perera, 2023; Song et al., 2023; Zeng et al., 2024). A fundamental aspect of this comprehension involves delving into the injection and withdrawal processes, which are pivotal for the ingress and egress of hydrogen into and from the storage reservoir.

During the injection phase, hydrogen permeates through the porous media via dedicated injection wells. This migration process is governed by diffusion, advection, and dispersion mechanisms, which are influenced by factors such as rock porosity, permeability, and fluid properties. Optimizing injection necessitates meticulous control over parameters like injection rates, pressures, and volumes to achieve maximal storage capacity while mitigating energy losses and operational hazards (Gbadamosi et al., 2023; Ugarte et al., 2022). Strategic positioning of injection wells beneath impermeable caprocks plays a crucial role in containing lateral spreading and preventing unintended interactions between hydrogen and existing fluids. This not only ensures efficient storage but also safeguards reservoir integrity. Maintaining a low and steady injection rate is essential to stabilize the injection front, reducing losses due to fingering, spreading, and fracturing, thereby ensuring safe operations within prescribed pressure limits. Utilizing cushion gases such as nitrogen or carbon dioxide during injection optimizes pressurization efficiency, minimizes residual hydrogen saturation, and facilitates smooth cyclic operations during injection and withdrawal phases (Heinemann, 2021; Kumar, 2023; Zivar et al., 2023). On the withdrawal front, extracting hydrogen from the reservoir

TABLE 14.2 Challenges, strategies, and recommendations for scientific and technical consideration (flow dynamics, physiochemical, geochemical, and microbial) at the hydrogen-rock interface (Carden et al., 1997; Hemme et al., 2018; Panfilov et al., 2006, 2016).

Aspect	Challenges and risks	Feasible strategies	Recommendations
Flow behavior	The movement of hydrogen in geological formations is influenced by physiochemical, geochemical, and microbial interactions, which can reduce storage efficiency. Challenges like capillary forces, viscous fingering, and gravity segregation need to be managed to ensure effective storage.	Control injection rates and pressures to prevent viscous fingering and capillary forces. Use hydrogen-miscible storage fluids to optimize distribution. Modify reservoir structure with gels or foams to enhance hydrogen spread.	Conduct numerical simulations and optimize injection strategies for uniform hydrogen distribution and minimal storage losses.
Loss mechanisms	Risk of hydrogen loss through cap rock integrity, well boreholes, and casing materials. Improper operating pressures or production rates can lead to water coning and storage instability.	Maintain optimal operating pressures and production rates within safe limits. Regularly monitor cap rock integrity and well borehole conditions.	Implement real-time monitoring systems for continuous storage parameter assessment and early detection of leakage risks.
Geochemical & microbial reactions	In-situ geochemical reactions and microbial activities can alter storage dynamics and reservoir conditions significantly.	Conduct detailed geochemical modeling studies to predict mineral effects accurately. Monitor microbial growth and its impact on hydrogen consumption.	Integrate advanced modeling tools with microbial monitoring systems for comprehensive reservoir assessment. Collaborate for joint research projects to enhance understanding and technology development.
Research needs	Limited understanding of complex hydrogen-geological interactions. Lack of advanced monitoring and modeling tools for real-time assessment.	Invest in interdisciplinary research and advanced technology development. Develop monitoring technologies for real-time data analysis.	Collaborate with experts for comprehensive research and technology implementation. Allocate funds for advanced technology development in UHS facilities for improved safety and efficiency.

involves specific production wells. Reservoir pressure, fluid properties, saturation levels, and withdrawal rates significantly impact the efficiency of this process. Deploying multiple shallow extraction wells strategically beneath impermeable caprocks enhances hydrogen recovery and withdrawal efficiency, particularly in high-permeability zones.

Addressing challenges like gas mixing, hysteresis effects, losses during injection and withdrawal, and the risk of caprock leakage demands meticulous planning and execution. Strategies encompass controlling gas mixing through appropriate cushion gases, managing injection and withdrawal rates, optimizing cycle durations, and conducting comprehensive site characterizations for efficient operations (Table 14.3). Understanding pressure

TABLE 14.3 Challenges, strategies, and recommendations for efficient injection and withdrawal processes (Feldmann et al., 2016; Hagemann et al., 2016; Katarzyna and Radosław, 2020; Pfeiffer and Bauer, 2015, 2019; Sainz-Garcia et al., 2017).

Challenges and risks	Feasible strategies	Recommendations
Injection strategies		
Well pattern: efficient storage relies on optimal well placement to mitigate lateral spreading and avoid unwanted interactions with existing fluids in the reservoir.	Strategically position injection wells to minimize lateral spreading and ensure effective interaction between hydrogen and reservoir fluids.	Select well locations carefully, particularly beneath caprocks, to prevent lateral spreading and enhance overall storage efficiency.
Injection rate: high injection rates can lead to fingering, lateral spreading, and significant hydrogen losses.	Maintain a controlled and steady injection rate within safe pressure limits to minimize losses and ensure stable storage conditions.	Implement strict pressure limits to prevent fracturing and maintain consistent injection rates for optimal storage and operational safety.
Cushion gas: the choice of cushion gas impacts pressurization speed, residual saturation, and gas mixing during cyclic operations.	Use cushion gases with density differences conducive to efficient water displacement and consider CO_2 for better control during cyclic operations.	Balance investment costs with operational benefits when selecting cushion gas types for long-term storage efficiency and safety.
Mechanism and monitoring		
Gas mixing: uncontrolled gas mixing reduces the quality of recovered gas due to interactions with cushion gas and preexisting gases.	Utilize appropriate cushion gas types and control injection/withdrawal rates to minimize gas mixing and maintain gas quality.	Monitor and adjust gas mixing to ensure consistent gas quality throughout storage cycles, optimizing overall storage performance.
Hysteresis: gas trapping due to hysteresis can reduce storage efficiency over time, affecting gas recovery rates.	Increase injection/withdrawal cycles strategically to mitigate hysteresis effects and improve overall gas recovery efficiency.	Implement a cycle management strategy to reduce hysteresis effects and enhance gas recovery rates during storage operations.
Losses: various factors contribute to irrecoverable hydrogen volumes during storage, impacting overall storage efficiency.	Optimize injection rates and consider geological factors to minimize losses below acceptable limits and enhance storage efficiency.	Aim for minimal hydrogen losses through strategic operational practices and continuous monitoring of storage systems.
Leakage: hydrogen leakage through caprock fissures poses safety risks and reduces storage efficiency over time.	Select caprocks with high threshold pressures and implement monitoring systems to detect and mitigate any hydrogen leakage.	Employ advanced monitoring technologies to detect and prevent hydrogen leakage, ensuring long-term storage integrity and safety.
Hydrogen in the gas mixture		
Injected amount: maintaining optimal hydrogen percentages in gas mixtures is crucial for system integrity, flow behavior, and gas quality.	Monitor and adjust hydrogen percentages to preserve system integrity, flow behavior, and gas quality during storage operations.	Maintain consistent hydrogen percentages to ensure stable gas behavior and minimize system-related issues during storage cycles.

(*Continued*)

TABLE 14.3 (Continued)

Challenges and risks	Feasible strategies	Recommendations
Withdrawal strategies		
Well pattern: efficient hydrogen recovery depends on well placement to maximize recovery rates and preserve reservoir integrity.	Position extraction wells strategically to maximize recovery beneath caprocks and high-permeability zones within the reservoir.	Optimize well locations for maximum hydrogen recovery rates and minimal impact on reservoir integrity during extraction operations.
Withdrawal rate: controlled withdrawal rates are necessary to prevent fluid conning, pressure drops, and gas property alterations.	Maintain constant pressure in wells and control extraction rates to prevent fluid conning and maintain stable gas properties.	Implement strict pressure control measures during extraction to avoid reservoir damage and maintain gas quality during withdrawal.
Cycle duration and purity: optimal cycle durations and gas purity levels are critical for long-term storage performance and gas quality.	Optimize cycle durations for fluid segregation and remove impurities from gas streams to maintain gas quality and storage efficiency.	Adjust cycle durations and monitor gas purity to ensure consistent gas quality and performance during storage and extraction processes.

dissipation processes in saline aquifers is crucial for optimizing injectivity and productivity. Precise cushion gas management is pivotal for maximizing storage capacity (Heinemann, 2021; Kumar, 2023; Zivar et al., 2023). Accurate estimations and cost-benefit analyses are indispensable for cushion gas utilization during injection and production phases. Furthermore, optimizing initial cycles in deep reservoir storage significantly enhances long-term efficiency and reduces undesired water production, ensuring the success and sustainability of storage operations. Advances in well design, reservoir engineering, and operational practices can enhance the performance of storage facilities and maximize storage capacity (Muhammed, 2022; Zivar et al., 2021). Additionally, the integration of advanced monitoring and control systems enables real-time monitoring of storage operations and facilitates proactive management of storage reservoirs.

14.3.4 Modeling requirements

A modeling study to explore UHS is challenging, primarily due to the limited availability of comprehensive literature on hydrogen storage in porous media. This lack of data significantly impedes our ability to develop a thorough understanding of the intricate processes involved and to optimize storage strategies effectively. Furthermore, the modeling complexities increase substantially when dealing with depleted gas fields instead of cavern systems. These fields demand consideration of multiple gases such as hydrogen, native natural gas, and supplementary gases (cushion gas) like CO_2 and N_2, in addition to the aquifer (Craig et al., 2022; Zivar et al., 2021). As a result, accurate modeling necessitates the use of compositional simulation techniques, which are beyond the capabilities of conventional black oil simulators commonly employed in the hydrocarbon industry.

The migration behavior and fluid dynamics of hydrogen pose notable challenges during modeling due to its lower density and higher mobility compared to natural gas. Accurately predicting its movement and interaction with other gases is crucial yet challenging. Uncertainties surrounding the estimation of relative permeabilities for hydrogen and natural gas in mixed gas systems further complicate modeling efforts, highlighting the need for extensive literature review and possibly core analysis to validate modeling assumptions (Craig et al., 2022). Moreover, the geochemical interactions of hydrogen with geological components like kerogen and clays raise concerns about potential hydrogen losses and the integrity of caprock formations. Rigorous laboratory tests and research are imperative to understand these interactions better and to assess their implications accurately. Operational risks associated with material integrity and safety in scenarios involving pure hydrogen storage also require thorough validation and testing protocols. This underscores the multidisciplinary nature of UHS modeling, necessitating collaboration among experts in geology, fluid dynamics, chemistry, and engineering disciplines. Addressing data adequacy and quality through comprehensive validation processes and potentially new data acquisition strategies is paramount to enhancing the accuracy and reliability of UHS modeling outcomes and ensuring robust and effective UHS solutions.

The successful study and implementation of UHS heavily rely on understanding and modeling a multitude of controlling factors and processes. These encompass physical, geo-hydraulic, geochemical, biochemical, and mineralogical aspects (Zivar et al., 2021). Mathematical and numerical modeling emerge as indispensable tools in comprehending how these processes influence reservoir properties, fluid dynamics, and storage efficiency during injection and withdrawal cycles. One significant consideration is the distinction between aquifers and depleted hydrocarbon reservoirs, as their storage mechanisms and interactions with injected hydrogen differ due to resident fluids. Aquifers are initially saturated with water or brine, whereas depleted reservoirs may contain residual oil and gas along with brine.

Numerical models such as those utilizing CMG, Eclipse, COMSOL, DuMu^X, OpenGeoSys, OPM, PHREEQC, GWB (the Geochemist's workbench), and TOUGHREACT software play a pivotal role in investigating UHS feasibility by considering hydrogeochemical, geophysical, and biological mechanisms (Hagemann, 2017; Hemme and van Berk, 2018; Zivar et al., 2021). These models can simulate complex processes like gravity-driven displacement, bio-reactive two-phase transport, dynamic system stability, and storage cycles. Notably, strategies like "selective technology" have been proposed to mitigate issues such as lateral gas spreading and viscous fingering, which can hamper storage efficiency.

Furthermore, geochemical modeling is crucial in assessing fluid-rock interactions, predicting hydrogen losses, and understanding reservoir rock property changes. Microbial presence is another critical aspect, as certain species can metabolize injected hydrogen, affecting energy content. Various models, like Monod's model, aid in predicting microbial growth and activity, which is essential for assessing hydrogen consumption rates over storage cycles.

Numerical simulations and modeling have also proven invaluable in evaluating injection schemes, storage medium characteristics, seasonal variations, and the effectiveness of extraction wells in maximizing hydrogen recovery while minimizing unwanted fluid production. Studies have shown that efficient UHS operation depends on factors such as injection rates, gas mixing, hydrodynamics, and reservoir heterogeneity, all of which can be effectively analyzed through mathematical and numerical models. Additionally, integrating geophysical monitoring techniques with numerical simulations provides detailed insights into storage dynamics and performance over time (Table 14.4). These modeling strategies and findings collectively contribute to the development of optimized UHS designs and operational strategies for sustainable energy storage.

14.3.5 Operational, deployment, and infrastructure

The operational challenges and prospects associated with deploying UHS projects, focusing on well drilling and injection strategies, significantly influence the storage prospects. The prospect of implementing innovative operational practices to enhance the efficiency and safety of UHS operations, considering advancements in drilling technologies, monitoring systems, and data analytics, should be considered while determining the total value of an UHS project. Through a critical examination of these challenges and prospects, it is aimed at identifying strategies for optimizing the operational deployment of UHS projects and accelerating the transition towards a sustainable energy future. The operational deployment of UHS projects presents a range of challenges and prospects that must be carefully considered to ensure the successful implementation of storage facilities. One of the primary operational challenges involves well drilling and completion, which are critical stages in the construction of storage reservoirs. Drilling wells in heterogeneous subsurface formations requires careful planning and execution to ensure optimal well placement and integrity (Iyer et al., 2022; Salah et al., 2017). Challenges such as formation damage, wellbore stability, and drilling fluid compatibility must be addressed to minimize operational risks and maximize reservoir performance. Injection strategies play a crucial role in the operational deployment of UHS projects, as they determine how hydrogen is introduced into the storage reservoir and distributed within the subsurface formations. Optimizing injection strategies involves selecting appropriate injection rates, pressures, and volumes to maximize storage capacity while maintaining reservoir integrity (Ampomah et al., 2017; Rasool et al., 2017). Challenges such as gas breakthrough, plume migration, and reservoir heterogeneity must be addressed to ensure efficient injection and distribution of hydrogen within the reservoir.

Despite these operational challenges, there are promising prospects for implementing innovative operational practices to enhance the efficiency and safety of UHS operations. Advances in drilling technologies, such as horizontal drilling and directional drilling, offer opportunities to improve well placement and reservoir access, thereby increasing storage capacity and performance. Additionally, the integration of advanced monitoring and control systems enables real-time monitoring of injection operations and facilitates proactive management of operational risks. Furthermore, the prospect of implementing enhanced operational practices, such as smart well technologies and real-time data analytics, offers opportunities to optimize injection strategies and enhance reservoir performance (Bello et al., 2017; Popa et al., 2015). By leveraging data-driven insights and predictive analytics, operators can optimize injection rates, pressures, and volumes to maximize storage efficiency and minimize operational risks.

TABLE 14.4 Insights on modeling requirements for underground hydrogen storage.

Objective	Storage media	Software(s)/ tool(s)/technique(s)	Findings	Authors
Investigating hydrogeochemical mechanisms to identify potential risks during storage.	Depleted gas fields.	1-D reactive mass transport (1DRMT) model, PHREEQC version 3, Monod model.	Hydrogen losses occurred due to gas-water-rock interactions and bacterial processes converting some hydrogen to H_2S and CH_4. Diffusion through the cap rock also contributed to losses. Storage duration, pressure, temperature, kinetic rate, and gas composition were key factors influencing overall loss.	Hemme & van Berk (2018)
Studying effect of Microorganism including methanogens, acetogens, and acetotrophs.	Porous media with residual water coating.	Pore-scale model, biofilm model, and viscous-fluid model.	The growth rates of methanogenic microorganisms were found to rely on biomass activity and the level of residual water containing biomass.	Ebigbo et al. (2013)
Assessing Deep Aquifer Feasibility.	2D model of layered sandstone and conglomerates.	Immiscible multiphase flow model using finite element method code, COMSOL Multiphysics.	The efficiency of storage was found to be influenced by various properties of both the gas and the aquifer. Factors such as fluid viscosity, injectivity duration, and regional groundwater flow were observed to impact how easily the gas moved, expanded, saturated, and behaved within the storage system.	Sainz-Garcia et al. (2017)
Investigating the impact of bacteria on the composition of stored gases.	Depleted gas (natural) reservoir in an aquifer.	Mathematical model and Monod's model alongside a multicomponent gas injection model.	The transport dynamics were notably influenced by methanogenic and bacterial activities, leading to the conversion of some H_2 and CO_2 (cushion gas) into CH_4 and H_2O.	Panfilov et al. (2006)
Exploring storage operations and geophysical monitoring through numerical simulation.	Synthetic, heterogeneous geological structure (field site).	Gravity modeling, seismic mapping, flow simulation model, facies modeling approach, petrophysical rock model, geoelectric resistivity model, Gassmann fluid substitution model, P-wave velocity model, Peng–Robinson equation of State, Eclipse E300.	The study offered comprehensive multidimensional insights into the storage operation. It found that H_2 storage increased with cycle numbers, leading to more availability of H2 at the final storage cycle. Detailed information on UHS dynamics was presented in this study.	Pfeiffer et al. (2016)
Investigating the hydrodynamic effects, including gas rising, lateral gas spreading, and other related phenomena.	2D reservoir (vertical slice through an anticlinal structure).	Box-method, $DuMu^x$, Darcy's law, Brooks–Corey correlation, and selective technology.	The study highlighted the significant influence of injection rates on gravitational and viscosity forces, impacting fluid displacement processes during hydrogen storage.	Hagemann et al. (2015)

Simulating hydrodynamic processes, gas mixing, and seasonal cyclic operations.	3D depleted gas reservoir and water-saturated anticline structures.	Peaceman's model, DuMu^x, COMSOL Multiphysics, and selective technology.	The study found significant energy storage potential in underground facilities but highlighted challenges like viscous fingering and gravity override in fluid displacement. Selective technology improved H_2 production, and storing H_2 in stratified aquifers reduced gas loss due to lateral spreading or viscous fingering.	Feldmann et al. (2016)
Investigating hydrodynamic behavior coupled with biochemical processes.	2D reservoir representing a depleted gas field through an anticlinal structure.	Mathematical model, biochemical model, hydro-dynamical model, DuMuX, double Monod model, Box-method, Darcy's, law, Fick's law, Brooks–Corey.	The study found that increased microbial activity partially converted injected H_2 to CH_4 and H_2O. It emphasized high injection rates causing unstable lateral spreading below the caprock. Notably, UHS showed more pronounced lateral fingers than natural gas storage.	Hagemann et al. (2016)
Investigating fluid-rock interactions through geochemical modeling.	Data from molasse basin in upper Austria.	Geochemist's workbench (GWB), Equilibrium and kinetic batch simulations, LLNL thermodynamic database (Lawrence Livermore National Laboratory).	The controlling factors were observed to include reaction rates, equilibrium assumptions among redox pairs, and the reduction of pyrite to pyrrhotite.	Hassannayebi et al. (2019)
Investigating subsurface H_2 storage behavior.	Saline aquifer model based on the existing Rhaetian anticlinal structure.	Eclipse 300, and Peng–Robinson EOS.	The simulation demonstrated that the storage system could meet 20% of the weekly electrical energy demand. Optimizing the injection scheme further enhanced storage performance, indicating potential efficiency improvements in meeting short-term energy needs.	Pfeiffer & Bauer (2015)
Modeling the unstable transport phenomena, such as gas rising, lateral spreading, and hydrodynamics.	Aquifer	DuMu^x, Darcy's law, Brooks–Corey, Box-method, and the selective technology.	The study found that controlling injection rates managed gravitational and viscosity forces during gas displacement. Stratified storage with selective technology delayed gas rise and reduced leaks. Barrier extent significantly affected gas movement, varying by gas type.	Hagemann et al. (2015)

(Continued)

TABLE 14.4 (Continued)

Objective	Storage media	Software(s)/ tool(s)/technique(s)	Findings	Authors
Comparing the seasonal storage capacities and deliverability.	Depleted gas reservoir.	PHREEQC, chemical stability study, volumetric analysis, and leakage studies.	The facility was able to store approximately 42% of the supplied energy capacity. Additionally, the study found that the loss of H_2 due to dissolution and diffusion was reduced to less than 0.1%.	Amid et al. (2016)
Examine dimensioning and induced hydraulic effects.	Existing Rhaetian anticlinal structure based on Saline aquifer model.	Eclipse 300, the Penge Robinson EOS, and facies modeling.	The study found that storage was primarily limited by the achievable extraction rates. It also provided detailed insights into how the distribution of heterogeneous parameters affected storage efficiency.	Pfeiffer et al. (2017)
Assessing feasible strategies for seasonal underground H_2 storage.	Saline aquifer.	COMSOL Multiphysics, 3D multiphase model, Brooks–Corey, and Darcy's law.	The study achieved a 78% maximum H_2 recovery rate, observed no viscous fingering, noted the suitability of steeply dipping structures for H2 storage without cushion gas (prone to upconing), and recommended positioning extraction wells beneath the caprock for optimal H_2 recovery.	Sainz-Garcia et al. (2017)
Comparing UHS using heterogeneous and homogeneous flow property models.	Existing Rhaetian anticlinal structure based on Saline aquifer model.	Eclipse 300 and analytical averaging methods.	This study provided detailed multidimensional information on how spatial averaging methods could be applied to approximate storage characteristics effectively.	Pfeiffer & Bauer (2019)

The backbone of UHS deployment lies in its infrastructure, which encompasses the generation, transportation, storage, and distribution of hydrogen. This complex network demands meticulous planning, design, and development to seamlessly integrate into global energy frameworks, aligning with sustainability goals (Kumar, 2023). However, substantial investments in energy and materials are required throughout infrastructure construction, with transportation presenting challenges due to its impact on hydrogen retail prices. Exploring viable transportation modes while considering existing infrastructure globally remains imperative. Safety considerations are paramount across all phases of UHS operations, necessitating comprehensive safety frameworks covering production, storage, transportation, and utilization. Rigorous hazard and accident analyses, coupled with standardized safety protocols, are indispensable for risk mitigation and stakeholder acceptance. Skills development is a critical aspect, requiring workforce training and skill enhancement for the safe handling of hydrogen across its lifecycle. Moreover, adherence to engineering standards and robust quality control measures throughout the value chain is essential, necessitating governmental support for standardization and harmonization of regulations. Deployment strategies must align with sustainability objectives while addressing public acceptance and engagement challenges. Local authorities play a pivotal role in managing infrastructure delivery and aligning with national strategies, ensuring seamless integration into evolving energy systems. Investment in digital technologies is imperative for enhancing operational efficiency and integrating the hydrogen value chain seamlessly.

Despite challenges, UHS presents significant opportunities for technological innovation, global collaboration, and market leadership in hydrogen technologies. Leveraging existing infrastructure, transitioning to low-carbon materials, and harnessing renewable energy resources position countries to play pivotal roles in the global hydrogen market. Investment in research, development, and operational technologies is crucial for unlocking the full potential of UHS in achieving sustainable energy futures globally.

14.3.6 Feasibility and economic considerations

The feasibility and economic viability of UHS projects, considering factors such as capital investment, operational costs, and revenue potential, are encountered in deep consideration as these directly influence the project. The prospects for reducing costs and improving the economic attractiveness of UHS through technological advancements and policy support are very important. Through a critical examination of these considerations, it is important to identify strategies for optimizing the economic performance of UHS projects and accelerating the transition towards a sustainable energy future. Assessing the feasibility and economic viability of UHS projects is essential for determining their potential to contribute to the transition towards sustainable energy systems. This evaluation involves analyzing various factors, including capital investment, operational costs, and revenue potential, to determine the overall economic attractiveness of UHS projects (Chen et al., 2022; Qiu et al., 2020; Quintos Fuentes & Santos, 2023). One of the key considerations in evaluating the feasibility of UHS projects is the capital investment required for project development and construction. This includes costs associated with site selection, drilling and completion of storage wells, the installation of surface facilities, and the implementation of monitoring and control systems. Additionally, factors such as geological complexity, regulatory requirements, and environmental considerations may impact capital investment and project timelines (Aftab et al., 2022; Kumar, 2023). Fig. 14.4 illustrates

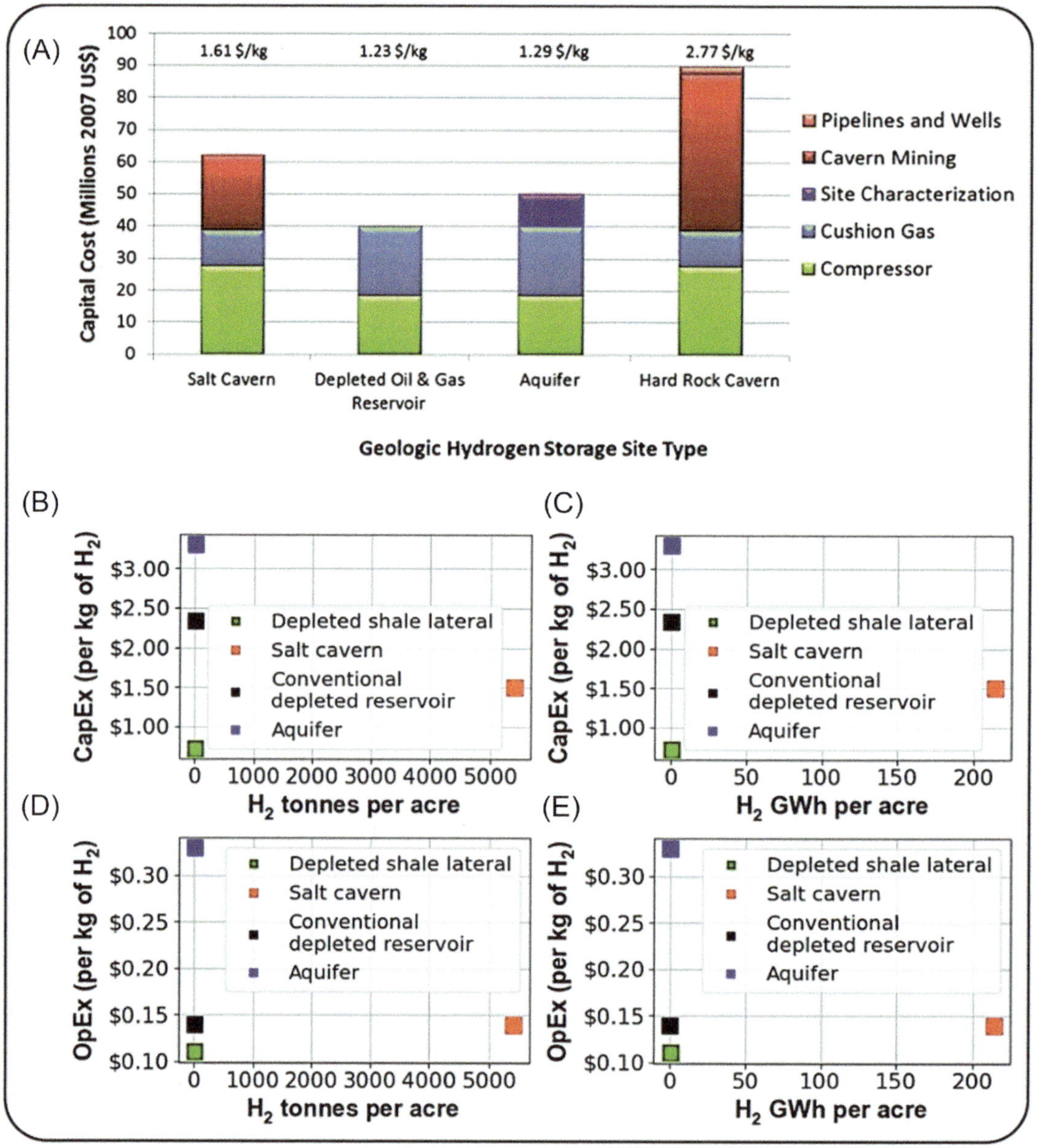

FIGURE 14.4 Cost of underground hydrogen storage versus storage density (per acre) for storage options (A) Capital cost of underground hydrogen storage, (B-C) Capital expenditure (D-E) Operational expenditure in differnt geological formations (Singh, 2022).

the capital cost and operation cost for different types of storage systems. It is found from the data that depleted reservoirs require the lowest cost for UHS.

Operational costs are another important factor to consider when assessing the economic viability of UHS projects. These costs include expenses associated with hydrogen production, injection and withdrawal operations, reservoir monitoring and maintenance, and safety and environmental compliance. Optimizing operational practices and leveraging technological advancements can help reduce operational costs and improve the overall economic performance of UHS projects (Epelle, 2022; Quintos Fuentes & Santos, 2023). Revenue potential is also a critical consideration in evaluating the economic viability of UHS projects. This includes potential revenue streams from hydrogen sales, ancillary services such as grid balancing and peak shaving, and revenue from carbon credits or other incentives. Assessing revenue potential requires a thorough understanding of market dynamics, including hydrogen demand and pricing trends, as well as regulatory frameworks and policy support for hydrogen storage and utilization (Clemens & Clemens, 2022; Santos et al., 2022). Despite the challenges associated with capital investment and operational costs, there are promising prospects for reducing costs and improving the economic attractiveness of UHS projects. Technological advancements, such as advanced drilling and completion techniques, reservoir modeling and simulation tools, and hydrogen production technologies, offer opportunities to enhance project economics and reduce overall costs (Epelle, 2022; Quintos Fuentes & Santos, 2023). Furthermore, policy support and regulatory incentives can play a significant role in improving the economic viability of UHS projects. This includes financial incentives, such as grants, loans, and tax credits, as well as regulatory frameworks that provide certainty and support for project development and operation. By fostering a supportive policy environment and leveraging technological advancements, stakeholders can work together to enhance the feasibility and economic attractiveness of UHS projects.

14.3.7 Legal and regulatory frameworks

Understanding the legal and regulatory landscape is crucial for the successful deployment of UHS projects. This section delves into the challenges and prospects associated with the legal and regulatory frameworks governing permitting, licensing, and compliance for UHS projects. One of the primary challenges in the legal and regulatory domain is navigating the complex permitting and licensing processes required for UHS projects. These processes involve obtaining approvals from various governmental bodies at the local, regional, and national levels and may include environmental impact assessments, land use permits, and water rights approvals. Delays or uncertainties in the permitting process can significantly impact project timelines and economics. Additionally, compliance with relevant regulations and standards is essential to ensure the safety, environmental sustainability, and social acceptance of UHS projects (Matos et al., 2022; Tchouvelev et al., 2019). This includes compliance with regulations related to underground storage, such as construction and integrity requirements, as well as regulations governing hydrogen production, transportation, and utilization. Ensuring compliance with these regulations requires comprehensive monitoring, reporting, and verification processes throughout the project lifecycle (Tchouvelev et al., 2019).

Despite these challenges, there are promising prospects for developing supportive regulatory frameworks and incentives to facilitate the deployment of UHS on a global scale. Governments and regulatory authorities are increasingly recognizing the importance of hydrogen as a clean and versatile energy carrier and are taking steps to support the development and deployment of hydrogen storage infrastructure. One approach to fostering supportive regulatory frameworks is to establish clear and streamlined permitting and licensing processes specifically tailored to UHS projects. This can involve the development of standardized guidelines, templates, and best practices for permitting and licensing, as well as the establishment of dedicated regulatory bodies or agencies responsible for overseeing UHS projects. By reducing regulatory uncertainties and streamlining approval processes, supportive regulatory frameworks can help accelerate the deployment of UHS projects (Jahanbakhsh et al., 2024; Watson et al., 2024). In addition to permitting and licensing, regulatory incentives such as financial incentives, tax credits, and grants can play a crucial role in incentivizing investment in UHS projects. These incentives can help offset capital costs, reduce project risks, and improve the overall economic viability of UHS projects. Furthermore, regulatory frameworks that prioritize the development of hydrogen infrastructure and support the integration of hydrogen into existing energy systems can create a favorable environment for UHS deployment. Overall, the development of supportive legal and regulatory frameworks is essential for unlocking the full potential of UHS and accelerating the transition towards a sustainable energy future. By addressing legal and regulatory challenges and capitalizing on opportunities for regulatory innovation and incentives, stakeholders can pave the way for the widespread adoption of UHS on a global scale.

14.3.8 Strategic implications

Understanding the strategic implications of UHS is essential for recognizing its potential role in addressing key energy and environmental challenges. This section examines the strategic implications of UHS for energy security, climate change mitigation, and economic development, while also exploring prospects for international cooperation and collaboration to accelerate its adoption.

14.3.8.1 Analysis of strategic implications

Understanding the strategic implications of UHS for energy security, climate change mitigation, and economic development is essential for recognizing its potential role in addressing key energy and environmental challenges.

14.3.8.1.1 Energy security

UHS has significant implications for enhancing energy security by providing a reliable and resilient energy storage solution. By enabling the storage of surplus renewable energy, such as wind and solar power, UHS can help mitigate the intermittency and variability of these energy sources, thereby reducing dependence on fossil fuels and enhancing energy independence (Blanchard, 2023). Furthermore, UHS can serve as a strategic reserve during periods of high energy demand or supply disruptions, contributing to energy security at the national and regional levels.

14.3.8.1.2 Climate change mitigation

UHS plays a critical role in advancing climate change mitigation efforts by enabling the large-scale deployment of renewable energy and facilitating the transition to a low-carbon energy system. By storing excess renewable energy as hydrogen, UHS helps decarbonize key sectors such as transportation, industry, and power generation, reducing greenhouse gas emissions and mitigating climate change impacts (Rosman et al., 2023). Additionally, UHS enables the utilization of renewable hydrogen in fuel cells for zero-emission electricity generation and transportation, further contributing to climate change mitigation objectives.

14.3.8.1.3 Economic development

UHS presents significant opportunities for economic development by fostering innovation, creating jobs, and driving investment in clean energy technologies. The deployment of UHS infrastructure requires investment in research and development, manufacturing, construction, and operation, leading to job creation and economic growth across various sectors (Iordache et al., 2014; Uliasz-Misiak et al., 2022). Furthermore, the development of a robust UHS industry can stimulate the growth of related supply chains and value-added services, generating additional economic benefits for local communities and economies.

14.3.9 Exploration of prospects for international cooperation

Given the global nature of energy and environmental challenges, international cooperation and collaboration are essential for accelerating the adoption of UHS on a global scale. Countries can leverage their respective strengths and resources to overcome common barriers and facilitate the development and deployment of UHS infrastructure (Amirthan & Perera, 2023; Jahanbakhsh et al., 2024; Le Duigou et al., 2017; Lemieux et al., 2020; Raza et al., 2024). This includes sharing best practices, technical expertise, and financial resources, as well as collaborating on research and development initiatives, pilot projects, and demonstration efforts. By working together, countries can maximize the benefits of UHS and achieve collective energy and climate objectives. International cooperation and collaboration can help address challenges related to technology transfer, financing, policy harmonization, and regulatory frameworks, enabling the rapid and widespread adoption of UHS as a key component of the global energy system (Amirthan & Perera, 2023; Raza et al., 2024). Thus, exploring prospects for international cooperation and collaboration is essential for unlocking the full potential of UHS and realizing its strategic implications for energy security, climate change mitigation, and economic development on a global scale.

14.3.10 Safety concerns and technological innovations

Safety is paramount in the development and operation of UHS facilities. While UHS offers significant benefits for energy storage and climate mitigation, it also presents unique safety challenges that must be addressed through technological innovations and rigorous safety protocols.

14.3.10.1 *Safety concerns*

UHS involves handling and storing large quantities of hydrogen, which is highly flammable and poses explosion risks if not managed properly. Additionally, the presence of hydrogen in confined underground spaces raises concerns about potential leaks, migration, and accumulation of hydrogen gas. Furthermore, interactions between hydrogen and reservoir fluids or geological formations may lead to unexpected reactions or hazards (Navaid et al., 2023). Addressing these safety concerns requires a comprehensive understanding of the potential risks and hazards associated with UHS operations.

14.3.10.2 *Technological innovations*

To mitigate safety risks and enhance the reliability of UHS infrastructure, ongoing technological innovations are crucial. Advanced design features and engineering solutions can help minimize the likelihood of accidents and mitigate the consequences of unforeseen events (Navaid et al., 2023; Tarkowski & Uliasz-Misiak, 2022). For example, innovative materials and construction techniques can enhance the integrity and resilience of storage reservoirs and containment systems. Similarly, advanced monitoring and detection systems can provide real-time data on gas concentrations, pressure levels, and other critical parameters to enable the early detection of potential safety hazards.

14.3.10.3 *Prospects for safety enhancement*

The prospect of integrating cutting-edge technologies and safety protocols offers promising avenues for enhancing the safety of UHS facilities. This includes the development of predictive modeling tools and risk assessment methodologies to identify potential safety hazards and evaluate their likelihood and consequences (Navaid et al., 2023; Tarkowski & Uliasz-Misiak, 2022). Furthermore, ongoing research on hydrogen behavior in porous media and geological formations can provide valuable insights into risk mitigation strategies and best practices for safe UHS operations.

14.3.10.4 *Integration of advanced design features*

Incorporating advanced design features and safety protocols into UHS infrastructure is essential to ensuring robustness and resilience against safety hazards. This may include the use of redundant safety systems, passive and active ventilation measures, and remote monitoring and control capabilities. Additionally, innovative materials such as high-strength alloys and corrosion-resistant coatings can enhance the durability and longevity of storage components, reducing the risk of structural failure or integrity loss. Addressing safety concerns through technological innovations and advanced design features is essential for realizing the full potential of UHS as a safe and reliable energy storage solution (Navaid et al., 2023; Tarkowski & Uliasz-Misiak, 2022). By integrating state-of-the-art safety protocols and engineering solutions, UHS projects can minimize risks and maximize the benefits of UHS for sustainable energy systems.

14.3.11 Advancements in mechanical components

The successful operation of UHS facilities relies heavily on the performance and reliability of various mechanical components such as pipes, valves, connectors, regulators, and measuring instruments. As UHS projects continue to evolve and expand, advancements in mechanical design and engineering play a crucial role in enhancing the efficiency, safety, and reliability of UHS infrastructure.

14.3.11.1 Role of advanced mechanical components

Advanced mechanical components form the backbone of UHS infrastructure, facilitating the safe and efficient transport, storage, and distribution of hydrogen. Pipes and valves are essential for conveying hydrogen from production facilities to storage sites and regulating the flow of gas within the storage reservoirs (Amirthan & Perera, 2023; Raza et al., 2024). Connectors and regulators ensure seamless integration and operation of different system components, while measuring instruments provide critical data on pressure, temperature, and gas composition for monitoring and control purposes.

14.3.11.2 Prospects for innovative designs

The prospect of developing and implementing innovative mechanical designs holds significant promise for improving the performance and functionality of UHS infrastructure. Advanced materials such as high-strength alloys and corrosion-resistant coatings can enhance the durability and longevity of mechanical components, reducing the risk of leaks, corrosion, and structural failure (Matos et al., 2022; Tchouvelev et al., 2019). Similarly, novel valve designs incorporating features such as fail-safe mechanisms and remote actuation capabilities can enhance operational safety and reliability.

14.3.11.3 Integration of smart technologies

The integration of smart technologies and digitalization solutions offers exciting prospects for optimizing the performance and efficiency of mechanical components in UHS infrastructure. Sensor-enabled valves and regulators can provide real-time data on system parameters, enabling proactive maintenance and troubleshooting to prevent downtime and ensure continuous operation (Kumar, 2023; Song et al., 2023). Furthermore, advanced monitoring and control systems can enhance the responsiveness and adaptability of UHS infrastructure to changing operational conditions and demand patterns.

14.3.11.4 Collaborative research and development

Collaborative research and development efforts involving industry stakeholders, research institutions, and government agencies are essential for driving innovation and advancing the state-of-the-art in mechanical components for UHS. By fostering partnerships and knowledge exchange, stakeholders can leverage collective expertise and resources to accelerate the development and deployment of innovative mechanical designs that meet the unique challenges and requirements of UHS projects (Kumar, 2023; Song et al., 2023). Advancements in mechanical components hold great potential for enhancing the efficiency, reliability, and safety of UHS infrastructure. By embracing innovation and collaboration, stakeholders can leverage cutting-edge technologies and engineering solutions to unlock the full potential of UHS as a key enabler of sustainable energy systems.

14.3.12 Expert manpower and training

The successful implementation and operation of UHS projects rely heavily on the availability of expert manpower equipped with the necessary skills and knowledge. As UHS technologies continue to evolve and expand, the importance of fostering talent development and specialized training programs becomes increasingly paramount to ensuring the effective management and execution of these complex projects. However, Table 14.5 illustrates the challenges, technical aspects, and opportunities of UHS projects.

14.3.12.1 Importance of expert manpower

Expert manpower plays a critical role in all stages of UHS projects, from site selection and design to construction, operation, and maintenance. Professionals with specialized expertise in geology, engineering, hydrogeology, reservoir management, and safety regulations are essential for navigating the technical, logistical, and regulatory challenges associated with UHS. Additionally, skilled technicians and operators are required to ensure the safe and efficient operation of UHS infrastructure and equipment (Chen et al., 2022; Qiu et al., 2020).

14.3.12.2 Prospects for talent development

The prospect of fostering talent development and knowledge exchange offers significant opportunities for building a skilled workforce capable of addressing the challenges and opportunities of UHS. Specialized training programs, workshops, and educational initiatives can provide professionals with the specialized skills and knowledge required to excel in various roles within the UHS industry. Furthermore, mentorship programs and knowledge-sharing platforms can facilitate the transfer of expertise and best practices from experienced professionals to the next generation of UHS practitioners (Kumar, 2023).

14.3.12.3 Knowledge exchange and collaboration

Collaboration among industry stakeholders, academic institutions, research organizations, and government agencies is essential for promoting knowledge exchange and talent development in the field of UHS. By leveraging collective expertise and resources, stakeholders can facilitate the development of comprehensive training curricula, certification programs, and professional development initiatives tailored to the specific needs and challenges of UHS projects. Additionally, international collaboration and partnership initiatives can enable the cross-border exchange of best practices and lessons learned, enriching the pool of talent and expertise available to support global UHS initiatives (Kumar, 2023; Song et al., 2023). The importance of expert manpower and specialized training programs cannot be overstated in the context of UHS projects. By investing in talent development

TABLE 14.5 Challenges, aspects, and opportunities in underground hydrogen storage (Heinemann, 2021; Kumar, 2023).

Challenges	Technical aspects	Opportunity/prospect
Technical challenges		
Fluid dynamics	Understanding fluid flow behavior in porous media.	Optimizing injection-reproduction cycles to enhance storage efficiency.
	Preventing gas migration.	Ensuring long-term storage integrity to maintain safety and reliability.
Geochemical reactions	Managing complex hydrogen-mineral interactions.	Preventing reservoir damage and unintended mineral alterations.
	Mitigating dissolution-precipitation reactions.	Maintaining cement integrity to avoid infrastructure damage.
Microbial impacts	Monitoring and mitigating microbial interactions.	Preventing biocorrosion, microbiological-induced fractures, and system contamination.
	Preventing biogeochemical changes.	Ensuring system and storage integrity through effective microbial management.
Geomechanical considerations	Predicting and managing stress-strain effects.	Preventing induced seismicity and infrastructure instability.
	Managing reservoir compaction.	Ensuring fault stability and long-term structural integrity.
Regulatory and policy constraints		
Infrastructure integration	Adapting existing infrastructure for hydrogen compatibility.	Ensuring safety, reliability, and seamless integration with current energy systems.
	Aligning safety standards and regulations.	Supporting sustainable hydrogen production, distribution, and utilization.
Energy import dependency	Reducing reliance on energy imports.	Promoting domestic energy security and sustainable energy practices.
	Aligning policies for sustainable hydrogen production.	Encouraging investment in local green energy solutions and reducing emissions.
Safety and monitoring	Developing robust safety protocols.	Implementing advanced safety measures for hydrogen production and usage.
	Implementing early detection technologies.	Ensuring continuous monitoring and risk mitigation across the hydrogen value chain.
Cost and efficiency challenges		
Production cost	Advancing electrolysis technologies for cost efficiency.	Achieving cost-competitive green hydrogen production through innovation.
	Reducing capital and operational expenditures for production.	Integrating renewable energy for sustainable and economical hydrogen production.

(Continued)

TABLE 14.5 (Continued)

Challenges	Technical aspects	Opportunity/prospect
Storage efficiency	Enhancing underground storage technologies for scalability.	Optimizing cushion gas strategies and minimizing energy losses during storage.
	Minimizing energy losses during storage and retrieval.	Scaling up storage capacities to meet future hydrogen demands.
Purity and energy efficiency	Ensuring high purity levels in hydrogen production.	Minimizing contaminants and maximizing energy efficiency in utilization.
	Maximizing energy efficiency in electrolysis processes.	Optimizing hydrogen utilization in end-use applications for maximum efficiency.
Global collaboration and standards		
International cooperation	Fostering global collaboration for standardized practices.	Accelerating global hydrogen technology adoption and interoperability.
	Developing common standards and regulations.	Promoting harmonized regulations and ease of technology transfer globally.
Knowledge sharing	Promoting research collaboration and public-private partnerships.	Accelerating innovation, technology diffusion, and expertise development globally.
	Building capacity through knowledge-sharing initiatives.	Facilitating international partnerships for sustainable hydrogen solutions
Long-term planning and investment	Planning sustainable hydrogen infrastructure projects.	Evaluating business models and ensuring profitable and scalable hydrogen initiatives.
	Evaluating investment strategies and risk-sharing mechanisms.	Attracting capital, ensuring long-term profitability, and supporting energy transition.
Environmental sustainability	Balancing energy security with environmental preservation.	Reducing carbon footprints, promoting ecosystem preservation, and ensuring sustainability.

and knowledge exchange, stakeholders can build a skilled workforce capable of driving innovation, ensuring operational excellence, and advancing the sustainable development of UHS technologies.

14.4 Conclusions

This chapter has provided a comprehensive examination of the challenges and prospects associated with UHS. From geological site selection to safety concerns, technological innovations, and the importance of expert manpower and training, various aspects critical to the successful implementation and operation of UHS projects have been explored.

Key prospects and challenges are demanding and critical for UHS projects. Throughout the chapter, we have identified and discussed a range of challenges and prospects related to UHS, including the complexities of geological site selection, the need for technological innovations to enhance safety and reliability, and the importance of fostering talent development and knowledge exchange. While UHS presents unique technical, operational, and regulatory challenges, it also offers significant opportunities for advancing energy security, mitigating climate change, and promoting economic development.

Timely action is required to be taken for the implementation of UHS projects. As we look to the future, it is imperative for stakeholders across the public and private sectors to collaborate and innovate towards realizing the full potential of underground hydrogen storage. By working together to address the challenges and seize the opportunities presented by UHS, we can accelerate the transition to a sustainable energy future and unlock the numerous environmental, social, and economic benefits associated with hydrogen as a clean and versatile energy carrier. In conclusion, the journey towards realizing the full potential of UHS requires concerted efforts, bold initiatives, and collaborative partnerships.

Acknowledgment

The first author, Sunil Kumar, gratefully acknowledges the support of RINOLYST, SciGlyph Exploration Private Limited, India, to complete the work. Another author, Achinta Bera, appreciates the support provided by the Department of Petroleum Engineering, School of Energy Technology, Pandit Deendayal Energy University, Gandhinagar-382426, Gujarat, India.

References

Aftab, A., Hassanpouryouzband, A., Xie, Q., Machuca, L. L., & Sarmadivaleh, M. (2022). Toward a fundamental understanding of geological hydrogen storage. *Industrial & Engineering Chemistry Research, 61*(9), 3233–3253.

Amid, A., Mignard, D., & Wilkinson, M. (2016). Seasonal storage of hydrogen in a depleted natural gas reservoir. *International Journal of Hydrogen Energy, 41*(12), 5549–5558. Available from https://doi.org/10.1016/j.ijhydene.2016.02.036.

Amirthan, T., & Perera, M. S. A. (2023). Underground hydrogen storage in Australia: A review on the feasibility of geological sites. *International Journal of Hydrogen Energy, 48*(11), 4300–4328. Available from https://doi.org/10.1016/j.ijhydene.2022.10.218.

Ampomah, W., Balch, R. S., Grigg, R. B., McPherson, B., Will, R. A., Lee, S. Y., Dai, Z., & Pan, F. (2017). Co-optimization of CO2-EOR and storage processes in mature oil reservoirs. *Greenhouse Gases: Science and Technology, 7*(1), 128–142.

Bello, O., Yang, D., Lazarus, S., Wang, X.S. and Denney, T., 2017. Next generation downhole big data platform for dynamic data-driven well and reservoir management. In *SPE Reservoir Characterisation and Simulation Conference and Exhibition?* (p. D031S014R002). SPE.

Blanchard, A., 2023. Strategic Investments: Electrolysis vs. Hydrogen Storage for Europe's Energy Security in the Hydrogen Era (No. hal-04297136).

Caglayan, D. G., et al. (2020). Technical potential of salt caverns for hydrogen storage in Europe. *International Journal of Hydrogen Energy, 45*(11), 6793–6805. Available from https://doi.org/10.1016/j.ijhydene.2019.12.161.

Calabrese, M., et al. (2024). Hydrogen safety challenges: A comprehensive review on production, storage, transport, utilization, and CFD-based consequence and risk assessment. *Energies, 17*(6). Available from https://doi.org/10.3390/en17061350.

Clemens, M., & Clemens, T. (2022). Scenarios to Decarbonize Austria's Energy Consumption and the Role of Underground Hydrogen Storage. *Energies, 15*(10), 3742.

Craig, A., Newman, S., Stephenson, P., Evans, C., Yancazos, S., & Barber, S. (2022). Hydrogen storage potential of depleted oil and gas fields in Western Australia. *The APPEA Journal, 62*(1), 185–195.

Capurso, T., et al. (2022). Perspective of the role of hydrogen in the 21st century energy transition. *Energy Conversion and Management*, 251. Available from https://doi.org/10.1016/j.enconman.2021.114898.

Carden, P., & Paterson, L. (1979). Physical, chemical and energy aspects of underground hydrogen storage. *International Journal of Hydrogen Energy*, *4*(6), 559–569.

Chai, M., et al. (2023). Numerical simulation of large-scale seasonal hydrogen storage in an anticline aquifer: A case study capturing hydrogen interactions and cushion gas injection. *Applied Energy*, 334. Available from https://doi.org/10.1016/j.apenergy.2023.120655.

Chen, F., Ma, Z., Nasrabadi, H., Chen, B., Mehana, M. and Van Wijk, J.W., 2022. Technical and Economic Feasibility Analysis of Underground Hydrogen Storage: A Case Study in Intermountain-West Region USA. *arXiv pr-print arXiv:2209.03239*.

Le Duigou, A., Bader, A. G., Lanoix, J. C., & Nadau, L. (2017). Relevance and costs of large scale underground hydrogen storage in France. *International Journal of Hydrogen Energy*, *42*(36), 22987–23003.

Du, Z., et al. (2024). Exploring hydrogen geologic storage in China for future energy: Opportunities and challenges. *Renewable and Sustainable Energy Reviews*, *196*, 114366. Available from https://doi.org/10.1016/j.rser.2024.114366.

Ebigbo, A., Golfier, F., & Quintard, M. (2013). A coupled, pore-scale model for methanogenic microbial activity in underground hydrogen storage. (in english) *Advances in Water Resources*, *61*, 74–85.

Elberry, A. M., Thakur, J., & Veysey, J. (2021). Seasonal hydrogen storage for sustainable renewable energy integration in the electricity sector: A case study of Finland. *Journal of Energy Storage*, 44. Available from https://doi.org/10.1016/j.est.2021.103474.

Epelle, E. I., et al. (2022). Perspectives and prospects of underground hydrogen storage and natural hydrogen. *Sustainable Energy and Fuels*, *6*(14), 3324–3343. Available from https://doi.org/10.1039/d2se00618a.

Ershadnia, R., et al. (2023). Impact of geological and operational conditions on underground hydrogen storage. *International Journal of Hydrogen Energy*, *48*(4), 1450–1471. Available from https://doi.org/10.1016/j.ijhydene.2022.09.208.

Feldmann, F., Hagemann, B., Ganzer, L., & Panfilov, M. (2016). Numerical simulation of hydrodynamic and gas mixing processes in underground hydrogen storages. *Environmental Earth Sciences*, *75*(16), 1165.

Gabrielli, P., et al. (2020). Seasonal energy storage for zero-emissions multi-energy systems via underground hydrogen storage. *Renewable and Sustainable Energy Reviews*, 121. Available from https://doi.org/10.1016/j.rser.2019.109629.

Gbadamosi, A. O., Muhammed, N. S., Patil, S., Al Shehri, D., Haq, B., Epelle, E. I., Mahmoud, M., & Kamal, M. S. (2023). Underground hydrogen storage: A critical assessment of fluid-fluid and fluid-rock interactions. *Journal of Energy Storage*, *72*, 108473.

Hagemann, B. "Numerical and analytical modeling of gas mixing and bio-reactive transport during underground hydrogen storage," Doctoral, Université de Lorraine, 2017.

Hagemann, B., Rasoulzadeh, M., Panfilov, M., Ganzer, L., & Reitenbach, V. (2015). Mathematical modeling of unstable transport in underground hydrogen storage, in English *Environmental Earth Sciences*, *73*(11), 6891–6898.

Hagemann, B., Rasoulzadeh, M., Panfilov, M., Ganzer, L., & Reitenbach, V. (2016). Hydrogenization of underground storage of natural gas, in English *Computat Geosci*, *20*(3), 595–606.

Harati, S., et al. (2024). Multi-criteria site selection workflow for geological storage of hydrogen in depleted gas fields: A case for the UK. *International Journal of Hydrogen Energy*, *51*, 143–157. Available from https://doi.org/10.1016/j.ijhydene.2023.10.345.

Hassannayebi, N., Azizmohammadi, S., De Lucia, M., & Ott, H. (2019). Underground hydrogen storage: application of geochemical modelling in a case study in the Molasse Basin, Upper Austria. (in english) *Environmental Earth Sciences*, *78*(5), 177.

Heinemann, N., Booth, M. G., Haszeldine, R. S., Wilkinson, M., Scafidi, J., & Edlmann, K. (2018). Hydrogen storage in porous geological formations e onshore play opportunities in the midland valley (Scotland, UK). *International Journal of Hydrogen Energy*, *43*(45), 20861–20874.

Heinemann, N., et al. (2021). Enabling large-scale hydrogen storage in porous media-the scientific challenges. *Energy and Environmental Science*, *14*(2), 853–864. Available from https://doi.org/10.1039/d0ee03536j.

Hemme, C., & van Berk, W. (2018). Hydrogeochemical Modeling to Identify Potential Risks of Underground Hydrogen Storage in Depleted Gas Fields. (in english) *Applied Sciences-Basel*, *vol. 8*(11), 2282.

Hydrogen council 2017. https://hydrogencouncil.com/wp-content/uploads/2017/11/Hydrogen-Scaling-up_Hydrogen-Council_2017.compressed.pdf.

Iordache, I., Schitea, D., Gheorghe, A. V., & Iordache, M. (2014). Hydrogen underground storage in Romania, potential directions of development, stakeholders and general aspects. *International Journal of Hydrogen Energy, 39*(21), 11071–11081.

Iyer, J., Lackey, G., Edvardsen, L., Bean, A., Carroll, S. A., Huerta, N., Smith, M. M., Torsæter, M., Dilmore, R. M., & Cerasi, P. (2022). A review of well integrity based on field experience at carbon utilization and storage sites. *International Journal of Greenhouse Gas Control, 113*, 103533.

Jahanbakhsh, A., Potapov-Crighton, A. L., Mosallanezhad, A., Kaloorazi, N. T., & Maroto-Valer, M. M. (2024). Underground hydrogen storage: A UK perspective. *Renewable and sustainable energy reviews, 189*, 114001.

Katarzyna, L., & Radosław, T. (2020). Numerical simulation of hydrogen injection and withdrawal to and from a deep aquifer in NW Poland. *International Journal of Hydrogen Energy, 45*, 2068–2083.

Kharel, S., & Shabani, B. (2018). Hydrogen as a long-term large-scale energy storage solution to support renewables. *Energies, 11*(10). Available from https://doi.org/10.3390/en11102825.

Kumar, S., et al. (2023). Underground hydrogen storage and its roadmap and feasibility in India toward net-zero target for global decarbonization. *Fuel*, 350. Available from https://doi.org/10.1016/j.fuel.2023.128849.

Lemieux, A., Shkarupin, A., & Sharp, K. (2020). Geologic feasibility of underground hydrogen storage in Canada. *International Journal of Hydrogen Energy, 45*(56), 32243–32259.

Lewandowska-Śmierzchalska, J., Tarkowski, R., & Uliasz-Misiak, B. (2018). Screening and ranking framework for underground hydrogen storage site selection in Poland. *International Journal of Hydrogen Energy, 43*(9), 4401–4414. Available from https://doi.org/10.1016/j.ijhydene.2018.01.089.

Liebscher, A., Wackerl, J., & Streibel, M. (2016). Geologic storage of hydrogen – Fundamentals, processing, and projects. *Hydrogen Science and Engineering: Materials, Processes, Systems and Technology*, 629–658. Available from https://doi.org/10.1002/9783527674268.ch26.

Luo, X., Tveit, S., Gholami, R., & Andersen, P. Ø. (2024). Underground hydrogen storage (UHS) in natural storage sites: A perspective of subsurface characterization and monitoring. *Fuel, 364*, 131038.

Matos, C. R., Silva, P. P., & Carneiro, J. F. (2022). Overview of compressed air energy storage projects and regulatory framework for energy storage. *Journal of Energy Storage, 55*, 105862.

Matos, C. R., Carneiro, J. F., & Silva, P. P. (2019). Overview of large-scale underground energy storage technologies for integration of renewable energies and criteria for reservoir identification. *Journal of Energy Storage, 21*, 241–258. Available from https://doi.org/10.1016/j.est.2018.11.023.

Muhammed, N. S., et al. (2022). A review on underground hydrogen storage: Insight into geological sites, influencing factors and future outlook. *Energy Reports, 8*, 461–499. Available from https://doi.org/10.1016/j.egyr.2021.12.002.

Navaid, H. B., Emadi, H., & Watson, M. (2023). A comprehensive literature review on the challenges associated with underground hydrogen storage. *International Journal of Hydrogen Energy, 48*(28), 10603–10635.

Olabi, A. G., et al. (2021). Large-vscale hydrogen production and storage technologies: Current status and future directions. *International Journal of Hydrogen Energy, 46*(45), 23498–23528. Available from https://doi.org/10.1016/j.ijhydene.2020.10.110.

Osman, A. I., et al. (2022). Hydrogen production, storage, utilisation and environmental impacts: A review. *Environmental Chemistry Letters, 20*(1), 153–188. Available from https://doi.org/10.1007/s10311-021-01322-8.

Ozarslan, A. (2012). Large-scale hydrogen energy storage in salt caverns. (in -english) *International Journal of Hydrogen Energy, 37*(19), 14265–14277.

Panfilov, M. (2016). *Underground and pipeline hydrogen storage. In: Gupta RB, Basile A, Veziroglu TN, editors. Compendium of hydrogen energy* (pp. 91–115). Woodhead Publishing.

Panfilov M., Gravier G., Fillacier S. Underground storage of H2 and H2-CO2-CH4 mixtures. In: ECMOR X-10th European conference on the mathematics of oil recovery; 2006.

Perera, M. S. A. (2023). A review of underground hydrogen storage in depleted gas reservoirs: Insights into various rock-fluid interaction mechanisms and their impact on the process integrity. *Fuel, 334*, 126677.

Pfeiffer, W. T., al Hagrey, S. A., Kohn, D., Rabbel, W., & Bauer, S. (2016). Porous media hydrogen storage at a synthetic, heterogeneous field site: numerical simulation of storage operation and geophysical monitoring. (in english) *Environmental Earth Sciences, 75*(16), 1177.

Pfeiffer, W. T., & Bauer, S. (2015). Subsurface porous media hydrogen storage e scenario development and simulation. *Energy Procedia, 76*, 565–572.

Pfeiffer, W. T., & Bauer, S. (2019). Comparing simulations of hydrogen storage in a sandstone formation using heterogeneous and homogenous flow property models, in English *Petrol Geosci*, *25*(3), 325–336.

Pfeiffer, W. T., Beyer, C., & Bauer, S. (2017). Hydrogen storage in a heterogeneous sandstone formation: dimensioning and induced hydraulic effects. *Petroleum Geoscience*, *23*(3), 315–326.

Popa, A.S., Grijalva, E., Cassidy, S., Medel, J. and Cover, A., 2015, September. Intelligent use of big data for heavy oil reservoir management. *SPE Annual Technical Conference and Exhibition?* (p. D021S011R003). SPE.

Qiu, Y., Zhou, S., Wang, J., Chou, J., Fang, Y., Pan, G., & Gu, W. (2020). Feasibility analysis of utilising underground hydrogen storage facilities in integrated energy system: case studies in China. *Applied Energy*, *269*, 115140.

Quintos Fuentes, J. E., & Santos, D. M. (2023). Technical and Economic Viability of Underground Hydrogen Storage. *Hydrogen*, *4*(4), 975–1000.

Raad, J., Mostafa, S., Yuri, L., & Hassan, H. (2022). Hydrogen storage in saline aquifers: Opportunities and challenges. *Renewable and Sustainable Energy Reviews*, 168. Available from https://doi.org/10.1016/j.rser.2022.112846.

Rasool, M. H., Ahmad, M., & Ayoub, M. (2023). Selecting geological formations for CO2 storage: A comparative rating system. *Sustainability*, *15*(8), 6599.

Raza, A., Arif, M., Glatz, G., Mahmoud, M., Al Kobaisi, M., Alafnan, S., & Iglauer, S. (2022). A holistic overview of underground hydrogen storage: Influencing factors, current understanding, and outlook. *Fuel*, *330*, 125636.

Raza, A., Mahmoud, M., Arif, M., & Alafnan, S. (2024). Underground hydrogen storage prospects in the Kingdom of Saudi Arabia. *Fuel*, *357*, 129665.

Rosman, A., Hamdi, Z., Ali, M. and Bataee, M., 2023, September. Mitigating Global Warming with Underground Hydrogen Storage: Impacts of H2S Generation. *SPE Offshore Europe Conference and Exhibition* (p. D031S013R005). SPE.

Sainz-Garcia, A., Abarca, E., Rubi, V., & Grandia, F. (2017). Assessment of feasible strategies for seasonal underground hydrogen storage in a saline aquifer. *International Journal of Hydrogen Energy*, *42*(26), 16657–16666.

Salah, M., Orr, D., ElSebaee, M., Kamar, A., Batmaz, T., ElDarawy, Y., & Salem, L. (2017). *Integrated advanced workflows and heterogeneity analysis for planning unconventional horizontal wells in Western Desert of Egypt. Offshore Mediterranean conference and exhibition*. OMC. (pp. OMC-2017).

Sambo, C., et al. (2022). A review on worldwide underground hydrogen storage operating and potential fields. *International Journal of Hydrogen Energy*, *47*(54), 22840–22880. Available from https://doi.org/10.1016/j.ijhydene.2022.05.126.

Santos, B. H., Lopes, J. P., Carvalho, L., Matos, M., & Alves, I. (2024). Public policies to foster green hydrogen seasonal storage: Portuguese study case model until 2040. *Energy Strategy Reviews*, *52*, 101354.

Scafidi, J., et al. (2021). A quantitative assessment of the hydrogen storage capacity of the UK continental shelf. *International Journal of Hydrogen Energy*, *46*(12), 8629–8639. Available from https://doi.org/10.1016/j.ijhydene.2020.12.106.

Schultz, R. A., et al. (2023). Underground energy-related product storage and sequestration: Site characterization, risk analysis and monitoring. *Geological Society Special Publication*, *528*(1), 37–59. Available from https://doi.org/10.1144/SP528-2022-66.

Singh, H. (2022). Hydrogen storage in inactive horizontal shale gas wells: Techno-economic analysis for Haynesville shale. *Applied Energy*, *313*, 118862.

Song, H., Lao, J., Zhang, L., Xie, C., & Wang, Y. (2023). Underground hydrogen storage in reservoirs: pore-scale mechanisms and optimization of storage capacity and efficiency. *Applied Energy*, *337*, 120901.

Tarkowski, R., & Uliasz-Misiak, B. (2022). Towards underground hydrogen storage: A review of barriers. *Renewable and Sustainable Energy Reviews*, 162. Available from https://doi.org/10.1016/j.rser.2022.112451.

Tarkowski, R., & Czapowski, G. (2018). Salt domes in Poland e potential sites for hydrogen storage in caverns. *International Journal of Hydrogen Energy*, *43*(46), 21414–21427.

Tchouvelev, A. V., de Oliveira, S. P., & Neves, N. P. (2019). *Regulatory framework, safety aspects, and social acceptance of hydrogen energy technologies. Science and Engineering of Hydrogen-Based Energy Technologies* (pp. 303–356). Academic Press.

Ugarte, E. R., & Salehi, S. (2022). A review on well integrity issues for underground hydrogen storage. *Journal of Energy Resources Technology*, *144*(4), 042001.

Uliasz-Misiak, B., Lewandowska-Śmierzchalska, J., Matuła, R., & Tarkowski, R. (2022). Prospects for the Implementation of Underground Hydrogen Storage in the EU. *Energies*, *15*(24), 9535.

Wallace, R. L., et al. (2021). Utility-scale subsurface hydrogen storage: UK perspectives and technology. *International Journal of Hydrogen Energy*, *46*(49), 25137–25159. Available from https://doi.org/10.1016/j.ijhydene.2021.05.034.

Watson, M., Ennis-King, J., Hortle, A., & Raab, M. (2022). Developing Australia's underground hydrogen storage through demonstration. *The APPEA Journal*, *62*(2), S196–S199.

Williams, J. D. O., et al. (2022). Does the United Kingdom have sufficient geological storage capacity to support a hydrogen economy? Estimating the salt cavern storage potential of bedded halite formations. *Journal of Energy Storage*, *53*. Available from https://doi.org/10.1016/j.est.2022.105109.

Zeng, L., Sander, R., Chen, Y., & Xie, Q. (2024). Hydrogen Storage Performance During Underground Hydrogen Storage in Depleted Gas Reservoirs: A Review. *Engineering*.

Zivar, D., Kumar, S., & Foroozesh, J. (2021). Underground hydrogen storage: A comprehensive review. *International Journal of Hydrogen Energy*, *46*(45), 23436–23462.

Zivar, D., Rezaei, A., Shad, S., & Khosravi, V. (2023). *Carbon dioxide injection in tight oil reservoirs. Gas Injection Methods* (pp. 317–338). Gulf Professional Publishing.

Zhang, L., et al. (2024). A comprehensive review of the promising clean energy carrier: Hydrogen production, transportation, storage, and utilization (HPTSU) technologies. *Fuel*, 355. Available from https://doi.org/10.1016/j.fuel.2023.129455.

Index

Note: Page numbers followed by "*f*" and "*t*" refer to figures and tables, respectively.

D

H

P

Q

R

S

V